Introduction to Human Biology
Second Edition

Front cover photographs

Top left   A mosquito sucking blood from a human arm. Some species of mosquito transmit malaria and other diseases. *See* pages 241–5. (Shell)

Top centre   Students at a WHO centre for research and training in the biology of immunity. *See* pages 252–4. (WHO)

Top right   A cast showing the air passages of the lung. It is made by pouring liquid plastic into the lung removed from a dead animal. The plastic hardens and the lung tissue is digested away, leaving a replica of the air passages. *See* pages 85–6. (D. G. Mackean)

Centre left   A group of nurses learning about the use of surgical instruments. (WHO)

Centre centre   A microscope section of a piece of cartilage which is becoming ossified, i.e. turning into bone. *See* pages 111–13 for further details. (Philip Harris Biological)

Centre right   A health worker in Kenya checks a baby's vaccination scar. A world-wide campaign of vaccination eliminated smallpox in 1978. *See* pages 252–6. (WHO)

Bottom left   Apples do not clean the teeth or prevent decay as was once thought, but if apples are eaten instead of sweets, dental decay will be reduced and there will be less likelihood of obesity. *See* pages 126–8 and page 37. (WHO)

Bottom right   Clean water for drinking, cooking and washing is taken for granted in the industrialized countries, but in many other countries the water is a source of diseases. *See* pages 275 and 278. (WHO)

Back cover photograph

Human blood smear. *See* pages 70–2. (Biophoto Associates)

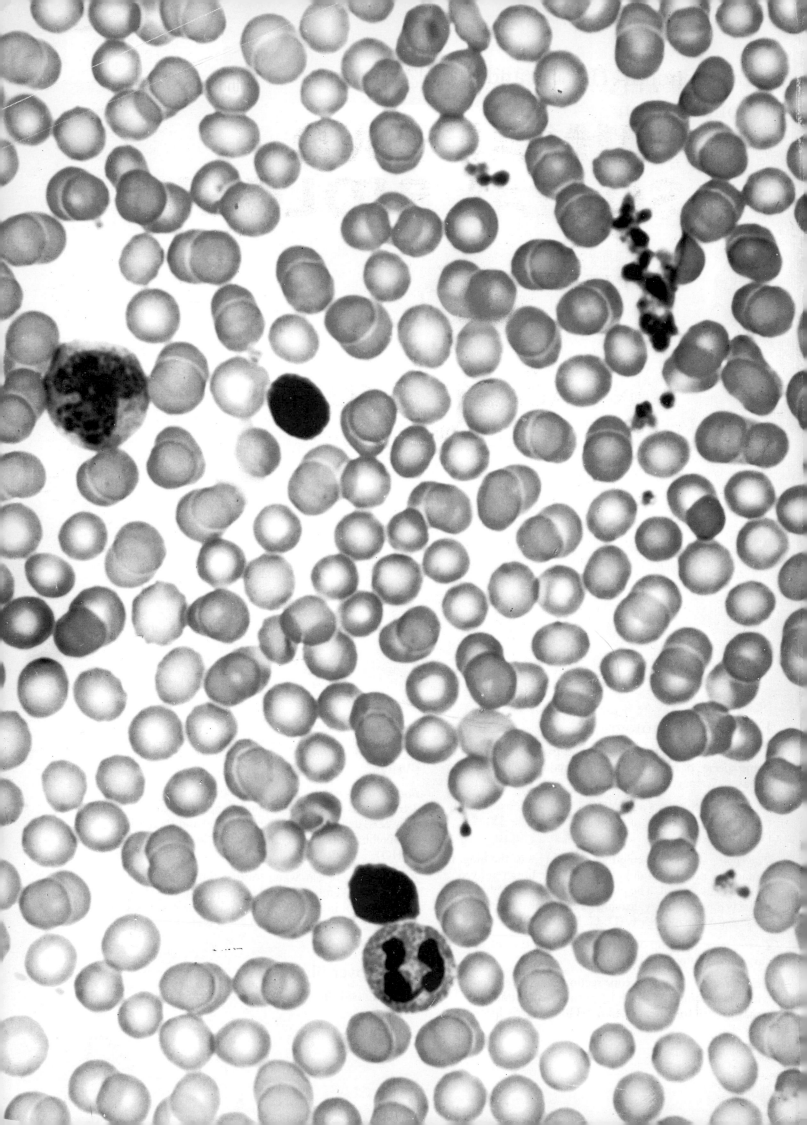

INTRODUCTION TO
HUMAN AND SOCIAL BIOLOGY

Second Edition

DON MACKEAN BA, FIBiol
BRIAN JONES BSc, FIBiol

John Murray

Other books by Don Mackean

Advanced Biology (with C. J. Clegg)

Introduction to Biology, 5th edition
Introduction to Biology: Third Tropical edition
Introduction to Biology: Second West African edition
Introduction to Genetics, 3rd edition
Life Study: A Textbook of Biology
Class Experiments in Biology (with C. J. Worsley and P. C. G. Worsley)
Class Experiments in Biology – Teachers' Book (with C. J. Worsley and
    P. C. G. Worsley)
Certificate Health Science for West Africa (with Brian Jones)
Human Life

GCSE Biology
GCSE Practical Assessment: Biology (with M. M. Oakes)
Biology Resource Pack for GCSE

Science for Today and Tomorrow (with M. A. Atherton and T. Duncan)
Science in Zimbabwe, 2nd edition (Ed. W. Stout: with G. Bethell,
    T. Duncan, I. Elliott and Brian Jones)

© Don Mackean & Brian jones, 1975, 1985

Published by John Murray (Publishers) Ltd,
a member of the Hodder Headline Group,
338 Euston Road, London NW1 3BH

First published 1975
Second edition 1985
Reprinted 1987, 1990, 1992 with revisions, 1993 (twice), 1995, 1996
1998, with revisions, 2000, 2002, 2004 (twice)

Printed and bound in Great Britain by CPI Bath.

British Library Cataloguing in Publication Data

ISBN 0 7195 4167 0

Preface to First Edition

Introduction to Human and Social Biology is primarily a textbook for students following courses leading to 'O' and 'AO' examinations in the General Certificate of Education. The contents cover the Human and Social Biology syllabuses of the Cambridge and the London Overseas examinations, and also the Human Biology syllabuses at 'O' and 'AO' level in Great Britain. It is expected that the book will also be useful for pre-nursing courses and as a general reference book for a variety of other social and biological studies.

Human Biology is of universal significance and though it is often necessary to quote examples peculiar to certain regions, the facts presented in this book are relevant to human society in all parts of the world. It is as valid for a student in Britain to learn about the problems of disease and food production in tropical countries as it is for a student in Malaysia to study the dangers of pollution and faulty nutrition in the Western Hemisphere.

Our objective has been to present the facts as clearly as possible, without advocating any particular order or method of study (although a knowledge of the facts and principles described in the early chapters is taken for granted in the later sections). The extensive use of cross-references and the presence of a glossary should make it relatively easy to use the book for reference at any point.

The questions at the ends of chapters are designed to make the reader use, reorganize or appraise the information in the text rather than simply replicate it. Some of the essay questions set by the main examining boards in recent years have been reproduced at the end of the book.

The suggestions for practical work have been kept as simple as possible so that the students can attempt the experiments with a minimum of apparatus.

Preface to Second Edition

The contents have been revised to take account of changes in national syllabuses from 1984. The opportunity has also been taken to introduce up-to-date information and improve some of the illustrations, and to introduce new chapters on Mental Health, Socially Significant Diseases, Pollution, and First Aid.

Don Mackean
Brian Jones

Acknowledgements

The authors would like to thank the following who read various sections of the manuscript and made many constructive criticisms. We greatly value the contribution they have made to the accuracy and clarity of the text;

    Mr S. W. Hurry (physiology)
    Dr C. O. Carter (genetics)
    Dr P. R. Travers (posture and exercise)
    Mr C. J. D. Sykes (teeth)
    Mr G. T. Creber

We are also extremely grateful to all those who have supplied photographs. They have taken a great deal of trouble to supply us with prints, some of which were specially prepared for this book. They are acknowledged individually on the pages where the illustrations appear. Cover photographs are acknowledged on p. (i).

Contents

1

Living Organisms

Characteristics of living organisms

Biology is the study of life (Greek *bios* = life, *logos* = knowledge) which, in practice, means the study of living things.

In most animals, the characteristics by which we know they are alive are self-evident: they move about, they feed, they have young, and they respond to changes in their surroundings.

These features are less obvious in plants and certain small animals; and when dealing with organisms like bacteria and viruses the distinctions between living and non-living can often be drawn only by trained scientists with the appropriate apparatus and techniques at their disposal. The main differences between living organisms and non-living objects can be summarized as follows:

1 **Respiration.** This is the process by which energy is made available as a result of chemical changes within the organism, the commonest of which is the chemical decomposition of food as a result of its combination with oxygen. This is not a particularly obvious occurrence in plants and animals; but it is fairly easy to demonstrate that living creatures take in air, remove some of the oxygen from it and increase the volume of carbon dioxide in it. More simply expressed it can be said that living organisms take in oxygen and give out carbon dioxide. Sometimes this takes place with obvious breathing movements. Respiration also results in a rise of temperature, which is more easily detectable in animals than in plants.

2 **Feeding.** This is an essential preliminary to respiration, since energy comes ultimately from food. The feeding of a tree by its leaves is less obvious than that of an animal, which moves actively in search of food. Feeding may also result in growth.

3 **Excretion.** Living involves a vast number of chemical processes, including respiration, many of which produce substances that are poisonous when moderately concentrated. The elimination of these from the organism is called excretion.

4 **Growth.** Strictly, growth is simply an increase in mass, but it usually implies also that the organism is becoming more complicated and more efficient. An illustration of this is an animal which changes its form from larva to adult, for example, a frog or a butterfly.

5 **Movement.** An animal can generally move its whole body, whereas the movements of the higher plants are usually restricted to certain parts such as the opening and closing of petals, or to movements of parts as a result of growth.

Even if an organism does not move as a whole, the contents of its cells (p. 6) may be seen to move. The cytoplasm of the plant cell shown in Fig. 2.3*c* (p. 7) and Fig. 7.2*d* (p. 41) exhibits streaming movements within the cell, and the white blood cells (p. 72) of animals can move about by flowing movements.

6 **Reproduction.** No organism has a limitless life, but although individuals must die sooner or later their life is handed on to new individuals by reproduction, resulting in the continued existence of the species.

7 **Sensitivity.** Sensitivity is the ability to respond to a stimulus. Obvious signs of sensitivity are the movements made by animals as a result of noises, on being touched or on seeing an enemy. Fully grown plants do not show such responses under casual observation, but during growth they respond to the direction of light, gravity and moisture.

The classification of living organisms

The earth is populated by enormous numbers of different kinds of organisms. When these organisms are studied and described it becomes apparent that they can be divided into groups. In each group of organisms the members show strong likenesses to each other. These similarities are not always immediately obvious but become clearer when the characteristics of the group are known. Bees and butterflies, for example, though differing considerably in appearance, size, colour and habits, belong to the same group, insects, because they both have hard outer skeletons and three distinct regions in their bodies, the middle region carrying six legs and two pairs of wings.

The largest division of organisms is into the *kingdoms*.

Living organisms can be grouped into five kingdoms, although not all biologists agree that this is the best, or the only way, to group or 'classify' them.

Viruses have not been included in this classification because they are not living organisms in the same sense as the other creatures. A virus can reproduce only in the cells of living organisms (see p. 200).

1 **Bacteria and blue-green algae.** These are single-celled organisms which do not have a nucleus (p. 8) in their cells. Most of them live in water or soil and play an important part in decay and recycling (p. 51).

2 **Single-celled organisms** which do have a nucleus. Most of these live freely in fresh water or sea water but some of them cause human diseases.

3 **Fungi.** This group includes toadstools and moulds as well as a number of fungi which cause diseases in crop plants.

4 **Green plants.** In their leaves, green plants can absorb sunlight and use its energy to build up the food they need for survival.

5 **Animals.** These organisms cannot make their own food. They have to take in food by eating plants, other animals or the dead remains of either. Man is classified as an animal.

Each of the kingdoms can be divided up into smaller groups called *phyla*. The *Crustacea* (crabs, lobsters, shrimps, water-fleas), constitute one animal phylum. The *Flowering plants* form one of the phyla in the plant kingdom. Phyla are further divided into *classes*.

The animal kingdom can also be divided into *vertebrates* (animals with 'backbones') and '*invertebrates*' (animals without 'backbones'). 'Invertebrates' is not really a proper biological group because it contains organisms which differ greatly from each other, e.g. worms, insects and snails. However, it is convenient to use the term 'invertebrate' to distinguish these animals from the vertebrates. The vertebrates are divided into five classes:

(i) Fish
(ii) Amphibia (frogs and toads)
(iii) Reptiles (snakes, lizards, tortoises)
(iv) Birds
(v) Mammals (Fig. 1.1)

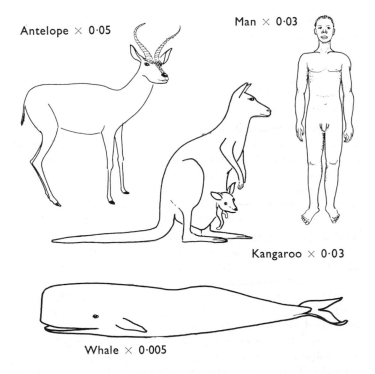

Antelope × 0·05

Man × 0·03

Kangaroo × 0·03

Whale × 0·005

Fig. 1.1   Mammals

The fish, amphibia and reptiles are all 'cold-blooded'. They cannot regulate their body temperature in the same way as mammals. Fish will have the same temperature as their surroundings. Amphibia and reptiles may warm up or cool down when their surrounding temperature changes. Birds and mammals keep their body temperatures at a steady level. They are usually warmer than their surroundings.

Fish, amphibia, reptiles and birds all reproduce by laying eggs. These may be fertilized externally as in most fish and amphibia, or internally as in reptiles and birds.

Characteristics of mammals

(a) Mammals, with the exception of the Australian platypus and spiny anteater, do not lay eggs. Their eggs are fertilized internally and are kept in the female's body for weeks or months while they develop into embryos. They are born, that is, delivered to the outside world, more or less fully formed (Fig. 1.2). Man is classified as a mammal.

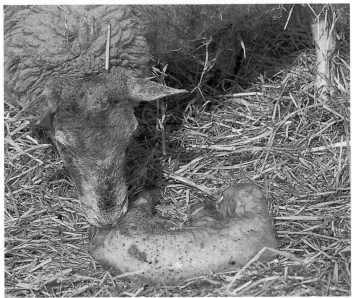

(ECOSCENE/Angela Hampton)

Fig. 1.2   Most mammals produce their young fully formed. This sheep has just given birth to a lamb

In addition to giving birth to their young, mammals are different from the other vertebrates in the following ways:

(b) Female mammals feed their young on milk produced from *mammary glands*, until the young animals can eat solid food.

(c) They have hair or fur on their bodies (Fig. 1.3). This may be an obvious dense covering as in cats and dogs, sparsely distributed bristles as in pigs or a simple tuft at the end of an elephant's tail.

(Zoological Society of London/M. Lyster)

Fig. 1.3   Mammalian characteristics. The leopard shows the covering of fur, the ear pinnae and the vibrissae (sensory whiskers)

(d) In all mammals, the internal body space is divided into two regions: the *thorax*, which contains the heart and lungs, and the *abdomen* which contains digestive organs, liver, kidneys and reproductive organs. The two cavities are separated by a sheet of tissue called the *diaphragm* (p. 89). Other vertebrates do not have a diaphragm.

(e) In most but not all mammals, the first set of teeth, the 'milk' teeth, is shed at an early stage in life and replaced with a set of permanent teeth.

(f) In most mammals, but not, for example, in dolphins, the teeth are of different shapes and sizes (Fig. 1.4) and are adapted for special functions, e.g. incisors for gripping or cutting, molars for crushing or grinding (p. 125).

In a reptile or a fish, the teeth are usually all the same shape, forming rows of simple sharp spikes or pegs which grip the prey.

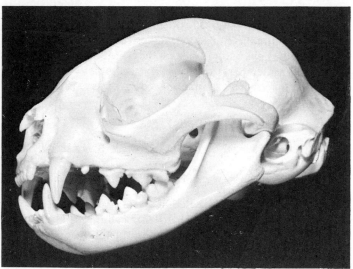

(Philip Harris Biological)

Fig. 1.4   Skull of cat showing the different shapes and sizes of teeth for holding prey, cutting flesh and crushing bones

(g) Sweat glands occur only in mammals, but not all mammals have them. In cats they are present only in the feet; in whales they are absent altogether.

(h) Although birds, reptiles and amphibians have ears (hearing organs), only mammals have the structures called ear *pinnae*, i.e. the flaps which project from the head and direct sound vibrations into the ear (Fig. 1.3). Most marine mammals such as whales and some seals do not have ear pinnae.

(i) One of the main features which separates mammals from all other vertebrates is the much greater development of their brains. Part of the front region of the brain has developed into two large *cerebral hemispheres* (see p. 155). It is thought that this region is responsible for learning and intelligence as well as many other activities which make mammals a successful group.

Primates. The vertebrate classes are each sub-divided into *orders*. Amongst the 25 mammalian orders are *Rodents* (rats, mice, voles), *Carnivores* (lions, wolves, stoats), *Artiodactyls* (deer, cows, pigs), *Chiroptera* (bats) and *Primates* (lemurs, monkeys, apes and man).

The primates differ from most others mammals in having their eyes directed forward, giving stereoscopic vision (p. 138), very large brains with highly developed cerebral hemispheres (p. 153), and small numbers of offspring. In addition, they have greater powers of learning and adaptability than most other mammals. Their hands are often unspecialized, i.e. not adapted as hooves or wings or to any specific purpose such as burrowing or climbing.

Man as a mammal

(a) **Birth.** Human eggs are fertilized internally and the young develop in about 38 weeks inside the mother's uterus, obtaining their food and oxygen through a structure called the *placenta* (see p. 105). The young are born fully formed but helpless and depend on their parents for food and shelter for many years.

(b) **Suckling.** The mammary glands of the females form the breasts. These contain fatty tissue and milk-producing glands. The young suck the milk from the breasts and it is normally their only source of food for several months (p. 107).

(c) **Hair.** Apart from the palms of the hands and soles of the feet, human skin is covered with hair (p. 96). This grows longer and more densely on the head, the face and chest (in males), in the armpits and pubic regions.

(d) **Diaphragm.** Humans have a diaphragm which separates the abdomen and thorax. The diaphragm plays an important part in breathing (p. 89).

(e) **Milk teeth.** Children have 24 milk teeth which they lose between the age of 6–12 years. The milk teeth are replaced by 32 permanent teeth (p. 124).

(f) **Teeth shape.** Although human teeth are not so varied in shape as, for example, those of a dog, the incisors have chisel-like crowns which cut off pieces of food, and the molars are broad and knobbly for crushing the food.

(g) **Sweat glands.** Sweat glands are present in all parts of human skin and play a part in temperature regulation (p. 96).

(h) **Ear pinnae.** Human ears have pinnae which project from the side of the head, but whether they are important in directing sound waves into the ear is not clear.

(i) **Brain.** The brain in humans is very large in proportion to the body size, and the cerebral hemispheres are particularly well developed. It seems likely that this gives man a learning ability and intelligence which is greater than that of the other mammals.

Man's stereoscopic vision, unspecialized hands, large cerebral hemispheres and small number of offspring further classify him as a primate.

Special human features

Upright posture. By standing upright instead of on all fours, man leaves his hands free to manipulate tools and weapons. This has enabled him to build shelters, defend himself against predators and develop writing and other technical skills (Fig. 1.5).

threatening cries and greeting signals but other mammals do not have the vocal equipment or brain capacity for passing on the detailed information which speech can convey.

Learning and intelligence. The combination of man's speech, writing, memory and ability to learn from experience helps him to survive all kinds of hazard. Also, by speaking and writing he can pass on information to his offspring so that they can benefit from years of accumulated learning. It may have taken our ancestors thousands of years to change from hunting to agriculture but now the information is available and can be used at once by anybody. We learn by watching others and by reading and listening. This is all part of human culture which is not shared by any other mammal.

Adaptability. Man is not a specialized mammal. His fore-arms are not flippers for swimming, wings for flying or hooves for running but by using his intelligence and memory he can use his simple, five-fingered hand to build machines and buildings, to reach and survive in almost any part of the world. By wearing clothes, building houses, storing food and passing on information humans can adapt themselves to be independent of their surroundings and to carry on the essential living processes in extremes of temperature and in inhospitable conditions that no other single species of mammal could endure (Fig. 1.6).

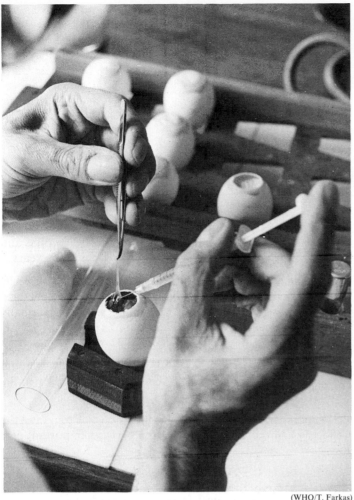

(WHO/T. Farkas)

Fig. 1.5 Delicate manipulation by human hands. Notice the opposable thumb and fore-finger holding the forceps. The egg is being used to culture a virus for the production of antibodies

Speech. Other mammals communicate with each other by sounds but, as far as we can tell, the amount of information they exchange is fairly limited. There may be warning sounds,

Questions

1   If acid is dropped on a piece of chalk or marble, it gives off carbon dioxide. Does this mean that the chalk is living? What other evidence would you need to convince you?
2   Make a list of the processes in your own body that need energy to make them work.
3   If a crystal of copper sulphate is placed in a strong solution of copper sulphate it will get larger. Why is this not 'growth' in the same sense that plants and animals grow?
4   Choose any animal you know well and describe briefly the ways in which it exhibits the seven characteristics of living organisms listed on p. 2.
5   What is the main difference between green plants and animals? Suggest some other differences.
6   Make a list of all the mammals you can think of. Choose one of these and say how it differs from man.
7   Man is described as an unspecialized or adaptable mammal. How does this apply to his choice of food?

(Canada House/D. Wilkinson)

(WHO/J. Mohr)

Fig. 1.6   Man can survive in a wide variety of climatic conditions

(a) An expedition in Northern Canada. The temperature may be as low as $-30\,°C$. Heavy outer clothing is needed

(b) A herdsman in Nigeria. The temperature may be $+35\,°C$. Little clothing is needed but the hat provides some shade

2
Cells

If almost any structure from a plant or an animal is examined microscopically it will be seen to consist of more or less distinct units—cells—which, although too small to be seen individually, in large numbers make up the structure or organ (Fig. 2.1*a* and *b*).

Methods of studying cells

Cells are too small to be seen with the naked eye so they must be magnified, at least × 100 and usually much more, in order to get any idea about their structure. To make a microscopic examination it is necessary to direct light through the tissue being studied and so the layer of tissue must be very thin. This may be achieved by squashing or smearing the tissue thinly on a glass slide, as in the case of cells from the lining of the mouth, or cells in the blood. In most cases, however, the tissue is cut into very thin slices, 10 μm* thick or less. The slices or *sections* are passed through one or more dyes (*stains*) which show up their outlines and their contents more clearly, then mounted on a glass slide and sealed under a thin glass cover slip (Fig. 2.2).

The kind of microscopes available in schools and colleges are

* A micrometre, symbol *μ*m, (often called a micron) is one-thousandth of a millimetre.

light microscopes. Tissues are studied by passing light through them or reflecting light off them. The light microscope will magnify up to × 1 500. Magnifications up to × 200 000 can be achieved by using the electron microscope, which passes a

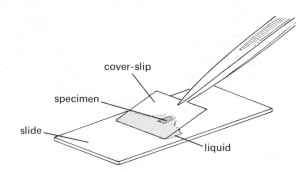

Fig. 2.2   Mounting a specimen for examination under the microscope

beam of electrons rather than rays of light through the object and takes a photograph of the image. Modern knowledge of cell structure is largely due to studies with the electron microscope.

Fig. 2.1   Plant and animal cells

(*a*) *Cells in plant stem. The photograph shows part of a very thin slice taken across the stem. The cell contents are not visible at this magnification* (× 70)

(G.B.I. Laboratories Ltd)

(*b*) *Cells from the human adrenal gland* (× 1 000)

(Brian Bracegirdle/Biophoto Associates)

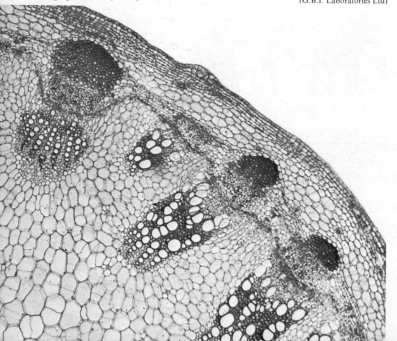

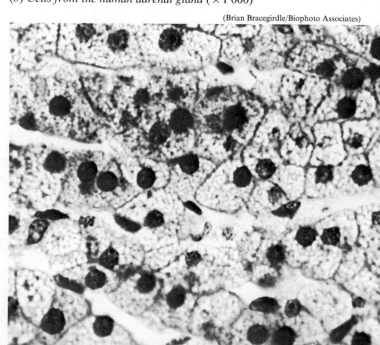

Cell structure

The size of human cells varies according to the type of cell being considered. One of the smallest is the red blood cell (p. 70) which is 7.5 μm in diameter and 2.2 μm thick. The largest human cell is probably the female egg, or ovum (p. 102), which is 100 μm across. The average diameter of human cells is therefore probably 10–30 μm.

The shape also depends on the type of cell. A cell of a gland such as the salivary gland may be more or less spherical, but a nerve cell in the leg having a diameter of only 10 μm can be over one metre long, because it runs from the foot to the spinal cord. Since the cells of any organ are usually specially developed in their size, shape and chemistry to carry out one particular function there is, strictly speaking, no such thing as a typical cell. Nevertheless, all animal cells have certain important features in common. They all consist of an outer membrane enclosing a mass of cytoplasm in which is contained a nucleus. It is these common features which are illustrated in Fig. 2.3a.

Cell membrane. This forms the outer boundary of the cell and keeps the cell contents intact, preventing them from mixing with the medium outside the cell or with the contents of neighbouring cells. Under the light microscope the cell membrane appears as little more than a line, but the electron microscope shows it to be a structure about 1/100 000 mm (10 nanometres) thick. One of its principal functions is to exercise control over which substances enter and leave the cell. The wrong type or quantity of a substance entering the cell could upset its delicately balanced chemistry.

Cytoplasm. Under the light microscope, this appears to be a uniform, semifluid, structureless substance containing a variety of particles and occupying most of the space inside the cell. The electron microscope, however, shows that it is by no means a structureless jelly but appears to consist of a variety of folded membranes forming tubes and passages, called the *endoplasmic reticulum*, which communicate with the external medium and the nucleus (Figs. 2.3b and 2.4). In addition to the internal system of membranes, there are other structures or *organelles* in the cytoplasm. Examples of organelles are the *mitochondria*, the *ribosomes* and the *Golgi apparatus*, all involved in essential chemical processes for the maintenance of life.

The ribosomes play a part in building up proteins, complex chemicals from which all cells and tissues are constructed (p. 166). In the mitochondria, food substances such as sugar are broken down chemically to release the energy that is needed to drive the reactions in the cell (p. 27).

The Golgi apparatus consists of a variable group of flattened sacs and vacuoles probably derived from the nuclear membrane or endoplasmic reticulum. It is thought to play a part in assembling the materials for making the cell membrane and to prepare proteins, e.g. enzymes, that are to be released from the cell. These functions probably vary according to the cell type.

In addition to these and other organelles in the cytoplasm there may be *inclusions* of non-living material. Depending on the cell there might be food reserves such as oil droplets or glycogen granules (p. 16). Sometimes droplets of fluids collect in the cytoplasm and form vacuoles.

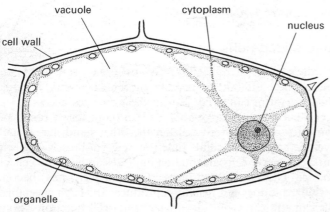

(a) Generalized animal cells as seen with the light microscope

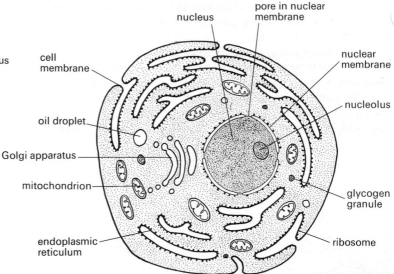

(b) Generalized animal cell as seen with the electron microscope. (This is a simplified, diagrammatic interpretation of the appearance of a cell section as it appears in the electron microscope. The cell has been treated with various chemicals and dehydrated before sectioning but it is assumed that the features shown do represent structures in the living cell)

(c) Typical plant cell. There is a cellulose wall outside the cell membrane, the cytoplasm is confined to a thin layer lining this wall and there is a large, fluid-filled vacuole

Fig. 2.3   Generalized cells

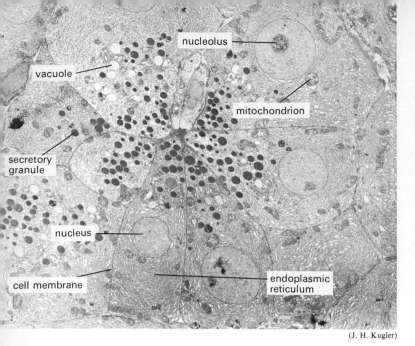

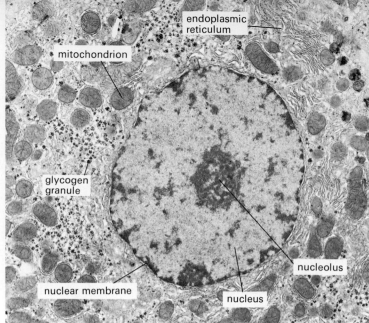

(a) Cells from the pancreas (× 22 000)

(J. H. Kugler)

(b) Part of a liver cell (× 50 000)

(J. H. Kugler)

Fig. 2.4 Human cells as seen under the electron microscope

The nucleus is a large spherical or ovoid body enclosed in the cytoplasm but separated from it by a membrane similar to the cell membrane, Fig. 2.4(b). The contents are more acid than the cytoplasm and so react differently to the stains used in making microscopical preparations. The stains are selected because they are taken up more strongly by the nucleus. Consequently in most microscopical preparations (and hence in most drawings and photographs of cells) the nucleus is seen as a dark object in the cytoplasm.

In the nucleus there are a number of fine thread-like bodies called *chromosomes* (p. 162). These cannot normally be seen with either the light microscope or the electron microscope unless the nucleus is dividing, but there is plenty of evidence that they are there all the time and produce substances that pass into the cytoplasm and control the chemical reactions going on there.

It is the nucleus that ultimately determines the shape and function of the cell. A cell may live for a time without its nucleus but it cannot divide and produce new cells.

Protoplasm. The cell membrane, cytoplasm and nucleus are sometimes collectively described as protoplasm.

Cell division

Animals and plants grow as a result of cell division and cell enlargement. Most animals begin their existence as a single cell, i.e. a fertilized egg. This cell divides into 2, 4, 8, 16, 32 and so on to produce a body consisting of millions of cells, specialized for particular functions. These become grouped into tissues, organs and systems. Cell division begins by the nucleus dividing into two, followed by the cytoplasm dividing, so forming two smaller cells which will then grow to the size of the parent (Fig. 2.5). In the early stages of development all the cells are able to reproduce, but as they become specialized to form bone, muscle, blood and so on, they lose the ability to divide. This power is retained by certain cells only; for example, there are cells in the bone marrow that constantly produce new blood cells, cells in the skin that continuously replace the outer layers as they wear away, and cells in the reproductive organs that produce sperms or eggs.

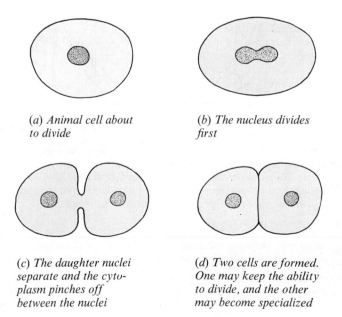

(a) Animal cell about to divide

(b) The nucleus divides first

(c) The daughter nuclei separate and the cyto-plasm pinches off between the nuclei

(d) Two cells are formed. One may keep the ability to divide, and the other may become specialized

Fig. 2.5 Cell division (animal cell)

Specialized cells

As already implied, cells have different functions and consequently different shapes and internal chemistry. A single-celled organism can carry out all the processes necessary for its existence; it can move about, obtain and digest food, and reproduce. A specialized cell, on the other hand, has usually developed one particular function: e.g. a muscle cell can contract, a nerve cell can conduct impulses and a gland cell can produce chemical substances. Once a cell has become specialized it does not usually reproduce, although specialized cells can still carry out all other normal cell functions. Some examples of specialized cells are shown in Fig. 2.6.

8

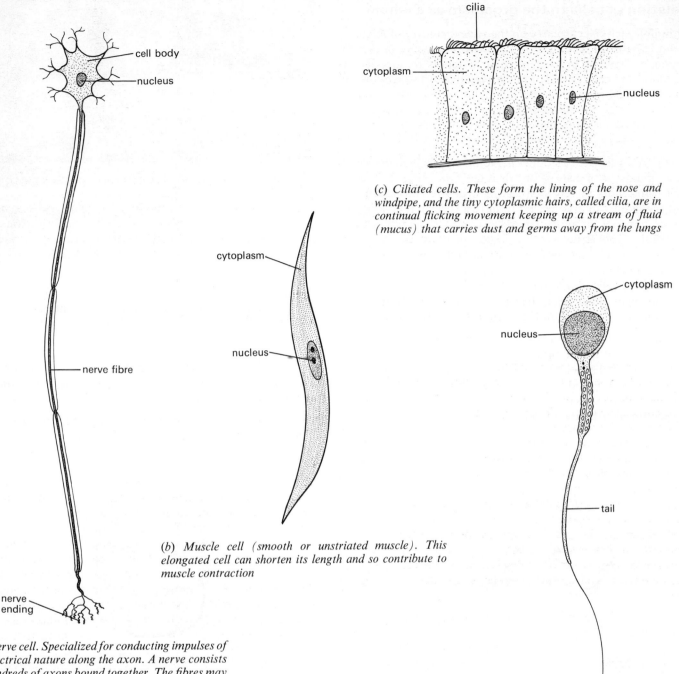

cilia

cytoplasm

nucleus

(c) Ciliated cells. These form the lining of the nose and windpipe, and the tiny cytoplasmic hairs, called cilia, are in continual flicking movement keeping up a stream of fluid (mucus) that carries dust and germs away from the lungs

cell body

nucleus

cytoplasm

nerve fibre

nucleus

cytoplasm

nucleus

tail

(b) Muscle cell (smooth or unstriated muscle). This elongated cell can shorten its length and so contribute to muscle contraction

nerve ending

(a) Nerve cell. Specialized for conducting impulses of an electrical nature along the axon. A nerve consists of hundreds of axons bound together. The fibres may be very long, e.g. from the foot to the spinal column

(e) Sperm cell. The movements of the tail help the sperm to reach the female's egg and fertilize it. (The shape of the sperm's head depends on whether it is seen from 'above' or from the 'side'. See Fig. 15.10, p. 103)

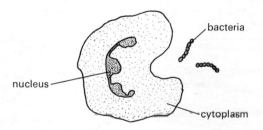

bacteria

nucleus

cytoplasm

(d) White blood cell. Occurs in the blood stream and is specialized for engulfing harmful bacteria. It is able to change its shape and move about, even through the walls of blood vessels into the surrounding tissues

Fig. 2.6   Specialized cells

Relation of cells to the organism as a whole

Although each cell can carry on the vital chemistry of living, it is not capable of existence on its own. A muscle cell cannot obtain its own food or oxygen. These materials are supplied by the blood and transported or made available by the activities of other specialized cells. Unless individual cells are grouped together in large numbers and made to work together by the co-ordinating mechanisms of the body, they cannot exist for long.

Tissue. A tissue such as bone, nerve or muscle is made up of many hundreds of cells of one or a few types (Fig. 2.7), each type being more or less identical in structure and activity so that the tissue can also be said to have a specific function, e.g. nerves conduct impulses, muscles contract, glands secrete chemicals. The structures of tissues such as blood, bone, cartilage, muscle and nerve are described more fully in the chapters dealing with human physiology.

The tissue called *epithelium* forms the lining of many organs. Various types of epithelia, their position in the body and their functions, are described in Fig. 2.10, p.12.

Organs consist of several tissues grouped together making a functional unit; for example, a muscle is an organ containing long muscle cells held together with connective tissue and permeated with blood vessels and nerve fibres. The arrival of a nerve impulse causes the muscle to contract, using the food and oxygen brought by the blood vessels to provide the necessary energy.

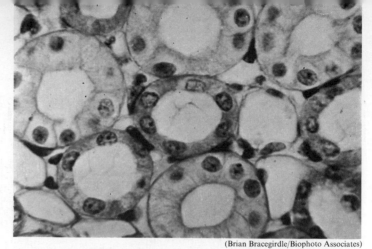

(Brian Bracegirdle/Biophoto Associates)

Fig. 2.8   Transverse section through kidney tubules (× 500)

System. A system is usually a series of organs whose functions are co-ordinated to produce effective action in the organism; for example, the heart and blood vessels constitute the circulatory system; the brain, spinal cord and nerves make up the nervous system (Fig. 2.9).

Organism. A multicellular organism is formed from a number of organs and systems whose working is efficiently co-ordinated. It is able to reproduce its own kind.

Plant cells

Although a plant cell has a nucleus, cytoplasm, organelles and cell membranes, it differs from an animal cell in having a cellulose wall outside its cell membrane and usually a large central vacuole in its cytoplasm (Fig. 2.3c).

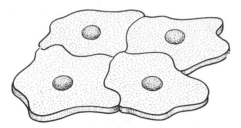

(a) *Cells forming an epithelium, a thin layer of tissue, e.g. that lining the mouth cavity*

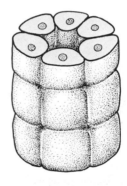

(b) *Cells forming a fine tube, e.g. a kidney tubule (see p. 93 and Fig. 2.8 above)*

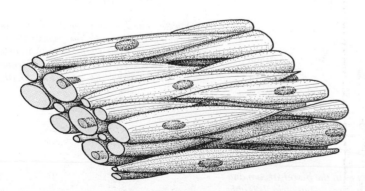

(c) *Unstriated muscle cells forming a sheet of muscle tissue. Blood vessels, nerve fibre and connective tissues will also be present*

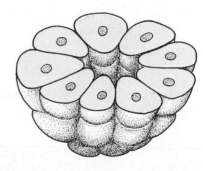

(d) *Cells forming part of a gland. The cells make chemicals which are released into the central space and are carried away by a tubule such as shown in (b)*

Fig. 2.7   How cells form tissues

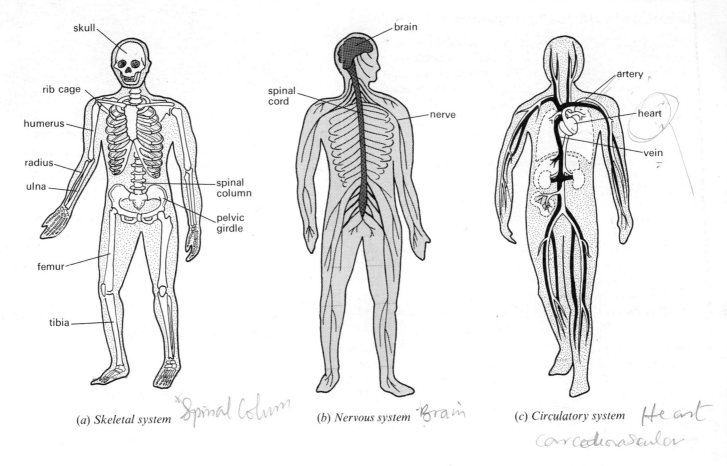

(a) Skeletal system *Spinal Column*

(b) Nervous system *Brain*

(c) Circulatory system *Heart*
Cardiovascular

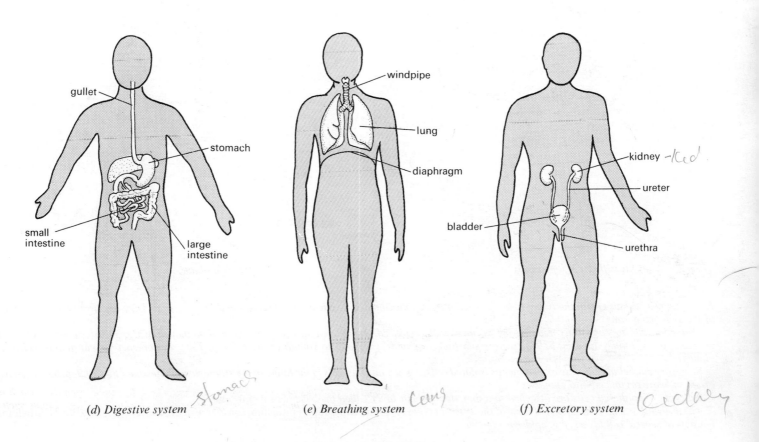

(d) Digestive system *Stomach*

(e) Breathing system *Lungs*

(f) Excretory system *Kidney*

Fig. 2.9   Some of the systems of the body

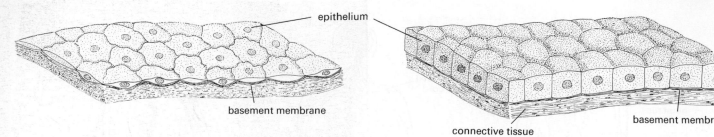

epithelium

basement membrane

(a) *Squamous*, e.g. lining of blood vessels

connective tissue

basement membrane

(b) *Cuboidal*, e.g. outer lining of ovary

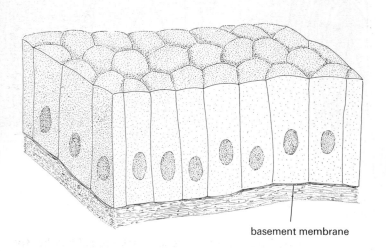

basement membrane

(c) *Columnar*, e.g. intestinal lining

cilia

mucous cell

(d) *Ciliated*, e.g. lining of brochioles

cells being shed

(e) *Stratified*, e.g. epidermis

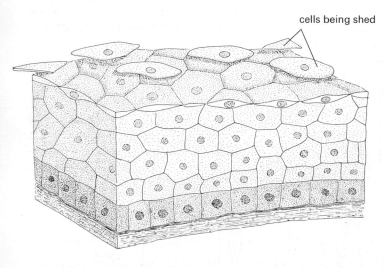

duct

connective tissue

secretory cell

(f) *Glandular*

Epithelial cells usually form the lining to body structures, e.g. they line the inside of blood vessels and the intestine. They also form the outer layer of the skin (epidermis).

Epithelial cells are constantly dying, being worn away and replaced. Apart from their protective function, they may also absorb substances (small intestine, p. 64, kidney tubules, p. 93), secrete substances (glands), respond to stimuli (taste buds, p. 132), or contract (sweat glands, salivary glands).

Secretory epithelial cells may exist either as isolated cells, e.g. the mucous cells in the lining of the stomach and intestine (Fig. 10.9, p. 65) or, more often, they are organized into glands.

All types of epithelial cells are attached to a thin, non-cellular layer called the basement membrane.

The drawings in Fig. 2.10 represent some of the different types of epithelium. Simple epithelium *consists of one cell layer only;* Stratified epithelium *consists of several cell layers, e.g. epidermis (p. 96)*

Fig. 2.10   Epithelia

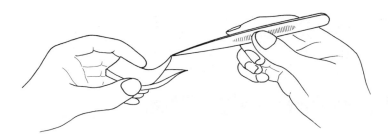

(a) Peel the epidermis from the inside of an onion scale

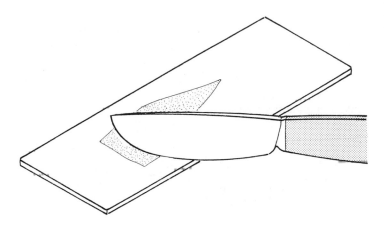

(b) Cut a small piece of epidermis

Fig. 2.11   Preparing plant cells for microscopic study

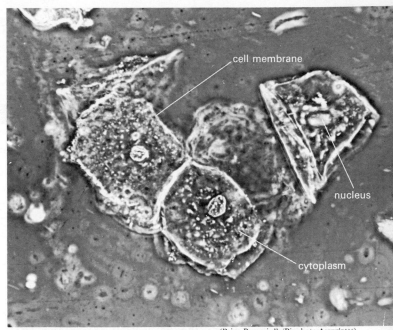

cell membrane

nucleus

cytoplasm

(Brian Bracegirdle/Biophoto Associates)

Fig. 2.12   Cells from the cheek lining ($\times 1\ 000$)

Practical Work

Experiment 1   **Plant cells**

Cut a piece of onion as shown in Fig. 2.11a. With forceps or your fingers you will find that you can peel off the epidermis from the inside of the scale. The epidermis is made up of a single layer of cells. Place this tissue on a microscope slide and cut a small piece from it with a scalpel or razor blade (Fig. 2.11b). Now cover the square of epidermis with a drop of iodine solution and carefully lower a cover slip over it as shown in Fig. 2.2. Examine the tissue under the microscope. Because the tissue is so thin, light passes through it and you can see the individual cells. The iodine will stain the cell nuclei pale yellow. If there are any starch grains present they will turn blue.

Experiment 2   **Human cells**

Note: The Department of Education and Science recommends that schools no longer use the technique which involves studying the epithelial cells that appear in a smear taken from the inside of the cheek because of the very small risk of transmitting the AIDS virus. The Institute of Biology suggests that if the following procedure is adopted the risk is negligible (*Biologist* **35** (4) p. 211, September 1988).

Cotton buds from a freshly opened pack are rubbed lightly on the inside of the cheek and gums. The buds are rubbed on to clean slides and then dropped into a container of absolute alcohol. The smear on the slide is covered with a few drops of methylene blue solution before being examined under the microscope (Fig. 2.12). The slides are placed in laboratory disinfectant before washing.

Various alternatives to cheek epithelial smears have been put forward. For example:

(a) Some 'Sellotape' is pressed on to a 'well-washed' wrist. When the tape is removed and studied under the microscope, cells with nuclei can be seen. A few drops of methylene blue solution will stain the cells and make the nuclei more distinct.

(b) Microscope slides are pressed against the corneas of fresh or refrigerated bullock's eyes. Corneal cells stick to the slides and can be stained with methylene blue.

Questions

1   What features are (a) possessed by both plant and animal cells, (b) possessed by plant cells only?

2   With what materials must cells be supplied if they are to live and grow?

3   In what ways would you say that the white blood cell (Fig. 2.6d) is less specialized than the nerve cell (Fig. 2.6a)?

4   In many microscopical preparations of animal tissues, it is difficult to make out the cell boundaries and yet the disposition and numbers of cells can usually be determined. Which cell structure makes this possible?

5   At one time it was thought that cytoplasm was a kind of structureless jelly. At the same time it was difficult to understand how a complicated series of ordered chemical reactions could take place in such an amorphous fluid. How do you think that electron microscope studies of the cell have helped to resolve this problem?

6   Most mature cells are incapable of dividing to produce new cells. In which parts of the human body in (a) a baby, (b) an adult, would you expect cells to retain their power of division?

7   The systems illustrated in Fig. 2.9 seem to be independent of each other but, in fact, they are closely interconnected. Say how the functions of any one system might influence or be influenced by any of the other systems.

3

The Chemicals of Living Cells

The preceding chapter described cells as the units which, in their thousands, go to make up the bodies of plants and animals. By suitably magnifying the cells they can be seen to consist of nucleus and cytoplasm, containing smaller units such as ribosomes and mitochondria. These subcellular structures are themselves built up from particles, which for the most part are too small to be seen even using the electron microscope. These particles are the molecules of the various chemical substances that contribute to the structures listed above. Chemical substances can conveniently be considered under the headings of *elements* and *compounds*.

Elements

An element is a substance that cannot be split up into other substances. Copper, iron, sulphur and carbon are examples of solid elements; oxygen and nitrogen are gaseous elements.

The smallest particle of an element is an *atom*, and so elements can be visualized as consisting of countless millions of atoms, all of the same kind, and with a good deal of space between them. When describing and explaining chemical reactions, the atom of an element is represented by a letter, often the initial letter of the element, e.g. C represents an atom of carbon, O represents an atom of oxygen, S sulphur, H hydrogen and N nitrogen.

Compounds

When two or more elements combine chemically they make a compound. If the elements carbon and oxygen combine they make the compound carbon dioxide. Since each element consists of atoms, it is assumed that the combination takes place between the atoms of the different elements (Fig. 3.1).

In many cases, one atom of an element will combine with more than one atom of another element. The atoms may be linked in small, discrete groups called *molecules* or continuously

throughout the material in a three-dimensional array described as a *giant structure*. A molecule of carbon dioxide thus consists of one carbon atom joined to two oxygen atoms, and for simplicity can be visualized as in Fig. 3.2. In silicon dioxide, there are two oxygen atoms to each silicon atom but they are linked in a giant structure (Fig. 3.4).

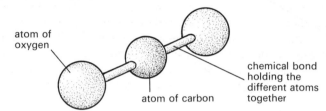

Fig. 3.2   Model of carbon dioxide molecule

In describing compounds and their reactions, the molecules are represented by the letters of their constituent atoms. Carbon dioxide is depicted as CO_2, the figure 2 after the O signifying that there are two atoms of oxygen in the molecule; CO_2 is called the *formula* of carbon dioxide. The formula for silicon dioxide, SiO_2, represents the simplest ratio of atoms present in the giant structure. It so happens that the smallest particle of many gases, including oxygen and nitrogen, consists of two atoms joined together, i.e. it is a molecule, so that the formula of oxygen is O_2 and for nitrogen is N_2 (i.e. N—N).

Gases, liquids and solids

Gases. The molecules of a gas are spaced very far apart and move about rapidly in all directions. If the gas is heated the molecules move faster and further apart, and so the gas expands (Fig. 3.3; Experiment 1). If the gas is cooled the molecules slow down and the gas contracts. Because the molecules are so far apart, it is not difficult to push them closer

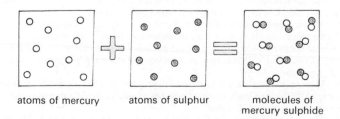

atoms of mercury   atoms of sulphur   molecules of mercury sulphide

Fig. 3.1   Combination of atoms to make molecules

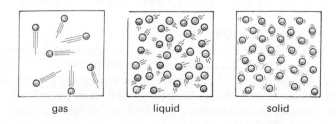

gas   liquid   solid

Fig. 3.3   Movement and spacing of molecules in solid, liquid and gas

14

together by applying pressure to the gas. Thus a gas is easily compressed (Experiment 2). Also, being so far apart, free to move and not greatly attracted to each other, the molecules of a small quantity of gas released into a large space will spread out until they are uniformly distributed through that space (*see* diffusion, p. 22).

Liquids. In liquids the molecules are closer together than they are in a gas but they are still free to move at random. Hence a liquid can flow, though it cannot be compressed.

Solids. The molecules of a solid are also close together (although there is still a great deal of space between them) but they are not free to move, apart from vibrating. Usually they are held in a three-dimensional pattern which gives the substance a crystalline structure (Fig. 3.4, Experiment 3).

Organic and inorganic chemicals

The distinction between these two classes of chemicals was made originally because it was thought that the organic chemicals were produced only by living organisms and could not be made artificially in a laboratory. Although this is not true, there are still significant differences between the two types of chemicals. Inorganic chemicals are substances such as the compounds of metals, for example copper sulphate or sodium chloride. Their molecules are usually small, consisting of up to ten atoms; or, in giant structures, the ratio of the different atoms is small.

Organic chemicals

Organic molecules are usually large, sometimes consisting of hundreds or thousands of atoms. Examples are sugar, starch, oil, fat and protein. The principal chemical feature that identifies organic compounds is that they consist for the most part of carbon atoms joined to each other in chains or rings.

The lines in the formulae represent the chemical bonds holding the atoms together. The 'spare' bonds sticking out from the two formulae above would normally be holding hydrogen, oxygen, nitrogen, phosphorus or other carbon atoms.

Note that each carbon atom has four chemical bonds for holding other atoms, including carbon atoms, in place. Each bond must be used in some way for combining with other

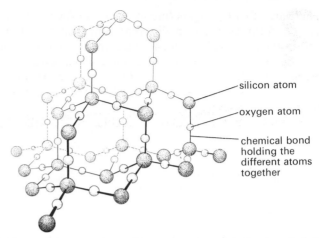

This shows the arrangement of atoms in a tiny part of a crystal of silica (silicon oxide). The atoms are geometrically spaced as shown and thus, in their millions, produce crystals

Fig. 3.4   Atoms in a crystal lattice

atoms. One, two or three bonds, depending on the compound, may be used for holding other carbon atoms. The chain compound, propane, has only single bonds between carbon atoms in its molecule. The benzene molecule, however, has three *double bonds*, shared, in effect, between the six carbon atoms.

The organic molecules which constitute the greater part of living cells are classified as carbohydrates, fats and proteins.

Carbohydrates

Carbohydrate molecules consist of chains of usually six or more carbon atoms combined with hydrogen and oxygen atoms only. The principal groups of carbohydrates are sugar, starch, glycogen and cellulose. Apart from cellulose in plant cell walls, the carbohydrates do not form permanent structures in cells but are used in chemical reactions to provide energy for driving other reactions.

Sugar. One of the simplest and most important sugars is *glucose*. Its formula is $C_6H_{12}O_6$; i.e. a molecule of glucose contains six carbon atoms, twelve hydrogen atoms and six oxygen atoms. It is sometimes represented structurally as

or more simply as

showing only the reactive parts of the molecule.

There are other C_6 sugars (monosaccharides) with the same $C_6H_{12}O_6$ formula but with different arrangements of these atoms in their molecules, which makes them react differently from glucose. Two examples are *fructose* and *galactose*.

Maltose has the formula $C_{12}H_{22}O_{11}$ and is made up from two glucose molecules. Some cells can build up a maltose molecule by combining two glucose molecules. The reaction

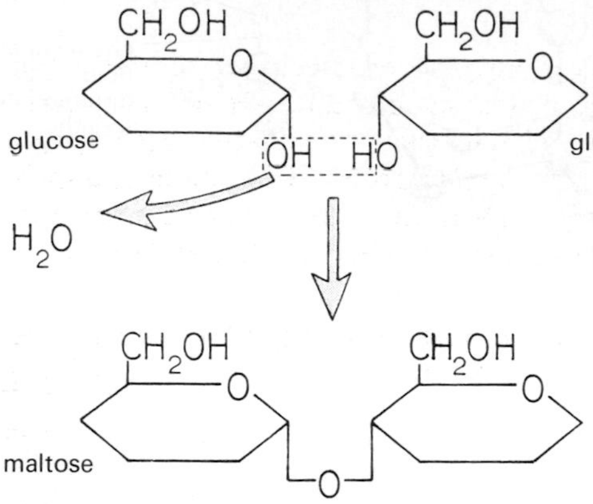

proceeds by removing the elements of water (H_2O) from the two glucose molecules (an —OH from one and an —H from the other). The reaction will not take place spontaneously; that is, when two glucose molecules meet they will not combine to form maltose of their own accord. Another type of chemical in the cell, called an *enzyme* (p. 26), is responsible for making this reaction occur.

There are other C_{12} sugars (disaccharides) of which the most important is *sucrose* (cane sugar), consisting of a glucose and a fructose molecule combined.

Starch. The starch molecule is very large, being made up of 300 or more glucose molecules linked together in straight and branching chains (Experiment 6). Starch is a common storage material in plant cells.

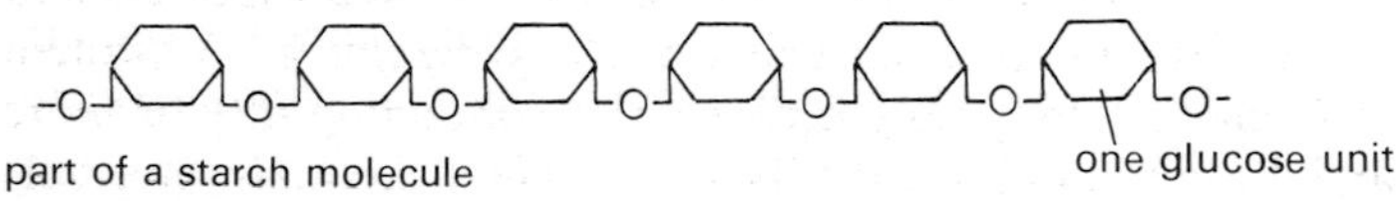

part of a starch molecule

one glucose unit

Glycogen. Animal cells cannot make or store starch. Their storage carbohydrate is usually glycogen which, like starch, has a large molecule built up from thousands of glucose molecules joined together in a branching chain. Glycogen granules occur as inclusions in animal cells.

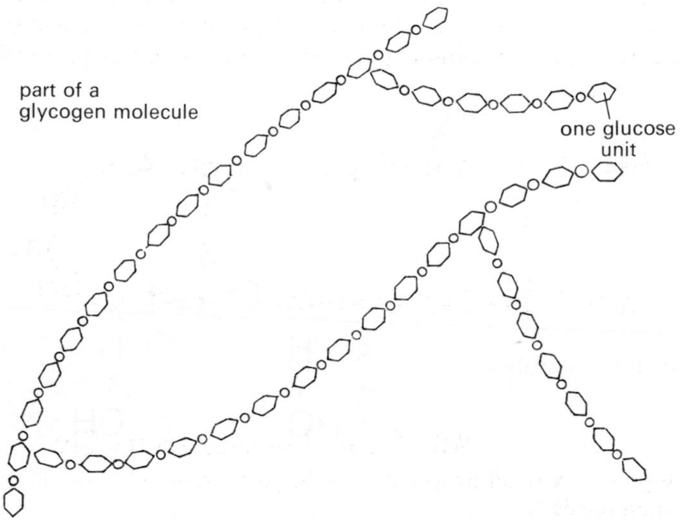

part of a
glycogen molecule

one glucose
unit

Cellulose is chemically similar to starch in that it consists of one thousand or more glucose units, but the units are joined together in a slightly different way so that the chemical and physical properties of cellulose are different from starch. Only plants can make cellulose and they incorporate it in their cell walls (p. 7), the long molecules being packed together in bundles to make tough micro-fibrils which give the cell wall its strength. Textile fibres such as cotton and flax, made from plant materials, consist largely of cellulose.

Starch, glycogen and cellulose are sometimes called *polysaccharides*.

Hydrolysis of carbohydrates. The disaccharides and polysaccharides are built up from glucose and fructose molecules joined together by eliminating the elements of water (—OH and —H). They can be broken down again by reacting them with water (Experiment 5). This process, called hydrolysis

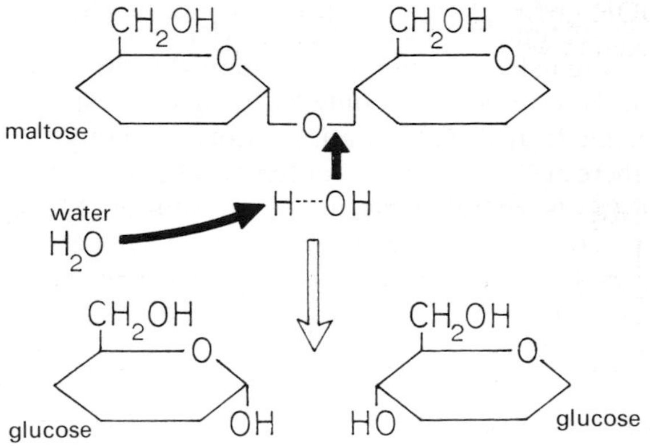

(Greek *hydro* = water, *lysis* = breakdown), does not take place spontaneously in cells but requires the intervention of a particular enzyme. In human saliva there is an enzyme which brings about the hydrolysis of starch to maltose.

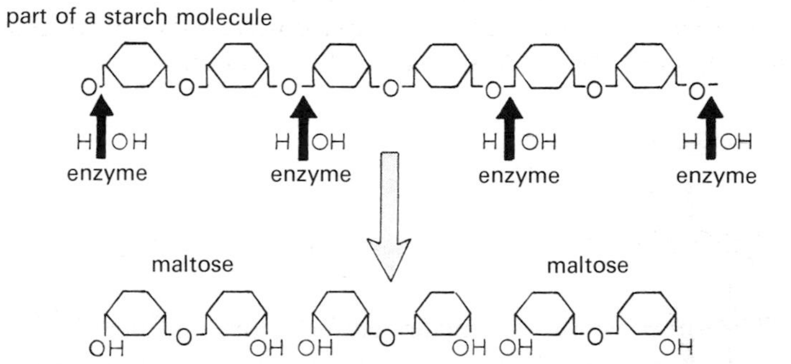

part of a starch molecule

Fats

Like carbohydrates, fats contain atoms of only carbon, hydrogen and oxygen, but there are four distinct parts to a fat molecule. One of these is *glycerol* and the other three are organic acids called *fatty acids*.

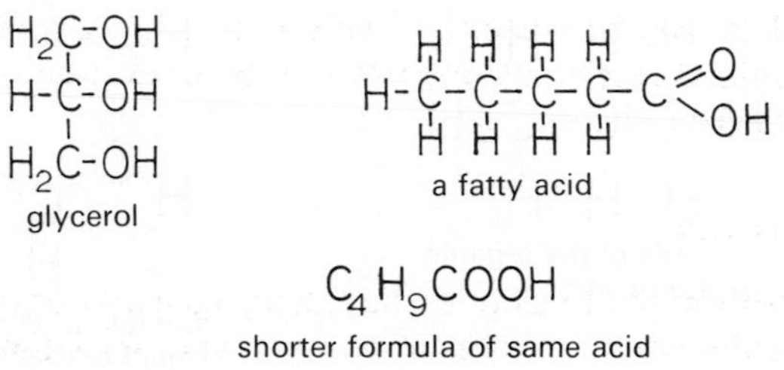

glycerol

a fatty acid

$$C_4H_9COOH$$
shorter formula of same acid

The short formula is written in this way and not as $C_5H_{10}O_2$ because there could be other compounds with this formula but with quite different properties, e.g.

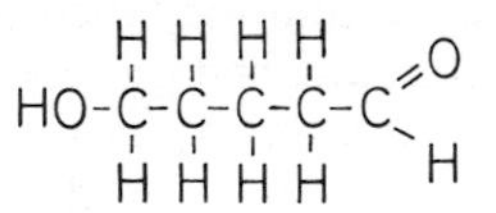

$$HO-\overset{\overset{\displaystyle H}{|}}{\underset{\underset{\displaystyle H}{|}}{C}}-\overset{\overset{\displaystyle H}{|}}{\underset{\underset{\displaystyle H}{|}}{C}}-\overset{\overset{\displaystyle H}{|}}{\underset{\underset{\displaystyle H}{|}}{C}}-\overset{\overset{\displaystyle H}{|}}{\underset{\underset{\displaystyle H}{|}}{C}}-C\overset{\displaystyle O}{\underset{\displaystyle H}{\diagdown}}$$

It is the —COOH group at the end of a fatty acid molecule that gives it its characteristic properties.

The reactive parts of the glycerol and the fatty acid are the —OH groups at the ends of the molecules. By removing the elements of water, an —H from the glycerol and an —OH from the fatty acid, the glycerol molecule can be made to combine with three molecules of a fatty acid such as stearic acid to make a *triglyceride*, one form of fat. In the formula below, the —COOH group of the fatty acid is written in reverse to show the reaction with glycerol more clearly.

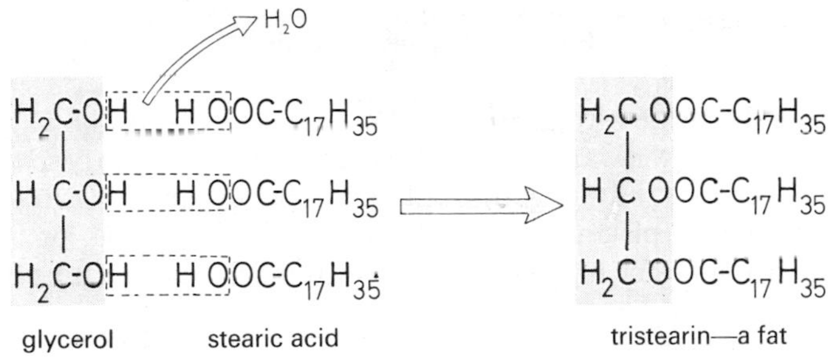

$$H_2C\text{-}OH \quad H \; OOC\text{-}C_{17}H_{35}$$
$$H \; C\text{-}OH \quad H \; OOC\text{-}C_{17}H_{35} \longrightarrow$$
$$H_2C\text{-}OH \quad H \; OOC\text{-}C_{17}H_{35}$$

glycerol   stearic acid

$$H_2C \; OOC\text{-}C_{17}H_{35}$$
$$H \; C \; OOC\text{-}C_{17}H_{35}$$
$$H_2C \; OOC\text{-}C_{17}H_{35}$$

tristearin—a fat

As in all the other examples, there is a specific enzyme in the cell that brings about this reaction. Most natural fats have more than one kind of fatty acid combined in the molecule, and whether they are hard fats or liquid oils will depend on which particular fatty acids are involved.

Fats can be hydrolysed by the appropriate enzymes to split them up into fatty acids and glycerol once again.

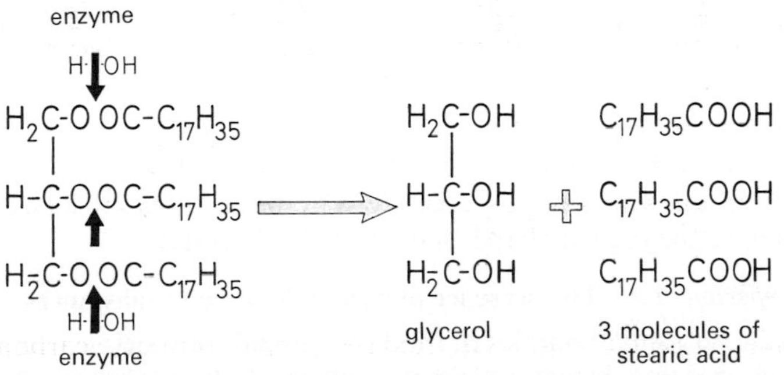

enzyme
H   OH

$$H_2C\text{-}O\;OC\text{-}C_{17}H_{35}$$
$$H\text{-}C\text{-}O\;OC\text{-}C_{17}H_{35} \longrightarrow$$
$$H_2C\text{-}O\;OC\text{-}C_{17}H_{35}$$

H   OH
enzyme

$$H_2C\text{-}OH \qquad C_{17}H_{35}COOH$$
$$H\text{-}C\text{-}OH \quad + \quad C_{17}H_{35}COOH$$
$$H_2C\text{-}OH \qquad C_{17}H_{35}COOH$$

glycerol   3 molecules of stearic acid

Fats form part of the permanent structures of the cell, particularly the cell membrane and the internal membranes. Oil droplets may be present as inclusions in cells, and both plant and animal cells break down fats in chemical reactions to obtain energy from them.

Proteins

In addition to atoms of carbon, hydrogen and oxygen, protein molecules contain atoms of nitrogen and, sometimes, sulphur.

The units from which protein molecules are built are called *amino acids* and the simplest of these is *glycine*.

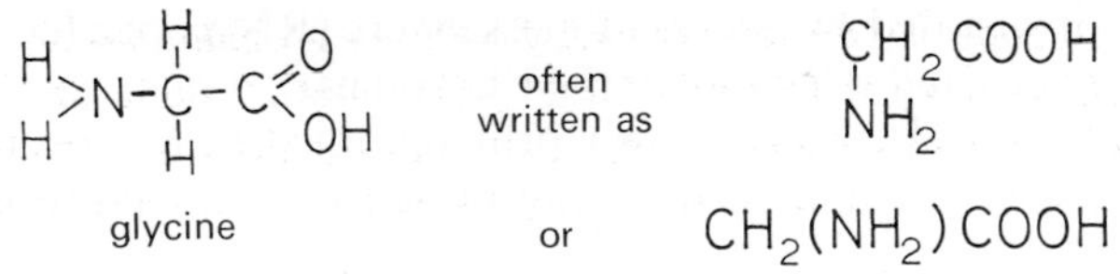

$$\overset{\displaystyle H}{\underset{\displaystyle H}{>}}N-\overset{\overset{\displaystyle H}{|}}{\underset{\underset{\displaystyle H}{|}}{C}}-C\overset{\displaystyle O}{\underset{\displaystyle OH}{\diagdown}}$$

often written as

$$\overset{\displaystyle CH_2COOH}{\underset{\displaystyle NH_2}{|}}$$

glycine   or $CH_2(NH_2)COOH$

As with fatty acids, the reactive —H of the —OH group confers acidic properties but the —H of the —NH₂ is also reactive. The —COOH group is called the *carboxyl* group and the —NH₂ is the *amino* group. The amino group is always attached to the carbon atom next to the carboxyl group, as shown in the formulae of alanine, valine and serine.

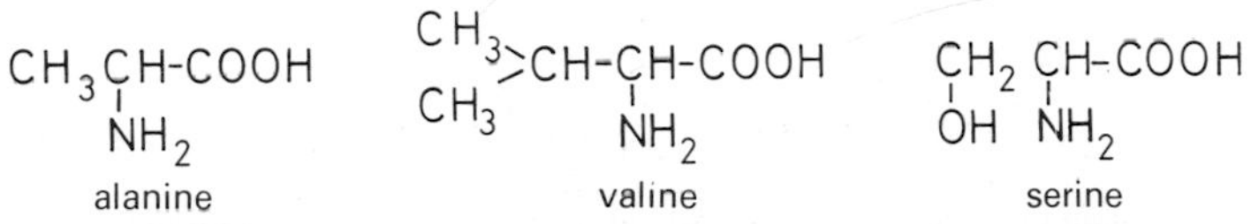

$$CH_3\overset{\displaystyle CH\text{-}COOH}{\underset{\displaystyle NH_2}{|}}$$

alanine

$$\overset{\displaystyle CH_3}{\underset{\displaystyle CH_3}{>}}\overset{\displaystyle CH\text{-}CH\text{-}COOH}{\underset{\displaystyle NH_2}{|}}$$

valine

$$\overset{\displaystyle CH_2\;CH\text{-}COOH}{\underset{\displaystyle OH\;\;NH_2}{|\quad\;\;|}}$$

serine

By removing the elements of water from two amino acid molecules, —OH from the carboxyl group and —H from the amino group, the amino acids combine to form a *dipeptide*.

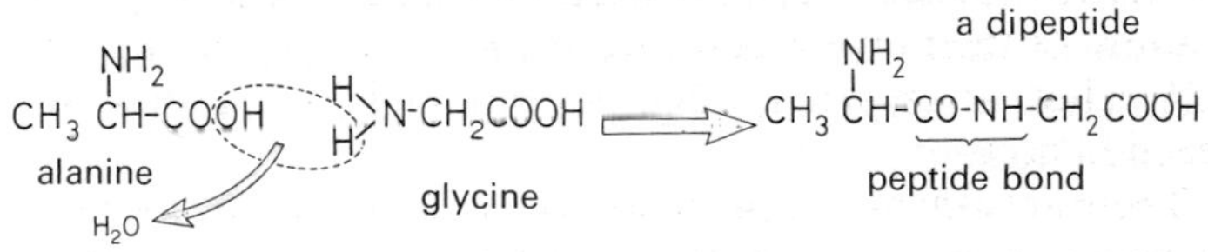

a dipeptide

$$CH_3\overset{\displaystyle CH\text{-}COOH}{\underset{\displaystyle NH_2}{|}} \quad \overset{\displaystyle H}{\underset{\displaystyle H}{>}}N\text{-}CH_2COOH \longrightarrow CH_3\overset{\displaystyle CH\text{-}CO\text{-}NH\text{-}CH_2COOH}{\underset{\displaystyle NH_2}{|}}$$

alanine   glycine   peptide bond

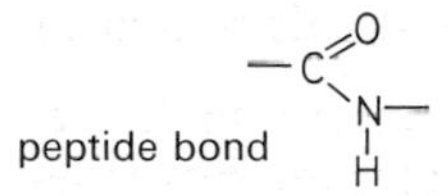

H_2O

$$-C\overset{\displaystyle O}{\underset{\displaystyle N-}{\diagdown}}$$
$$\underset{\displaystyle H}{|}$$

peptide bond

Three amino acids joined in this way will form a *tripeptide* and numerous amino acids will form a *polypeptide*. A protein molecule is a long polypeptide chain containing fifty or more amino acids, but instead of being a straight chain, the molecule is often branched and folded or coiled up because of cross-linkages which form between amino acids at different parts of the chain. Some amino acids such as *cysteine* contain sulphur atoms which are important in forming these cross links. There are twenty different commonly occurring amino acids, and the relative numbers of each amino acid, the order in which they are joined up and the folding and cross-linkages which subsequently form, determine the type of protein and its properties (Fig. 3.5).

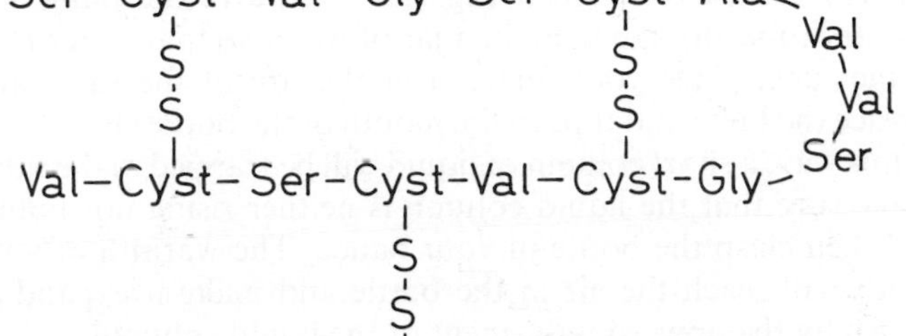

Ser—Cyst—Val—Gly—Ser—Cyst—Ala
        | | Val
        S   S |
        | | Val
        S   S |
        | | Ser
Val—Cyst—Ser—Cyst—Val—Cyst—Gly
                    |
                    S
                    |
                    S
                    |
Val—Cyst—Ala—Ala—Ser—Gly

Fig. 3.5 A small imaginary protein made from only five different amino acids

In cells, proteins are built up by the ribosomes. Amino acids in the cytoplasm surrounding the ribosomes are assembled in the 'correct' order to make a particular protein. The correct order is dictated by chemical messengers (RNA) produced in the nucleus and taken up by the ribosomes.

Cells can also break down proteins into their constituent amino acids by hydrolysis in the presence of the appropriate enzymes.

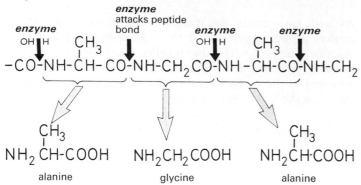

Although proteins and amino acids can be broken down to provide energy, their principal value in cells is as the material from which the cell is constructed. They contribute very largely to the cytoplasm and its membranes, the mitochondria, ribosomes, the nucleus and its chromosomes, and the enzymes themselves which control the direction and rate of all of the chemical reactions in the cell.

Proteins are adversely affected by quite small rises of temperature. Sugar, starch and fat can be subjected to temperatures of 100 °C without their decomposing, other than by a slow hydrolysis, but some proteins exposed to temperatures even above 50 °C for any length of time are irreversibly altered. The folds and cross-linkages of the molecule are disarranged and the protein is said to be *denatured*. In this case, the structural proteins of the cell membranes are damaged and the protein of enzymes rendered ineffective. As a result, the cell can no longer function properly and will die. This is the prime reason why continued exposure to temperatures above 50 °C ultimately proves lethal to most living organisms.

Practical Work

Experiment 1   Expansion of air and water

Fit a small bottle or 250 cm³ flask with a bung and glass delivery tube, as shown in Fig. 3.6a. Remove the bung and delivery tube, dip the bung in a jar of water so that water rises in the tube, place your finger over the top of the tube, and replace the bung and tube in the mouth of the bottle (Fig. 3.6b). In this way, a short column of liquid will be trapped in the tube. Make sure that the liquid column is neither rising nor falling and then clasp the bottle in your hands. The warmth of your hands will reach the air in the bottle and make it expand as shown by the upward movement of the liquid column.

Remove the bung and tube and fill the bottle with cold water. When the bung is replaced, the water will rise some way up the tube. Make a mark on the tube at the water level and again

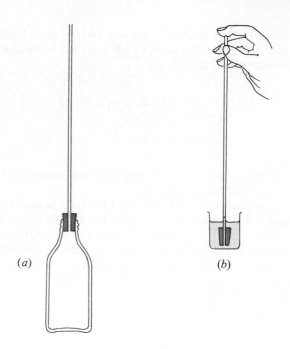

Fig. 3.6   Expansion of air

warm the bottle with your hands. As the heat expands the water in the bottle, the water will rise up the tube but not nearly so far as when the bottle contained air. The expansion can be increased by placing the bottle in water at 40–50 °C, or the expansion can be magnified by substituting a capillary tube for the delivery tube.

Experiment 2   Compressibility of gases and liquids

Select a large plastic syringe, 5 or 10 cm³, without a needle and in which the plunger moves freely in the barrel. Withdraw the plunger, place a finger firmly over the short tube at the end and push the plunger in. The gas can fairly easily be compressed in this way to half its original volume and will return to its first volume when the plunger is released.

Now draw up water with the syringe to fill it. Expel air bubbles by pointing the syringe upwards and depressing the plunger slightly. If the syringe opening is now blocked with a finger or thumb, it will be found impossible to compress the water.

Experiment 3   Crystals

Examine a few grains of salt and sugar with a × 10 hand lens or, better still, with a microscope. Many of the particles will be seen to have the regular shape characteristic of crystals.

Experiment 4   The presence of carbon in organic substances

Heat very small samples of food on a tin lid from below with a bunsen flame. Steam and smoke will be produced, but in each case there will be a black residue of carbon. This experiment produces strong smells and is best conducted in a fume cupboard, outside the laboratory or at the end of a school session when the laboratories can be cleared.

Experiment 5   The hydrolysis of starch by acid

Place 5 cm³ of 3 per cent starch solution in each of three test-tubes labelled 1 to 3. Add 1 cm³ dilute (2M) hydrochloric acid to each and place all three in a water bath of boiling water (Fig. 3.7). Remove tube 1 after 5 minutes in the water bath, tube 2 after 10 minutes and tube 3 after 15 minutes. Cool the

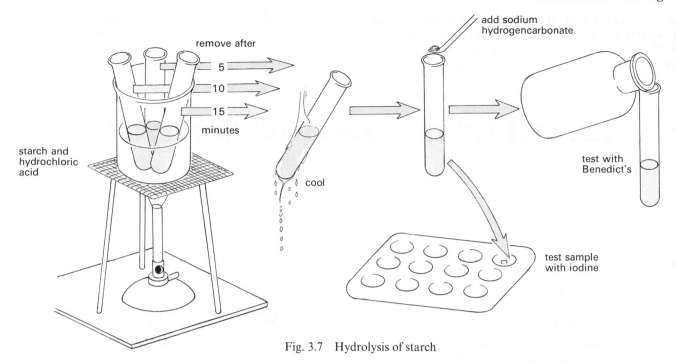

Fig. 3.7   Hydrolysis of starch

tubes under the cold tap or by dipping in a jar of cold water, and then add solid sodium hydrogencarbonate, a little at a time until all fizzing stops and the acid is neutralized. With a dropping pipette take a sample of the liquid from each tube and add it to a little iodine solution in a test-tube or on a tile. Look for the presence or absence of a blue colour. Now add 3 cm³ Benedict's solution to each tube and return them to the boiling water bath for 5 minutes. After this time examine the tubes for evidence of a sugar.

Result. A red precipitate after heating with Benedict's solution indicates that sugar is present (*see* p. 38). A blue colour with iodine shows that starch is present. There should be evidence of an increasing quantity of sugar in tubes 1 to 3 and a decrease in the amount of starch.

Interpretation. Heating with dilute hydrochloric acid converts starch to sugar by hydrolysis (*see* p. 16). The longer the heating is maintained the further the reaction proceeds.

(*Note.* Experiments involving hydrolysis by enzymes can be found on p. 30 and p. 69.)

Experiment 6   **Building up starch from glucose**

The glucose used in this experiment is glucose-l-phosphate, a reactive compound of glucose. The reaction is brought about by an enzyme (*see* p. 27) extracted from potatoes.

A cube of potato of side about 2 cm is crushed in a mortar with 10 cm³ distilled water and a little clean sand. The liquid is filtered into a clean test-tube through a filter paper and will contain, amongst other substances, the enzyme *starch phosphorylase*. Half the solution is poured into another test-tube and boiled over a low bunsen flame. A dropping pipette is used to place two rows of four single drops of a 5 per cent solution of glucose-l-phosphate on a cavity tile. One drop of unboiled potato extract is added to each drop of glucose-l-phosphate in the top row and one drop of boiled extract to each drop of glucose-l-phosphate in the bottom row. After 5 minutes one drop of iodine solution is placed on the first drop in each row. Five minutes later the second drop in each row is similarly tested, and so on at 5-minute intervals until all the drops have

been tested. A sample of the potato extract and the glucose-l-phosphate solution is tested separately to show that no starch is present to begin with.

Result. In the top row, the iodine test should produce first a mauve and then a blue colour which becomes more intense with successive samples. The bottom row should give no blue colour.

Interpretation. Since no starch was present to start with in either solution, it is reasonable to infer that starch molecules have been built up from glucose-l-phosphate units (*see* p. 16). This inference is strengthened by the fact that when the liquid thought to contain the enzyme is boiled, it fails to produce starch with glucose-l-phosphate. This supports the idea that it is an enzyme that causes the glucose units to combine.

The formulae for the chemicals and reagents mentioned in the practical work will be found on p. 319. For other experiments involving enzymes, giving fuller practical details, see Experimental Work in Biology No. 2, *Enzymes* (*see* p. 320).

Questions

1   Which of the following are elements and which are compounds: sugar, carbon dioxide, silver, water, iodine, calcium, benzene?

2   What are the differences between a molecule and an atom?

3   The molecule of sodium carbonate contains two sodium atoms, one carbon atom and three oxygen atoms. Write a formula for this compound (sodium = Na, carbon = C, oxygen = O).

4   Give three general differences between organic and inorganic compounds. Why is there no such thing as an organic element?

5   In what way do maltose, starch, glycogen and cellulose resemble each other?

6   The formula on p. 17 represents a molecule of one kind of fat. Write the formula for a different fat using palmitic acid, $C_{15}H_{31}COOH$, instead of stearic acid.

7   What is the final product of hydrolysis of glycogen, starch and cellulose?

8   A molecule of starch consists of hundreds of glucose units joined together. A molecule of protein consists of perhaps hundreds of amino acids joined together. In what way is the joining-up process similar, and in what way do the final products differ?

9   Write the formula for a dipeptide made from joining valine and serine.

4
Solutions, Diffusion and Osmosis

Before a substance can enter a cell it must dissolve. Moreover, all the chemical reactions in a cell take place in solution. For these and many other reasons it is desirable for a biologist to understand some of the chemical and physical properties of solutions.

When sugar is thoroughly mixed with water the solid crystals of sugar disappear and samples from all parts of the liquid will taste equally sweet. The sugar molecules are evenly dispersed throughout the liquid, and the sugar–water mixture is called a *solution*. The sugar is said to have *dissolved* in the water (Fig. 4.1).

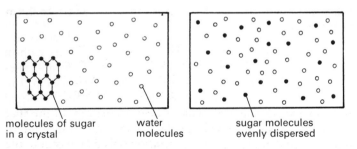

molecules of sugar in a crystal   water molecules   sugar molecules evenly dispersed

Fig. 4.1 Sugar dissolving

The liquid part of the mixture, in this case water, is called the *solvent* and the solid which dissolves is called the *solute*. Because the sugar will dissolve in water it is said to be *soluble*, while a substance like sand which does not dissolve is described as *insoluble* in water (Experiment 1a).

Although the terms 'soluble' and 'solubility' nearly always refer to water, it is important to realize that other liquids can act as solvents. Sugar is insoluble in petrol; fats and oils, however, dissolve readily in this solvent though they are insoluble in water (Experiment 1b).

Since living organisms consist of about 70 per cent water, it is the solubility of substances in water that is of prime importance. The cell membrane, however, contains fats, and consequently substances that can dissolve in fats and oils will sometimes penetrate the membrane more readily than water-soluble substances. Also there are a number of vitamins which are fat-soluble and therefore present only in natural fats and oils such as butter and fish-liver oils.

The process of digestion in animals is generally concerned with turning a variety of insoluble compounds into other compounds that are soluble in water and can therefore be carried in solution in the blood stream.

Solubility. There is a limit to the quantity of solute that will dissolve in a given volume of water. If too much solute is added, it will remain undissolved. A solution that can dissolve no more solute is called *saturated*. Some substances are more soluble than others; sodium nitrate, for example, is much more soluble than lead chloride. In 100 g water at 20 °C it is possible to dissolve 87 g sodium nitrate, but in the same conditions only 2 g lead chloride will dissolve before the solution is saturated. In general, the higher the temperature of the solution, the larger the amount of solid that will dissolve; i.e. solubility increases with temperature (Experiment 1c).

The quantity of solute dissolved in a solution is referred to as its *concentration*. A solution of 80 g sodium nitrate in 100 g water is more concentrated than one containing only 8 g sodium nitrate.

Gases in solution. It is not only solids that can dissolve in water; so can liquids and gases. The gases which concern the biologist are those present in the atmosphere—oxygen, carbon dioxide and nitrogen—and their solubilities are given in the table below.

	cm³ gas in 1 cm³ water	
	at 0 °C	*at 15 °C*
carbon dioxide	1.8	1.0
oxygen	0.05	0.035
nitrogen	0.023	0.017

The table shows that carbon dioxide is the most soluble and nitrogen the least soluble of the atmospheric gases, and also that the solubility of these gases *decreases* with rise in temperature. It is the carbon dioxide dissolved in water which enables aquatic plants to make their food by photosynthesis (p. 40) and the oxygen dissolved in water which is used by all aquatic organisms for their respiration (p. 27). In fact, the oxygen and carbon dioxide entering and leaving all cells in the body must be in solution, irrespective of whether the animal lives in water or on land.

If a gas is in contact with a liquid and the pressure of the gas is increased, more gas will dissolve. The carbon dioxide in fizzy drinks has been dissolved under high pressure so that when the stopper is removed the gas escapes from solution as bubbles. The air supplied to a diver has to be delivered at high pressure in order to force it down to him against water pressure. At high pressure, more of all the atmospheric gases will dissolve in the blood. The oxygen and carbon dioxide are taken out of solution by chemical reactions but not the nitrogen. Consequently, if the diver surfaces too rapidly, the nitrogen that has dissolved in his blood comes out of solution and the bubbles get trapped in small blood vessels in the limbs giving intense pain known as the 'bends' or decompression sickness. The symptoms can be avoided by bringing the diver to the surface slowly, or counteracted by placing him in a pressure tank when he surfaces and reducing the pressure slowly so that the surplus nitrogen can escape from his blood as it passes through the lungs rather than coming out of solution in the body tissues and blood vessels.

Ions in solution. In most cases where a compound has a giant structure the linking arises because the atoms develop electrical charges when they combine. In the salt sodium chloride, the sodium atoms become positively charged and the chlorine atoms negatively charged. Such charged atoms are called *ions*. Because positive and negative charges strongly attract one another, the ions are held by the attractive forces in a continuous regular array. When such compounds are dissolved in water the giant structure breaks down and the ions become free to move independently in solution. The compound is said to have *dissociated*. The properties of ions are very different from atoms. For example, sodium atoms react violently with water but sodium ions are quite stable in water. Chlorine atoms are combined in pairs and constitute a gas; chloride ions, on the other hand, exist singly and remain in solution.

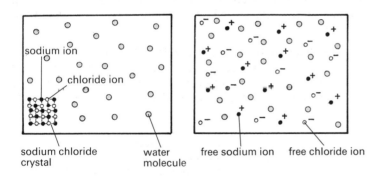

Fig. 4.2 Sodium chloride dissolving

Ions are not necessarily single atoms but may themselves be discrete groups of atoms that carry electrical charges. Potassium nitrate, for example, has the formula KNO_3 and is composed of potassium and nitrate ions.

$$KNO_3 \longrightarrow K^+ + NO_3^-$$
potassium nitrate   potassium ion   nitrate ion

Nitrate, NO_3^-, cannot exist as a compound on its own but it is perfectly stable as an ion in solution.

Most inorganic salts dissociate into free ions when dissolved in water. If both sodium chloride and potassium nitrate are dissolved in water, the solution will contain ions of sodium, potassium, chloride and nitrate (Fig. 4.3). In such a case, it would be impossible to say whether the salts originally dissolved had been sodium chloride and potassium nitrate or sodium nitrate and potassium chloride. Since the ions formed from a substance behave more or less independently in solution, the biologist is more concerned with, say, the reactions involving chloride ions and sodium ions in the blood than with the reactions of sodium chloride as a compound.

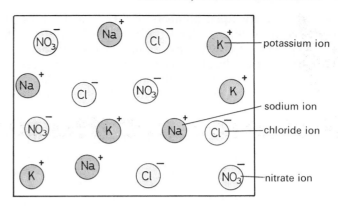

Fig. 4.3 Two salts ionized in solution (water molecules not represented)

When some compounds composed of molecules dissolve in water, the process of dissolving may split at least some of the molecules into ions, even though the atoms in the molecules were originally uncharged and not present as ions. The dissociation of the molecules in a compound may be complete or only partial so that not all the molecules dissolved form ions.

Acids and alkalis

When acids are dissolved in water, they dissociate to form hydrogen ions, e.g.

$$HCl \longrightarrow H^+ + Cl^-$$
hydrochloric acid   hydrogen ion   chloride ion

$$H_2SO_4 \longrightarrow H^+ H^+ + SO_4^{2-}$$
sulphuric acid   two hydrogen ions   sulphate ion

With an acid such as hydrochloric acid, 90 per cent of the HCl molecules in solution will be dissociated into ions, but in sulphuric acid of the same concentration only 50 per cent of the molecules will form ions. Because, at a given concentration, hydrochloric acid produces more ions than does sulphuric acid, the former is said to be the stronger acid.

Organic acids such as fatty acids and amino acids are very weak acids. Acetic (ethanoic) acid, one of the simplest fatty acids, will have only 6 out of every 1 000 molecules dissociated in a solution of the same concentration as the hydrochloric and sulphuric acids mentioned above.

Water is very slightly dissociated to H^+ and OH^- ions which enables it to take part in reactions such as the hydrolysis described on p. 16.

Alkalis such as sodium hydroxide, NaOH, dissociate into a positive ion and a negative OH^- (hydroxyl) ion.

$$NaOH \longrightarrow Na^+ + OH^-$$
sodium hydroxide   sodium ion   hydroxyl ion

pH

The acidity of a compound is determined by how many hydrogen ions it produces. The strong acids such as hydrochloric acid, which dissociate almost completely, produce a high concentration of H^+ ions. Similarly, the alkalinity of a compound depends on the concentration of OH^- ions that it can produce. The degree of acidity or alkalinity of a compound is expressed as a *pH value* (Experiment 7).

When there are equal numbers of OH^- and H^+ ions, the solution is said to be *neutral* and its pH value is 7; e.g. pure water $(H_2O \rightarrow H^+ + OH^-)$ has this pH value. As the proportion of OH^- ions increases, the pH value rises (up to 14). With a rise in the proportion of H^+ ions, the pH value falls (to 1). The scale is logarithmic, i.e. there are ten times more H^+ ions in a solution of pH 3 than there are in a solution of pH 4, and one hundred times more than in a solution of pH 5. A solution of pH 2 is strongly acid and one of pH 6 is weakly acid. The pH of a solution can be measured approximately by use of dyes called *indicators*. A familiar pH indicator is litmus, which is red in solutions of pH 5 or less (acid) and blue at pH 8 or more.

The reactions which take place in cells, particularly those involving enzymes, are very sensitive to changes in pH (*see* p. 27). A marked change of pH in a cell can have very harmful effects, and one function of many organ systems in the body is to keep the pH of the blood, tissue fluid and cells within very narrow limits (*see* homeostasis, p. 92).

Diffusion

If a lump of sugar is placed in a bowl of water and left without stirring or disturbing in any way, the sugar will dissolve. At first, the sugar solution in the immediate neighbourhood of the lump will be very concentrated and the concentration will lessen as the distance from the lump increases. Eventually, however, the sugar molecules will distribute themselves evenly through the water so that all samples of the liquid will have the same concentration (Fig. 4.4).

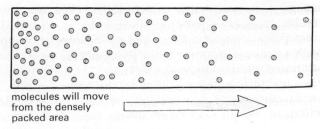

molecules will move from the densely packed area

Fig. 4.5   Diffusion gradient

extent, for the exchange of dissolved gases between the blood and tissue fluid (*see* p. 83) and the movement of substances within cells.

Osmosis

Osmosis can be regarded as a special case of diffusion: the diffusion of water from a weaker to a stronger solution.

When a substance dissolves in water, each ion or molecule attracts a small number of water molecules around itself. In Fig. 4.2, for example, each sodium ion would be surrounded by three water molecules. The ions are said to be *hydrated*; it is the hydrated ions which move about in the solution.

The hydration process means that some of the water molecules in a solution are no longer free to move independently. This has the effect of reducing the concentration of free water molecules in a solution. A concentrated solution of sugar will contain fewer free water molecules than the same volume of a dilute solution. If two such solutions were in contact, the

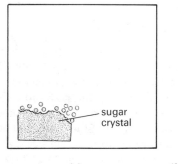 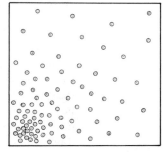 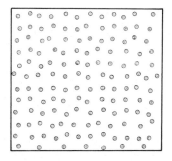

sugar crystal

(*a*)   (*b*) *Sugar dissolves; molecules diffuse*   (*c*) *Sugar molecules evenly spaced*

Fig. 4.4   Diffusion (water molecules not represented)

This kind of directional movement of molecules is called *diffusion* and occurs because the molecules of solute are in constant motion (Experiment 3). Although molecules move in random directions, each molecule will continue in a straight line until it is deflected from its path by colliding with other molecules or the wall of the container. This will go on until the solute molecules are evenly distributed.

The same phenomenon occurs with gases (Experiment 2). If a small quantity of oxygen is introduced into a container full of nitrogen, the oxygen molecules will diffuse until they are uniformly spaced throughout the container.

The difference in concentration which brings about diffusion is called a *diffusion gradient*; the greater the difference in the two concentrations the 'steeper' is the gradient and the more rapidly will diffusion occur (Fig. 4.5).

Diffusion can account for the movement of oxygen and carbon dioxide within the alveoli of a lung (p. 86) and, to some

hydrated sugar molecules would diffuse from the concentrated to the dilute solution, but the free water molecules would diffuse from the dilute to the concentrated solution.

Fig. 4.6 shows the two solutions separated by a thin membrane that stops them from mixing but allows individual molecules to pass through pores in the membrane. More water molecules will pass through the membrane from left to right because there are more free water molecules in the dilute solution. More hydrated sugar molecules will pass from right to left because there are more of them in the concentrated solution.

The small water molecules diffuse faster than the large sugar molecules and so the most obvious effect is that water diffuses from the dilute to the concentrated solution.

The pores in the membrane which permit the passage of molecules and ions are very small but quite big enough to allow both sugar and water molecules through. Nevertheless, because water molecules diffuse through the membrane more rapidly than sugar molecules do, the membrane is called 'selectively permeable' (sometimes 'semi-permeable').

A definition of osmosis, therefore, is the diffusion of water through a 'selectively permeable' membrane from a dilute to a concentrated solution (Experiment 6).

If you find the explanation of osmosis difficult or confusing, you need only remember this definition in order to understand the effects of osmosis in living organisms.

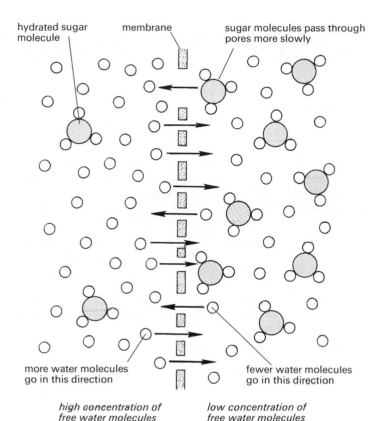

high concentration of free water molecules   low concentration of free water molecules

Fig. 4.6   Explanation of osmosis

Osmotic pressure. If a concentrated solution of sugar is enclosed in a bag made from 'selectively permeable' material such as cellophane and immersed in water, molecules of water will pass through the membrane into the sugar solution faster than they will pass out (Fig. 4.7). Consequently the volume of water inside

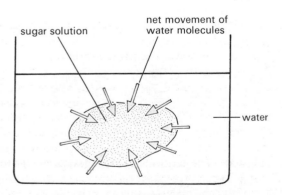

Fig. 4.7   Osmotic pressure

the membrane will increase, and this increase in volume will be accompanied by an increase in pressure which may swell the bag or burst it. The pressure that does this is called the *osmotic*

pressure of the sugar solution. The more concentrated the solution, the higher is its osmotic pressure (Experiment 5).

If the bag of sugar solution is removed from the water, osmosis cannot take place and pressure will not build up. Even when separated, however, the water and the sugar solution are said to have an *osmotic potential* because they can cause osmosis in the appropriate situation. The water has a higher osmotic potential than the sugar solution because the diffusion gradient for water molecules is from the pure water to the solution.

Osmosis in cells. The cell membrane has 'selectively permeable' properties, and the cytoplasm inside it contains many substances in solution. The cell thus has a low osmotic potential and if it is surrounded by a solution more dilute than that in the cytoplasm, water will pass into it by osmosis. If surrounded by a solution more concentrated than that in the cytoplasm, the cell will lose water by osmosis through its membrane and shrink. Loss or gain of water in this way will distort the cell and may damage its internal structure, affecting the important chemical reactions taking place. It is vital, therefore, that the tissue fluid which surrounds the cells does not become appreciably more or less concentrated than the cytoplasm. The composition of the tissue fluid depends on the composition of the blood, and the latter is kept within narrow limits by the combined action of the brain, kidneys and liver (*see* pp. 72 and 92). The osmotic potential of the blood is also thought to play a part in the exchange of water between the capillaries and the tissue fluid (p. 83).

Practical Work

Experiment 1   Solubility of solids

(a) *Salt and sand.* Make marks on three test-tubes, 1 cm and 7 cm from the base. Pour salt (sodium chloride) into two of the tubes up to the 1-cm mark and the same depth of sand into the third tube. In the tube with sand and one of the tubes with salt pour water up to the 7-cm mark. In the remaining tube of salt pour ethanol (industrial methylated spirit will do) up to the 7-cm mark. Cork each tube or cover the mouth with your thumb, shake them vigorously for about 30 seconds and then let them settle down. Examine the contents of each tube and record your conclusions on the solubility of salt and sand in water and the solubility of salt in ethanol.

(b) *Fats.* Select two clean, dry test-tubes. Into one pour about 5 cm water and into the other a similar quantity of isopropyl alcohol (propan-2-ol). Use a dropping pipette to add 10 drops of vegetable oil to each tube. Observe the effects in each tube both before and after shaking them to mix the contents. What conclusions do you reach about the solubility of oil in (i) water, and (ii) isopropyl alcohol?

(c) *Temperature and saturation.* Select two clean, dry test-tubes. In one tube place some potassium nitrate to a depth of 1 cm. In the other, place potassium nitrate to a depth of 2 cm. Add water to both to a depth of 5 cm. Shake the tubes for a few seconds to mix the contents and notice that the smaller quantity of nitrate dissolves entirely in the water, while in the other tube some salt is left undissolved, i.e. the solution is saturated at room temperature. Heat the second tube with a bunsen flame or place it in a water bath at 100 °C and observe that all the solid dissolves when the temperature of the solution is increased.

Experiment 2   **Diffusion of a gas**

Squares of wetted red litmus paper are pushed with a glass rod or wire into a wide glass tube that has been corked at one end, so that they stick to the side and are evenly spaced out (Fig. 4.8). The open end of the tube is closed with a cork carrying a plug of cotton wool saturated with a strong solution of ammonia. The alkaline ammonia vapour diffuses along inside the tube at a rate which can be determined by observing the time when each square of litmus paper turns completely blue. If the experiment is repeated using a more dilute solution of ammonia the rate of diffusion is seen to be slower.

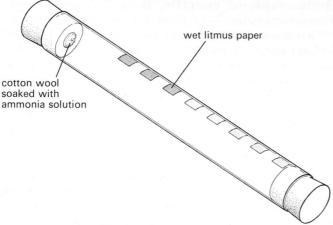

Fig. 4.8   Diffusion of ammonia

Experiment 3   **Diffusion in a liquid**

Diffusion in a liquid is slow and liable to be affected by convection currents or other physical disturbances in the liquid. In this experiment the water is 'kept still' so to speak, by dissolving gelatin in it. 10 g gelatin is dissolved in 100 g hot water and the solution is poured into test-tubes to half fill them. Some of the liquid gelatin remaining is coloured with methylene blue and when the first layer of gelatin in the test-tube has set firmly, a narrow layer of blue gelatin is poured into it. When the blue layer of gelatin is cold and firm, the test-tube is filled with cool but liquid gelatin and cooled quickly so that the blue gelatin is sandwiched between two layers of clear gelatin (Fig. 4.9). After a week, the blue dye is seen to have diffused into the clear gelatin, upwards and downwards to equal extents.

Experiment 4   **Dialysis**

A 15-cm length of 6 mm dialysis tubing (cellophane or Visking tubing) is cut, soaked in water for a few minutes, and a knot tied tightly near one end. Using a dropping pipette, the tubing is partly filled with 1 per cent starch solution, placed in a test-tube and held in place by an elastic band as shown in Fig. 4.10. The test-tube and dialysis tube are now washed with water from a running tap to remove any starch solution that may have escaped from the dialysis tubing. Fill the test-tube with water, add one or two drops of iodine solution, sufficient to colour the water yellow, and leave the tube in a rack for 10–15 minutes. After this time it will be seen that the starch inside the dialysis tube has turned blue but the iodine outside remains yellow.

When iodine mixes with starch or starch mixes with iodine, a blue colour results. In this experiment, the simplest interpretation of the result is that the dialysis tubing allows iodine through

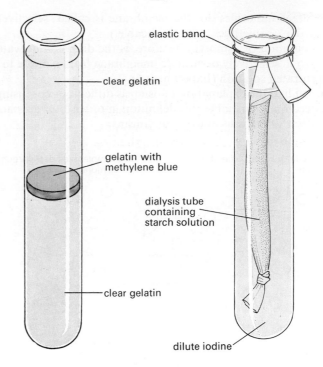

Fig. 4.9   Diffusion in liquid

Fig. 4.10   Dialysis

to reach the starch but does not allow starch out to reach the iodine. Since the starch molecules are hundreds of times larger than the iodine molecules this might be the reason for their failure to pass through the membrane of the dialysis tubing.

The separation of large and small molecules by this method is called dialysis. It should not be confused with osmosis even though the dialysis tubing is used for osmotic experiments.

Experiment 5   **Osmotic pressure**

A 20-cm length of 6 mm dialysis tubing is cut, soaked in water, and securely knotted near one end. The tube is then partly filled with a strong solution of syrup or sugar and the open end tightly knotted. The tube should be flabby and easily bendable (Fig. 4.11). Immerse the dialysis tube in a test-tube full of water

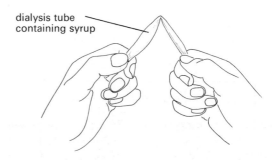

Fig 4.11   The partly filled tube is flexible enough to bend

and leave it for 30–45 minutes. When it is removed from the water the dialysis tubing will be taut and full. Water has entered by osmosis and has thus increased the pressure inside the tubing.

Experiment 6 **Osmosis**

A length of dialysis tubing (6 mm cellophane or Visking tubing) is soaked in water, securely knotted at one end, filled with a strong solution of syrup or sugar, and fitted over the end of a capillary tube with the aid of an elastic band (Fig. 4.12). The dialysis tube is lowered into a beaker or jar of water and clamped vertically. In a few minutes, the level of liquid is seen to rise up the capillary tube and may continue to do so for a metre or more according to the length of the tube.

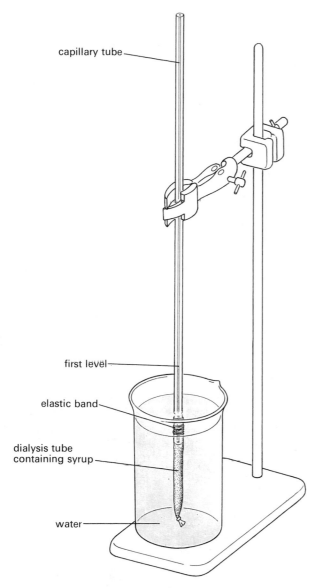

Fig. 4.12   Demonstration of osmosis

Interpretation. The most plausible interpretation is that water molecules have passed through the cellophane tubing into the sugar solution, increasing its volume and forcing it up the capillary tube. This movement should theoretically continue until the pressure of the column of syrup in the capillary is equal to the diffusion pressure of the water entering the dialysis tube.

In practice, the dialysis tubing does allow sugar molecules to pass through into the water but more slowly than it allows water molecules in. Eventually, the concentrations of sugar in the beaker and the dialysis tubing would become the same.

Experiment 7 **pH and universal indicator**

Fill three test-tubes to a depth of about 5 cm with tap water. To each add 10 drops of universal indicator. Using a dropping pipette add dilute (M/10) hydrochloric acid, two drops at a time, to one of the tubes. Shake the tube after each addition and note any change in colour by comparing it with the other tubes. Continue adding the acid until there is no further colour change. Wash the pipette and use it to add dilute (M/20) sodium carbonate solution (an alkali), two drops at a time, to one of the remaining tubes. Again, compare the colours with the third tube and go on adding the alkali until no further colour change occurs. If the colours produced are compared with a special colour chart for this indicator, the pH represented by each colour can be seen. Approximately, red is pH 4, orange is pH 5, yellow 6, green 7, blue 8–9 and purple 10.

Experiment 8 **Carbon dioxide and pH**

Two test-tubes are half filled with tap water, and carbon dioxide from a cylinder or carbon dioxide generator is bubbled through one of them for a minute. Ten drops of universal indicator are added to each tube and the colours compared with the results of Experiment 7. The experiment may also be tried by bubbling carbon dioxide through hydrogencarbonate indicator (*see* p. 319). It will be seen that carbon dioxide makes an acid solution when it dissolves in water.

Further experiments on diffusion and osmosis, with more detailed instructions, can be found in the 'Experimental Work in Biology' series (*see* p. 320).

Questions

1   Rubber dissolves in benzene; chloroform dissolves Perspex; salt and water make a solution. In each case, say which substance is the solvent and which is the solute.
2   Which of the following solutions is the more concentrated: 10 g potassium chloride in 50 g water, or 25 g potassium chloride in 130 g water?
3   The dissolved air in the water of a pond and the air in the atmosphere above it are in equilibrium, i.e. the gases are escaping from the water at the same rate as they are dissolving in it. Nevertheless, the proportions of oxygen, carbon dioxide and nitrogen in the air and pond water are quite different. Suggest a reason for this.
4   What ions will form when the following salts dissolve in water: potassium chloride, KCl; copper nitrate, $CuNO_3$; zinc sulphate, $ZnSO_4$; nitric acid, HNO_3; potassium hydroxide, KOH?
5   Which is the more acid, a solution of pH 6 or a solution of pH 4?
6   As a general rule diffusion takes place in gases and liquids but not in solids. Why is this so?
7   Which system offers the steeper diffusion gradient: (a) a solution containing 200 g salt per litre in contact with one containing 20 g per litre, or (b) a solution containing 20 g per litre in contact with one containing 2 g per litre?
8   A strong solution of sodium chloride is sometimes used as an antiseptic, i.e. it destroys certain bacteria. Suggest, in terms of osmosis, how it might achieve its effect.
9   What activities in man are likely to (a) increase and (b) decrease the osmotic potential of his blood and body fluids?

5 Reactions in the Cell

Chapters 3 and 4 dealt with the structures of the cell and the chemicals of which it is composed or which take part in reactions inside it. These reactions are necessary (a) to enable the cell to grow and reproduce, (b) to produce energy for driving other reactions and (c) to carry out special functions of the cell such as contraction or conduction. Some of these reactions will be considered in more detail but first it is necessary to describe the chemicals which control and direct all the reactions in a cell, the *enzymes*.

Enzymes

Enzymes are proteins made in the cell and they have the function of accelerating chemical reactions. Chemicals that affect the speed of reactions but are not themselves used up in the reaction are called *catalysts*, and enzymes are organic catalysts in cells (Fig. 5.1).

Catalytic action. The reactions that can take place between the substances in cells are usually slow; for example, it would take days for even a small amount of starch to be hydrolysed to maltose (p. 16) by simply mixing the starch with water. In the laboratory, the reaction can be speeded up by adding acid and raising the temperature of the mixture to boiling point. High temperatures and acids are harmful to living cells, and one of the distinctive features of enzymes is that they accelerate reactions at low temperatures and without extreme chemical conditions. The enzyme in saliva will convert starch to sugar at room temperature in a minute or two, a reaction which would otherwise need fifteen minutes or more of boiling with hydrochloric acid.

Specificity. Any one enzyme will act as a catalyst for only one kind of reaction. An enzyme that builds proteins out of amino acids (p. 17) will not catalyse the hydrolysis of maltose. Thus for each type of reaction there is one specific enzyme, and they are often given names related to the reaction in which they participate, the name usually ending in -*ase*. An enzyme that catalyses the removal of hydrogen from a compound is called a *dehydrogenase*. Enzymes that break down proteins are *proteases*, and those that hydrolyse fats (sometimes called *lipids*) are *lipases*. Starch is a mixture of two substances, *amylose* and *amylopectin*, so enzymes that act on starch are called *amylases*. The substance on which the enzyme acts is called its *substrate*.

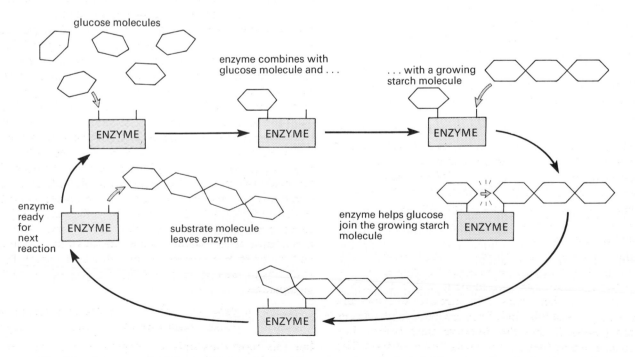

Fig. 5.1 An enzyme building a starch molecule. The enzyme is a catalyst and is not used up during the reaction

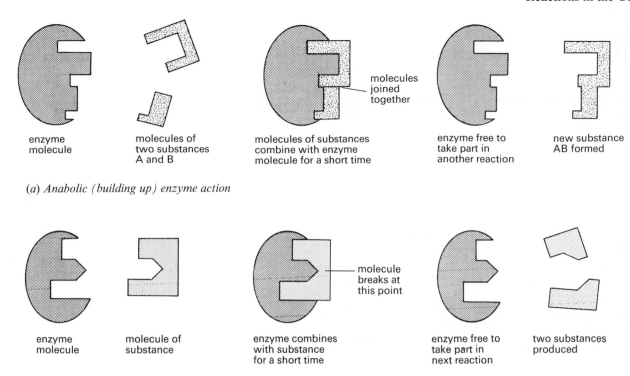

(a) Anabolic (building up) enzyme action

enzyme
molecule

molecules of
two substances
A and B

molecules of substances
combine with enzyme
molecule for a short time

molecules
joined
together

enzyme free to
take part in
another reaction

new substance
AB formed

enzyme
molecule

molecule of
substance

enzyme combines
with substance
for a short time

molecule
breaks at
this point

enzyme free to
take part in
next reaction

two substances
produced

(b) Katabolic (breaking down) enzyme action

Fig. 5.2   Simple models of enzyme action. Note that the shape of the enzyme molecule allows it to combine with only one kind of substrate. This accounts for its specificity

Optimum temperature and pH. Chemical reactions are speeded up by an increase in temperature; a rise of 10 °C will double the rate of most reactions. The reactions in cells also are accelerated by temperature rise, but because enzymes are proteins they are adversely affected by high temperatures; above 45 °C they are probably denatured (see p. 18). As the temperature of a cell is raised the reactions go faster, but as the temperature approaches 50 °C the enzymes are progressively inactivated so that they can no longer catalyse the reaction. The highest temperature at which reactions are speeded up without at the same time inactivating the enzyme is called the *optimum* temperature. Temperatures above or below the optimum will slow down the reaction. In man, the optimum temperature for most of the enzymes in his cells is between 30 and 40 °C (Experiment 1).

The structure and reactions of proteins, including enzymes, are affected by the acidity or alkalinity (pH) of the medium in which they are working. Just as there is an optimum temperature for enzymes there is also an optimum degree of acidity, but for somewhat different reasons. The acidity and alkalinity of the cytoplasm in cells varies very little but the optimum conditions for different digestive enzymes range from neutral to strongly acid (Experiment 2).

Extracellular enzymes. The majority of enzymes are *intracellular*, that is, they catalyse reactions taking place inside the cell. A few enzymes, however, are made in cells but released from the cells to do their work. The digestive enzymes are extracellular enzymes of this kind. They are made in gland cells but secreted (released) into the digestive tract before they become active, and here they start to break down the food. The digestive enzymes are considered in more detail on p. 62.

Respiration

The chemical reactions in the cell involve building up large molecules from small ones, or breaking down large molecules to smaller ones and reassembling them in a different way. There may also be other molecular changes in specialized cells; for instance, when a muscle fibre shortens to produce muscle contraction, there is a temporary rearrangement of its molecules. The movements of substances into and out of a cell may also involve chemical reactions. Many of these reactions involving enzymes need a supply of energy to make them take place. This energy is obtained from food substances such as sugars and fats. The molecules of the food substance are broken down to smaller molecules and the energy that was in their chemical bonds is transferred to other molecules in the cell. The food molecules are broken down in stages by enzymes (e.g. dehydrogenases) removing hydrogen and carbon atoms one at a time and eventually combining these atoms with oxygen to form water (the oxide of hydrogen, H_2O) and carbon dioxide (an oxide of carbon, CO_2). The carbon dioxide and water are eventually eliminated from the cell (Experiments 3 and 4). This transfer of energy from food to other chemicals in the cell is called *respiration*. In its simplest terms, respiration is the breakdown of carbohydrates and fats to form carbon dioxide and water with a corresponding release of energy for other reactions in the cell. It is sometimes expressed in the form of an equation.

$$C_6H_{12}O_6 + 6O_2 = 6CO_2 + 6H_2O + 2\,830 \text{ kilojoules}$$

glucose   oxygen   carbon dioxide   water   energy

but this represents only the beginning and the end of the process, because the glucose is actually broken down in small steps involving many intermediate compounds with energy being produced at each step. The 2 830 kilojoules (kJ) represent the maximum energy to be obtained from one gram-molecule (180 g) of glucose if it is completely oxidized to water and carbon dioxide.

A distinction is usually made between two forms of (or stages in) respiration called *aerobic* and *anaerobic*. Aerobic respiration involves the use of oxygen in the breakdown of carbohydrates or fats which are eventually oxidized completely to carbon dioxide and water. Anaerobic respiration is the breakdown of carbohydrates to release energy without the use of oxygen (*see* below). Each step in the chemistry of respiration is catalysed by a specific enzyme.

The enzymes for aerobic respiration are contained in the mitochondria, while those for the anaerobic stage are in the cytoplasmic fluid.

The term 'respiration' is also often used loosely in reference to breathing, as in 'artificial respiration', 'pulse and respiration rate', or in connection with gaseous exchange as in 'the respiratory surface of the lungs', 'organs of respiration'. For this reason, the respiration described in this chapter is sometimes called *tissue respiration* or *internal respiration* to distinguish it from either the breathing movements (ventilation) or the intake of oxygen and the output of carbon dioxide (gaseous exchange).

Anaerobic respiration. This is the release of energy from food material by a process of chemical breakdown that does not require oxygen. The food, e.g. carbohydrate, is not broken down completely to carbon dioxide and water but to intermediate compounds such as lactic acid or alcohol.

$$C_6H_{12}O_6 = 2CO_2 + 2C_2H_5OH + 118 \text{ kJ}$$

glucose   carbon dioxide   alcohol   energy

The incomplete breakdown of the food means that less energy is released in anaerobic respiration than in aerobic respiration.

Both processes may take place in cells at the same time. Indeed, the first steps in the breakdown of glucose in respiration are anaerobic.

glucose $\xrightarrow{\text{anaerobic}}$ lactic acid $\xrightarrow{\text{aerobic}}$ carbon dioxide and water

During vigorous activity, the oxygen supply may not be sufficient to completely oxidize the food required to meet the energy demands of the body, so that the products of the initial, anaerobic, stages (e.g. lactic acid) accumulate in the cell. These products have to be oxidized or converted back to carbohydrate so that even after the vigorous activity has ceased the uptake and use of oxygen continues at a high rate. The organism is said to have incurred an '*oxygen debt*' as a result of its excess of anaerobic respiration.

Some bacteria and fungi derive all or most of their energy from anaerobic respiration, and the end products are frequently alcohol and carbon dioxide; the process in this case is called *fermentation* (Experiment 5).

Adenosine triphosphate, ATP. The energy in the chemical bonds of carbohydrate and fat molecules is largely associated with hydrogen atoms. As the molecules are broken down in respiration, the hydrogen atoms and their energy are passed to a series of chemicals called *hydrogen acceptors*. As the hydrogen atom is passed from one hydrogen acceptor to the next, it releases a proportion of its energy and this is trapped by a substance called *adenosine diphosphate* (ADP). The chemical structure of ADP need not concern us except for the fact that it contains two phosphate groups, i.e. adenosine—phosphate—phosphate. The energy from the transferred hydrogen atom allows ADP to combine with a third phosphate group to make adenosine triphosphate (ATP), i.e. adenosine—phosphate—phosphate ~ phosphate. The chemical bond between the second and third phosphate is more easily broken than the bond between the first and second. When the third phosphate bond is broken it releases energy which can be used for other reactions (*see* Fig. 5.3).

ATP is a kind of universal and readily available 'energy currency'. It is used to drive a wide variety of chemical reactions in the cell, from the conversion of glycogen to glucose to the contraction of a muscle. In suitable chemical conditions, ATP breaks down to ADP and phosphate, releasing energy for some vital reaction. The ADP will eventually be used to rebuild ATP once more.

The energy from food molecules cannot be used directly for cell reactions but must first be transferred to ATP; the ATP is then available as an energy source, acceptable to a wide variety of different reactions. Moreover, the transfer of energy from food is a relatively slow process requiring many steps and is therefore unlikely to meet the demand for sudden and continued energy expenditure. The release of energy from ATP, however, is a rapid, single step.

ATP thus acts as a kind of energy store. Food is broken down steadily in the cell and a reserve of ATP molecules built up from ADP. One molecule of glucose has enough energy to build 38 molecules of ATP. When the cell needs energy it takes it from the ATP stock which is subsequently replenished by respiration of more food.

Protein synthesis

Protein synthesis has been described briefly in Chapter 3. On the ribosomes amino acids are assembled in a certain order to make particular proteins (p. 166). The sequence of amino acids is dictated by molecules of a substance called *ribonucleic acid*

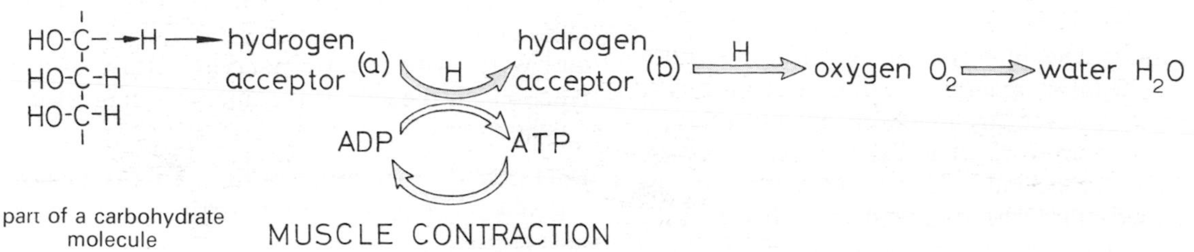

Fig. 5.3   ATP in respiration

(RNA) which is made in the nucleus but released into the cytoplasm where it combines temporarily with a ribosome. This 'messenger' ribonucleic acid 'calls up' amino acids which are present in the cytoplasm and, with the aid of enzymes, joins them together in a chain to make a protein. In this way the cell makes enzymes and all the proteins necessary for the structures in its protoplasm. Various stages in·this process need energy which is supplied by ATP.

Transport

If a cell is to carry out the process of respiration and protein synthesis it must obtain supplies of sugar, oxygen, amino acids and other substances, and it must be able to get rid of the waste products such as carbon dioxide and water which accumulate as a result of respiration and other chemical changes. Cells in the bodies of animals are bathed in a fluid, *tissue fluid*, which is derived from the blood. The blood keeps the tissue fluid supplied with food molecules and oxygen in ways described in more detail on p. 83. The cells extract the substances they need from the tissue fluid and release into it the substances they do not want. The methods by which cells take up substances through the cell membrane are not well understood but can be divided into two main processes, active and passive transport.

Passive transport includes processes such as diffusion and osmosis which are described more fully on p. 22. Diffusion occurs when molecules of a gas or a dissolved substance are unevenly distributed. The molecules are moving at random but eventually they will be evenly distributed.

If the cell membrane is permeable to the molecules (i.e. will let them through) the concentrations on each side of the membrane will eventually become equal whatever the inequality of the initial distribution (Fig. 5.4).

Diffusion could thus account for the movement of oxygen, glucose and amino acids into a cell and carbon dioxide and water out. If rapid respiration is taking place inside a cell, the concentrations of glucose and oxygen will fall as these substances are used up. The concentrations of glucose and oxygen in the tissue fluid outside the cell will be greater than the concentration inside and so these two substances will diffuse into the cell to restore the balance. Similarly, carbon dioxide and water will accumulate inside the cell during respiration and diffuse out. Such a difference in concentration of a particular substance with respect to two locations is called a *diffusion gradient* (p. 22).

If diffusion were the only method whereby substances entered or escaped from cells, the cell would have little control

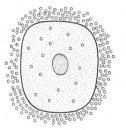 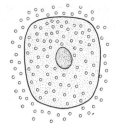

(a) *Greater concentration outside cell* (b) *Concentrations equal on both sides of cell membrane*

Fig. 5.4   Molecules entering a cell by diffusion

over the process. Any substance, harmful or not, that was more concentrated outside the cell would tend to diffuse in. Useful substances, such as glucose, would tend to diffuse out. In fact, the cell membrane exercises almost complete control over what gets in and out. The methods by which it does this are grouped under the heading *controlled diffusion*, *active transport* and *endo-* or *exo-cytosis*.

Controlled diffusion. The rate of diffusion through a cell membrane depends on the concentration gradient of the substance concerned, but the rate is often faster or slower than expected. Water diffuses more slowly and amino acids diffuse more rapidly through a membrane than might be expected. In some cases this is thought to happen because the ions or molecules can pass through the membrane only by means of special pores. These pores may be few in number or they may be open or closed in different conditions.

Active transport. In some cases, substances are taken into or expelled from the cell against the concentration gradient. For example, sodium ions may continue to pass out of a cell even though the concentration outside is greater than inside. The processes by which such reverse concentrations are produced are not fully understood and may be different for different substances, but are all generally described as active transport.

Anything which interferes with respiration, e.g. lack of oxygen or glucose, prevents active transport taking place. Also, during active transport, ATP is broken down to ADP. Thus it seems that active transport needs a supply of energy.

Fig. 5.5 offers a theoretical 'model' to explain how active transport might work. A 'carrier' is activated by the transfer of energy from ATP. In this activated form it combines with the substance, carries it across the cell membrane and releases it into the cell. The carrier, probably an enzyme, is then re-activated and can be used again.

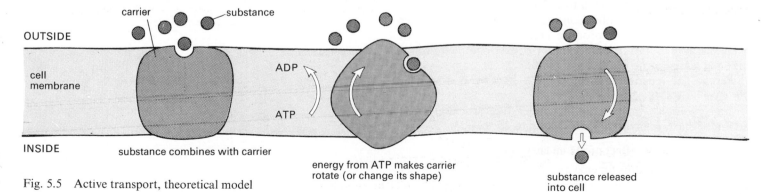

Fig. 5.5   Active transport, theoretical model

Exo- and endo-cytosis. Some cells can take in (*endo*cytosis) or expel (*exo*cytosis) solid particles or drops of fluid through the cell membrane. Endocytosis occurs in some single-celled organisms when they feed, and in certain white blood cells (phagocytes, p. 72) when they engulf bacteria; a process called *phagocytosis*. Exocytosis takes place in the cells of some glands. A secretion, e.g. a digestive enzyme, collects in vacuoles in the cytoplasm, and these are then expelled through the cell membrane to do their work outside the cell.

Metabolism

All the chemical activities which take place in cells and therefore in the organisms to which they contribute are described by the collective term *metabolism*. Metabolism is subdivided into *anabolism* and *katabolism*. Anabolism refers to the chemical reactions that build up molecules, e.g. protein synthesis, and katabolism to the breakdown reactions such as respiration.

Practical Work

Experiment 1   **Effect of temperature on enzymes**

Diastase is an enzyme extracted from germinating cereal grains. It acts on starch and hydrolyses it to maltose.

Place 5 cm³ of 1 per cent starch solution in each of three test-tubes, and 5 cm³ of 5 per cent diastase solution in each of three others. Prepare three water baths from beakers or cans containing (a) cold water and ice at about 10 °C, (b) cool water from the tap at about 20 °C, and (c) warm water at about 35 °C. Place a tube of starch solution and a tube of diastase solution in each water bath and leave them for five minutes to acquire the temperature of the water. Meanwhile, use a dropping pipette to place several rows of drops of dilute iodine solution on a tile or dish. After five minutes, note the time and pour the diastase into the corresponding tube of starch solution, starting with the cold one. Wash the pipette in the warm water and use it to take a small sample from the test-tube in the warm water bath. Place this sample on the first iodine drop, which will probably go blue indicating the presence of starch. Rinse the pipette again and use it to take a sample from the tube in the cool water and place it on another iodine drop. Repeat the procedure for the tube in the ice water.

Continue taking samples in this way at intervals of about one minute. When a sample ceases to give a blue colour with the iodine the reaction is complete, i.e. all the starch has been changed by the enzyme to maltose. Note the time and take no more samples from this tube. Continue sampling from the other two tubes until they fail to produce a blue colour with iodine. In each case note the time for the reaction to reach completion, and take the temperature of the water bath.

Result. The reaction is most rapid at 35 °C and slowest at 10 °C.

Interpretation. An increase in temperature accelerates the action of diastase on starch, at least up to 35 °C.

Experiment 2   **The effect of pH on enzymes**

Human saliva contains an enzyme, amylase, which hydrolyses cooked starch to maltose (*see* Experiment 1, p. 69).

Place 5 cm³ of 1 per cent starch solution in each of five tubes, labelled 1 to 5. Add acid or alkali as follows: to tube 1, 1 cm³ M/20 sodium carbonate solution; to tube 2, 0.5 cm³ M/20 sodium carbonate; tube 4, 2 cm³ M/10 acetic acid; tube 5, 4 cm³ M/10 acetic acid. Tube 3 is left without either acid or alkali. Using a dropping pipette, place rows of drops of dilute iodine solution on a tile or dish. Collect about half a test-tube full of saliva; note the time, then add 1 cm³ saliva to each tube and shake them to mix the contents.

With a clean pipette take up a little of the mixture from tube 1 and place a drop or two on the first iodine spot. The blue colour indicates the presence of starch. Rinse the pipette in clean water and test a sample from each tube in turn, rinsing the pipette between each test. Test samples from each tube at approximately one-minute intervals. Eventually one of the samples will not go blue with iodine, which means that the reaction in that tube is complete and all the starch has been changed to maltose. Note the time and take no more samples from that tube. Continue sampling the other tubes, noting the time in each case when the starch has disappeared. After about fifteen minutes from the start of the experiment, stop testing and find the pH in each tube by placing a drop of liquid from each on a piece of pH paper and comparing its colour with that on the colour chart supplied.

The pH values will be approximately as follows.

Tube	1	2	3	4	5
pH	9	7	6–7	6	3

Result. Starch will disappear first from tube 3, more slowly from tubes 2 and 1, and more slowly still from tube 4. In tube 5, starch may still be present after fifteen minutes or more.

Interpretation. The optimum pH for salivary amylase is between 6 and 7. Acid conditions (pH 3) slow down the reaction very markedly. Alkaline conditions also retard the reaction but to a lesser extent.

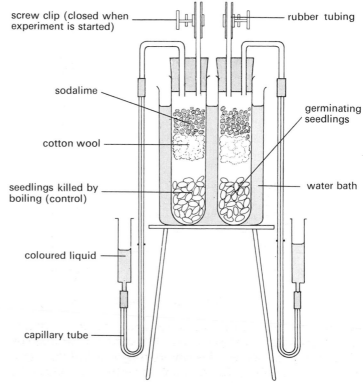

Fig. 5.6   Uptake of oxygen in respiration

Experiment 3 **Uptake of oxygen by living organisms**

20 g seeds are soaked for 24 hours, the water poured away and the seeds left to germinate for 3 days in a closed container. Half of the seedlings are killed by boiling for 5 minutes and then cooled in tap water.

In Fig. 5.6 the manometers are made by fitting 2 cm³ plastic syringe barrels to capillary tubing. The liquid in the manometers is water with colouring and a little detergent.

Put 10 g live seedlings into one tube and 10 g dead seedlings into the other. Place 5 g sodalime in each tube, supported on a wad of cotton wool. Place both tubes, with screwclips open, in the water bath and leave for 5 minutes to reach the water temperature. Close both clips and, after 10–15 minutes, note any change of level of liquid in the capillary tubes.

Result. The level of liquid should rise in the manometer connected to the living seeds and remain the same in the manometer connected to the dead seeds.

Interpretation. If the seeds are respiring they will be taking in oxygen and giving out, probably, almost an equal volume of carbon dioxide so that there would be no change in volume of the gas in the tube. Since the sodalime, however, absorbs all the carbon dioxide given out, the reduction in volume leading to the rise in liquid level is therefore most likely due to uptake of oxygen by the living seeds.

Temperature changes are reduced to a minimum by the water bath and any changes that do occur affect both tubes equally. So a *difference* in liquid level can be attributed to the living seeds.

Experiment 4 **Production of carbon dioxide by seedlings**

Some seeds which germinate quickly are soaked for 24 hours and allowed to germinate on moist cotton wool in a closed container. When they begin to germinate, half are killed by boiling for 5 minutes. A large test-tube is half filled with the living seedlings and a similar tube with the dead ones. The mouth of each tube is covered with aluminium foil and they are

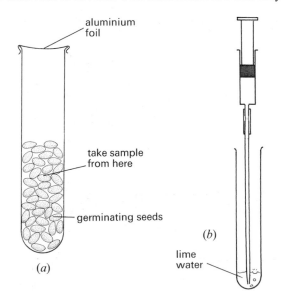

Fig. 5.7   Production of carbon dioxide by seeds

left for 30 minutes. A length of glass tubing, fitted to a 10 cm³ syringe, is pushed through the foil and a sample of air taken by withdrawing the plunger. The air is slowly expelled through a little lime water (Fig. 5.7b). A sample of air from the other tube is tested in the same way.

Result. The air from the living seeds should turn lime water milky indicating the presence of carbon dioxide. The air from the dead seeds should not affect the lime water.

Interpretation. The living seeds have produced carbon dioxide. This is reasonable evidence that respiration is taking place.

Experiment 5 **Anaerobic respiration in yeast**

Water is boiled to expel all dissolved oxygen and when cool is used to make a 5 per cent solution of glucose and a 10 per cent suspension of dried yeast. 5 cm³ of the glucose solution and 1 cm³ of the yeast suspension are placed in a test-tube and covered with a thin layer of liquid paraffin to exclude atmospheric oxygen. A delivery tube, fitted as in Fig. 5.8, dips into clear lime water. A control is set up in the same way but with a yeast suspension that has been boiled to kill the yeast cells.

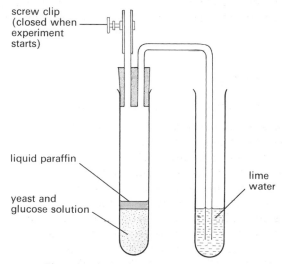

Fig. 5.8   Anaerobic respiration in yeast

In 15 minutes, fermentation should begin in the living yeast-glucose mixture, with bubbles escaping through the lime water. If this does not happen, both tubes with the yeast should be warmed in a beaker or jar of water at 30 °C.

Result. The bubbles of gas from the tube with living yeast will turn the lime water milky showing that carbon dioxide is being produced. The dead yeast should not produce any gas.

Interpretation. The fact that the yeast is producing carbon dioxide is evidence that respiration is taking place. Since air was expelled and excluded, this respiration, in the absence of oxygen, must be anaerobic.

Questions

1   (a) Where does respiration occur? (b) What is the importance of respiration? (c) What materials are used up during respiration, and what are the products?
2   What methods do organisms have for storing energy?
3   List the differences between aerobic and anaerobic respiration. Are these two forms of respiration mutually exclusive? Explain.
4   Which aspects of respiration can readily be measured or demonstrated?
5   In the mammal, which classes of food can be used to provide energy? Which ones provide the most energy? (*See* pp. 32–3.)
6   If a biologist wants to demonstrate that a reaction is controlled by enzymes and will not take place without these enzymes, how, in principle, does he design his experiment?
7   What evidence would you look for in deciding whether a substance was entering a cell by active transport or passive diffusion?

6
Nutrition

The food taken in by man is used in three principal ways, namely for energy, growth and replacement.

(a) It is broken down in cells during respiration (p. 27) to provide energy for movement and essential chemical reactions.

(b) It provides the raw materials for making protoplasm and so contributes (i) to the growth of cells and tissues, (ii) to the replacement of cells which die, e.g. red blood cells and the cells of the epidermis, and (iii) to the repair of wounds and other damage to body structures.

The substances that can meet all these requirements are chemical materials with food value, namely proteins, fats and carbohydrates. In addition to these, the diet must contain water, salts, vitamins and fibre (roughage) because although some of these substances have no energy value they do participate in chemical reactions or bodily functions.

Carbohydrates

Carbohydrates contain the elements carbon, hydrogen and oxygen; their chemistry is outlined on p. 15. The principal forms of carbohydrate in man's diet are starch, sugar and cellulose. Foods relatively rich in starch are the cereals such as wheat, maize, rice, millet and their products (e.g. bread), and peas, beans and certain 'root' crops (e.g. potatoes, yams and cassava). Sugar is abundant in confectionery, cakes and jam. Cellulose occurs in all the whole-plant material eaten. Though man has no enzymes which digest cellulose, it plays an important part in the diet as dietary fibre.

Carbohydrates are principally of value as energy-giving food, providing on average 17 kJ per gram. They are usually the cheapest and most abundant foods and therefore the main source of energy. In man, excess carbohydrate is stored as glycogen in the liver and the muscles, or converted to fat and stored beneath the skin in other fat depots.

Proteins

Proteins contain the elements carbon, hydrogen, oxygen, nitrogen and usually sulphur (see p. 17 for their chemistry). Examples of foods containing protein are lean meat, eggs, beans, fish, milk and milk products such as cheese.

Proteins are broken down by digestion to amino acids (see p. 64) which are absorbed into the blood stream and eventually reach the cells of the body. In the cells, the amino acids are assembled to form the structural proteins of the protoplasm and its constituent enzymes (p. 26). There are only about 20 different kinds of amino acids but there are millions of different proteins. The difference between one protein and another depends on which amino acids are used to build it, how many of each there are, their sequence and their arrangement (Fig. 6.1).

Plants can build all the amino acids they need from carbohydrates and nitrates but animals cannot. They must therefore obtain their amino acids from proteins already made by plants or present in the flesh of other animals, and so the diet must include a minimum quantity of protein of one sort or another. A diet with a sufficient energy content of fats and carbohydrates and rich in vitamins and salts but lacking protein will lead eventually to illness and death. Proteins are particularly important during periods of pregnancy and growth when new protoplasm, cells and tissues are being made. Growing children and pregnant women need more than average supplies of protein; manual workers, contrary to popular belief, probably do not.

Essential amino acids. Although animals cannot make amino acids they can, in some cases, convert one amino acid into another. There are, however, ten or more amino acids which man cannot produce in this way and these *essential amino acids* must be obtained directly from proteins in the diet. Proteins

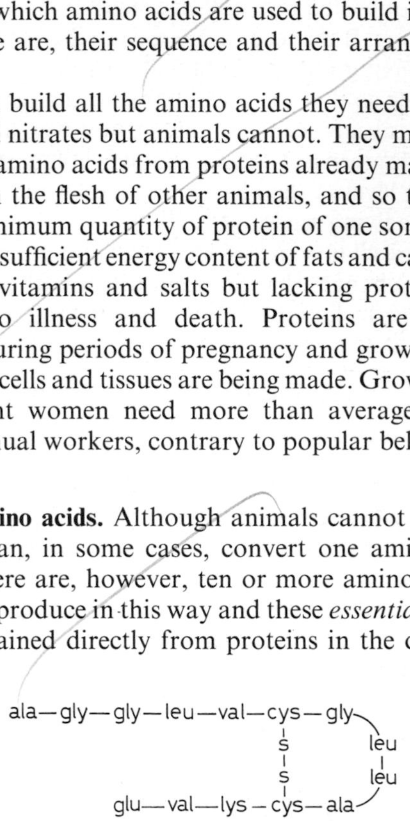

(a) *Plant polypeptide of 14 amino acids*

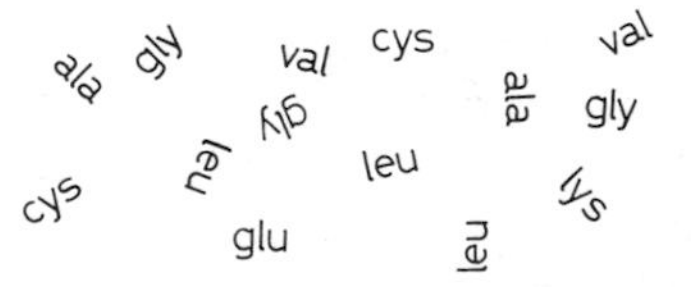

(b) *Digestion breaks up polypeptide into its amino acids*

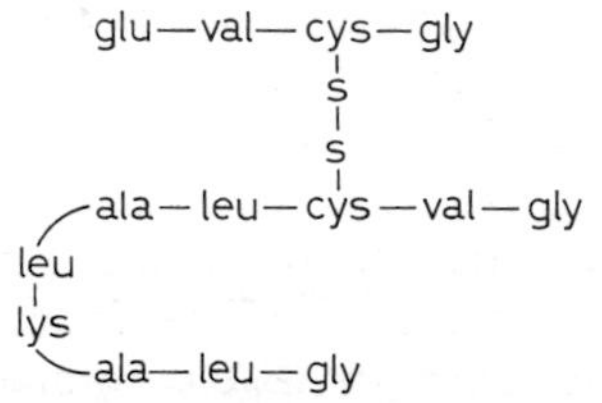

(c) *Our body builds up the same 14 amino acids into a polypeptide that it needs*

Fig. 6.1. Digestion and use of a polypeptide (part of a protein molecule)

which contain the essential amino acids in the right proportions are sometimes called *first class proteins*. Most animal proteins come into this category. Vegetable proteins often lack one or more of the essential amino acids (they are second class proteins) or contain them in the wrong proportions. To obtain the essential amino acids from plant protein alone, a very mixed vegetable diet in large quantities is needed. Vegetarians who include milk, cheese and eggs in their diet will not lack essential amino acids since they occur in these foods in very high proportions.

Protein needs. Estimates of the daily quantity of protein needed by man have been changing in recent years but it seems to be no more than about 0.6 g per kg body weight, i.e. a 70 kg man needs about 42 g protein each day assuming that the protein contains all the essential amino acids. Thus on average, a man needs about 40 g and a woman 30 g good quality protein per day. A pregnant woman will need at least 40 g to supply herself and her embryo which is building up its tissues, and a woman breast-feeding a baby should take in 60 g protein per day to compensate for the loss of protein in the milk. Babies need 1.5 g, young children 1 g and adolescents about 0.7 g per kg body weight.

If proteins are eaten in excess, there will be more amino acids in the body than are needed to produce or replace cells. The excess amino acids are converted in the liver to carbohydrates, which are then oxidized for energy or converted to glycogen and stored. The energy value of protein when oxidized is 17 kJ per gram. Amino acids cannot be stored.

Fats

Like carbohydrates, fats contain only carbon, hydrogen and oxygen but in different proportions (p. 16). They are present in foods such as animal fat, butter, margarine, cheese, milk and groundnuts.

Fats are used in making cell structures, particularly the membrane systems of the cell. They are also an important source of energy.

Although fats are less easily digested and absorbed than carbohydrates, they have more than double the energy value, providing 39 kJ per gram. They can be stored in the body and thus provide an important way of storing energy. About 25 per cent of our energy requirements should be supplied by fats. Europeans eat 80–150 g per day, people in tropical countries somewhat less. Eskimos eat up to 300 g.

Although man can convert carbohydrates to fatty acids there are probably one or two that he cannot make, and it is necessary to take in some fats with the diet in order to supply these *essential fatty acids*. Fat, particularly animal fat, is needed for the fat-soluble vitamins A, D, E and K which it contains. In addition the use of fat and oil in the process of cooking usually makes the food more palatable.

On pages 16 and 17 it was explained that fats are formed from glycerol and fatty acids. The fatty acids which make up many vegetable oils are often called *unsaturated*. This means that they have a double bond (p. 15) in their molecule. If there is more than one double bond, the fatty acid is said to be *poly-unsaturated* (Fig. 6.2).

The presence or absence of double bonds affects the properties of the fats. There is some evidence that poly-unsaturated fatty acids are less likely to contribute to heart and arterial disease than are saturated fatty acids. this view is not held by all doctors or nutritionists.

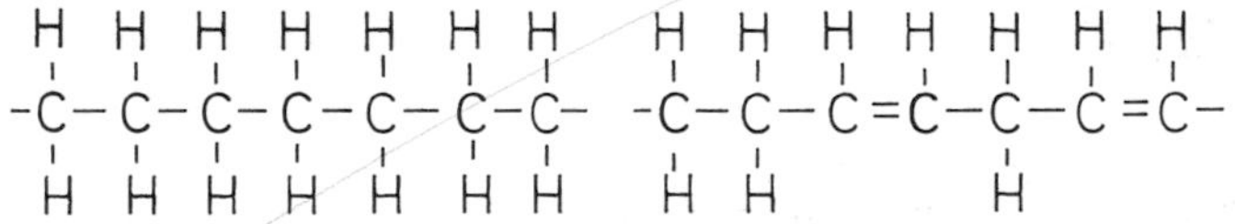

(a) Part of a saturated fatty acid

(b) Part of a polyunsaturated fatty acid

Fig. 6.2 Saturated and unsaturated fatty acids

Mineral salts and elements

In addition to the carbon, hydrogen, oxygen, nitrogen and sulphur present in carbohydrates, proteins and fats, a variety of other elements is needed for the chemical activities of the body and for the construction of certain tissues. These elements are obtained from food where they are usually combined in organic molecules.

Sodium, potassium, magnesium and phosphates are normally present in adequate quantities in the diet. Sodium and chlorine are taken in mostly as sodium chloride, which is used in cooking or added to the food. It is important in maintaining the osmotic concentration of the blood (pp. 23 and 72) in providing sodium ions for the transport of carbon dioxide as sodium hydrogencarbonate (bicarbonate) and chloride ions for the production of hydrochloric acid in the stomach. Potassium is present in most living cells, particularly red blood cells and muscles, and is important in growth. Cereals contain good supplies of potassium. Magnesium is a component of bone and is found in a wide variety of vegetable material. Phosphates are obtained in sufficient quantities from the protein part of the diet, if this is adequate. They are needed in the body for the formation of bone, nucleic acids (p. 166), ATP (p. 28) and enzymes, and they form part of most cell membranes. The elements mentioned above are most unlikely to be missing from any diet. On the other hand, calcium, iron and iodine are frequently in short supply.

Calcium. This element enters into the composition of bones as calcium phosphate and into the dentine of teeth. It also plays a part in the clotting of blood (p. 73), contraction of muscles, permeability of cells and conduction of nerve impulses. It is abundant in cheese and milk and in the latter form is most readily absorbed. Beans and soya flour contain appreciable quantities. In districts where the water supply comes from chalk or limestone areas, it will contain quantities of calcium hydrogencarbonate (bicarbonate). The absorption of calcium in the ileum depends on the presence of bile salts and adequate supplies of vitamin D.

Adults need about 800 mg calcium each day, children about 1 400 mg and pregnant or nursing mothers 1 600–2 000 mg. The demand of the growing foetus for calcium which is contributing to its skeleton depletes the mother's blood of its calcium ions. The milk secreted after birth also drains the maternal calcium resources. If there is insufficient calcium in the diet it will be withdrawn from the mother's bones to meet the requirements of the foetus.

Iron. Iron is another element which may be deficient in the diet. It is essential for the formation of haemoglobin in the red blood cells and the hydrogen acceptor system (p. 28) in nearly all living cells. Adults need about 15 mg per day. Red meat, particularly liver, is the best source of iron, but eggs, spinach and other vegetables contain appreciable quantities. Iron is conserved in the body, very little being excreted. The liver is mainly responsible for the storage of iron from disintegrated red cells. Shortage of iron in the diet leads to one form of anaemia.

Anaemia is a shortage of red cells or haemoglobin in the blood. It may result from the failure of the red bone marrow to make enough cells or pigment, from an excessive rate of destruction of red cells, or simply from a shortage of iron in the diet. In the latter case, called iron-deficiency anaemia, the condition can be cured by eating meat, liver or green vegetables such as the leaves of spinach, cocoyam or groundnuts, all of which contain iron, or by taking tablets containing suitable compounds of iron.

Iodine. This is needed in small quantities, 0.05–0.1 mg per day, for the synthesis of thyroxine, the hormone of the thyroid gland (p. 157). A deficiency of iodine results in 'simple' goitre, a gross swelling of the thyroid gland in the neck. This is most likely to occur in regions where iodides are absent from drinking water, such as Switzerland and parts of Africa and America. Five million people in India are thought to suffer from iodine deficiency to the extent of goitre or cretinism (p. 157). The disease can be prevented by adding minute traces of iodides to drinking water or to table salt (iodized salt). After a certain age, goitre cannot be cured by administration of iodine. Seafood such as fish is the best source of natural iodine.

Other 'trace' elements. Very small quantities of cobalt, manganese, zinc, copper and fluoride are known to be essential but are far less likely to be missing from the diet than calcium, iron and iodine. Cobalt is involved in the formation of vitamin B_{12}, zinc in insulin, and fluorides in the teeth.

It is important to realize that although certain foods and salts may be rich in essential elements, they may not be in a form which the body is able to use or absorb. The phosphates in cereals cannot be utilized in the body; calcium taken in with fats or an excess of phosphates or alkalis is not likely to be efficiently absorbed. Doses of essential elements in organic or inorganic compounds will not cure a deficiency if absorption or metabolism of that element is faulty.

Vitamins

Vitamins are complex, chemically unrelated compounds which, although they have no energy value, are essential in small quantities for the normal chemical activities of the body. It was not until about 1900 that their importance was realized by Gowland Hopkins and other workers.

If a diet is deficient in one or more vitamins it results in a breakdown of normal bodily activities and produces some symptoms of disease. Such diseases can usually be effectively remedied by including the necessary vitamins in the diet.

Plants can build up their vitamins from simple substances but animals must obtain them 'ready-made' directly or indirectly from plants.

Fifteen or more vitamins have been isolated and most of them seem to act as catalysts in essential chemical changes in the body, each one influencing a number of vital processes; e.g.

some of the B group vitamins form the hydrogen acceptors involved in respiration (p. 27).

Vitamins A, D, E and K are the *fat-soluble vitamins*, occurring mainly in animal fats and oils and absorbed along with the products of fat digestion. Vitamins B and C are the *water-soluble vitamins*.

Some of the important vitamins are set out in the table on p. 35, together with their properties. It must be emphasized that nearly all mixed diets will include adequate amounts of vitamins, and deficiency diseases are only likely to occur where the bulk of the diet consists of one or two kinds of food such as rice or maize.

Apart from ascorbic acid, vitamins are not seriously affected by cooking or canning processes. Vitamin C is not so much affected by boiling as by oxidation when exposed to the air. If vegetables are plunged into boiling water, the enzymes that help to oxidize the ascorbic acid are destroyed and its loss is minimized. Also, if only small quantities of water are used for boiling vegetables, relatively little ascorbic acid is dissolved out. Canning destroys only a small proportion of vitamin C, but prolonged storage of fruit and vegetables reduces the quantity of the vitamin quite markedly.

Water

Water makes up 70 per cent by weight of the body tissues and is a fundamental constituent of protoplasm. Man can live for 2–3 weeks or more without food, if relatively inactive, but may die in two or three days if deprived of water. Water plays an important part in the chemistry of digestion (p. 62) and transports digested food and many other substances in solution to all parts of the body. All the chemical reactions in the body take place in solution. Excretion of urine, elimination of faeces and evaporation from the skin and lungs result in a loss of 2–3 litres per day. About 1.5 litres of this is replaced by drinking liquids, and the remainder comes from the food.

The regulation of water in the body is described on p. 94.

Dietary fibre (roughage)

This consists largely of the cellulose in the cell walls of plants. Humans have no enzymes for digesting cellulose but bacteria in the colon (p. 61) digest part of the fibre to fatty acids which the colon can absorb. Vegetable fibre, therefore, may supply some useful food material, but it has other important functions.

The fibre itself and the bacteria which multiply from feeding on it, add bulk to the contents of the colon and help it to retain water. This keeps the faeces soft (p. 66) and reduces the time needed for the undigested residues to pass out of the body. Both these effects help to prevent constipation and keep the colon in a healthy condition.

Milk

Milk is the sole component of diet during the first weeks or months of a mammal's life. It is an almost ideal food since it contains proteins, fats, carbohydrates, mineral salts, particularly those of calcium and magnesium, and vitamins. For adults, however, it is less satisfactory because of its high water content and lack of iron. Large volumes would have to be consumed if it were the principal component of diet for an adult, and serious blood deficiencies would result from the lack of iron. In the body of the embryo mammal, iron is stored while the embryo develops inside the body of its mother, and this supply must suffice until the young mammal begins to eat solid food.

The composition of the milk varies greatly in different species of mammal. Whale's milk is very rich in fat; rabbit's milk contains ten times more protein than human milk; cow's milk contains too much protein, too little sugar and the wrong proportion of minerals for human babies. Fresh cow's milk can be diluted and sweetened or it can be dried and processed to make it more suitable but it is still far less satisfactory than human breast milk. One reason for this is that human milk contains antibodies which help protect the baby against disease (p. 254). Another is that bottle-feeding using manufactured milk powders introduces a significant risk of infection, unless the mixture, the bottle and the teat are sterile.

Balanced diet

A balanced diet must contain enough carbohydrates and fats to meet our energy needs. It must contain enough protein of the right kind to provide the essential amino acids to make new cells and tissues for growth or repair. The diet must also contain vitamins and mineral salts, plant fibre and water.

Vitamins and their characteristics

NAME AND SOURCE OF VITAMIN	DISEASES AND SYMPTOMS CAUSED BY LACK OF VITAMIN	NOTES
Retinol (vitamin A; fat-soluble) Liver, cheese, butter, margarine, milk, eggs. **Carotene** (vitamin A precursor; water-soluble) Spinach, red palm oil, carrots.	Reduced resistance to disease, particularly those which enter through the epithelium. Poor night vision. Cornea of eyes becomes dry and opaque leading to *keratomalacia* and blindness.	The yellow pigment, carotene, present in green leaves, carrots, red peppers and palm oil, is turned into retinol by the body. Modified retinol forms part of the light-sensitive pigment in the retina (page 135). Retinol is stored in the liver.
Vitamin B complex Ten or more water-soluble vitamins usually occurring together. Four are described here.	The B vitamins are present in most unprocessed food. Deficiency diseases usually arise only in populations living on restricted diets.	Many of the B vitamins act as catalysts in the oxidation of carbohydrates during respiration. Absence of these catalysts upsets the body chemistry and leads to illness.
Thiamine (vitamin B$_1$) Whole grains of cereals, beans, groundnuts, green vegetables, meat, yeast and 'Marmite'.	Wasting and partial paralysis, or water-logging of the tissues and heart failure. These are symptoms of the two forms of *beriberi*.	Rice husks contain both thiamine and tryptophan, so highly milled rice is deficient in both. Populations living largely on milled rice are very prone to beriberi. Maize lacks tryptophan, so a diet which consists mainly of maize can lead to pellagra.
Nicotinic acid or **Niacin** (vitamin B$_3$) Beans, lean meat, liver, yeast and 'Marmite'. Nicotinic acid can be made in the body from the essential amino acid *tryptophan* which is present in most cereals.	Skin eczema on exposure to sunlight, diarrhoea, wasting and mental degeneration; all symptoms of *pellagra*.	
Cobalamin (vitamin B$_{12}$) From animal products only, e.g. meat, milk, eggs, cheese, fish.	*Vitamin deficiency anaemia* (pernicious anaemia).	This vitamin contains cobalt and is made by the bacteria in the intestines of herbivorous animals
Folic acid Liver, spinach, fish, beans, peas.	*Vitamin deficiency anaemia*. Not enough red blood cells are made.	May affect pregnant women on poor diets, and people with malaria or hookworm.
Ascorbic acid (vitamin C; water-soluble) Oranges, lemons, guava, pawpaw, mango, tomatoes, fresh green vegetables.	Vitamin C is needed for the formation of collagen fibres in connective tissues, e.g. in the skin (p. 96). If ascorbic acid is deficient, wounds do not heal properly; bleeding occurs under the skin, particularly at the joints; the gums become swollen and bleed easily. These are all symptoms of *scurvy*.	Possibly acts as a catalyst in cell respiration. Scurvy is only likely to occur when fresh food is not available. Milk contains little ascorbic acid so babies need additional sources. Cannot be stored in the body; daily intake needed.
Calciferol (vitamin D; fat-soluble) Butter, milk, cheese, egg-yolk, liver, fish-liver oil.	Calcium is not deposited properly in the bones, causing *rickets* in young children because the bones remain soft and are deformed by the child's weight. Deficiency in adults causes *osteomalacia*; fractures are likely.	Calciferol helps the absorption of calcium from the intestine and the deposition of calcium salts in the bones. Natural fats in the skin are converted to a form of calciferol by sunlight.

There are several other substances classed as vitamins, e.g. **riboflavin** (B$_2$), **tocopherol** (E), **phylloquinone** (K). but these are either (a) unlikely to be missing from the diet, or (b) not known to be important in the human diet. Vitamin K is synthesized by bacteria in the colon (p. 66). If it is absorbed, this might explain why dietary deficiency is unimportant.

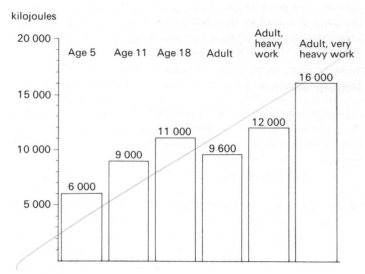

Fig. 6.3   The daily energy requirements as they change with age and occupation

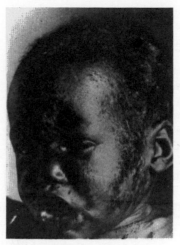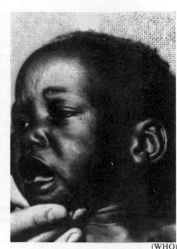

(WHO)

(a) *On admission to hospital*   (b) *After 15 days on a good diet with adequate protein*

Fig. 6.4   Kwashiorkor can often be cured by improving the quality of the diet

Energy requirements. Energy can be obtained from carbohydrates, fats and proteins. The cheapest energy-giving food is usually carbohydrate; the greatest amount of energy is available in fats; proteins give as much energy as carbohydrates but are usually more expensive. Whatever mixture of carbohydrate, fat and protein makes up the diet, the total energy must be sufficient (a) to keep internal body processes working (e.g. heart beating, breathing action), (b) to maintain the body temperature and (c) to meet the needs of work and other activities. The bodily activities (a) and (b), needed just to stay alive, are called *basal metabolism*.

The amount of energy that can be obtained from food is measured in calories or joules. One gram of carbohydrate or protein can provide up to 17 kJ (kilojoules). A gram of fat can give 39 kJ. We need to obtain about 12 000 kJ of energy each day from our food. The table below shows how this figure is obtained. However, the figure will vary greatly according to our age, occupation and activity (Fig. 6.3). It is fairly obvious that a person who does hard manual work, such as digging, will use more energy than someone who sits in an office.

Energy requirements in kJ

8 hours asleep		2 400
8 hours awake; relatively inactive physically		3 000
8 hours physically active		6 600
	Total	12 000

If the diet includes more food than is needed to supply the energy demands of the body, the surplus food is stored either as glycogen in the liver (*see* p. 69) or as fat below the skin and in the abdomen.

Pregnancy. The embryo growing in the uterus of a pregnant woman will need supplies of amino acids for building the proteins of its tissues, calcium for its bones, iron for its blood, and energy to carry out all these building-up processes. A pregnant woman, therefore, needs to increase her intake of protein, calcium and iron. She should also ensure that her diet has sufficient vitamin D to help the calcification of the embryo's bones. Increased intake of vitamins A, B₁, C and folic acid is also recommended.

If the woman's diet is already adequate, the amount of additional protein, minerals and vitamins needed is quite small, e.g. only 2–10 g of extra protein per day, as the pregnancy proceeds. However, if her diet is already barely adequate, more effort will be needed to supply the extra nutrients demanded by the growing embryo.

Lactation. 'Lactation' means the production of breast milk for feeding the baby. The production of milk, rich in proteins and minerals, makes a large demand on the mother's resources. She needs to take in even more protein than during pregnancy and to maintain at least the same level of vitamins and minerals (particularly calcium) as when she was pregnant.

Growing children. Most children up to the age of about 12 years need less food than adults, but they need more in proportion to their body weight. For example, an adult may need 0.57 g protein per kg body weight, but a 6–11 month baby needs 1.53 g per kg, and a 10-year-old needs 0.8 g per kg. The extra protein is needed for making new tissues as the child grows. Large amounts of starchy foods such as cassava, potatoes or rice should be avoided because the child soon feels 'full-up' and yet may not have taken in sufficient protein.

In addition to protein, children need extra calcium for their growing bones, iron for their red blood cells, vitamin D to help calcification of their bones and vitamin A for disease resistance. Children's eyesight is particularly at risk if vitamin A is lacking in the diet (*see* p. 35).

It is also desirable to give young babies extra vitamin C to make up for the shortage of this vitamin in milk. If extra vitamins A and D are given in the form of fish-liver oil, the dose rate must be strictly observed. Both these vitamins can be poisonous if too much is taken.

Old age. Elderly people generally need a smaller intake of food as they become less active. As their appetite gets less or they lose their teeth, their food intake may become so low that they lack essential amino acids, vitamins and minerals. It is important to offer them diets which, though they may be easy to chew and swallow, still contain a good level of vitamins and minerals, particularly iron.

Unbalanced diets

Tropical diets are often unbalanced for the following reasons.

(i) They contain too much yam, cassava, banana and maize. These foods are easy to grow but deficient in protein. In pregnant mothers, babies and young children a protein shortage can lead to malnutrition. Rice, millet, guinea corn and beans are a far better source of vegetable protein than cassava and yams. Maize contains some protein but does not provide all the amino acids needed by the body and, moreover, one of the B vitamins, nicotinic acid, cannot be adequately obtained from maize.

(ii) They contain too little vitamin A. Special care needs to be taken to include green vegetables or red palm oil in the diet, in addition to tomatoes and other sources of carotene. Alternatively the vitamin A in the diet may be supplemented with fish-liver oil.

(iii) In areas where the diet is mainly rice or maize, there is likely to be a shortage of vitamins B_1 and B_3. Different methods of milling and cooking rice could reduce this deficiency, while a maize diet needs to be supplemented with other cereals to avoid pellagra (*see* table on p. 35).

(iv) Insufficient joules, or in other words, not enough food. About half the people in the world, principally in Asia, South America and Indonesia, have an average daily intake of less than 9 240 kJ. Even allowing for children whose needs may be less than this, these figures suggest that many people have only enough food to stay alive and not enough to do a day's work.

Kwashiorkor. This is a Ghanaian term meaning 'the sickness the old baby gets when the new baby comes'. For a long time it was thought that kwashiorkor was the result of a shortage of protein in the diet. It is when babies are weaned from protein-rich breast milk to protein-deficient yam and cassava that they tend to develop kwashiorkor. They become listless and miserable; their skin cracks and becomes scaly (Fig. 6.4); their abdomens swell and their hair takes on a reddish colour. Many children affected by kwashiorkor die before the age of five. Because kwashiorkor occurs mainly in children whose diets are deficient in quantity as well as in protein, the disease has also been described as protein-calorie, or protein-energy malnutrition (PEM). In many cases it has been possible to cure the disease by providing more food and a better balance in the diet.

However, there are many cases where some children develop kwashiorkor and others do not, even though they all have the same low-protein, low-energy diet. Also, kwashiorkor occurs mainly in hot, humid regions rather than in hot, dry regions despite the fact that diets in the two regions differ very little.

Recent studies suggest that the disease might result from various causes, including protein deficiency. Other causes might be the after-effects of measles or even the poisons produced by a mould which grows in certain types of food. These poisons, called *aflatoxins*, have been found in ground-nuts, chick peas and dried okra, and affected children have higher levels of aflatoxins in their blood than normal. Aflatoxins are known to cause liver damage.

It is in the hot, humid regions, and in the rainy season that mould grows most rapidly on food. This would explain the distribution and time of occurrence of kwashiorkor.

It seems likely that some children are able to break down the aflatoxins in their bodies better than others, which would explain why they do not develop kwashiorkor even though they eat the same food. If this theory is correct, one way to combat the disease would be to improve the methods of storing and marketing the suspect food, to prevent the growth of mould.

Western diets. In the affluent societies of the USA, USSR and Europe, there is no general shortage of food and most people can afford a diet with an adequate energy and protein content. Consequently there are few people who suffer from malnutrition, but it seems very likely that many people eat too much food of the wrong kind. This causes illnesses in middle-age and old-age.

Some nutritionists believe that the high level of animal fats in the meat, butter and cheese of western diets contributes to the increased frequency of heart attacks. (Vegetable oils, other than coconut oil and palm oil, may be safer in this respect.)

Westerners also eat a great deal of refined sugar (sucrose). This is the white sugar in the sugar bowl, or the sugar added to a great many processed foods. Like animal fats, it is suspected of contributing towards heart disease and there is little doubt that it causes tooth decay (page 126) and fatness (obesity). People who are overweight are more likely to suffer from heart disease, high blood pressure and diabetes (page 157).

Western diets usually contain too little vegetable fibre and this may lead to constipation and diseases of the large intestine.

As some tropical countries become wealthier, there is a danger, at least in the towns, of a change to western diets with their low-fibre and high-energy content and an excess of animal fats. There is plenty of evidence to show that a totally vegetarian diet, even one which excludes milk, butter, cheese and eggs, can be as healthy as, if not healthier than, a diet which contains meat.

It should also be remembered that meat production may involve a very inefficient use of food resources. If cereals and beans are fed to cattle, only about 10 per cent of their energy is converted to meat. The rest is used by the animal for its life processes. It is far more efficient for humans to eat the cereal and beans themselves (*see* p. 56).

Cooking. Cooking food improves its taste and flavour and makes it easier to chew, swallow and digest. The starch grains in cereals and root crops are almost indigestible unless they have been cooked. The processes of boiling, frying and baking food also destroy bacteria and other harmful organisms which may be present.

However, there are some disadvantages in cooking food, mainly due to the loss of vitamins. The fat-soluble vitamins, A and D, are not destroyed by heat or dissolved out in the cooking water. Vitamins B and C, on the other hand, are dissolved by boiling water and affected by heat. Cooking may destroy 30–50 per cent of thiamine (vitamin B_1). Vitamin C is also affected by boiling, partly because it is destroyed, but partly because in the early stages of cooking, the enzymes in the plant cells attack the vitamin. Plunging green vegetables straight into boiling water destroys these enzymes quickly and so helps to retain the vitamin C.

To avoid losing the water-soluble vitamins, the smallest possible quantity of water should be used, and boiling should not be prolonged more than is needed for adequate cooking. If the fluids used for boiling meat and vegetables are kept for soups or gravy, some of the dissolved vitamins will be retained.

Mineral salts are not much affected by cooking, though meat and fish may lose 40 per cent of their salts when boiled.

Practical Work

Experiment 1 Determination of balanced diet

Draw up a table such as that given below and, using the foods listed on p. 39 (note that all the values in the table are for 10 g of food), fill in the details of what you consider a reasonable series of meals for one day. Work out the total energy value of the day's food intake and compare it with the figures given on p. 36. Also calculate the total protein content and indicate which vitamins are present.

The table is completed for one meal to show how to work out the values, but it will be necessary to weigh samples of the food selected in order to complete the calculations for the rest of the day. If your results do not conform to the criteria given on p. 35 for a balanced diet, suggest ways in which your day's menus could be adjusted.

When the sugar has finished burning and cannot be ignited again, gently stir the water in the can with the thermometer and record its new temperature. Calculate the rise in temperature by subtracting the first from the second temperature. Work out the quantity of energy transferred to the water from the burning sugar as follows.

4.2 joules raise 1 g water 1 °C
100 cm³ cold water weighs 100 g
Let the rise in temperature be T °C

To raise 1 g water 1 °C needs 4.2 joules
∴ To raise 100 g water 1 °C needs 100×4.2 joules
∴ To raise 100 g water T °C needs $T \times 100 \times 4.2$ joules
∴ 1 g burning sugar produced $420 \times T$ joules

Meal	Menus	Weight (in grams)	Energy (in kilojoules)	Vitamins	Protein (in grams)
Breakfast	1 fried egg	65	$6.5 \times 67 = 436$	A, B, D	$6.5 \times 1.2 = 8$
	(cooking fat)	6	$0.6 \times 334 = 200$	A, D	
	1 slice bread	35	$3.5 \times 97 = 340$	B	$3.5 \times 0.8 = 2.8$
	(butter)	5	$0.5 \times 334 = 167$	A, D	
	1 cup coffee		0		
	(¼ cup milk)	60	$6 \times 28 = 168$	A, B, C, D	$6 \times 0.3 = 1.8$
	(teaspoon sugar)	7	$0.7 \times 162 = 113$		
Totals					

Experiment 2 Energy from food

Arrange the apparatus as shown in Fig. 6.5. Use a measuring cylinder to place 100 cm³ cold water in the can. With the thermometer, find the temperature of the water and make a note of it. In the nickel crucible or tin lid place 1 g sugar and heat it with the bunsen flame until it begins to burn. As soon as it starts burning, slide the crucible under the can so that the flames heat the water. If the flame goes out, *do not* apply the bunsen burner to the crucible while it is under the can, but return the crucible to the bunsen flame to start the sugar burning again and replace the crucible beneath the can as soon as the sugar catches light.

The experiment may now be repeated using 1 g vegetable oil instead of sugar and replacing the warm water in the can with 100 cm³ cold water.

(*Note*. The experiment is very inaccurate because much of the heat from the burning food escapes into the air without reaching the water, but since the errors are about the same for both samples, the results can at least be used to compare the energy released from sugar and oil.)

Experiment 3 Food tests

(a) *Starch*. A little starch powder is shaken in a test-tube with some cold water and then boiled to make a clear solution. When the solution is cold, 3 or 4 drops of iodine solution* are added. The dark blue colour that results is characteristic of the reaction between starch and iodine.

(b) *Glucose*. A little glucose is heated in a water bath (*see* Fig. 7.11, p. 47) with some Benedict's solution* in a test-tube. The solution changes from clear blue to opaque green, yellow, and finally a brick-red precipitate of copper(II) oxide appears.

(c) *Sucrose* is recognized by its failure to react with Benedict's solution until after it has been boiled with dilute hydrochloric acid and neutralized with sodium hydrogencarbonate.

(d) *Protein*. To a 1% albumen solution is added 5 cm³ dilute sodium hydroxide (CARE: this solution is caustic) and 5 cm³ 1% copper sulphate solution. A purple colour indicates protein. (This is the *biuret* reaction.)
*See p. 319.

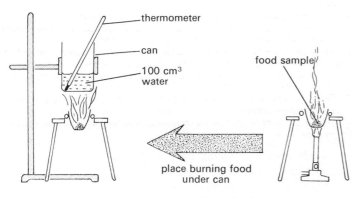

thermometer

can

100 cm³ water

food sample

place burning food under can

Fig. 6.5 Energy from food

Table of food values

Food 10 grams	Energy value kilojoules	Protein grams	Fat grams	Carbo-hydrate grams	Calcium milligrams	Iron milligrams	Vitamin A micrograms	Thiamine (B_1) micrograms	Ascorbic acid (C) micrograms	Vitamin D micrograms
Bread, wholemeal	101	0.96	0.3	4.7	2.8	0.3	0	24	0	0
Potato	32	0.2	0	1.8	0.8	0.07	0	11	2000	0
Rice	150	0.6	0.1	8.7	0.4	0.04	0	8	0	0
Beans	107	2.1	0	4.5	18.0	0.67	0	45	0	0
Groundnuts	245	2.8	4.9	0.8	6.1	0.2	0	23	0	0
Soya flour	181	4.0	2.4	1.3	21.0	0.7	0	74	0	0
Beef	131	1.5	2.8	0	1.0	0.4	0	7	0	0
Pork	171	1.2	4.0	0	1.0	0.1	0	100	0	0
Liver	58	1.7	0.8	0	0.8	1.4	600	30	2000	0.07
Chicken	60	2.1	0.7	0	1.1	0.15	0	4	0	0
White fish	30	1.6	0.05	0	2.5	0.1	0	6	0	0
Herring	80	1.6	1.4	0	10.0	0.15	4.5	3	0	2.2
Milk, whole	27	0.3	0.4	0.5	12.0	0.01	4.4	4	130	0.005
Dried milk, whole	206	2.7	2.8	3.8	81.0	0.07	24.6	31	1100	0.03
Butter	312	0.05	8.3	0	1.5	0.02	99.5	0	0	0.125
Cheese	173	2.5	3.5	0	81.0	0.06	42.0	4	0	0.04
Eggs	66	1.2	1.2	0	5.6	0.25	30.0	10	0	0.15
Sugar	165	0	0	10.0	0.1	0	0	0	0	0
Carrots	9	0.07	0	0.5	4.8	0.06	200	6	600	0
Spinach	9	0.3	0	0.3	7.0	0.3	100	12	6000	0
Orange	15	0.08	0	0.8	4.1	0.03	0.8	10	5000	0
Banana	32	0.1	0	1.9	0.7	0.04	3.3	4	1000	0

Note that: (a) the weights quoted are for 10 of the edible parts of the natural food, (b) water and indigestible matter are not included, (c) tea and coffee have no food value apart from any milk and sugar added, (d) fried food will have about 1 g extra fat and 40 kJ extra for every 10 g food cooked. From McCance and Widdowson, *The Composition of Foods*, by permission of the Controller of Her Majesty's Stationery Office.

(e) *Fats*. Two drops of cooking oil are thoroughly shaken with about 5 cm³ ethanol in a dry test-tube until the fat dissolves. The alcoholic solution is poured into a test-tube containing a few cm³ water. A cloudy white emulsion is formed. This shows that the solution contained some fat or oil.

(f) *Application of the food tests*. The tests can be used on samples of food such as milk, pineapple, yam, onion, beans, egg-yolk, groundnuts, to find what food materials are present. The solid samples are crushed in a mortar and shaken with warm water to extract the soluble products. Separate samples of the watery mixture of crushed food are tested for starch, glucose or protein as described above. To test for fats, the food must first be crushed with ethanol, not water, and then filtered. The clear filtrate is poured into water to see if it goes cloudy, indicating the presence of fats.

Experiment 4 **Comparison of the vitamin C content of fruit juices**

A blue dye, phenolindo-2-6-dichlorophenol (P.I.P. for short), is turned colourless by vitamin C (ascorbic acid). Place 1 cm³ of a 0.1 per cent solution of P.I.P. in a test-tube. Draw up exactly 2 cm³ fresh orange juice in a syringe (without the needle). Add the juice one drop at a time to the P.I.P. solution until it goes colourless. Note the volume of juice added (i.e. the difference between the volume left in the syringe and the original 2 cm³). Repeat the experiment with lemon juice and grapefruit juice and a 0.1 per cent solution of ascorbic acid.* Bottled or canned fruit squashes or cordials can also be tested.

Draw up a table to show the juices in order of vitamin C content on the assumption that the more vitamin C is present, the less juice is needed to decolourize the blue dye.

For further experiments, see the 'Experimental Work in Biology Series', p. 320.

Questions

1 What sources of protein are available to a vegetarian who (a) will eat animal products but not meat itself, (b) will eat only plants and their products? Why must all diets contain some protein?
2 What principles must be observed when working out a diet designed to produce a reduction in weight (a slimming diet)? What dangers are there if such diets are not scientifically planned?
3 Which tissues of the body need (a) iron, (b) glucose, (c) calcium, (d) protein, (e) iodine?
4 List the ways in which green leaves contribute to our diet.
5 It is sometimes believed that a person who does hard physical work needs to eat a lot of protein. Try to explain why this is not true.
6 From the table on p. 36 work out the approximate minimum amount of energy needed each day to keep an adult alive, e.g. lying quietly in bed.
7 Eating a large amount of protein at one meal is wasteful. It is better to take in a little protein at each meal. Why do you think this is the case?
8 What nutritional problems are likely to be experienced by people living mainly on rice or maize?
9 Why might it be better to cook vegetables in the steam from boiling water rather than in the boiling water itself?

* See p. 319.

7
Sources of Food

Man obtains his food from plants and their products, and from animals which feed on plants. The food substances that provide him with energy or body-building materials are made, in the first instance, by plants. Obvious plant products in the diet are seeds, fruits, leaves and roots; but equally, flour and bread are derived from cereal grains, and sugar from stems and roots. Beef, milk, butter and cheese all come from cattle, but the raw materials for their manufacture were in the grass eaten by the cattle. No matter what article of diet is considered, the source of its matter and energy can be traced back to green plants. Green plants, therefore, are called primary producers and it is important to understand the process by which green plants make their food. This process is called *photosynthesis*.

Photosynthesis

Green plants *make* their food. Although they need and use proteins, fats, carbohydrates and vitamins in the same way as do animals, they have to build up these complex organic food substances from simple inorganic materials which they obtain from the air and the soil (Experiment 1). Carbohydrates and fats are built up from carbon dioxide in the air and water from the soil (Experiment 2). Amino acids and proteins are made by combining the carbohydrates with nitrogen obtained in the form of nitrate ions in solution from the soil. The chemical reactions that build up all these molecules need a supply of energy, and this energy comes in the first instance from sunlight.

In the cells of the green parts of plants, particularly in the leaves, there are organelles called *chloroplasts* (Fig. 7.2). The chloroplasts contain a green chemical called *chlorophyll*. When light strikes a chlorophyll molecule some of the energy from the light is absorbed by the chlorophyll, making it temporarily unstable. The energy thus trapped in the chlorophyll molecule is used to split a water molecule into hydrogen and oxygen ions, and the chlorophyll molecule reverts to its stable condition.

The oxygen ions combine to make oxygen molecules, O_2, and consequently oxygen gas is produced by a green plant when photosynthesis is going on (Experiment 3). The hydrogen ions now contain the energy that was trapped by the chlorophyll and they are passed to a series of hydrogen acceptors, giving up their energy and forming ATP from ADP and phosphate in the same way as described on p. 28. The store of energy-rich ATP so produced is used firstly to make molecules of carbon dioxide and hydrogen combine to produce glucose and then to build up other molecules such as starch, cellulose and protein from the glucose molecules as indicated on pp. 15–17.

A simple definition of photosynthesis is the production of food by a green plant from carbon dioxide and water using the energy from sunlight which is trapped by chlorophyll. An equation is often given which sums up the process,

$$6CO_2 + 6H_2O \xrightarrow{\text{energy from sunlight}} C_6H_{12}O_6 + 6O_2$$

carbon dioxide   water   glucose   oxygen

but it is not a very accurate representation of the many chemical steps which actually take place.

A consideration of our dependence on plants for our food, and the plants' dependence on sunlight for the energy to make their food, leads to the conclusion that man and all other living organisms obtain the energy for their living processes indirectly from the sun. When a molecule of glucose in the body is broken down by respiration, the energy released from it is the same energy that the plant put into making it, and that energy came from sunlight.

$$\text{Chlorophyll} \xrightarrow{\text{light}} \text{energized chlorophyll} \longrightarrow \text{stable chlorophyll}$$

$$\underset{\text{water}}{H_2O} \quad \underset{\text{2 hydrogen ions}}{2H^+} + \underset{\text{oxygen ion}}{O^{2-}}$$

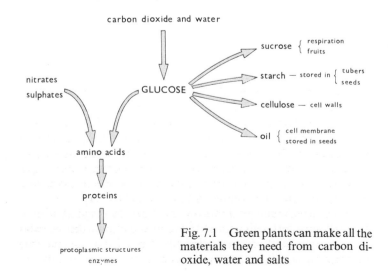

Fig. 7.1 Green plants can make all the materials they need from carbon dioxide, water and salts

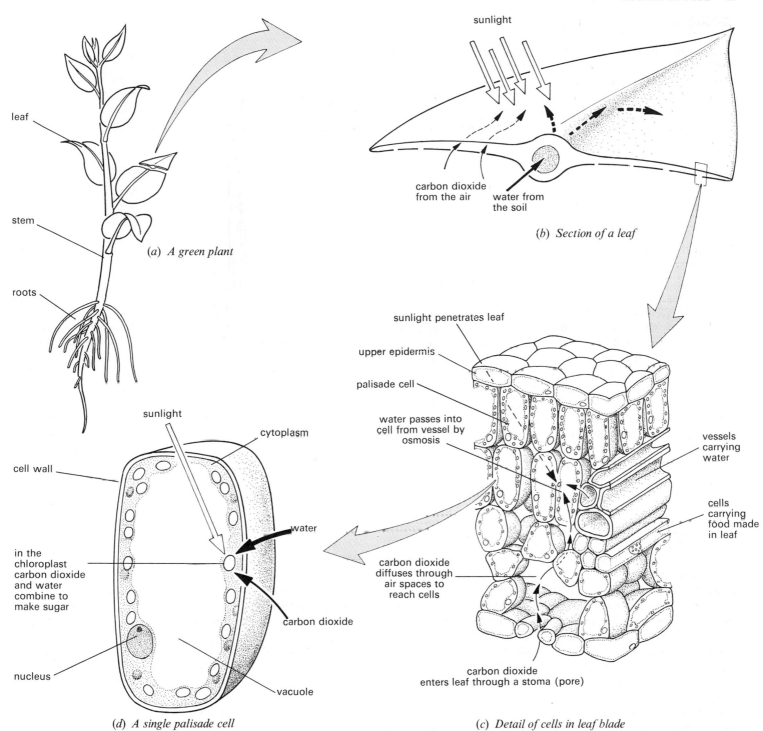

leaf

stem

roots

(a) *A green plant*

sunlight

carbon dioxide from the air

water from the soil

(b) *Section of a leaf*

sunlight penetrates leaf

upper epidermis

palisade cell

water passes into cell from vessel by osmosis

vessels carrying water

cells carrying food made in leaf

carbon dioxide diffuses through air spaces to reach cells

carbon dioxide enters leaf through a stoma (pore)

(c) *Detail of cells in leaf blade*

sunlight

cytoplasm

cell wall

in the chloroplast carbon dioxide and water combine to make sugar

water

carbon dioxide

nucleus

vacuole

(d) *A single palisade cell*

Fig. 7.2   Photosynthesis in a leaf

Synthesis of other materials by plants (Fig. 7.1). Starting from glucose, a green plant can make all the carbohydrates, fats, proteins and vitamins it needs for its growth and its chemical reactions. Sucrose is made by combining two $C_6H_{12}O_6$ molecules (p. 16). Starch and cellulose are made by joining hundreds of glucose molecules in long chains (*polymerization*). Glucose is converted to amino acids using nitrates from the soil. The amino acids are combined to make proteins (p. 17).

Some substances need metallic elements for their synthesis; for instance, the chlorophyll molecule includes the element magnesium, the hydrogen acceptors need iron, the material between the cell walls incorporates calcium. These metals and the non-metallic elements are obtained by the green plant from the soil, where they occur as ions (p. 21) in solution in soil water (Experiment 4). Nitrogen, phosphorus and sulphur occur as nitrates, $-NO_3$, phosphates, $-PO_4$, and sulphates, $-SO_4$, respectively. These inorganic ions in the soil came originally from the rocks which gave rise to the soil but as the plants use them up, or as they are washed out of the soil by rain, they are replaced by the decay of plant and animal remains as described on p. 50. In agricultural practice, the supply of essential elements in the soil is increased by adding artificial 'fertilizers', such as ammonium nitrate (for nitrogen) and compound NPK fertilizers (for nitrogen, phosphorus and potassium).

Principles of agriculture

Arable farming. The principles of arable farming involve the removal of the natural vegetation from the soil in the first place. Then (a) the soil is planted with seeds of a single plant species, (b) their growth is encouraged by improving the soil fertility, e.g. adding water and mineral salts, and (c) plant competitors and parasites are eliminated by hoeing or spraying. In this way, a limited area of soil is made to produce a much larger quantity of edible plant material than if it were left to develop its natural vegetation such as grass or forest trees.

(a) The plant species cultivated today must have had their origins in wild plants, but by careful selection and cross-breeding over many hundreds of years they have become plants with large seeds (e.g. maize), edible fruits (e.g. pineapple) or substantial root or stem tubers (e.g. yam and potato). The process of breeding and selection continues even more intensively today on a scientific basis to try and find plants that yield more food, are resistant to disease or can survive in adverse conditions such as drought.

The agricultural practice of growing large colonies of a single species of plant year after year is called *monoculture.*

(b) The soil used for agriculture is 'improved' mechanically by ploughing. This loosens the soil, admits air, allows water to drain and makes penetration by plant roots easier. The nutritive value of the soil is increased by adding natural fertilizers such as farmyard manure, or artificial fertilizers such as ammonium sulphate (Fig. 7.3). Soil structure and fertility are both improved by the practice of crop rotation (p. 53). If water is scarce it can be added to the soil by various methods of irrigation, and if the soil holds too much water it can be drained by digging ditches or placing pipes beneath the soil.

(c) In the favourable conditions provided by cultivation, plants other than the crop plants will grow and compete with them for water, root space, nutrients and light. Such plants are collectively called weeds. The weeds are kept in check firstly by ploughing the soil before planting and then by hoeing between the rows as the crops grow. Certain chemicals, *herbicides*, can be used to reduce weeds both before and after planting.

Fungus diseases of the crop plants are suppressed by selecting resistant strains of plant and by spraying chemical *fungicides* on the crop. Insect pests which eat or damage crop plants or their seeds are reduced by spraying the crops with *insecticides* and by dipping the seeds in similar chemicals before planting.

Animal husbandry. Animal protein contains more essential amino acids than does plant protein (p. 33) and man, generally, finds animal products particularly palatable. In consequence, man uses some of his crops to feed animals and then eats the animals or their produce such as milk and eggs.

The principles of animal husbandry are much the same as those applied in growing crops. A limited number of species is used. One such is the cow, which after years of selective breeding has high milk yield, produces much meat or is resistant to disease or adverse conditions . The animals are allowed to grow in as near ideal conditions as possible and pests and competitors are eliminated. Only about 10 per cent of the food given to animals appears as flesh, or eggs or milk. The rest is used for energy to keep the animal alive, maintain its temperature and enable it to move about. The logical step in modern animal husbandry is therefore to reduce the energy losses to a minimum. In 'factory' farming, the animals are kept indoors at a constant temperature to reduce heat losses, and their

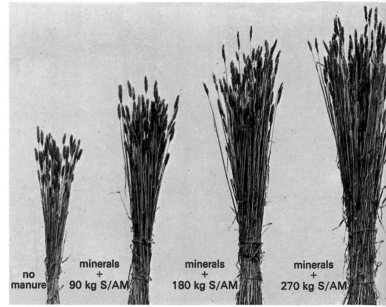

(Rothamsted Experimental Station)

Fig. 7.3   Wheat harvested from equal areas of soil (S/AM = sulphate of ammonia)

movements are restricted so that the food they eat is used for making flesh rather than being used for energy production. If the world population continues to increase at its present rate, it will become necessary to use more plant products directly for food instead of losing 90 per cent of their energy value by feeding them to animals.

Food chains and energy transfer

Food chains. A mouse eats the seed of a plant; a fox eats the mouse. This sequence: plant→mouse→fox is called a *food chain.* The plants which bear the leaves and seeds are called *producers.* Animals (herbivores) which eat the plants are called *first order consumers.* Animals (carnivores) which eat other animals are *second* or *third order consumers.* Man can be a first, second or third order consumer.

When plants or animals die, their remains are broken down by bacteria and fungi. These organisms are called *decomposers.* Food chains are discussed more fully on p. 56.

Energy transfer. With the exception of atomic energy and tidal power, all the energy released on Earth is derived from sunlight. The energy released by animals is derived ultimately from plants that they or their prey eat and the plants depend on sunlight for making their food. The energy in organic fuels also comes ultimately from sunlight trapped by plants. Coal is formed from fossilized forests and petroleum comes from the cells of ancient marine plants.

To try and estimate just how much life the Earth can support it is necessary to examine how efficiently the sun's energy is utilized. The amount of energy from the sun reaching the Earth's surface in one year ranges from 2 million to 8 million kilojoules per square metre ($2–8 \times 10^9$ J m^{-2} y^{-1}) depending on the latitude. When this energy falls on to grassland, about 20 per cent is reflected by the vegetation, 39 per cent is used in evaporating water from the leaves (transpiration), 40 per cent warms up the plants, the soil and the air, leaving only about 1 per cent to be used in photosynthesis for making new organic matter in the leaves of the plants (Fig. 7.4a). This figure of 1 per

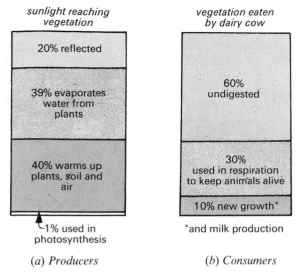

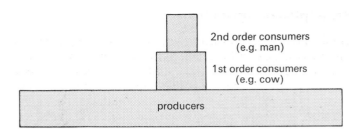

Fig. 7.5   Energy transfer between producers and consumers

Fig. 7.4   Utilization of energy by producers and consumers

cent will vary with the type of vegetation being considered and with climatic factors such as availability of water and the soil temperature. Sugar cane grown in ideal conditions can convert 3 per cent of the solar energy into photosynthetic products, and sugar beet at the height of its growth has nearly a 9 per cent efficiency. Tropical forests and swamps are far more productive than grassland but it is difficult, and probably undesirable, to harvest and utilize their products.

In order to allow crop plants to approach their maximum efficiency they must be provided with sufficient water and mineral salts. This can be achieved by irrigation and the application of fertilizer. In some cases the small amount of carbon dioxide in the air may limit the rate of photosynthesis but little can be done about this except in an artificially enclosed system such as a glass-house.

Having considered the energy conversion from sunlight to plant products the next step is to study the efficiency of transmission of energy from plant products to first order consumers. On land, first order consumers eat only a small proportion of the available vegetation. In a deciduous forest only about 2 per cent is eaten; in grazing land, 40 per cent of the grass may be eaten by cows. In open water, however, where the producers are microscopic plants (phytoplankton) and swallowed whole by the first order consumers in the zooplankton (p. 56), 90 per cent or more may be eaten. In the land communities, the parts of the vegetation not eaten by the first order consumers will eventually die and be used as a source of energy by the decomposers.

A cow is a first order consumer; of the grass it eats over 60 per cent passes through its alimentary canal (p. 60) without being digested. Another 30 per cent is used in the cow's respiration to provide energy for its movement and other life processes. Less than 10 per cent of the plant material is converted into new animal tissue to contribute to growth (Fig. 7.4b). This figure will vary with the diet and the age of the animal. In a fully grown animal all the digested food will be used for energy and replacement and none will contribute to growth. Economically it is desirable to harvest the first order consumers before their rate of growth starts to fall off.

The transfer of energy from first to second order consumers is probably more efficient since a greater proportion of the animal food is digested and absorbed than is the case with plant

material. The transfer of energy at each stage in a food chain may be represented by classifying the organisms in a community as producers, first, second and third order consumers and showing their relative masses in a pyramid such as shown in Fig. 7.5. The width of the horizontal bands is proportional to the masses (dry weight) of the organisms.

In human communities, the use of plant products to feed animals which provide meat, eggs and dairy products is very wasteful because only 10 per cent of the plant material is converted to animal products. Although a proportion is undigested whether it is a man or a chicken which eats wheat, it is far more economical for the man to eat bread made from the wheat than to feed the wheat to hens and eat the eggs and chicken meat because it avoids using any part of the energy in the wheat to keep the chicken alive and healthy.

It is estimated that about 50 g protein is needed each day by an adult. Since animal protein contains more essential amino acids (p. 32) than plant protein, about half this amount should come from animal sources. However, animal protein eaten in excess of about 40 g per day is wasteful of food resources.

In Europe, much of the animal protein harvested from a dwindling fish population is used not to feed man but to provide fish meal for animal feed stuffs.

Consideration of the energy flow in a modern agricultural system reveals other sources of inefficiency because energy from the sun is not the only energy used to produce a crop. To produce 1 tonne of nitrogenous fertilizer takes energy equivalent to burning 5 tonnes of coal. Calculations show that if the energy needed to produce the fertilizer is added to the energy used to produce a tractor and to power it, the energy derived from the food so produced is less than that expended in producing it.

Hazards of agriculture

Agricultural practice is essential for feeding the present population of the world. If man were to revert to collecting and hunting his food, only a tiny fraction of the present world population could survive. Some authorities suggest that for each family of hunters and gatherers, five square kilometres of hunting territory is needed. The world population in this case could be about 10 million instead of its present 4 700 million.

Although agriculture is so essential to our continued survival, its widespread use and increased intensity carry certain dangers. Removal of natural vegetation and ploughing in some areas make the soil much more liable to be washed away by heavy rain or blown away by high winds (see Erosion, p. 53). Excessive use of some artificial fertilizers has led to streams and rivers becoming polluted with the chemicals which are washed out of the soil (see p. 52). The practice of growing plants of the same species very close to each other enables the diseases of these species to spread easily and rapidly from one plant to the next so that whole fields of crops can be infected and destroyed. The use of chemicals to control weeds, fungal diseases and insect pests is a cause for concern because the chemicals used are often harmful to man and other animals (*see* p. 58).

Fishing

Fish and other sea food make up about 6 per cent of man's total protein intake and 17 per cent of his protein from animal sources.

The principal methods of catching fish are seine-netting, drift-netting and trawling. Trawling catches fish which feed at the bottom of the sea by pulling a bag-like net along the sea bottom. It can be used only in relatively shallow water. There are also mid-water trawls.

Drift-nets are long, straight nets up to 4 km long. They are suspended by floats at the surface and hang down in the top 15 metres of the sea. Migrating shoals of fish, swimming near the surface to feed on the plankton (p. 56), become trapped in the mesh.

In seine-netting, a floating net is pulled around and under a shoal of fish near the surface.

Although sophisticated electronic aids may be used to locate shoals of fish, and scientific data help to predict their movements, fishing is still little more than organized hunting.

There is evidence to suggest that we are not exploiting the potential fish stocks effectively, mainly because the most fruitful areas are being over-fished. For example, in 10 years the North Sea herring catch declined from 700 thousand tonnes to 160 thousand tonnes per year. If so many fish are removed that too few are left to breed and replace the stock, the fisheries are destroyed. There is a 'maximum sustainable yield' for any fish population but it is very difficult to establish just what this is. A 4-year-old fish increases its weight by 50 per cent in a year, while a 6-year-old fish increases by only 35 per cent and a 20-year-old by 1 per cent. A population of rapidly growing small fish may well convert food into flesh at a far more economic rate than a smaller population of slowly growing large fish.

A carefully regulated fishing industry harvests only those fish above a certain size (by regulating the net mesh size) and ensures that a strong breeding population is left (by putting an upper limit on the total catch).

World food production

The world population is increasing at the rate of about 50 million people per year. Whether this rate of increase will continue is impossible to say but at the moment, the rate of increase is itself accelerating. United Nations experts suggest that by A.D. 2000 the world population may have increased from its present 4 700 million to 6 000 million. Such a rate of increase introduces many problems but one of the most urgent is the supply of food. Whether increase in food production can match this increase in numbers is another controversial matter, both in theory and in practice. Theoretically a ten- or twenty-fold increase in world food production seems possible, and yet the

Food and Agriculture Organization of the United Nations (FAO) estimated that, in the least developed countries, food production increased by only 1.6 per cent per year compared with an increasing population of 2.6 per cent per year. In Africa and the Near East, food production had not kept up with population growth and in Latin America it had barely kept pace. Clearly, a large proportion of the world population receives inadequate food, though often this is not so much a shortage in quantity as a lack of a small number of essential amino acids and vitamins. The situation gives little cause for complacency, but fatalistic gloom is of no value. The following paragraphs outline some of the ways in which food production can be increased.

Intensification of agriculture. The scientific principles of agriculture are mentioned on p. 42 and some of the consequences of bad agricultural practices discussed on p. 43. Not all land is suited to conventional agriculture, but land that is should be made to yield greater quantities of food while conserving its productive capacity.

In some areas, more intensive agriculture has been very effective. New varieties of wheat and rice in India, Pakistan and Sri Lanka, for example, with the aid of fertilizers and irrigation, have doubled or trebled the yields of grain. Crop varieties can be bred not only to raise their yields but to increase resistance to disease and adverse conditions. In this way, plants resistant to frost could be developed and grown in winter and spring in temperate zones so that the soil is productive all year round. Any land without green cover wastes valuable energy from sunlight.

Human sewage, often discharged wastefully into the sea, could be used, after treatment, as an agricultural fertilizer. Irrigation can increase productivity in arid regions. However, intensified production in agricultural regions and the conversion of new areas to agriculture need to be accompanied by conservation measures, as discussed on p. 53.

More nutritious crops, better extraction and different feeding habits. To raise the total energy value of man's diet to the arbitrary minimum of 12 000 kilojoules per day, though essential, is not enough. The most acute shortage is of proteins containing essential amino acids (p. 32). Peas, beans, groundnuts and green leaves contain a higher proportion of protein than many grains and far more than most roots such as yam and cassava. A great deal of work has to be done to find protein-rich crops which grow well in particular localities. European and North American garden vegetables do not grow well in tropical conditions, but new varieties of vegetable could be bred from existing wild types, increasing the leaf protein available in these regions.

Some methods of harvesting and extraction waste useful parts of plants. For example, extraction of oil from groundnuts, soya beans, coconuts and cotton seed tends to waste their high protein content when the residues are fed to animals. The world demand may be for oil but the need is for protein, and more efficient methods need to be devised for its extraction. The quantities of protein involved in these plants could supply an estimated third of the world's needs. In addition, methods of extracting protein from poisonous or unpalatable leaves could be sought.

As explained on p. 43 there is considerable loss of energy (up to 90 per cent) in the conversion of plant material to animal

(Radio Times Hulton Picture Library)

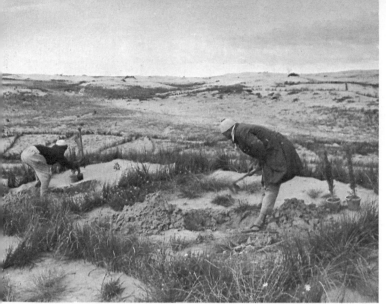

Fig. 7.6 Dunes being stabilized by planting grass in 5 m squares in the desert in Tripolitania

products, and the most efficient use man can make of food plants is to eat them directly. A suitable combination of plant material can supply all the essential amino acids and it may be that we shall have to forgo, to some extent, the more palatable animal protein in order to make the best use of available food. We can afford to feed to animals only those plant products or residues that cannot be used directly by man.

Reclamation of land. Thoughtless conversion of marginal land to conventional agriculture may lead to severe erosion, but new farming methods, coupled with appropriate conservation measures, may enable some marginal areas to produce food. Unproductive areas may be reclaimed, as is happening in North Africa and Israel today. Although some areas of natural desert may never be productive in the conventional agricultural sense, there is abundant evidence that many desert margins were once far more productive than they are today. On the nothern border of the Sahara and in the Negev and Kara Kum deserts of Israel and Russia, reclamation is being carried out with encouraging results.

The methods adopted fall roughly into two categories: (a) supplying water, and (b) establishing vegetation.

(a) *Water supply.* The water supply can be increased and controlled by replanting forests in the catchment areas and by carrying water in pipes from regions of high rainfall to arid regions. Dams are constructed to conserve water from winter floods which previously poured down the wadis and into the sea from the Kara Kum and Negev, or over the desert in the Sahara. Attempts have been made to remove the salt from sea water and brackish desert water. In Israel this has been carried out experimentally by distillation using the sun's energy, by deionization using chemical methods, or by using condensed water from power stations. In many cases, irrigation is all that is needed to make the soil productive.

(b) *Recolonization of arid areas by plants.* Firstly, coarse grasses are planted in rectangular grids. Their roots bind the soil and their foliage reduces wind erosion. In the squares of the grid are planted trees such as acacia, which can survive in conditions of water shortage. The roots and foliage of the trees enhance the soil-holding properties of the grasses. Finally, olive trees and date palms can be established and what was once shifting sand can be converted to stable, productive soil (Figs. 7.6–7.8).

(Radio Times Hulton Picture Library)

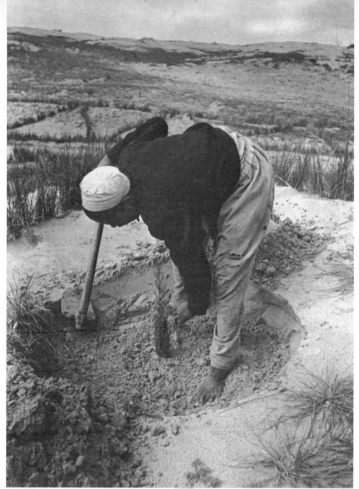

Fig. 7.7 Trees are planted in the squares

All these measures must be introduced with due consideration for the indigenous human population. It is one thing to see the solution to a problem but another to enlist the support of people who are reluctant to change their way of life. The nomads' goats are a menace to the recolonization schemes, but employing the nomads themselves to guard the plantations from wandering herds is more likely to achieve success than repressive measures.

Successful desert reclamation has its attendant problems. Irrigation increases the chances of locust infestation by making conditions suitable for their breeding, and also helps to spread the snails that carry schistosomiasis (p. 228). Similarly, fungus diseases which are suppressed by a hot, dry desert climate may flourish in irrigated areas. One thing is certain: the plants

Fig. 7.8 Aerial photograph showing desert reclamation

(Hulton Educational Publications)

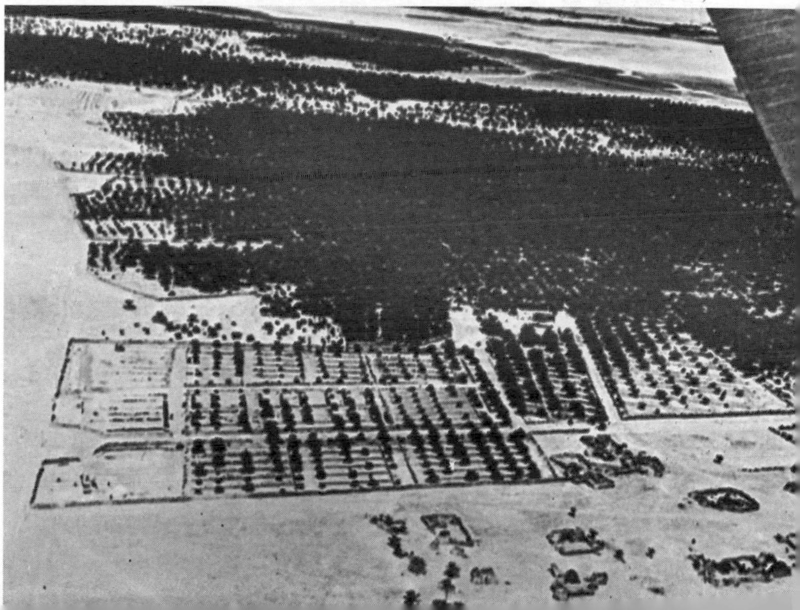

and animals and even the people who farm them must be well adapted to the environment. It is useless to introduce European pedigree cattle into tropical areas where they are unlikely to do well. The farm animals and plants must be bred from existing types that are already well adapted to arid conditions.

In some cases the large sums of money which might be spent on reclaiming desert areas, where the final yield may be small, would be better spent in improving productive or partially productive land.

Apart from the reclamation of eroded land, it may be possible to make areas such as the African swamps productive. Straightforward drainage seems to increase the hazards of malaria. The swamps may, however, be better suited to rice growing or fish farming than to conventional agriculture.

Pest control. 10 000 million kilograms of produce are lost each year to pests and diseases. Eradication or control of these pests and diseases would make this food available for human consumption. Modern pesticides have played an enormous part in reducing such losses but they are not without their hazards (*see* p. 58).

New methods of obtaining food. Most of the methods described below are in the experimental stages and are not in themselves likely to solve a world food shortage. Nevertheless, in certain regions they may make an important contribution.

(Dr A. J. Cooper, Glasshouse Crops Research Institute)

Fig. 7.9   Hydroponic culture. In this experimental situation tomato plants are growing in a nutrient solution circulating through the black plastic tubing. No sand or other material is used for rooting the plants, which are supported by taut strings. Since soil is not used, water losses by surface evaporation are greatly reduced

(a) *Game cropping*. As a result of work carried out in parts of Africa, it seems that the natural communities of wild animals, 'game', could if efficiently controlled or mixed with cattle provide more meat than cattle alone and with less damaging effects on the environment. The reasons for this are as follows:

(i)   Game animals are resistant to the trypanosomes carried by tsetse flies, which bring the disease ngana to domestic cattle.
(ii) Game use a wide range of vegetation, from the browsing of leaves by giraffes to the grazing of grass by the antelope. Cattle, however, feed almost exclusively on grass and are less interested in other vegetation.
(iii) The equilibrium established between the various wild mammals and the diverse plant population has become stable over millions of years of evolution. Some farming communities are more concerned with the numbers of cattle than with their quality and so tend to overstock and overgraze, thus destroying the habitat and its food resources.

Judicious protection of game and its habitat, with controlled slaughter, can be made to yield meat in valuable quantities. For example, hippos in Uganda provide 3 000 tonnes of meat per annum. Game cropping has also a psychological value in offering an outlet for 'hunting' other than indiscriminate poaching which often involves cruelty and destruction when the brush is burned to drive out game, killing many other species and ruining the habitat.

(b) *Culture of algae*. Agricultural crops make efficient use of only 0.3 per cent of incident sunlight. Many microscopic algae use as much as 2 per cent. One such unicellular alga, *Chlorella*, can be grown in solutions containing salts, if suitably exposed to sunlight. No soil is needed and less water than rooted plants since transpiration does not occur. At 26 °C, *Chlorella* divides every 14 hours and its cells contain 30 to 50 per cent protein, including most of the essential amino acids, and vitamins, particularly carotene, the precursor of vitamin A. Chickens have been shown to digest *Chlorella* powder and thrive on it, though rats are unable to do so. *Chlorella* thus seems to be an unlikely source of food for man but may be of value as fodder for cattle or chickens.

In Israel, an experimental station produces about 550 grams of *Chlorella* per square metre per day using tanks of nutrient solutions exposed to sunlight. Other systems have been tried to avoid overcrowding, shading and overheating. In America the culture solution has been circulated in transparent tubes with good results so far. Theoretically 1 250 kilograms per hectare of *Chlorella* could be produced each day. This is no better than the rate of production under conventional agriculture in good conditions, but these culture methods may have something to offer in arid zones with little prospect of irrigation and very little cloud cover throughout the year.

(c) *Hydroponics*. Experiments were started in 1957 in the Sahara where nutrient solutions were fed to plants in otherwise infertile desert sand. The technique adopted is similar to that described for solution 1 of Experiment 4, p. 48, but the plants are rooted in sand or other inert material (Fig. 7.9). The water requirements are much less and the yield per square metre much greater than that of conventional palm-grove oasis cultivation.

(d) *'Synthetic' food.* Chemical synthesis of food from materials such as coal or petroleum, even if possible, is unlikely to equal the efficiency of green plants, but the synthesis of proteins from organic substances and salts by micro-organisms is possible on a large scale. A leading research worker in this field has pointed out that 'In 24 hours half a tonne of bullock will make 500 grams of protein; half a tonne of yeast will make 50 tonnes and needs only a few square yards to do it on.' Urea, glycine, ethanoic (acetic) acid, succinic and malic acids can all be synthesized, and might be used by animals as energy sources. Concentrated mixtures of amino acids and vitamins might help to reduce the incidence of other deficiency diseases, without any vast increase in the demand on agricultural productivity.

Micro-organisms are a source of 'single cell protein'. Single-celled organism such as yeast can make protein from sugar and mineral salts. Bacteria are used, on a commercial scale, to produce protein from methanol (methyl alcohol) which comes from petroleum. The bacteria reproduce and grow rapidly. Their cells are separated from the medium and dried to form a powder which is used to supplement animal feedstuffs. For a variety of reasons, single cell protein has not been accepted, to any extent, for human diets.

(e) *Education.* Before any progress can be made in applying new ideas, governments and people must be made aware of the problems and prospects, the dangers and the hopes of intensified or novel agricultural practices. Farming methods and feeding habits may have to undergo drastic changes, and the sooner the facts are presented to the people, the sooner they will become acclimatized to the new ideas. In this way, opposition of the conservative and traditional variety, if not exactly eliminated, might at least become informed and constructive.

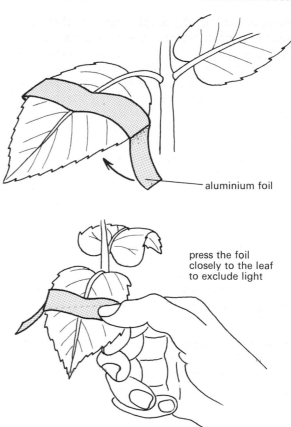

aluminium foil

press the foil closely to the leaf to exclude light

Fig. 7.10   Starch production in leaves

Practical Work

Experiment 1   **The production of starch by a green plant**

A potted plant is watered and placed in a dark cupboard for 48 hours during which time any starch present in its leaves is removed. After two days, the plant is taken from its cupboard and one of its leaves tested for starch as described below. If the test leaf proves to contain no starch, one or more of the leaves on the plant are partly covered with a strip of aluminium foil (Fig. 7.10) and the whole plant left in sunlight or strong artificial light for three hours or more. After this time the leaf with the foil strip is detached, the foil removed and the leaf tested for starch.

Result. Only those parts of the leaf that received light will go blue with iodine.

Interpretation. The blue colour indicates that starch has been made in the leaf which was previously free from starch. The fact that the area under the foil strip has produced no starch suggests that light is essential for starch production by photosynthesis.

Starch test. Prepare a beaker or can of boiling water. Hold the leaf in forceps and dip it momentarily in the water. Place the leaf in a test-tube and cover it with ethanol (industrial methylated spirit). Make sure all bunsen burners or spirit lamps are extinguished and place the test-tube in the boiling water (Fig. 7.11). The alcohol will boil and extract the chlorophyll from the leaf. When the alcohol is green and the leaf is white or

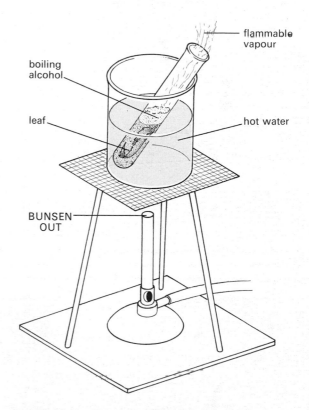

flammable vapour

boiling alcohol

leaf

hot water

BUNSEN OUT

Fig. 7.11   Extracting chlorophyll from a leaf

very pale green pour off the alcohol, take out the leaf, dip it once again in the hot water to soften it and spread it flat on a tile or dish. Cover the leaf with iodine solution and if any starch is present in it, a blue colour will appear.

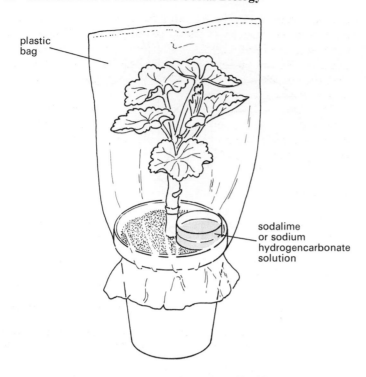

Fig. 7.12   The need for carbon dioxide

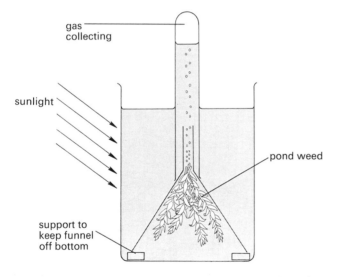

Fig. 7.13   Collecting oxygen from pond weed

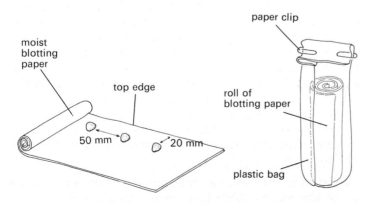

Fig. 7.14   Growing seedlings

Experiment 2   **The plant's need for carbon dioxide**

Two potted plants are watered and left in a dark cupboard for 48 hours so that all starch is removed from their leaves. After this period, one of the leaves from each plant is removed and tested for starch as described above. There should be no blue colour with iodine. Both plants are then watered and enclosed in clear plastic bags, as shown in Fig. 7.12, but one contains a dish of sodalime which absorbs all the carbon dioxide in the enclosed area, while the other has a dish of sodium hydrogen-carbonate solution which will release carbon dioxide into the air. Both plants are placed in sunlight for three hours or more after which a leaf from each is detached and tested for starch.

Result. The leaf deprived of carbon dioxide will not turn blue, while that from the carbon dioxide-enriched atmosphere will turn blue.

Interpretation. The fact that no starch is made in the leaf deprived of carbon dioxide suggests that the latter must be necessary for photosynthesis and starch production. The control (the plant with added carbon dioxide) rules out the possibility that high humidity or temperature in the plastic bag prevents normal photosynthesis.

Experiment 3   **Oxygen from pond weed**

A short-stemmed funnel is placed over some pond weed (*Elodea* or *Ceratophyllum*) in a beaker of water, preferably pond water, and a test-tube filled with water is inverted over the funnel stem (Fig. 7.13). The funnel is raised above the bottom of the beaker to allow free circulation of water. The apparatus is placed in sunlight and bubbles of gas soon appear from the cut stems, rise and collect in the test-tube. When sufficient gas has collected, the test-tube is removed and a glowing splint inserted in it. A control experiment should be set up in a similar way but placed in a dark cupboard. Little or no gas should collect.

Result. The glowing splint bursts into flame.

Interpretation. The relighting of a glowing splint does not prove that the gas collected is *pure* oxygen but it does show that, in the light, this particular plant has given off a gas which is considerably richer in oxygen than is atmospheric air.

(*Note.* Pond weed is used for this experiment because it is easier to collect the gas given off than from a land plant. It is, however, possible to demonstrate that all green plants give off oxygen in the light and it is this which replenishes the oxygen used up by combustion and respiration.)

Experiment 4   **The plant's need for mineral elements**

Some small seeds (e.g. wheat, sorghum, vigna) are soaked in water for 24 hours and then rolled up in moist newsprint or blotting paper. The paper rolls are stood on end in polythene bags (Fig. 7.14) for a few days until the seeds have germinated. Four seedlings at about the same stage of development are selected and each is wrapped in cotton wool and placed in the mouth of a test-tube, its roots dipping into one of four solutions (Fig. 7.15). Solution 1 contains all the elements thought to be necessary for the production of proteins, chlorophyll, enzymes, etc.; solution 2 contains the same elements except for calcium; solution 3 lacks only nitrates, and 'solution' 4 lacks all salts, i.e. it is distilled water. The plants are allowed to grow for two or three weeks in the light, the solutions being topped up with distilled water when the level drops. After this time the seedlings are compared for colour, size, number of leaves, length of roots, etc.

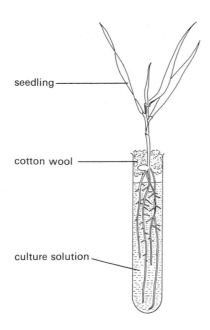

seedling

cotton wool

culture solution

Fig. 7.15   Seedling in water culture

Result. When nitrates are absent, the plant lacks the nitrogen needed to make proteins, so it is likely that the plant in solution 3 will be small and pale in colour. Calcium is needed for proper development of cell walls and its absence will have a bad effect on the growth of the seedling in solution 2. The seedling in distilled water is deprived of all mineral salts though it is still able to make carbohydrates by photosynthesis. When it has exhausted the food reserves in its seeds, it may cease growing and die. The reason for using small seeds in this experiment is that they quickly use up their own food reserves and become dependent on the solutions.

The formulae for the reagents and solutions for these experiments are given on p. 319. These experiments and many others are described more fully in the 'Experimental Work in Biology' series (*see* p. 320).

Questions

1   A group of explorers is stranded on a barren island where there is no soil and no vegetation. From their stores they have salvaged some living hens and some wheat. To make these resources last as long as possible should they
   (a) eat the wheat and when it is finished, kill and eat the hens, or
   (b) feed the wheat to the hens, collect and eat the eggs laid and when the wheat is gone, kill and eat the hens, or
   (c) kill and eat the hens first and when they are finished, eat the wheat? Justify your answer.
2   What are the requirements for photosynthesis? How are these requirements met in (a) a land plant, (b) an aquatic plant?
3   It can be claimed that the sun's energy is used indirectly to produce a muscle contraction in your arm. Trace the steps in the transfer of energy which would justify this claim.
4   Proteins contain carbon, hydrogen, oxygen, nitrogen and often sulphur. Name the source from which a green plant obtains each of these elements.
5   What factors do you think will limit the quantity of food that can be produced from a given area of soil? Suggest ways in which these limits could be raised by artificial means.

8
Soil

Components

Soil consists of a mixture of (a) mineral particles such as sand and clay, (b) humus, (c) water, (d) air, (e) dissolved salts, and (f) bacteria and fungi.

(a) Inorganic particles are formed from rocks which have been weathered and broken down. Particles from 2 to 0.02 mm diameter are classified as sand, 0.02 to 0.002 mm as silt and less than 0.002 mm as clay (Experiment 1). Chemically, sand is silicon oxide while clay may be various complexes of aluminium and silicon oxides. Iron oxide may give a red or brown coating to the particles.

Aggregates of these inorganic particles together with humus produce *crumbs* up to about 3 mm in diameter, which form the 'skeleton' of the soil. The crumb structure of a soil depends on the proportion of clay, sand and humus and the activities of plant roots; a good crumb structure is one of the most important attributes of a soil.

(b) Humus is the finely divided organic matter incorporated into the crumbs; it originates mainly from decaying plant remains (Experiment 2). The presence of humus in the crumbs affects the colour and physical properties of the soil. Humus is black, structureless, often forms a coating around sand particles, and may be important in glueing particles together to form soil crumbs. A sandy soil deficient in humus tends to have a poor crumb structure and is easily blown away if exposed by ploughing. The exclusive use of chemical fertilizers on certain soils in dry climates may lead to the formation of dust bowls or the advance of desert margins.

The bacterial decay of humus and the organic matter from which it originates produces the nitrates and other mineral salts needed for plant growth.

(c) Water is spread over the sand particles or clay aggregates as a thin film which adheres by capillary attraction. It may also penetrate the aggregates and be held to the clay particles by chemical forces. When a soil contains as much water as it can hold by capillary and chemical attraction (i.e. any more would drain away by the force of gravity), it is said to be at *field capacity*. Capillary attraction will tend to distribute water from regions above field capacity to drier regions. The forces holding water in the soil also set up considerable opposition to the 'suction' of plant roots when the soil begins to dry out.

(d) **Air** occurs in the spaces between the aggregates or sand particles unless the soil is waterlogged, in which case the air spaces are blocked up. A supply of oxygen is essential for the respiration of roots and some soil organisms, e.g. bacteria.

(e) **Salts.** Salts in the soil water are dissolved out from either the surrounding rock or from humus and organic matter in the soil. They make a very dilute solution in the soil water but are vital for plant growth, as explained on p. 41.

(f) **Bacteria.** Many microscopic plants, fungi and animals live in the soil, but among the most important to plant life are the bacteria which break down the organic matter and humus to form soluble salts which can be taken up in solution by roots. Other bacteria convert atmospheric nitrogen to organic compounds of nitrogen (*see* p. 51).

Types of soil

Heavy soils. A soil in which clay particles predominate and which has a poor crumb structure will be sticky and difficult to plough or dig. This results partly from chemical and capillary forces acting on the very large surface area of the minute clay particles, making them difficult to separate. When dry, the soil forms hard clods which do not break up readily during cultivation.

The small distances between particles (Fig. 8.1*a*) tend to produce poor aeration and drainage, but the large surface presented by the particles retains a high proportion of water in conditions of drought. There is also less tendency for soluble minerals to be washed away because they are held chemically to the clay particles.

A heavy soil can be made lighter, more workable and permeable to water and air by adding lime or organic matter, e.g. farmyard manure. The lime makes the particles clump together or *flocculate*, the clumps of particles behaving like the larger particles of a light soil. The crumb structure of a clay soil can be improved by growing grass on it for a year or two.

Light soils. The large, inorganic particles of a light soil give it its sandy texture (Fig. 8.1*b*). The wider separation of the particles leads to better aeration and drainage but there is a smaller surface for the water film. Such a reduced surface lessens the surface forces (surface tension) of the water and makes it easier to separate the particles in ploughing and digging, and the clumps break up easily when dry.

The mineral salts are more liable to be washed out from a lighter soil and it loses water rapidly in dry conditions. Its water-holding properties and nitrogen content can be improved by adding farmyard manure or compost (rotted vegetable matter).

Loam. A soil with a balanced mixture of particle sizes, a good humus content and stable crumb structure is called a loam. Loams are the most productive soils in agriculture.

Laterite soils. A laterite soil is formed in tropical conditions, where high temperatures cause very rapid decomposition of organic matter and the heavy rain washes out many of the minerals including much of the silica. The soil consists mainly of alumina particles coloured red with iron oxide, and is deficient in humus and mineral nutrients. When it dries, the alumina and iron oxide particles stick together forming a hard layer which is poorly aerated and difficult to plough or dig.

Soil pH (acidity). The pH of soils varies. A soil on limestone or chalk may be alkaline, up to pH 8. Some clay soils and soils containing much organic matter may be acid, down to pH 4.5. Acid conditions in the soil often lead to a deficiency of minerals by making them more soluble and easily washed out by rain. Alkaline conditions, on the other hand, may make some minerals so insoluble that they cannot be taken up in solution by the plant.

A pH of about 6.5 is considered favourable for most crops and cultivation methods can be used to adjust the pH. For example, application of lime (calcium hydroxide) will raise the pH of an acid soil, while addition of ammonium sulphate will lower the pH of an alkaline soil.

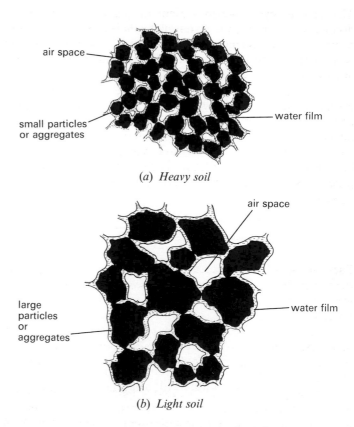

(a) *Heavy soil*

(b) *Light soil*

Fig. 8.1   Structure of light and heavy soils

Soil fertility

In natural conditions soil fertility is maintained by the activity of the organisms living on it or in it. For example, the plant roots maintain the soil's crumb structure and the burrows of earthworms enhance its drainage. Although plants remove mineral salts they are replaced by the death and decomposition of plant and animal bodies. This cycle of uptake from, and return of minerals to the soil applies to all mineral elements in the soil but can be studied in detail by reference to nitrogen, the element needed by plants to make their proteins.

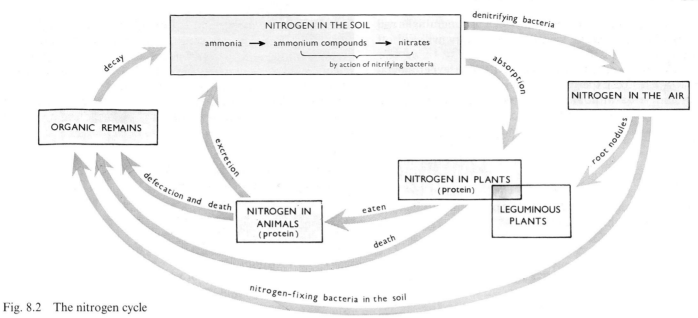

Fig. 8.2   The nitrogen cycle

The nitrogen cycle. When a plant or animal dies its tissues decompose, largely as a result of the action of enzymes and bacteria. One of the important products of this decomposition is ammonia, which is washed into the soil where it forms ammonium compounds (Fig. 8.2).

Nitrifying bacteria. In the soil there are many bacteria and certain of these, called *Nitrosomonas*, oxidize the ammonium compounds to nitrites. Others, *Nitrobacter*, further oxidize nitrites to nitrates which are taken up in solution by plants. The faeces of animals contain organic matter which is similarly broken down, and their urine is rich in nitrogenous waste products such as ammonia, which can be oxidized to nitrates by soil bacteria.

$$NH_3 \longrightarrow -NO_2 \longrightarrow -NO_3$$

ammonia nitrite nitrate
ion ion

Each step involves an increase in the proportion of oxygen in the molecule. Such reactions are called oxidations, and need well-aerated soil with a good supply of oxygen if they are to proceed effectively.

The bacteria derive energy from these oxidative processes in much the same way as plants and animals derive energy from respiration by oxidizing carbohydrates to form carbon dioxide and water (p. 27).

Nitrogen-fixing bacteria. Although green plants cannot utilize the nitrogen in the atmosphere, there are bacteria in the soil which absorb and combine it with other elements, so making nitrogen compounds such as amino acids. This is called the *fixation of nitrogen*. Such nitrogen-fixing bacteria, as well as existing free in the soil, are found in special root swellings or *nodules* (*see* Fig. 8.3) in plants of the pea family such as soya bean, groundnuts and lucerne, and these plants, if part of their vegetation is ploughed back into the soil, increase its nitrogen content. For this reason they are included in the system of crop rotation used in agricultural practice (*see* p. 53).

Denitrifying bacteria. Also present in the soil are bacteria that obtain energy by breaking down compounds of nitrogen to gaseous nitrogen which consequently escapes to the atmosphere.

Lightning. The high temperature of lightning discharge through the atmosphere causes nitrogen and oxygen to combine, forming oxides of nitrogen. These gases dissolve in the rain making weak acids which, when washed into the soil, combine with other elements there to form nitrates. This is said to add some millions of tonnes of nitrogen to the soil every year.

To sum up, the processes that remove nitrates from the soil are dissolving out by rain, absorption by plant roots and denitrification by bacteria. Processes that increase the nitrate content of the soil are the activities of nitrifying and nitrogen-fixing bacteria and lightning. When these two sets of processes are in balance, the nitrate content of the soil remains more or less constant.

Fertilizers. The practice of agriculture interrupts natural cycles by removing the crops at harvest but not returning to the soil the dead remains of either the plants or the animals which eat them.

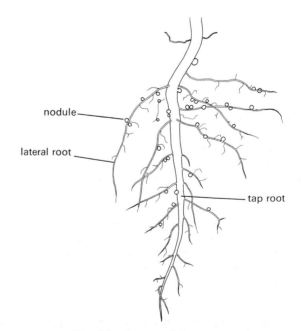

Fig. 8.3   Root nodules on groundnut

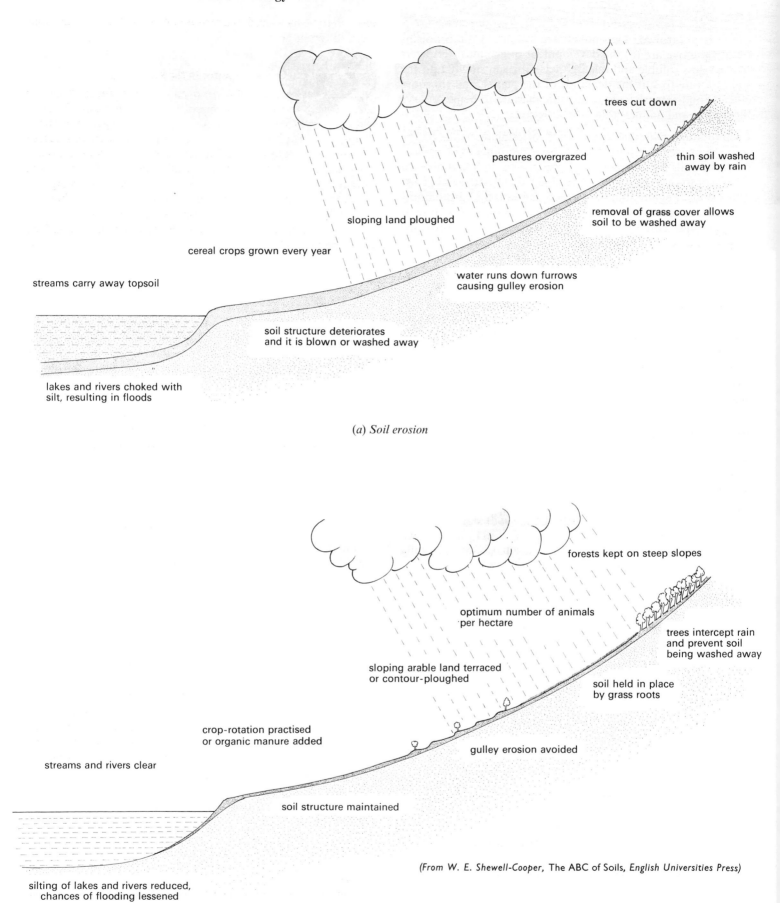

trees cut down

pastures overgrazed

thin soil washed
away by rain

sloping land ploughed

removal of grass cover allows
soil to be washed away

cereal crops grown every year

water runs down furrows
causing gulley erosion

streams carry away topsoil

soil structure deteriorates
and it is blown or washed away

lakes and rivers choked with
silt, resulting in floods

(a) Soil erosion

forests kept on steep slopes

optimum number of animals
per hectare

trees intercept rain
and prevent soil
being washed away

sloping arable land terraced
or contour-ploughed

soil held in place
by grass roots

crop-rotation practised
or organic manure added

gulley erosion avoided

streams and rivers clear

soil structure maintained

(From W. E. Shewell-Cooper, The ABC of Soils, *English Universities Press)*

silting of lakes and rivers reduced,
chances of flooding lessened

(b) Soil conservation

Fig. 8.7

(Agricultural Information Section, Ministry of Agriculture, Enugu, Eastern Nigeria)

Fig. 8.8 Contour strip cropping in the University farm, Nsukka, Eastern Nigeria

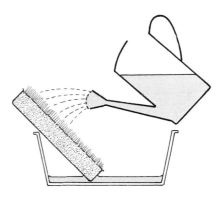

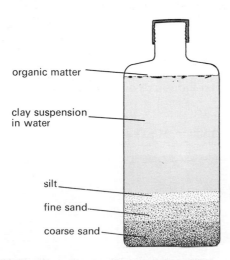

enough to be measured, the results can be used to compare soils from different areas.

Experiment 2 **Organic matter in the soil**

A sample of soil is dried in an oven at 100 °C for 24 hours or alternatively spread out on newspaper to dry in the air. The soil is then sieved to remove stones and particles larger than about 3 mm, 50 g of it is placed in a metal tray or tin can and heated strongly from below with a bunsen flame for 15 minutes. If the soil blackens and smoke is given off, this is evidence for the presence of organic matter. When the soil is cool, it is weighed. The loss in weight will represent partly the quantity of organic matter burnt away and partly the evaporation of water that had not completely escaped from the soil during drying. The experiment should be repeated to compare, for example, a poor laterite soil with a fertile garden soil.

Experiment 3 **Demonstration of erosion**

Obtain two metal or plastic trays about 5 cm deep. Fill one with soil and press it down tightly. Pour water on it carefully till it is saturated and then leave the tray tilted on end to let the surplus water drain away. Cut a rectangle of turf to fit exactly into the second tray. Water this in the same way as the soil.

Place the tray with the turf at an angle of about 45° in a plastic bowl or similar container and water the soil from a watering can with a fine 'rose' fitted, for 30 seconds (Fig. 8.10). Empty the bowl and repeat the experiment with the soil.

Fig. 8.10 Demonstration of erosion

Compare the quantity of soil washed into the bowl in each case. A further comparison can be made by cutting all the grass blades from the turf and watering it again. If a quantitative comparison is wanted, the soil washed into the basin can be collected by allowing it to settle, decanting the excess water, collecting, drying and weighing the soil.

Questions

1 What do you suppose is the biological significance of the following agricultural practices: (i) ploughing farmyard manure (animal faeces and straw) into the soil, (ii) adding lime to the soil, (iii) spreading sulphate of ammonia on the land?

2 Outline the ways in which careless agricultural practices can lead to rapid erosion of a light sandy soil.

3 Study the diagram of the nitrogen cycle (Fig. 8.2, p. 51) and suggest ways in which civilized man interferes with this natural chain of events.

4 List the ways in which soil fertility can be improved and indicate any disadvantages of these practices.

(e) **Strip cropping.** This consists of alternate bands of tilled and untilled soil following the contours (Fig. 8.8). Grass and cover-crop strips, between strips of ploughed land carrying grain, prevent the soil being washed away from the tilled portions. By alternating the grass and grain each year, the soil is allowed to rebuild its structure while under grass. Strip cropping is also effective against wind erosion if the strips are planted at right angles to the direction of the prevailing wind.

Practical Work

Experiment 1 **Observation of mineral particles in soil**

Some dry soil is sieved to remove stones and particles larger than about 3 mm. The soil is crushed lightly to break up the crumbs and 50 g is placed in a small, flat-sided bottle. The bottle is filled with water almost to the top, the cap screwed on and the bottle shaken for at least 30 seconds to disperse the soil throughout the water. The bottle is then allowed to stand for 10 to 15 minutes so that the soil settles down (Fig. 8.9). The large particles will fall most rapidly and form the bottom layer.

organic matter

clay suspension in water

silt

fine sand

coarse sand

Fig. 8.9 Mineral particles in the soil

Smaller particles will collect in the higher layers and some of the clay will remain suspended in the water. The larger particles of organic matter will float to the top. If the layers are distinct

9

Food Chains and the 'Balance of Nature'

All animals derive their food either directly or indirectly from plants. Carnivorous animals feed on other animals which themselves may feed on smaller animals, but sooner or later in such a series we come to an animal that feeds on vegetation. For example, small fish eat water fleas, and water fleas feed on microscopic plants in the pond or lake. This kind of relationship is called a *food chain*. The basis of food chains on land is vegetation in general, but particularly grass and other leaves. In water, the basis is *phytoplankton* (Fig. 9.1): the millions of microscopic plants living near the surface of the sea, ponds and lakes. These need only the water round them, the dissolved carbon dioxide, salts and sunlight to make all their vital substances (p. 40). Feeding on these microscopic plants are tiny animals, *zooplankton* (Fig. 9.2), such as the crustacea and the larvae of many kinds of animal. The small animals of the zooplankton are eaten by the surface-feeding fish; the fish, in turn, form part of the diet of other animals, including man.

This description of a food chain is inadequate in two respects, since it seems to suggest (a) that there are similar numbers of organisms at each stage in the chain, and (b) that one organism feeds exclusively on another. An alternative way of expressing the relationships of a food chain which overcomes the first objection, is by a pyramid of numbers, or pyramid of mass. Fig. 9.3 expresses this diagrammatically. At the beginning of a food chain, the organisms are usually small and very numerous. The plants of the phytoplankton are microscopic and very abundant. The animals of the zooplankton which feed on the phytoplankton, though still microscopic, are much larger than, for example, the diatoms. The fish which eat the zooplankton

are many times larger and far less numerous. Finally, it takes a large number of fish to feed one man.

These numerical relationships reflect the losses involved in the transference of energy and matter at each stage of a food chain. For every 100 kJ of energy present in the grass eaten by a cow, only about 5 kJ is used for making new flesh and bone. About 60 kJ is lost by not being wholly digested and utilized, and 35 kJ is used in respiration to meet the animal's energy needs. In food chains in general, about 10 per cent of the mass of plant material eaten is converted to new living matter, the other 90 per cent is undigested or used for energy. This shows why a large mass of living organisms at the beginning of a food chain will support only a small number of carnivorous animals at the end.

The second objection to the early description of the food chain is that any given animal, particularly a predator, does not usually live exclusively on one type of food. For example, snakes eat frogs and insects as well as lizards; small fish are eaten by herons as well as by larger fish. These more complex relationships can be shown as a *food web* (Fig. 9.4), but even this is greatly simplified and generalized.

If the population of one of the animals in a food web is altered, all the others are affected. In 1906 a game reserve was created in the Grand Canyon, U.S.A. Killing of deer was prohibited and hundreds of mountain lions, wolves and other predators of deer were killed. The deer multiplied rapidly in the absence of predators, so that by 1923 the population of deer had increased from 4 000 to 100 000. The range was depleted by overgrazing and thousands of deer were dying from starvation.

Fig. 9.1 Phytoplankton. These microscopic plants are diatoms (× 100)
(Dr D. P. Wilson)

Fig. 9.2 Zooplankton. Mostly adult and larval crustacea from the sea (× 15)
(Dr D. P. Wilson)

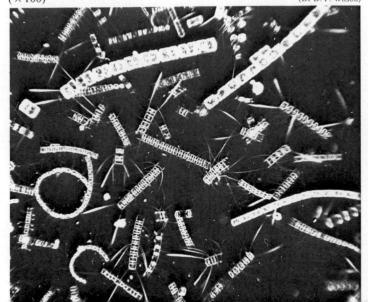

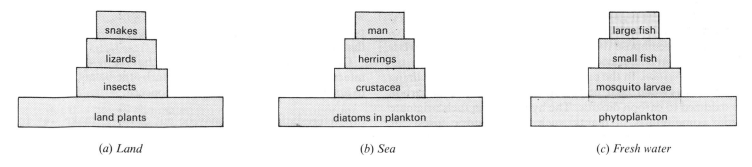

(a) Land   (b) Sea   (c) Fresh water

Fig. 9.3   Examples of pyramids of numbers in food chains

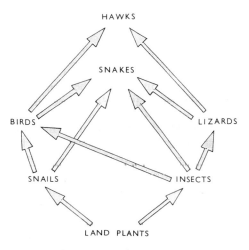

Fig. 9.4   A food web

In 1954–5, when the rabbits in Britain were almost exterminated by the disease myxomatosis, the vegetation in what had been rabbit-infested areas changed, and sheep could graze where rabbits had previously eaten all the available grass; trees which had hitherto been nipped off as seedlings began to grow to maturity, with the result that what had once been grassland, e.g. chalk downs, started to become scrub and eventually woodland. Foxes ate more voles, beetles and blackberries than before and attacked more lambs and poultry.

Ultimate sources of energy

All the energy released on the Earth, apart from atomic energy and tidal power, comes from the sun, via food oxidized in animals and plants, coal from fossilized forests, petroleum from deposits of marine algae, or hydro-electric power from rainwater in lakes. There are ways of using the sun's energy directly and making it heat water to produce steam, but, at the moment, one of the best ways of trapping sunlight is to grow plants from which food and other energy-rich products can be collected.

Of the sunlight that reaches an area of grassland, only about 1 or 2 per cent is used for photosynthesis. The rest escapes the leaves, is reflected from their surface or is used to evaporate water from them in transpiration. When the vegetation is eaten by animals, only about 10 per cent of the food is converted to milk, meat or eggs; the other 90 per cent is used as a source of energy by the animal, or lost in faeces and urine. It follows that the most efficient use is made of plants when they are eaten directly by man and that their conversion to animal products is a wasteful process. (See also pp. 42–3.)

The carbon cycle

Food chains and food webs are but one link in the constant use and re-use of the Earth's chemical resources. The nitrogen cycle described on p. 51 is one example of this, and the carbon cycle described below is another. The carbon cycle consists, in essence, of the processes that produce or use up the carbon dioxide in the environment (Fig. 9.5).

Removal of carbon dioxide from the atmosphere. Green plants, by their photosynthesis (p. 40), remove carbon dioxide from the atmosphere or water in which they grow (Experiment 3). The carbon of the carbon dioxide is incorporated at first into carbohydrates such as sugar or starch, and then eventually into the cellulose of cell walls and the proteins, pigments and other organic compounds comprised in living organisms. When the plants are eaten by animals the organic plant matter is digested, absorbed and built into compounds making the animals' tissues. Thus the carbon atoms from the plant become an integral part of the animal.

Addition of carbon dioxide to the atmosphere. (a) *Respiration.* Plants and animals obtain energy by oxidizing carbohydrates in their cells, breaking the compounds down to carbon dioxide and water (*see* p. 27). These products are excreted and the carbon dioxide returns once again to the environment (Experiment 2).

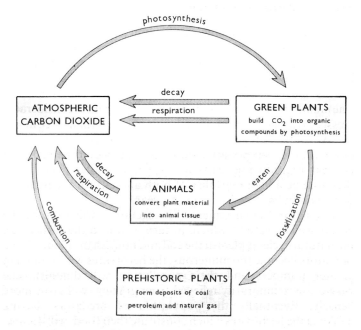

Fig. 9.5   The carbon cycle

three are placed in a rack in direct sunlight or a few centimetres from a bench lamp for about forty minutes.

Result. The indicator (which was originally orange) should not change colour in tube 3, the control; that in tube 1, with the leaf in darkness, should turn yellow; and in tube 2, with the illuminated leaf, the indicator should be scarlet or purple.

Interpretation. Hydrogencarbonate indicator is a mixture of dilute sodium hydrogencarbonate solution with the dyes cresol red and thymol blue. It is a pH indicator in equilibrium with the atmospheric carbon dioxide, i.e. its original colour represents the acidity produced by the cabon dioxide in the air. Increase in atmospheric carbon dioxide makes it more acid and it changes colour from orange to yellow. Decrease in atmospheric carbon dioxide makes it less acid and causes a colour change to red or purple.

Thus the results provide evidence that in darkness (tube 1) leaves produce carbon dioxide (from respiration), while in light (tube 2) they use up more carbon dioxide in photosynthesis than they produce in respiration. Tube 3 is the control, showing that it is the presence of the leaf which causes a change in the atmosphere in the test-tube.

For experiments on gaseous exchange in man *see* p. 90. The formulae for hydrogencarbonate indicator and other reagents are given on p. 319.

Questions

1   Trace the food chains involved in the production of the following articles of man's diet: eggs, cheese, bread, meat, wine. In each case show how the energy in the food originates from sunlight.
2   Discuss the advantages and disadvantages of man's attempting to exploit a food chain nearer to its source, e.g. the diatoms of Fig. 9.1.
3   Construct a diagram, on the lines of the carbon cycle (Fig. 9.5), to show the cycling process for hydrogen.
4   How do you think evidence is acquired in order to assign animals such as leopard and pigeon to their position in a food web?
5   What would be the desirable qualities of an insecticide to control an insect pest of a crop plant? What might be the disadvantages of total eradication of such an insect pest?
6   Why are the animals at the end of a food chain likely to be more seriously affected by a harmful pesticide than the animals at the beginning?

10

The Digestion, Absorption and Metabolism of Food

To be of any value to the body, the food taken in through the mouth must enter the blood stream and be distributed to all living tissues.

Digestion is the process by which insoluble food consisting of large molecules is broken down into soluble compounds with smaller molecules. These smaller molecules, in solution, pass through the walls of the intestine and eventually enter the blood stream. Digestion and absorption take place in the alimentary canal (Figs. 10.1 and 10.2), digestion being brought about by chemical compounds called enzymes (*see* p. 27). The *alimentary canal* is a muscular tube running from mouth to anus with a glandular lining called the *mucous membrane* or *mucosa*. In

general, the alimentary canal has an outer tough coat over layers of longitudinal and circular unstriated muscle (p. 119) which themselves enclose a *submucosa* containing networks of blood capillaries and a plexus of nerve fibres. Lining the inside of the alimentary canal is the mucous membrane or mucosa, consisting of an epithelium beneath which is connective tissue and a thin layer of unstriated muscle (Fig. 10.5). The outer muscular layers propel the contents of the alimentary canal while the mucosa secretes (a) enzymes which digest the food and (b) mucus which lubricates and protects the internal surface of the gut.

Certain regions of the alimentary canal are specialized for particular functions and have differing structures. Digestive juices are secreted into the alimentary canal from glands in the

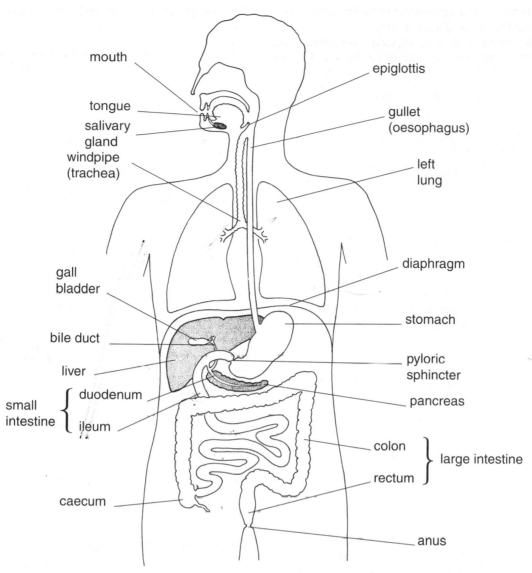

Fig. 10.1   The alimentary canal

Fig. 10.2   Alimentary canal of a rat unravelled

(Dissection by Griffin and George, Gerrard Biological Centre.)

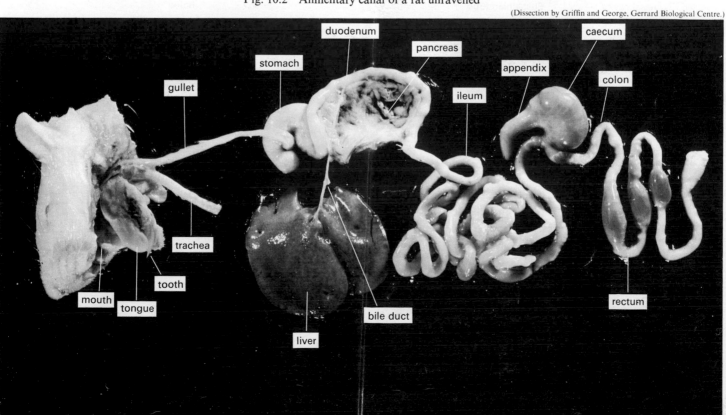

mucosa or through ducts from glandular organs outside it. The food is broken down in stages as it passes through the alimentary canal until the digestible material is dissolved and absorbed. The indigestible residue is expelled along with other products through the anus.

Enzymes are chemical compounds, protein in nature, made in the cells of living organisms. They act as catalysts and accelerate the rate of chemical changes in the organism without altering the end products. They occur in great numbers and varieties in all protoplasm and without them the chemical reactions would be too slow to maintain life. The vast majority of enzymes are *intracellular*, that is, they carry out their functions in the protoplasm of the cell in which they are made. Some enzymes, however, are secreted out of the cell in which they are made, to be used elsewhere. These are called *extracellular* enzymes. Bacteria (p. 199) and fungi secrete such extracellular enzymes into the medium in which they are growing. The higher organisms secrete extracellular enzymes into the alimentary tract to act on the food taken into it.

The chemical process of digestion is basically *hydrolysis*, in which the protein, carbohydrate and fat molecules have certain of their chemical bonds broken by adding the elements of water, —H and —OH, at these points. The process is discussed more fully on p. 16. Hydrolysis of food substances takes place on simply adding water, but the reactions are too slow to meet the needs of the body. The rates of the reactions are, however, greatly increased by enzymes, which speed up the hydrolysis and produce soluble products within minutes. Each type of food and each stage in its digestion needs a particular enzyme; those acting on starch are called *amylases*, those hydrolysing proteins are *proteinases*, and *lipases* accelerate the digestion of fats. The properties of enzymes are discussed more fully on p. 27, and the generalizations about them apply equally to digestive enzymes, namely (a) they speed up chemical reactions but cannot alter the end products, (b) they catalyse only one type of reaction, (c) their rate of working is influenced by temperature and pH, and (d) they are inactivated by temperatures above 45 °C.

Movement of food through the alimentary canal

Ingestion is the act of taking food into the alimentary canal through the mouth.

Swallowing (*see* Fig. 10.3). In swallowing, the following actions take place: (a) the tongue presses upwards and back against the roof of the mouth, forcing the pellet of food, called the *bolus*, to the back of the mouth or *pharynx*; (b) the soft palate closes the opening between the nasal cavity and the pharynx; (c) the laryngeal cartilage round the top of the trachea, or windpipe, is pulled upwards by muscles so that the opening of the larynx lies beneath the back of the tongue, and the opening of the trachea is constricted by the contraction of a ring of muscle; and (d) the *epiglottis*, a flap of cartilage, directs food over the laryngeal orifice. In this way food is able to pass over the trachea without entering it. The beginning of this action is voluntary, but once the bolus of food reaches the pharynx, swallowing becomes an automatic or reflex action. The food is forced into and down the *oesophagus*, or gullet, by *peristalsis* (*see* below). This takes about six seconds with relatively solid food and then the food is admitted to the stomach. Liquid travels more rapidly down the gullet. A small volume of air is inevitably swallowed with the food and accumulates in the upper part of the stomach.

Peristalsis (Fig. 10.4). The walls of the alimentary canal contain circular and longitudinal muscle fibres. The circular muscles, by contracting and relaxing alternately, urge the food in a wave-like motion through the various regions of the alimentary canal.

Egestion. The expulsion from the alimentary canal of the undigested remains of food is called *egestion*.

Digestion in the mouth

In the mouth the food is chewed and mixed with saliva. Chewing reduces the food to a suitable size for swallowing and

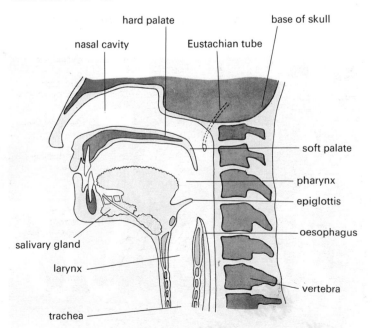

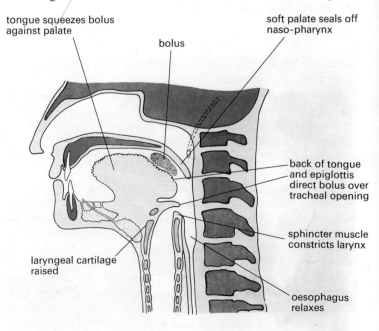

Fig. 10.3   Section through head to show swallowing action

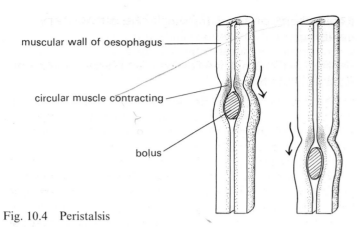

muscular wall of oesophagus

circular muscle contracting

bolus

Fig. 10.4   Peristalsis

also increases the available surface for enzymes to act on. Saliva is a digestive juice secreted by three pairs of glands whose ducts lead into the mouth (Fig. 10.3). The glands include two types of cell, one of which produces mucus and the other an enzyme, salivary amylase. The resulting fluid, which may be secreted in quantities up to 1.5 litres per day by an adult, is watery and neutral or slightly acid. The mucus helps to lubricate the food and stick the particles together, while the amylase acts on cooked starch and begins to break it down to maltose (p. 16), a soluble sugar (*see* Experiment 1). The longer that food is retained in the mouth, the further this starch digestion proceeds and the more finely divided does the food become as a result of chewing; but the digestive action of the salivary amylase does not seem particularly important since even well-chewed food does not remain long enough in the mouth for much digestion of starch to occur, although saliva continues to act for a time even when food is passed to the stomach.

Small quantities of saliva are secreted continually, keeping the lining of the mouth moist and permitting clear speech. In addition, dissolving some of the food taken into the mouth enables the taste-sensitive cells of the tongue to be stimulated by the chemicals present (*see* p. 132).

Digestion in the stomach

This part of the alimentary canal has elastic walls and so can be extended as food accumulates in it. This enables food from a particular meal to be stored for some time and released slowly to the rest of the alimentary canal.

There is no evidence to indicate that absorption, other than that of water, salts, alcohol and a little glucose, takes place in the stomach, but its mucosa produces gastric juice. There are numerous tube-like glands in the mucous membrane, which open in small groups into pits (Figs. 10.5 and 10.6). Lining the glandular tubes are three types of cell which produce *pepsinogen*, hydrochloric acid and mucus respectively, making up gastric juice which is secreted through the pits into the stomach. Gastric juice may also contain, in young children, an enzyme called *rennin*. When pepsinogen and hydrochloric acid meet in the stomach the former is activated to produce the enzyme *pepsin* which acts on the long chain molecules of proteins and breaks them down to shorter chains of more soluble compounds called *peptides* (*see* p. 17 and Experiment 2). Rennin, if present, clots the protein in milk, but pepsin alone has a similar effect and it is thought that the acidity in an adult's stomach would be too great for the effective action of rennin. The hydrochloric acid makes a solution of about 0.5 per cent by

weight in the gastric juice, providing the optimum pH (*see* p. 21) for pepsin to work in and also probably killing many of the bacteria taken in with the food. The salivary amylase swallowed with the food cannot digest starch in acid conditions, but it seems likely that it continues to act within the bolus of food until this is broken up and hydrochloric acid reaches all its contents.

The rhythmic, peristaltic movements of the stomach, about every twenty seconds, help to mix the food and gastric juice to a creamy liquid called *chyme*. Each wave of peristalsis also pumps a little of the chyme from the stomach into the first part of the small intestine, called the *duodenum*. The pyloric sphincter is usually relaxed but contracts at the end of each wave of peristalsis, so limiting the amount of chyme which escapes. Even when relaxed, the pyloric opening is narrow and only liquid is allowed through. When the acid contents of the stomach enter the duodenum, they set off a reflex action (p. 148), which closes the pyloric sphincter until the duodenal contents have been partially neutralized.

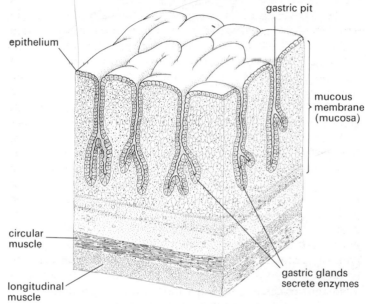

gastric pit

epithelium

mucous membrane (mucosa)

circular muscle

longitudinal muscle

gastric glands secrete enzymes

Fig. 10.5   Stereogram of section through stomach wall

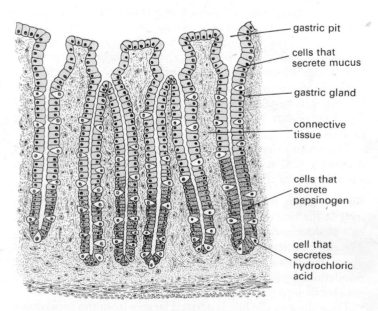

gastric pit

cells that secrete mucus

gastric gland

connective tissue

cells that secrete pepsinogen

cell that secretes hydrochloric acid

Fig. 10.6   The mucous membrane of the stomach

The length of time for which food is retained in the stomach depends to a large extent on the contents of the meal. Water may pass through in a few minutes, a meal of starch such as porridge may be retained less than an hour, one containing meat will be held longer while the presence of fat considerably retards the emptying of the stomach. Excitement may accelerate emptying, and fear may stop the gastric movements and cause the stomach to retain its contents up to twelve hours.

Digestion in the duodenum

An alkaline juice from the pancreas and bile from the liver are poured into the duodenum. The pancreas is a cream-coloured gland lying below the stomach (Fig. 10.7). Its cells make about

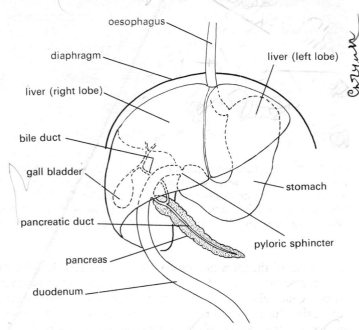

Fig. 10.7   Relationship of stomach, liver and pancreas

1 500 cm³ pancreatic juice per day, containing water, sodium hydrogencarbonate and three principal enzymes—an amylase, a lipase and *trypsinogen*—which act on starch, fats and proteins respectively. Trypsinogen is inactive until it meets a chemical activator called *enterokinase* secreted by the intestinal mucosa. The enterokinase activates the trypsinogen which becomes the enzyme *trypsin*; this enzyme breaks down proteins to peptides, which another enzyme converts to amino acids (*see* p. 18). Starch is broken down to maltose by the amylase, while fats are split into fatty acids and glycerol by the lipase (*see* p. 17 and Experiment 3). Glands in the duodenal mucosa produce mucus.

The sodium hydrogencarbonate in pancreatic juice partly neutralizes the acid chyme from the stomach and so creates a suitable pH for the pancreatic and intestinal enzymes.

Bile is a green, watery, alkaline fluid made continuously in the liver at the rate of about 500–1 000 cm³ per day, stored in the gall bladder and conducted to the duodenum by the bile duct. The gall bladder can hold about 30 cm³ bile and concentrates it about ten times by absorbing water. When food enters the duodenum from the stomach, the gall bladder contracts and expels bile into the duodenum.

The colour of the bile is derived largely from the breakdown products of the red pigment *haemoglobin* from decomposing red blood cells. Bile also contains complex organic salts which assist the digestion and absorption of fats, principally by reducing the surface tension of the fat drops. This results in their forming an emulsion of tiny droplets whose increased surface area permits rapid digestion by lipase. Most of the bile salts are reabsorbed in the small intestine and returned by the circulatory system to the liver where they stimulate further secretion of bile. Any disorder which impairs the absorption of fats also reduces the uptake of the fat-soluble vitamins, A, D and K.

Peristaltic waves pass down the duodenum and ileum at about 2 cm per second, moving the semidigested food along. In addition, there are muscular movements which churn the fluids back and forth.

Digestion in the ileum

The remaining three metres of small intestine after the duodenum are called the ileum. The glands in the lining produce mucus and the enzyme enterokinase, which activates pancreatic trypsinogen. These glands appear to produce no digestive enzymes but the pancreatic enzymes continue to act on the food, forming soluble maltose, peptides, amino acids, fatty acids and glycerol which can enter the epithelial cells of the villi (p. 65). Intracellular enzymes in these cells convert the disaccharides, maltose and glucose, to glucose and fructose. Other intracellular enzymes change peptides to amino acids.

Mammals have no enzymes for digesting cellulose, but in the paunch (rumen) of cows and goats, the large intestine of the horse and the caecum of the rabbit live a vast number of bacteria and other micro-organisms which are able to break down cellulose and are themselves eventually digested.

Most of the digestible material by this stage is reduced to soluble compounds which can pass through the intestinal lining and into the blood stream.

Prevention of self-digestion. The glandular lining of the alimentary canal secretes mucus continually and independently of enzyme secretion. This mucus helps to lubricate the passage of food but also prevents the digestive juices from reaching and digesting the alimentary canal itself. The cells which make the protein-digesting enzymes would themselves be digested by these chemicals were it not for the fact that the enzymes are made in an inactive form and cannot work until they reach the cavity of the alimentary canal, where they are activated by the chemicals present. Trypsin, for example, is made and secreted as an inactive substance, trypsinogen. When trypsinogen is set free in the duodenum, the enterokinase present converts it to active trypsin. This trypsin cannot now digest the duodenal walls because of their protective coating of mucus.

Absorption in the ileum

Nearly all the absorption of digested food takes place in the ileum. The following characteristics of the ileum greatly facilitate its absorbing properties:

(a) it is very long, and so presents a large absorbing surface to the digested food;

(b) this surface is greatly increased by thousands of tiny, finger-like projections called *villi*, 20–40 to each square millimetre (Figs. 10.8 and 10.10);

(c) the lining epithelium is very thin and the fluids can pass rapidly through it;

(d) there is a dense network of blood capillaries in each villus (Fig. 10.9).

The molecules of amino acids and glucose pass through the epithelium and capillary walls to enter the blood plasma. They are then carried away in the capillaries, which unite to form veins and eventually join up to form one large vein, the *hepatic portal*, which carries all the blood from the intestine to the liver. The liver may retain or alter some of the absorbed substances (*see* p. 69). After this, the digested food reaches the general circulation.

Glycerol, glycerides and fatty acids from the digestion of fats may also enter the capillaries of the villus, but for the most part the products of fat digestion are built up into fats again in the

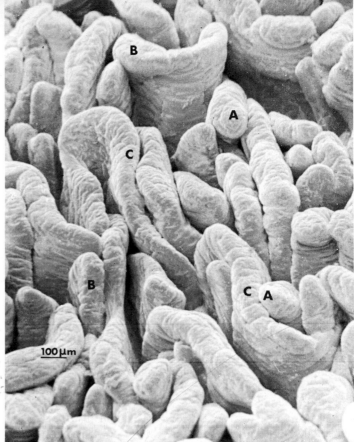

(K. E. Carr/P. G. Toner)

Fig. 10.10 Structure of the small intestine. In this scanning electron micrograph of the human small intestine (×60), the villi are about 0·5 mm long. In the duodenum they are mostly leaf-like (C), but further towards the ileum they become narrower (B), and in the ileum they are mostly finger-like (A). This micrograph is of a region in the duodenum

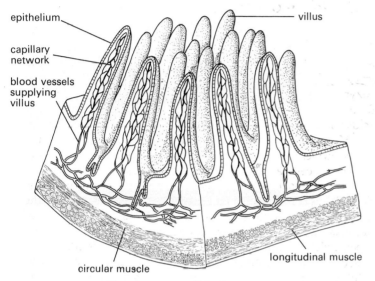

Fig. 10.8 Part of the ileum wall

epithelium of the intestinal mucosa and then enter the *lacteals* to be carried away in the lymph. Large fatty acid molecules may enter the lacteals, while the smaller molecules are carried in the capillaries together with amino acids, glucose and related sugars. Total digestion of fats to fatty acids and glycerol is not essential for absorption, but the emulsification by bile salts is

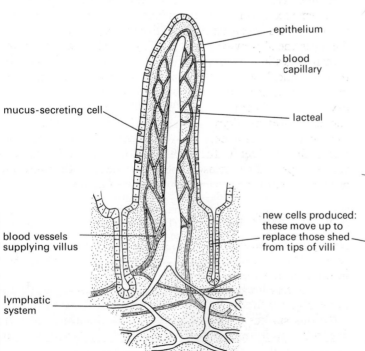

Fig. 10.9 Structure of villus

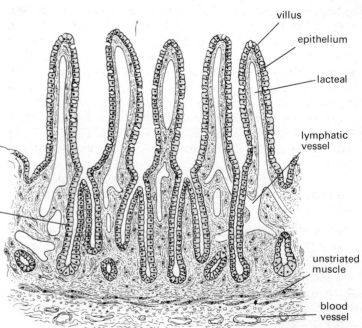

Fig. 10.11 Section through mucosa of ileum

65

very important and there is evidence that partial production of fatty acids and glycerol greatly accelerates emulsification of undigested fats. The fat-soluble vitamins A, D and K are absorbed along with the fats. The lacteals open into the lymphatic system (p. 83) which forms a network all over the body and eventually empties its contents into the blood stream.

The movements of the dissolved substances through the intestinal wall cannot be explained by diffusion alone. The soluble molecules may enter into chemical combination with substances in the epithelial cells in the course of their journey, for example the combination of glucose with phosphate, but the detailed mechanisms have not been worked out.

The food may spend from three to four hours in the small intestine before it passes into the large intestine through the ileo-colic sphincter.

The large intestine (colon and rectum)

The material passing into the large intestine consists of water and undigested matter, largely cellulose and vegetable fibre (roughage), mucus and dead cells from the lining of the alimentary canal. The colon secretes no enzymes but the bacteria in it digest part of the fibre. The colon absorbs much of the water from the undigested residues and other waste. The copious secretion of digestive juices extracts a great deal of water from the blood and pours it into the alimentary canal. If this water were not returned to the blood stream by the colon, the loss of fluid from the body would be excessive and lead to dehydration. This is one of the dangers of acute and prolonged diarrhoea (see p. 67). In normal conditions the partially dried and semi-solid wastes, the *faeces*, are passed into the rectum by peristalsis and expelled at intervals through the anus.

The entry of the contents of the colon into the rectum distends it and sets off a sensory impulse which produces the urge to defaecate, i.e. expel the faeces. Habit and opportunity control this urge and defaecation may take place with a frequency varying from three times a day to every third day according to the individual, there being no special physiological value in any particular frequency.

The components of faeces are largely the product of the alimentary tract itself, e.g. dead cells from its mucosa, dead bacteria and mucus, and are therefore produced whether food is taken or not, though the composition will vary with the food. The residue of vegetable fibre helps to retain enough water in the faeces to keep them soft and easily moved by peristalsis in the colon and rectum. The added bulk, by distending the colon, also stimulates this reflex contraction. For these reasons, dietary fibre helps prevent constipation.

Digestive action

Region of alimentary canal	Digestive gland	Digestive juice produced	Enzymes in the juice	Class of food acted upon	Substances produced	Notes
MOUTH	Salivary glands	Saliva	Salivary amylase	Starch	Maltose	Slightly acid or neutral. Mucus helps form bolus. Water lubricates food.
STOMACH	Gastric glands (in stomach lining)	Gastric juice	Pepsin	Proteins	Peptides	0.5% hydrochloric acid also secreted, provides acid medium for pepsin and kills most bacteria.
STOMACH	Gastric glands (in stomach lining)	Gastric juice	(Rennin)	(Milk protein)	(Clots it)	0.5% hydrochloric acid also secreted, provides acid medium for pepsin and kills most bacteria.
DUODENUM	Pancreas	Pancreatic juice	Trypsin	Proteins and peptides	Amino acids	Two other protein-digesting enzymes are present. Bile emulsifies fats and aids their absorption. Duodenum contents slightly acid.
DUODENUM	Pancreas	Pancreatic juice	Amylase	Starch	Maltose	Two other protein-digesting enzymes are present. Bile emulsifies fats and aids their absorption. Duodenum contents slightly acid.
DUODENUM	(Liver)	(Bile)	Lipase	Fats	Fatty acids and glycerol	Two other protein-digesting enzymes are present. Bile emulsifies fats and aids their absorption. Duodenum contents slightly acid.
ILEUM	The glands between the villi produce mucus and enterokinase but few, if any, digestive enzymes	Pancreatic enzymes still active		Peptides	Amino acids	The final stages of digestion take place in the ileum with the aid of pancreatic enzymes and the epithelial cells of the villi. The main function of the ileum is the absorption of the digested products.
ILEUM	The glands between the villi produce mucus and enterokinase but few, if any, digestive enzymes	Pancreatic enzymes still active		Fats	Fatty acids and glycerol	The final stages of digestion take place in the ileum with the aid of pancreatic enzymes and the epithelial cells of the villi. The main function of the ileum is the absorption of the digested products.
ILEUM	The glands between the villi produce mucus and enterokinase but few, if any, digestive enzymes	Pancreatic enzymes still active		Maltose	Glucose	The final stages of digestion take place in the ileum with the aid of pancreatic enzymes and the epithelial cells of the villi. The main function of the ileum is the absorption of the digested products.
ILEUM	The glands between the villi produce mucus and enterokinase but few, if any, digestive enzymes	Pancreatic enzymes still active		Sucrose	Glucose and fructose	The final stages of digestion take place in the ileum with the aid of pancreatic enzymes and the epithelial cells of the villi. The main function of the ileum is the absorption of the digested products.
ILEUM	The glands between the villi produce mucus and enterokinase but few, if any, digestive enzymes	Pancreatic enzymes still active		Lactose	Glucose and galactose	The final stages of digestion take place in the ileum with the aid of pancreatic enzymes and the epithelial cells of the villi. The main function of the ileum is the absorption of the digested products.
COLON	Bacterial enzymes produce fatty acids from vegetable fibre					Absorption of water

Constipation. Retention of the contents of the lower colon for long periods may lead to constipation. The urge to defaecate is depressed and the faeces are difficult to expel. Constipation may accompany or cause other symptoms of ill-health, not on account of the toxic products of bacterial decay in the colon but because of the physical distension of the rectum. Laxatives are chemicals which act by irritating the lining of the colon and in this respect are harmful on account of the likelihood of developing a dependence on these chemicals to produce regular emptying of the colon.

Diarrhoea usually results from inflammation of the large or small intestine during bacterial or virus infection. The irritation brings about frequent and violent peristalsis which expels the contents of the colon before they have been retained long enough for adequate absorption of water. Normally the residues should spend from 12 to 24 hours in the large intestine, being moved on by vigorous waves of peristalsis three or four times a day.

Caecum and appendix. These are relatively small structures in humans and their digestive functions, if any, are not clear. The appendix has many lymph nodes and may exert some control over the bacteria in the colon.

In herbivorous animals the caecum and appendix are large and it is here that most cellulose digestion occurs, largely as a result of bacterial activity. In humans, bacteria colonise the large intestine soon after birth and play a part in producing vitamin K and some vitamins of the B group.

Control of secretion

Digestive juices are only secreted when food is present in the appropriate part of the alimentary canal. In this way, wastage of enzymes is avoided. Secretion is co-ordinated by both the nervous and endocrine systems (p. 145).

Saliva is secreted continuously but the presence of food in the mouth sets off a nervous reflex, initiated in the sensory cells of the taste buds (p. 132), resulting in not only a greater volume of salivary secretion but also an increase in its enzyme content. Salivation is also subject to conditioned reflexes (p. 150).

Secretion in the stomach is initiated by a reflex set off by the presence of food in the mouth, and maintained by the presence of food, especially meat, in the stomach. A hormone, *gastrin*, is liberated by the pyloric end of the stomach and, on returning to the gastric glands in the blood circulation, stimulates the continued production of enzymes. The first reflex, as in salivation, is subject to conditioning and is inhibited by fear or disgust.

Pancreatic secretion takes place when the acid contents of the stomach reach the duodenum causing the production of two hormones, *secretin* and *pancreozymin*, which on being conveyed in the circulation to the pancreas stimulate it to produce enzymes. Secretin, however, is produced before food reaches the duodenum and its release is therefore likely to be initiated by nervous reflex. The pancreatic fluid resulting from stimulation by secretin is watery and alkaline, while that resulting from the effect of pancreozymin is rich in enzymes.

Hunger and thirst. The causes of the sensation of hunger are not understood but it is apparent that the sensation is an essential preliminary to adequate feeding. The depression of the sensation of hunger could lead to inadequate intake in respect of both dietetic balance (p. 37) and energy value. Smoking, chewing and the action of certain drugs will reduce hunger.

The empty stomach has been shown in some cases to contract slowly for periods of about 20 seconds over half an hour and may thus produce the 'pangs' of hunger, though it is more likely that there are centres of hunger and satiety in the brain. Hunger is a physiological need for food, whereas appetite is a psychological anticipation of food. It appears that appetite is an aid to digestion and that one's psychological state when eating can influence the course of digestion, probably by its effect on the muscular action of the alimentary canal and the secretion of enzymes. As has already been pointed out, fear and disgust can inhibit the secretion of saliva and gastric juice. Tension and disharmony at meal times therefore may produce indigestion.

Indigestion is the term applied in general to all sensations of discomfort in the stomach or duodenum. The causes vary from simple over-distension by eating too much or the accumulation of gases, to ulceration and excessive production of hydrochloric acid. 'Acid indigestion' is a fruitful source of profit for the manufacturers of patent alkalis, but only a doctor can diagnose whether the pain is the result of excessive hydrochloric acid in the stomach. Although in many cases antacids (alkalis) relieve the symptoms of indigestion by neutralization of hydrochloric acid, they can be harmful if taken over long periods. The sodium hydrogencarbonate (bicarbonate) mixtures can cause a disturbance in the salt balance in the body, and by suppressing the symptoms they may postpone a medical consultation by which a serious condition such as a stomach ulcer would be diagnosed.

Utilization of digested food

The products of digestion are carried round the body in solution in the blood. From the blood, most living cells are able to absorb and metabolize glucose, fats and amino acids.

(a) **Glucose.** During respiration in the protoplasm, glucose is oxidized to carbon dioxide and water (*see* p. 27). This reaction releases energy to drive the many chemical processes in the cell, and in specialized cells produces, for example, contraction (muscle fibres) and electrical changes (nerve cells).

(b) **Fats.** Fats are incorporated into cell membranes and other structures in cells. Fatty acids are oxidized in muscles to provide energy for muscle contraction. Twice as much energy can be obtained from fats as from glucose.

(c) **Amino acids** are absorbed by cells and reassembled to make proteins (p. 17). These proteins may form visible structures such as the cell membrane and other components of the protoplasm or the proteins may be enzymes which control and co-ordinate the chemical activity within the cell.

Amino acids not required for building proteins are de-aminated in the liver, that is, their nitrogen is removed and the residue is used in the same way as carbohydrate, namely oxidized, or converted to glycogen and stored.

Storage of digested food

If the quantity of food taken in exceeds the energy requirements of the body or the demand for structural materials, it is stored in one of the following ways:

(a) **Glucose** (Fig. 10.13). The concentration of glucose in the blood of a person who has not eaten for eight hours is usually between 90 and 100 mg/100 cm³ blood. After a meal containing carbohydrate, the blood sugar level may rise to 140 mg/100 cm³ but two hours later, the level returns to about 95 mg.

The sugar not required immediately for the energy supply in the cells is converted in the liver and in the muscles to glycogen. The glycogen molecule is built up by combining many glucose molecules in a long branching chain rather similar to the starch molecule. About 100 g of this insoluble glycogen is stored in the liver and about 300 g in the muscles. When the blood sugar level falls below 80 mg/100 cm³, the liver converts its glycogen back to glucose and releases it into the circulation. The muscle glycogen is not normally returned to the circulation but is used by active muscle as a source of energy in much the same way as glucose.

The glycogen in the liver is a 'short-term' store, sufficient for about only six hours if no other glucose supply is available. Excess glucose not stored as glycogen is converted to fat and stored in the fat cells of the fat depots. (*See* below.)

(b) **Fats.** Certain cells can accumulate drops of fat in their cytoplasm. As these drops increase in size and number, they join together to form one large globule of fat in the middle of the cell, pushing the cytoplasm into a thin layer and the nucleus to one side (Fig. 10.12). Groups of fat cells form *adipose tissue* beneath the skin and in the connective tissue of most organs (Fig. 14.3, p. 97).

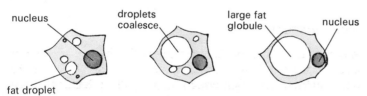

(a) *Accumulation of fat in a fat cell*

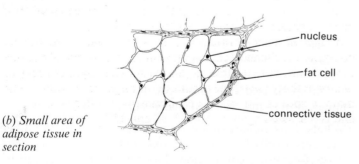

(b) *Small area of adipose tissue in section*

Fig. 10.12   Adipose tissue

Unlike glycogen, there is no limit to the amount of fat stored and because of its high energy value it is an important reserve of energy-giving food.

(c) **Amino acids** (Fig. 10.14). Amino acids are not stored in the body. Those not used in protein formation are deaminated. The protein of the liver and tissues can act as a kind of protein store to maintain the protein level in the blood, but absence of protein in the diet soon leads to serious disorders.

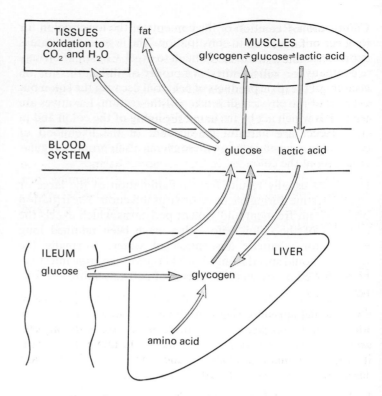

Fig. 10.13   Carbohydrate metabolism

(Figs. 10.13 and 10.14 by kind permission of Bell, G. H., Davidson, J. N., Scarborough, H., *Textbook of Physiology and Biochemistry*, 4th edn, Edinburgh, Livingstone, 1959)

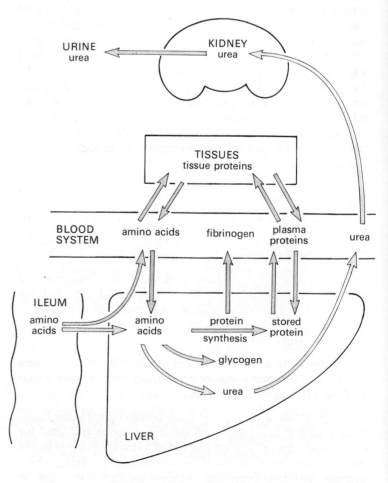

Fig. 10.14   Protein metabolism

The rate of oxidation of glucose and its conversion to glycogen or fat is controlled by hormones (p. 156). When intake of carbohydrate and fat exceeds the energy requirements of the body, the excess will be stored mainly as fat. Some people never seem to get fat no matter how much they eat, while others start to lay down fat when their intake only marginally exceeds their needs. Putting on weight is unquestionably the result of eating more food than the body needs but it is not clear why individuals should differ so much in their reaction. The explanation probably lies in the hormonal balance which, to some extent, is determined by heredity. A slimming diet designed to reduce calorific intake must, nevertheless, always include the essential amino acids, vitamins, mineral salts and certain essential fatty acids.

The liver

The liver is a large, reddish-brown organ which lies just below the diaphragm and partly overlaps the stomach (Fig. 10.7). In addition to a supply of oxygenated blood from the *hepatic artery*, it receives all the blood that leaves the alimentary canal. It has a great many important functions, some of which are described below. (*See* Figs. 10.13 and 10.14.)

1   **Regulation of blood sugar.** The liver is able to convert glucose, amino acids and other substances to an insoluble carbohydrate, *glycogen*. Some of the glucose so converted may be taken from the *hepatic portal* vein carrying blood rich in digested food from the ileum to the liver. About 100 g glycogen is stored in the liver of a healthy man. If the concentration of glucose in the blood falls below about 80 mg/100 cm³ blood, some of the glycogen stored in the liver is converted by enzyme action into glucose and it enters the circulation. If the blood sugar level rises above 160 mg/100 cm³, glucose is excreted by the kidneys. A blood glucose level below 40 mg/100 cm³ affects the brain cells adversely, leading to convulsions and coma. By helping to keep the glucose concentration between 80 and 150 mg the liver prevents these undesirable effects and so contributes to the homeostasis (*see* below) of the body. (*See* Fig. 11.9 for circulatory supply to liver.)

2   **Formation of bile.** Green and yellow pigments are formed when the red blood cells break down. These pigments are removed from the blood by the liver and excreted in the bile. The liver also produces bile salts which play an important part in the emulsification and subsequent absorption of fats (p. 64).

Bile is produced continuously by the liver cells, but stored and concentrated in the gall bladder. It is discharged through the bile duct into the duodenum when the acid chyme arrives there. Bile salts are reabsorbed with the fats they emulsify and eventually return to the liver.

3   **Storage of iron.** Millions of red blood cells break up every day. In the liver their decomposition is completed and the iron from the haemoglobin is stored.

4   **Deamination.** Excess amino acids are not stored in the body. Amino acids that are not built up into proteins and used for growth and replacement are converted to carbohydrates by the removal from the molecule of the *amino group*, —NH₂, which contains the nitrogen. The residue can be converted to glycogen, being stored or oxidized to release energy. The nitrogen of the amino group is converted in the liver to *urea*, an excretory product that is constantly eliminated by the kidneys.

5   **Manufacture of plasma proteins.** The liver makes most of the proteins found in blood plasma, including fibrinogen which plays an important part in the clotting action of the blood (p. 73).

6   **Body heat.** *NOTE:* Contrary to statements made in previous impressions, the liver is not a net exporter of heat. Although many of its chemical reactions release heat energy, many require an input of energy.

7   **Use of fats in the body.** When fats stored in the body are required for use in providing energy, they travel in the blood stream from the fat depots. Some are converted to fatty acids and used directly by the muscles and some are changed by the liver into substances which can be oxidized for energy by other tissues.

8   **Detoxication.** Poisonous compounds, produced in the large intestine by the action of bacteria on amino acids, enter the blood, but on reaching the liver are converted to harmless substances, later excreted in the urine. Many other chemical substances normally present in the body or introduced as drugs are modified by the liver before being excreted by the kidneys, e.g. hormones are converted to inactive compounds in the liver so limiting their period of activity in the body.

9   **Storage of vitamins.** The fat-soluble vitamins A and D are stored in the liver. This is the reason why animal liver is a valuable source of these vitamins in the diet. The liver also stores a product of the vitamin B₁₂. This product is necessary for the normal production of red cells in the bone marrow.

Homeostasis

A complete account of the functions of the liver would involve a very long list. It is most important, however, to realize that the one vital function of the liver, embodying all the details outlined above, is that it helps to regulate the concentration and composition of the body fluids, particularly in the blood.

Within reason, a variation in the kind of food eaten will not produce changes in the composition of the blood.

If this *internal environment*, as it is called, were not so constant, the chemical changes that maintain life would become erratic and unpredictable so that with quite slight changes of diet or activity the whole organization might break down. The maintenance of the internal environment is called *homeostasis* and is discussed again on pp. 72, 92 and 158.

Practical work

Experiment 1   **The digestive action of saliva on starch**

Saliva is collected in a test-tube after rinsing the mouth to remove traces of food, and about 1 cm³ is added to each of two test-tubes, A and B, containing approximately 2 cm³ of 2 per cent starch solution. A control is set up by preparing two more tubes, C and D, with the same volume of starch solution but with boiled saliva. After 5 minutes, tube A is tested with iodine solution while tube B is boiled with Benedict's solution (*see* p. 38). The controls C and D are similarly tested.

Result. Failure to obtain a blue colour with iodine indicates that starch is no longer present, while a red precipitate with Benedict's reagent shows that a sugar has been produced. Control C should give a blue colour with iodine while D should not give a red precipitate with Benedict's solution.

Interpretation. The disappearance of starch and the appearance of sugar when unboiled saliva is present suggests that the latter contains an enzyme that promotes this change.

Experiment 2 **The action of pepsin on egg-white (albumen)**

The white of one egg is stirred into 500 cm³ tap water. The mixture is boiled and filtered through glass wool to remove large particles. About 2 cm depth of the cloudy suspension is poured into each of four tubes labelled A to D. To A is added 1 cm³ of 1 per cent pepsin solution; to B is added 3 drops of dilute hydrochloric acid; to C, 1 cm³ of pepsin solution and 3 drops of acid; and to D, 3 drops of acid and 1 cm³ boiled pepsin solution. All four test-tubes are placed in a beaker of water at 35–40 °C for 5–10 minutes.

Result. The contents of tube C only will become clear; the others will remain cloudy.

Interpretation. The change from a cloudy suspension to a clear solution suggests that the solid egg-white particles have been digested to soluble products. This result, and those of the controls A, B and D, support the idea that pepsin is an enzyme that digests albumen (a protein) in acid conditions.

Experiment 3 **The action of lipase**

5 cm³ milk and 7 cm³ dilute (M/20) sodium carbonate solution are placed in each of three test-tubes labelled 1 to 3, and six drops of phenolphthalein are added to each tube to colour the contents bright pink. To tubes 2 and 3 is added 1 cm³ of 3 per cent bile salts solution. To tubes 1 and 3 is added 1 cm³ of 5 per cent lipase solution, and to tube 2 an equal volume of boiled lipase solution.

Result. In 10 minutes or less, the colour of the liquids in tubes 1 and 3 will change to white, tube 3 changing first. The liquid in tube 2 will remain pink.

Interpretation. Lipase is an enzyme that hydrolyses fats to fatty acids and glycerol. When lipase acts on milk fats, the fatty acids so produced react with the alkaline sodium carbonate and make the solution more acid. In acid conditions the pH indicator, phenolphthalein, changes from pink to colourless. The presence of bile salts in tube 3 seems to accelerate the reaction, although bile salts with the denatured enzyme in tube 2 cannot bring about the change on their own.

For experiments investigating the effect of temperature and pH on enzyme action *see* p. 30.

(*Note:* these experiments and several others are fully detailed in the laboratory manual *Enzymes*—*see* p. 320.)

Questions

1 List the chemical changes undergone by (a) a molecule of starch from the time it is placed in the mouth to its ultimate use in providing energy; (b) a molecule of protein from the time it is swallowed to the time when its components are used in a cell (other than the liver).
In each case, state where the changes are taking place.
2 Write down the menu for your breakfast and lunch (or supper); indicate the principal food substances present in each component of the meal and state the final digestion product of each and the use your body is likely to have made of them.
3 Suggest reasons to explain why digestion takes place in stages, e.g. in the mouth, stomach and duodenum, rather than at the same time and in one place.
4 Study the chemistry of carbohydrates, proteins and fats (or any *one* of these) on pp. 16–17. Write the structural formula, as simply as possible, of starch, a protein and a fat and show how the molecule is broken down by digestive enzymes to the products which are absorbed in the ileum. Indicate the point at which the enzymes attack the food molecules.

11

Blood, its Composition, Function and Circulation

Composition

Blood consists of a suspension of red cells, white cells and platelets in a liquid called plasma. In an adult man there are five to six litres of blood in the body while a woman, on average, has one litre less.

Cells

Red cells (erythrocytes) (Figs. 11.1a and 11.2). The red cells are minute, biconcave discs consisting of spongy cytoplasm in an elastic cell membrane. During their formation they have nuclei, but these are lost before they enter the circulation. In the cytoplasm of the erythrocytes is a red pigment called *haemoglobin*, which constitutes about 95 per cent of the solids in the cell, the remaining 5 per cent consisting of enzymes, salts and proteins.

Haemoglobin is a protein with iron in its molecule. It has an affinity for oxygen and readily combines with it in conditions of high oxygen concentration and alkalinity. It forms an unstable compound called *oxyhaemoglobin*, which equally readily breaks

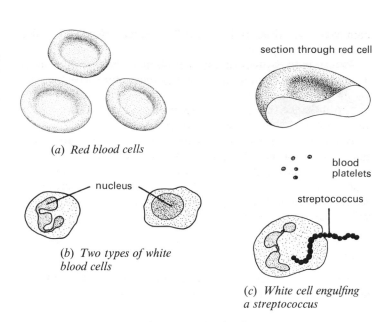

(a) Red blood cells

section through red cell

nucleus

(b) Two types of white blood cells

blood platelets

streptococcus

(c) White cell engulfing a streptococcus

Fig. 11.1   Blood cells

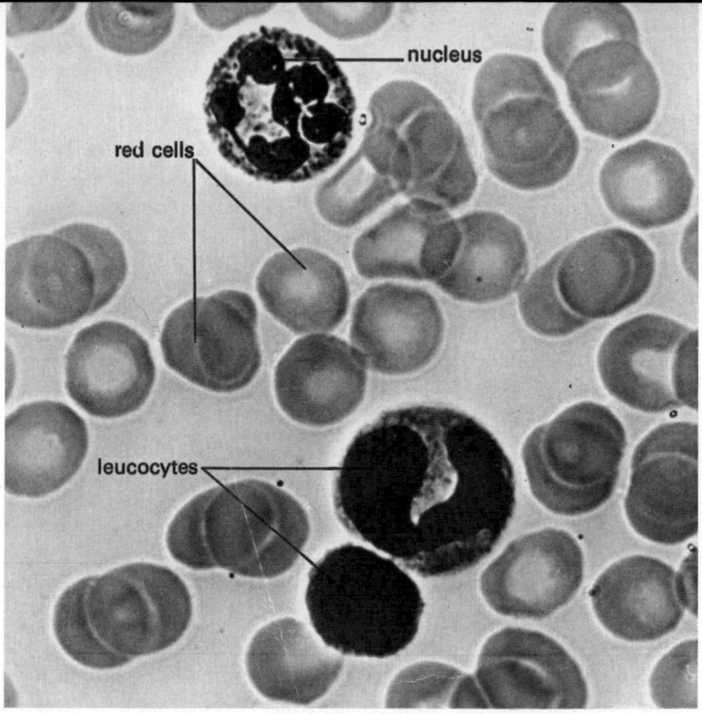

nucleus

red cells

leucocytes

Fig. 11.2   Red and white cells in the human blood ($\times 1\,500$)

down and releases the oxygen in conditions of low oxygen concentration and raised acidity. This property makes it very efficient in transporting oxygen from the lungs, where the oxygen concentration is high, to the tissues, where it is lower. The total absorbing surface presented by the membranes of all the red cells is about 3 000 square metres.

The red cells are made principally in the red bone marrow of the short bones such as the sternum, ribs and vertebrae. The cells lining the blood vessels and blood spaces in the red marrow divide repeatedly, budding off new cells which themselves divide again. Red cell formation takes about seven days, after which the nucleus disappears, the haemoglobin is formed and the cell is released into the circulation.

There are about five and a half million cells in a cubic millimetre of blood. One cell survives for about four months after which it breaks down and is disintegrated in the liver and spleen, partly as a result of ingestion by phagocytes (*see* p. 72). Part of the haemoglobin from the spent red cells is changed to *bilirubin* and *biliverdin*, green pigments which eventually form part of the bile. The iron from the haemoglobin is retained in the liver and spleen cells and probably used again for new erythrocytes. About 200 000 million red cells are formed and destroyed each day, representing about one per cent of the total. To examine the condition of the cells in the blood-forming marrow, a sample is usually taken from the sternum (breastbone), a procedure called *sternal puncture*.

At high altitudes atmospheric pressure is low, making it more difficult for the lungs to extract oxygen from the air. In these conditions, the red marrow makes greater numbers of red cells. There is a steady increase in the number of red cells in the blood of mountain climbers who spend long periods at high altitudes. This *acclimatization* takes several weeks. People living permanently at high altitudes have more red cells in their blood than people at sea level but there are other long-term physiological changes as well, e.g. a rise in blood pressure in the pulmonary artery and a thickening of the right ventricle (p. 79).

Anaemia is a shortage of red cells in the blood which may arise through a large loss of blood, insufficient iron taken in with the diet, failure of the red marrow to make enough haemoglobin or erythrocytes, or excessive destruction of the cells. The anaemic person usually feels weak, breathless and tired.

Loss of blood as a result of injury can usually be made good by transfusion. In diseases like malaria, however, the red cells are destroyed by the disease organism and only by killing the parasites can the anaemia be stopped. Lack of iron in the diet can be corrected by increasing its meat and liver content or by taking iron-containing tablets, but if absorption of iron in the ileum is deficient the iron compounds may not be taken up or used. Increasing the intake of iron is also of little value if the formative chemistry of haemoglobin is defective.

Extensive destruction of red cells follows an incompatible transfusion (p. 75) or occurs sometimes in newborn babies if the mother and father are not of the same Rhesus blood group (p. 76). The pigments from the rapidly destroyed red cells lead to a yellow discolouration of the skin called jaundice.

There are two forms of vitamin-deficiency anaemia. In both cases the bone marrow does not produce enough red cells. One form is caused by a shortage of folic acid in the diet and can be remedied by increasing the intake of this vitamin. The other form (sometimes called pernicious anaemia) is caused by a failure to absorb vitamin B_{12}. Vitamin B_{12} is widespread in animal products and unlikely to be missing from the diet except in the case of strict vegetarians (vegans). Cells in the stomach stop producing a substance which is needed for the absorption of B_{12}, so the condition has to be treated by regular injections of the vitamin.

Sickle-cell anaemia is an inherited condition. The haemoglobin molecules have a slightly different composition from normal which causes them to form rod-like structures at low oxygen concentrations. These structures distort the red cells and hasten their destruction (*see* p. 168).

White cells (leucocytes) (Figs. 11.2 and 11.3). There is about one white cell to every six hundred erythrocytes, the actual numbers varying from 4 000 to 13 000 cells per cubic millimetre. Accurate estimation is made difficult by the fact that their distribution in different parts of the body is very variable.

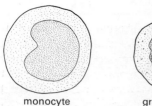

monocyte　　　granulocyte　　　lymphocyte

Fig. 11.3　Three types of leucocyte

red cell
(for comparison)

About 70 per cent of the white cells are of a type called *granulocytes*. They are larger than the red cells, irregular in shape and can change their form. They have a nucleus characteristically divided into two or three lobes, and there are many granules in their cytoplasm. These cells can move by a flowing action of their cytoplasm and, in certain situations, can pass out of the capillaries by squeezing between the cells of the capillary wall. They are made in the red marrow but outside the blood vessels. When mature they enter the circulation by passing through the capillary walls. They can ingest and destroy bacteria and dead tissue cells by flowing around, engulfing and digesting them. This action is called *phagocytosis* and these leucocytes are said to be *phagocytic*. The granulocytes accumulate at the site of an injury or infection and devour invading bacteria and damaged tissue, so preventing a spread of harmful bacteria as well as accelerating the healing of the infected region. Granulocytes survive for about a week.

Of the white cells, 23 per cent are *lymphocytes*, which are smaller than the granulocytes and have a round nucleus which occupies most of the cell. Lymphocytes are made in the lymph nodes and spleen (p. 83) and they make some of the chemicals called antibodies which are effective against foreign organisms and proteins in the blood. Lymphocytes survive on average for 100 days but some probably live for years. *Monocytes* are white cells with a simple or slightly lobed nucleus. They are made in the lymph nodes and spleen and are able to ingest almost any type of foreign particle. The life span for monocytes is not known for certain.

Platelets. These are pieces of special blood cells budded off in the bone marrow and they play an important part in the clotting action of the blood (p. 73). There are between 250 000 and 500 000 of them in each cubic millimetre of blood and they appear under a high-powered microscope as tiny oval structures.

Plasma

The liquid part of the blood is called plasma. It is a solution or suspension in water of many compounds. Some of the most important compounds are the plasma proteins, including albumin, globulin, fibrinogen and the antibodies. Other important constituents of plasma are the salts, or more strictly the ions, of sodium, potassium, calcium, chloride, phosphate and hydrogencarbonate. Varying in quantity with the area of the body and its activity, plasma will also contain food substances (e.g. amino acids, glucose and fats), hormones,

pigments (e.g. bilirubin, a bile pigment), urea and other nitrogenous compounds.

The sodium hydrogencarbonate (bicarbonate) acts as a chemical 'buffer', that is, it allows the blood to absorb hydrogen ions without effectively increasing its acidity (*see* p. 21). This is important since the chemical reactions in living cells are very sensitive to the slightest change of pH.

Serum is the name given to blood plasma from which the fibrinogen has been removed.

Functions of the blood

It is convenient at this point to distinguish between the functions of (a) the blood, which provides and controls the internal fluid surrounding the cells, (b) the circulation, which distributes food, oxygen, etc. to all parts of the body, and (c) defence against bacterial invasion.

Blood and the internal environment

All the living cells of the body are bathed by a fluid derived from blood plasma. This fluid, called *tissue fluid*, supplies them with the food and oxygen necessary for their living chemistry and removes the products of their activities which, if they accumulated, would poison the cells. The composition of the plasma is very precisely regulated by the liver and kidneys, so that the living cells are soaked in a liquid whose composition varies only within narrow limits. This provides them with the environment they need and enables them to live and grow in the most favourable conditions.

A single-celled animal living in a pond is able to adapt itself to variations of temperature, acidity, food and oxygen concentration, or move to a situation where these are favourable, but the cells of a complex organism have become specialized and unless the conditions are more or less right they will not work properly. The regulation of the composition of the tissue fluid is an aspect of *homeostasis* (*see* p. 92).

Circulation

The movement of blood in the vessels around the body constantly changes the fluid surrounding the living cells so that fresh supplies of oxygen and food are brought in as fast as they are used up and poisonous end products are not allowed to accumulate. The following account is concerned mainly with the circulation as a transport system rather than with the chemical properties of blood fluid as an internal environment.

On average, a red cell would complete the circulation of the body in 45 seconds.

1　**Transport of oxygen from the lungs to the tissues.** In the lungs, oxygen dissolves in the blood and is thus carried to all parts of the body. If the oxygen simply made a physical solution with the blood plasma, its solubility is such that it could not constitute more than 0.36 per cent by volume of the oxygenated blood. In fact there is 20 per cent by volume of oxygen present in the blood. This is because the combination of oxygen with haemoglobin in the red cells effectively takes up the oxygen from physical solution in the plasma and carries it in a chemically combined form.

As the blood passes through the lung capillaries it loses some of its carbon dioxide, so becoming more alkaline, and it is exposed to the relatively high concentration of oxygen in the air sacs of the lung. In these conditions, the haemoglobin combines with oxygen to make the unstable oxyhaemoglobin. Active tissues produce carbon dioxide and use oxygen. When blood reaches these tissues the conditions of raised acidity and low oxygen concentration cause the oxyhaemoglobin to break down, releasing its oxygen. The oxygen diffuses out of the cell membrane into the plasma, through the capillary wall and into the respiring cells.

Oxyhaemoglobin is bright red, while haemoglobin is a darker red colour.

2   **Transport of carbon dioxide from the tissues to the lungs.** Carbon dioxide produced from actively respiring cells diffuses through the capillary walls and dissolves in the plasma. As in the case of oxygen, a solution of carbon dioxide in blood would not exceed 2.7 per cent by volume and yet up to 60 per cent is found in the blood. Most of the carbon dioxide must therefore be carried in a combined form, usually as hydrogencarbonate (bicarbonate) ions ($—HCO_3^-$) which are transported in both the plasma and the red cells. The less oxygen there is present in the haemoglobin, the more carbon dioxide can the red cells carry. This favours the exchange of carbon dioxide and oxygen in the lungs and tissues. In the lungs, the hydrogencarbonate breaks down to carbon dioxide and water and the carbon dioxide diffuses through the capillary walls into the alveoli of the lungs (p. 86). Both the formation and breakdown of hydrogencarbonate are accelerated by an enzyme in the red cells called *carbonic anhydrase.*

3   **Transport of nitrogenous waste from the liver to the kidneys.** When the liver, in the course of deamination (p. 92), changes amino acids into glycogen, the amino ($—NH_2$) part of the molecule is changed into the nitrogenous waste product, *urea.* Similarly, the DNA bases (p. 166) from the nuclei of dead cells are deaminated to produce *uric acid.* These substances are carried away in the blood circulation. When the blood passes through the kidneys, most of the urea and uric acid are excreted (p. 94).

4   **Transport of digested food from the ileum to the tissues.** The soluble products of digestion enter the capillaries of the villi lining the ileum (p. 65). They are carried in solution by the plasma and after passing through the liver, enter the general circulation. Glucose and amino acids pass out of the capillaries and into the cells of the body. Glucose may be oxidized, in a muscle for example, and provide the energy for contraction; amino acids will be built up into new proteins and make new cells and tissues.

5   **Distribution of hormones.** Hormones are chemicals that affect the rate of vital processes in the body (p. 156). From the glands that make them, they are carried in the blood plasma all around the body. Each hormone has one or more 'target' organs and affects the rate at which the organ works.

6   **Distribution of heat and temperature control.** Muscular and chemical activity releases heat. These processes are going on far more rapidly in some parts of the body than in others, for example the abdomen, brain and active muscles. The heat produced locally is distributed all round the body by the blood, and in this way an even temperature is maintained in all regions.

The diversion of blood to or away from the skin also plays a part in keeping the temperature constant (*see* p. 98).

Defence

Formation of clots. When a blood vessel is cut open or its lining damaged, the platelets adhere to the damaged area. The protein, fibrinogen, in the plasma is converted to fibrin which appears to come out of solution as filaments radiating from the platelets. The network of fibres across the wound makes a plug which stops any more blood leaking out and prevents the entry of bacteria and poisons. Eventually the dried blood clot shrinks, hardens and forms a scab which protects the damaged area while new tissue is forming.

The details of the coagulation of the blood are very complicated and not all biologists agree about the different stages. It is vital that blood in undamaged vessels should not coagulate, and this may be the reason for the complex series of changes that must take place before clotting occurs.

The fibrinogen in the plasma is converted to fibrin by the enzyme-like action of *thrombin,* provided sufficient calcium ions are present in the plasma. In normal plasma, thrombin is in an inactive form known as *prothrombin.* In the region of the wound the damaged tissues and the platelets are thought to produce a substance called *thrombokinase* (thromboplastin) which converts prothrombin to thrombin.

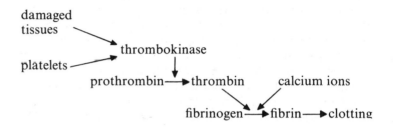

Prothrombin is made in the liver and a substance called vitamin K (p. 35) is essential for its production. Bile salts are necessary for the absorption of this fat-soluble vitamin, which is made by intestinal bacteria.

After coagulation, clot shrinkage pulls the tissues closer together and new cells made at the margins of the wound begin to spread over the inner surface of the scab, forming a new layer of skin at the rate of about 0.5 mm per day. New capillary branches grow from others in the area, and cut nerves grow out into the tissues. When the scab falls off a deep cut, a scar may be left due to the thinness of the epidermis and the extra white fibrous tissue (collagen) in the dermis.

Haemophilia is an inheritable disease in which a person's blood clots only very slowly owing to the lack of a substance needed for the formation of thrombokinase. Such people are liable to lose disproportionate quantities of blood from even minor wounds.

Prevention of infection. (a) *Infected wounds.* Normally the skin or the lining of the alimentary canal is a barrier to the entry of any bacteria. The layer of dead cells on the skin provides a mechanical barrier while the mucus and chemicals of the alimentary canal offer a chemical defence. If the skin is broken, however, and bacteria enter the cut, certain of the white cells migrate out of the capillaries in that region and begin to engulf and digest damaged tissues or any bacteria which have invaded the tissues. Sometimes large quantities of dead white cells together with digested and liquefied tissues accumulate at the site of the wound and form pus. The ingestion of bacteria by the white cells and the clot formation which prevents free circulation localize the site of infection, and most of the bacteria are destroyed before they can enter the general circulation. Those bacteria which do escape into the blood stream are trapped by stationary cells, like granulocytes, in the lymph nodes, spleen or liver. These stationary white cells are called *macrophages.*

Certain virulent strains of bacteria cannot be ingested by the white cells until they have been acted upon by chemicals called *antibodies* which are secreted into the blood. If these antibodies are not already present in the blood or are not made quickly enough, the virulent bacteria may invade the blood system and give rise to *septicaemia* (blood poisoning).

Other defence mechanisms come into action in the region of a wound, which either facilitate those already described or directly increase the chances of successful healing.

(i) A chemical similar in its action to—if not identical with —*histamine* is set free by the damaged tissues and causes the capillaries in that region to expand. The arterioles supplying the skin in that region also dilate so that more blood reaches the surface of the skin and the rate of flow is increased. This gives rise to the redness and warmth in the vicinity of a wound and also increases the supply of antibodies and leucocytes.

(ii) Another chemical from the damaged tissues makes the capillaries more permeable. This allows the leucocytes to squeeze out of the capillaries more easily but in addition permits some proteins to escape into the tissues. The osmotic influence of these proteins extracts water from the blood vessels and the fluid accumulates in the region of the wound giving rise to the familiar swelling. The accumulated fluid drains into the lymphatics and this direction of flow may be effective in carrying bacteria away from the blood circulatory system and into the lymphatic system where the lymph nodes quickly remove bacteria. In fact, the clotting of blood and the clumping of groups of bacteria by antibodies often prevent bacteria from entering either the blood or the lymphatic system.

(iii) The pain-sensory endings in the wounded area become more sensitive and tend to reduce movement of or interference with a damaged area. This may have beneficial effects in healing.

Asepsis. Cleanliness in dealing with wounds reduces the chances of bacterial infection. The dressing of a wound should include first a thorough washing of the area to remove, mechanically, as much bacteria-containing material as possible, and finally a sterile dressing to cover the wound and prevent further bacteria from reaching it. It follows that the instruments, dressings and hands of any assistants must be as free from bacteria as possible (*see* p. 293).

Antitetanus. With deep flesh wounds, an antitetanus injection is often given since, if tetanus bacteria enter the wound, by the time typical symptoms appear the tetanus toxin is firmly attached to the nervous system and almost unassailable by antibodies. The injected antitetanus antibody inactivates the bacteria before they leave the blood stream.

(b) *Disease and immunity.* Many diseases are caused by the presence of bacteria or viruses in the body, and the symptoms may be due to one or more of the following: the foreign proteins of the bacteria themselves; the poisonous chemicals (usually proteins) called *toxins,* which are produced by the bacteria;

and the breakdown products of the infected tissue. Recovery from the disease and the development of immunity to further attacks, in some cases, depend to a large extent on the production in the blood of antibodies. These antibodies are proteins, produced by cells such as lymphocytes and released into the plasma. They may affect bacteria or their products in a number of ways:

(i) *opsonins* adhere to the outer surface of bacteria and so make it easier for the phagocytic white cells to ingest them;
(ii) *agglutinins* cause bacteria to stick together in clumps; in this condition, the bacteria cannot invade the tissues;
(iii) *lysins* destroy the bacteria by dissolving their outer coats;
(iv) *antitoxins* combine with and so neutralize the poisonous toxins produced by the bacteria.

The foreign bodies that stimulate the blood to produce antibodies are called *antigens.* Antibodies are very specific, that is, they will act against and neutralize only one particular protein. An antibody against a *Streptococcus* bacterium, for example, would be ineffective against a *Staphylococcus.* Measles antibodies are ineffective against chickenpox.

When the organism recovers from the disease, the antibodies remain for only a short time in the circulatory system, but the ability to produce them is greatly increased, so that a further invasion by bacteria or viruses is likely to be stopped at once and the person is said to be *immune* to the disease. People may acquire immunity after recovering from an attack, as in measles, or it may be induced in them by *vaccination* or *inoculation.* Naturally acquired immunity may occur because disease bacteria are present in the body without being sufficiently numerous or suitably placed to produce disease symptoms. A person might thus experience a mild form of disease without even noticing it, but his blood will have made antibodies which render him immune to any further attacks. A population that has never experienced a particular disease is likely to suffer acutely from it when it is first encountered. When the white man came first to the Fiji islands he also introduced the disease measles. This was a minor ailment for the European but a fatal and devastating disease to the native population who had no natural immunity.

A *vaccine* is a collection of disease bacteria or viruses, killed or inactivated by heat, chemicals or other methods to prevent their reproduction, or it may consist of relatively harmless strains of the disease organisms. When a vaccine is introduced by injection or by mouth we undergo a mild form of the disease and our cells manufacture antibodies. In this way immunity is artificially acquired. The period of immunity during which antibodies can be produced rapidly, varies from a few months to many years, according to the nature of the infection. Immunity induced by an infection or by a vaccine is sometimes called *active immunity* because the person makes the antibodies himself. *Passive immunity* can be acquired by the injection into the blood of antibodies already made by a donor in the form of a serum.

Serum. The blood of a person (or animal) who has recently recovered from a disease will contain antibodies. If the cells and fibrinogen are removed from a sample of this blood, a serum is obtained which, when injected into other people, may give them temporary (passive) immunity or cure them if they already have the disease. Sera for treating tetanus and chickenpox are

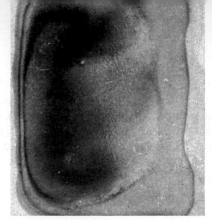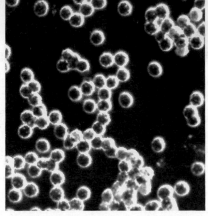

(a) *The cells and serum are matched* (b) *Effect of a mis-match between cells and serum* (c) *The red cells are being agglutinated* (× 800) (Philip Harris Biological Ltd)

Fig. 11.4 Agglutination; *a* and *b* show the appearance, to the naked eye, after compatible and incompatible cells and serum are mixed; *c* shows the red cells being agglutinated under the microscope

prepared from the blood of donors who have recently been inoculated or have recovered naturally from the disease. Blood is taken from the donor, the red cells are returned to his or her circulation and the antibodies are prepared from the plasma. For further examples of active and passive immunity, vaccines and sera, *see* Chapter 32.

Blood groups and transfusion

Excessive loss of blood from the body may cause a condition known as 'shock' which often produces more serious effects on the patient than the direct effect of his injuries. This loss of blood can be made good by a transfusion in which blood from another person, the donor, is fed into the patient's veins.

Early experiments on transfusion often produced either spectacular successes or very distressing symptoms due to the coagulation of the donor's red cells in the blood vessels of the recipient. When this happens, the kidneys have difficulty in eliminating the products of the clumped and disintegrating red cells and in extreme cases they may cease to function normally.

In 1901, Dr Landsteiner mixed samples of blood serum with suspensions of red cells taken from different people. In some cases the serum and cells made satisfactory mixtures while in others the red cells were clumped together (Fig. 11.4). It was discovered that there were antagonistic substances on the surface of the red cells and in the serum. These behave in much the same way as antigens and antibodies in the blood. The substance on the cell is called the *agglutinogen* and that in the plasma is the *agglutinin*. Antagonistic agglutinogens and agglutinins do not occur in any one person's blood.

There are four main blood groups of this kind, called A, B, AB and O. In group A, the cells carry agglutinogen A on their surface while in the plasma is the agglutinin anti-B. In group B the cells carry the B agglutinogen and the plasma contains the anti-A agglutinin. The cells of group AB carry both A and B agglutinogens and there are no agglutinins in the plasma. Group O cells have only a weak agglutinogen, not attacked by either anti-A or anti-B agglutinins but both of these antibodies are present in the plasma.

Group	Agglutinogen on cells	Agglutinin in plasma
A	A	anti-B
B	B	anti-A
AB	A and B	none
O	none	anti-A and anti-B

If group A blood is transfused into a group B person, the anti-A agglutinin in the recipient's plasma reacts with the agglutinogen A on the donor's cells and causes them to stick together. Similarly, group B cannot receive transfusions from group A. Group AB having no agglutinins in the plasma can receive blood from any group, while group O with both agglutinins in the plasma can receive blood only from group O donors whose cells have neither A or B agglutinogens. Conversely, group AB with both agglutinogens can donate blood only to their own group, whereas group O can donate to all groups since their cells cannot be agglutinated.

It is only the agglutination of the donor's cells that causes the pathological symptoms. Although the 'wrong' agglutinin in the donor's serum might be expected to react with the recipient's red cells, this does not appear to happen, or at least does not cause any ill effects. One possible reason is that the donor's serum with its antagonistic agglutinin is quickly diluted when it enters the recipient's blood stream, or that the foreign agglutinin does cause some cells to clump but in small, isolated and widely scattered groups.

To determine the blood group of a person, a little of his blood is mixed separately with sera from group A and group B. If the cells clump in A serum, he is group B; if they clump in group B serum he is group A. Agglutination in both indicates that he is group AB and if his cells react with neither, his group is O.

Group	Can donate blood to	Can receive blood from
A	A and AB	A and O
B	B and AB	B and O
AB	AB	all groups
O	all groups	O

A donor gives 420 cm³ of his blood from a vein in his arm. The blood is led into a sterilized container with sodium citrate which prevents clotting. The blood is then stored at 5 °C for 10 days, or longer if glucose is added. Before blood is transfused, even though both groups are known, it is carefully tested against the recipient's blood to make sure that no unforeseen factors produce agglutination. Then it is fed into one of the patient's arm veins at the correct rate and temperature. In a few hours, the donor's blood volume will have returned to normal, and the red cells are replaced in the next week or two.

The ABO blood groups are subject to a fairly straightforward pattern of inheritance as described in Chapter 21. There are known to be subdivisions of the ABO blood groups as well as several other groups but these seem to be relatively unimportant in transfusions.

The Rhesus factor. In 1940 it was found that 85 per cent of white Americans carried an agglutinogen similar to one previously investigated in Rhesus monkeys. The remaining 15 per cent had no such factor and were called Rhesus negative (Rh −), while those possessing the factor were called Rhesus positive (Rh +). There is no natural Rh agglutinin in the plasma so that the first transfusion from Rh + to Rh − is not likely to cause any harm. Exchange from Rh − to Rh + is safe at all times. The Rh + agglutinogen, however, stimulates the production of agglutinins in the plasma of a Rh − recipient so that a subsequent Rh + transfusion could be as incompatible as a faulty ABO transfusion. The Rh − individual is said to have been *sensitized* by the first Rh + transfusion.

If a Rh + man marries a Rh − woman there is a chance that some of their children will be Rh +. In most cases this is of no consequence but sometimes the red cells of the Rh + embryo get into the mother's circulation, possibly as a result of a fault in the placenta. The embryo's Rh + cells in the mother's blood act as antigens and she makes antibodies (agglutinins) against them, i.e. she has been sensitized to the Rh + factor. This will not affect the first embryo but if the mother conceives a second Rh + child, whose cells enter her circulation, the antibody reaction is much more vigorous and if her antibodies reach the embryo they will destroy its red cells, causing it to be born jaundiced or even stillborn.

Usually, if expectant mothers are found to be Rh −, the husband's blood is tested so that if he proves to be Rh + the precaution can be taken of suppressing the production of antibodies in the mother at the birth of each Rh + child. About one in every ten marriages is between Rh + men and Rh − women, but only about one in forty of these marriages is affected by the Rh incompatibility. It follows that Rh − women of child-bearing age should not be given Rh + blood in a transfusion in case it sensitizes them.

The circulatory system

The blood is distributed round the body in vessels, mostly (but not all) tubular and varying in size from about 10 mm to 0.01 mm in diameter (Fig. 11.6). They form a continuous system, communicating with every living part of the body. Blood flows in them always in the same direction, passing repeatedly through the heart whose muscular contractions maintain the circulation (Fig. 11.9).

There are three types of blood vessel connected to form a continuous system, the arteries, veins and capillaries (Fig. 11.7), and also the lymphatic vessels which return lymph from the tissues to the heart.

Arteries (Figs. 11.5 and 11.7a). These are fairly wide vessels that carry blood from the heart to the limbs and organs of the body. They consist of a fibrous outer layer, a middle layer of elastic fibres and circular muscle, and an inner layer of smooth endothelium. In the big arteries there is more elastic tissue than circular muscle, a feature that enables them to absorb the energy of the pressure surges caused by the contractions of the ventricles of the heart. The stretched elastic tissue contracts in between heartbeats, and this squeezing further aids the flow of blood along the arteries. The arteries divide into smaller vessels

called *arterioles*. In these there is more circular muscle than elastic tissue. The circular muscle is under the control of the nervous and endocrine systems. When stimulated, the muscle contracts, constricting the arterioles and reducing the flow of blood in them. The arterioles themselves divide repeatedly until they form a dense network of microscopic vessels permeating between the cells of every living tissue. These final branches are called capillaries.

Capillaries. Capillaries are tiny blood vessels with walls that are only one cell thick (Figs. 11.7c and 11.8). Although the blood appears confined within the capillary walls, the latter are permeable, with the result that water and dissolved substances pass in and out, exchanging oxygen, carbon dioxide, dissolved food and excretory products with the tissues around the capillary (Fig. 11.10).

The capillary network is so dense that no living cell is far from a supply of oxygen and food. In the liver, every cell is in direct contact with a capillary. Some capillaries are so narrow that the flexible red cells are squashed and distorted in passing through them. The diameter of capillaries can be altered by nervous stimulation which tends to close them, and by chemicals like histamine which dilate them. The change in diameter is brought about by a change in the shape of the cells constituting the walls.

Although the branching arterioles and capillaries are much smaller vessels than the arteries, their total volume is much greater. For example, the capillaries may be able to hold as much as 800 times the volume of blood in the aorta. Dilation or constriction of arterioles and capillaries controls the distribution of blood to different parts of the body. Sudden dilation of the abdominal vessels, as a result of shock, reduces the blood supply to the brain and causes fainting.

The friction between the blood and the capillary walls tends to reduce the intermittent surges of blood from the arteries to a steady flow. Eventually all the capillary branches join up again to form first *venules* and then veins.

Veins. Veins return blood from the tissues to the heart. The blood pressure in them is steady and much lower than in the arteries. They are wider and have thinner walls than arteries,

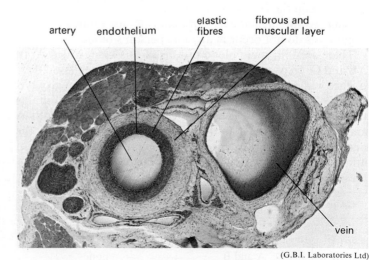

artery   endothelium   elastic fibres   fibrous and muscular layer

vein

(G.B.I. Laboratories Ltd)

Fig. 11.5   Transverse section through an artery and vein (× 20)

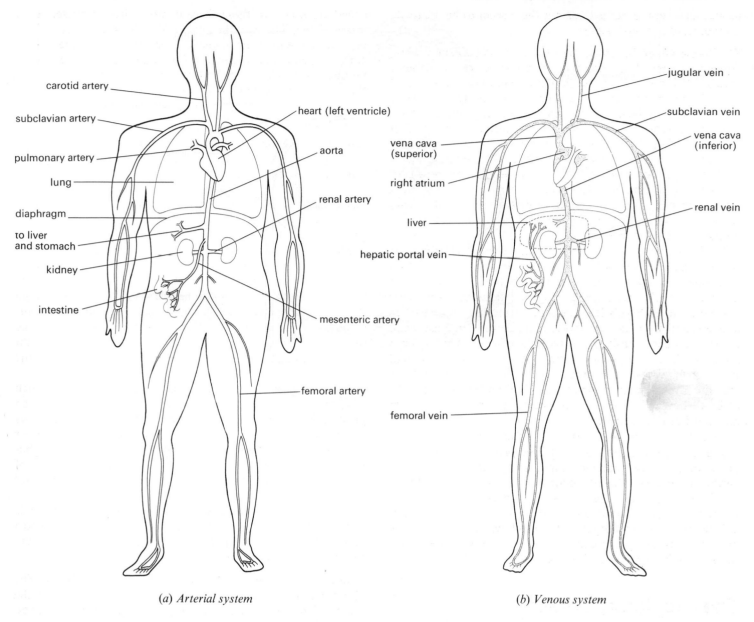

carotid artery

subclavian artery

pulmonary artery

lung

diaphragm

to liver
and stomach

kidney

intestine

heart (left ventricle)

aorta

renal artery

mesenteric artery

femoral artery

(a) Arterial system

jugular vein

subclavian vein

vena cava
(inferior)

vena cava
(superior)

right atrium

liver

hepatic portal vein

renal vein

femoral vein

(b) Venous system

Fig. 11.6   Human arterial and venous systems

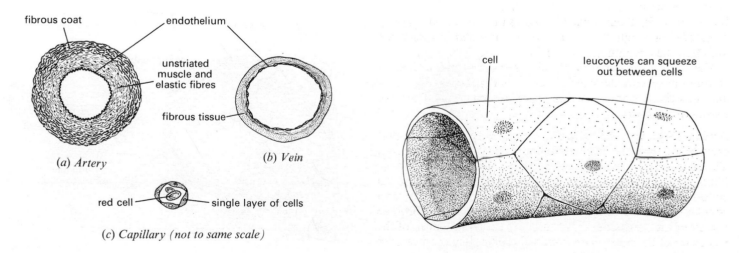

fibrous coat

endothelium

unstriated
muscle and
elastic fibres

fibrous tissue

(a) Artery

(b) Vein

red cell

single layer of cells

(c) Capillary (not to same scale)

Fig. 11.7   Transverse sections of blood vessels

cell

leucocytes can squeeze
out between cells

Fig. 11.8   Diagram of capillary (greatly enlarged)

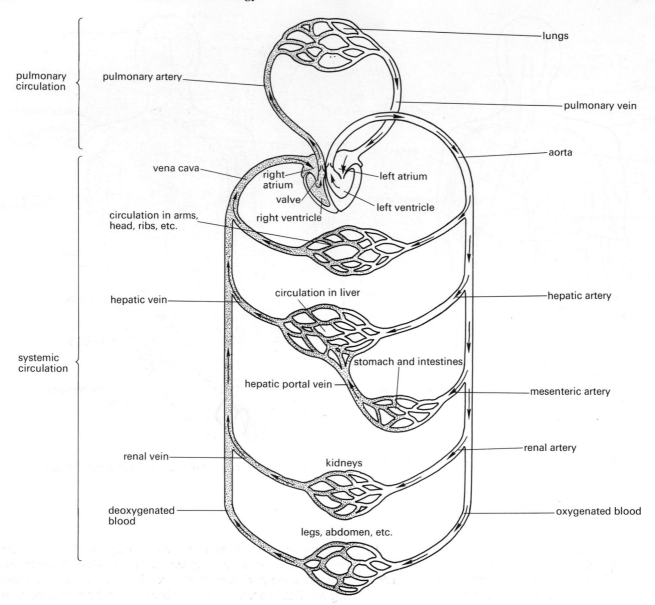

Fig. 11.9 Diagram of circulatory system

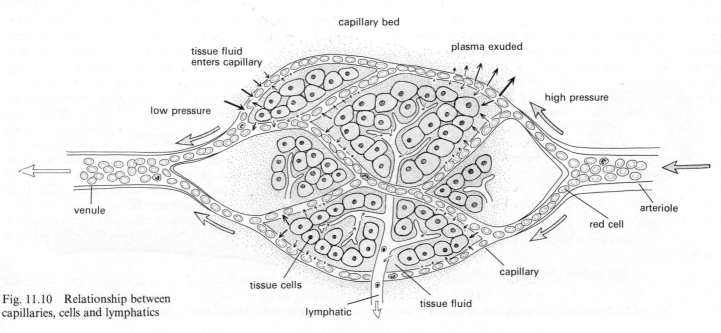

Fig. 11.10 Relationship between capillaries, cells and lymphatics

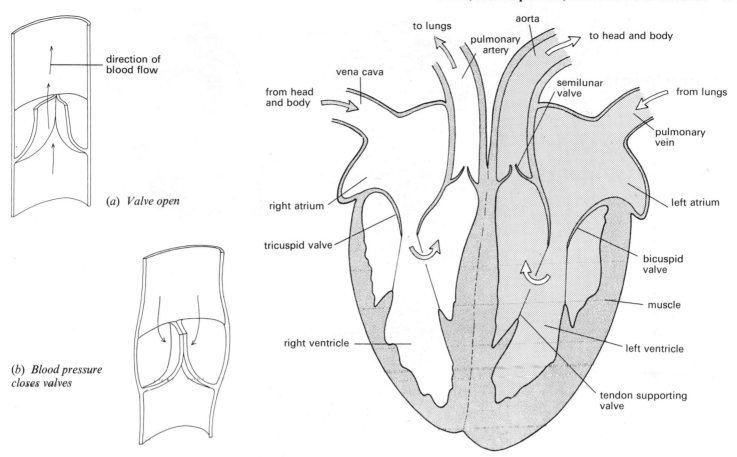

Fig. 11.11   The action of valves in the semilunar valves, or in a vein

(a) Valve open

(b) Blood pressure closes valves

direction of blood flow

to lungs
aorta
pulmonary artery
to head and body
vena cava
from head and body
semilunar valve
from lungs
pulmonary vein
right atrium
left atrium
tricuspid valve
bicuspid valve
muscle
right ventricle
left ventricle
tendon supporting valve

Fig. 11.12   Diagram of heart in longitudinal section

with an inextensible fibrous tissue replacing the elastic tissue of the latter (Fig. 11.7b and 11.5). Some veins also have valves in them (Fig. 11.11) which prevent blood flowing backwards, away from the heart, e.g. the veins of the limbs.

The pressure of surrounding muscles when they contract during activity tends to squash the veins and, since the valves prevent backward flow, assists the return of blood to the heart from the arms and legs. The positive pressure of the abdominal organs during inspiration of air also helps to keep the blood moving in the veins as does the simultaneous 'negative' (i.e. less than atmospheric) pressure in the thorax.

The blood in the veins usually contains less oxygen and food, and more nitrogenous waste and carbon dioxide, than arterial blood. The exceptions to this are (i) the pulmonary artery which carries deoxygenated blood to the lungs and (ii) the pulmonary vein which returns oxygenated blood from the lungs to the heart. Similarly, after a meal the hepatic portal vein from the intestines to the liver will have a higher concentration of glucose and amino acids than is present in any arteries. The renal vein leaving the kidney will have a reduced level of certain salts, nitrogenous waste and water.

The heart

The heart is a muscular pumping organ situated in the thorax. It develops in the embryo from a specialized region of an artery. The heart is enclosed in the *pericardium*, a tough membrane attached at its lower end to the diaphragm and at its upper region to the veins entering the heart. The fluid between the heart and the pericardium reduces friction with surrounding organs. The inextensible pericardium is also thought to prevent the heart becoming over-distended with blood in certain extreme conditions.

The heart itself consists of four chambers (Fig. 11.12); the left and right sides do not communicate. The upper chambers, the *atria*, are relatively thin-walled and receive blood from the veins. Oxygenated blood from the lungs enters the left atrium via the pulmonary vein, and deoxygenated blood from the body enters the right atrium from the vena cava. Each atrium opens into its corresponding *ventricle* through a large aperture guarded by a non-return valve (Fig. 11.11). Both ventricles are thick-walled and very muscular. They have the same capacity and expel the same volume of blood, i.e. about 70 cm³ per stroke. The walls of the left ventricle, however, are three to four times thicker than those of the right, reflecting its function of pumping blood all round the body via the aorta. The right ventricle pumps blood to the lungs through the pulmonary artery. The muscle of the atria and ventricles is supplied with oxygenated blood from the *coronary arteries* (Fig. 11.13) which branch from the aorta.

Coronary heart disease. If a coronary artery becomes obstructed by the formation of an internal blood clot, the blood flow is reduced and the heart muscle is deprived of oxygen.

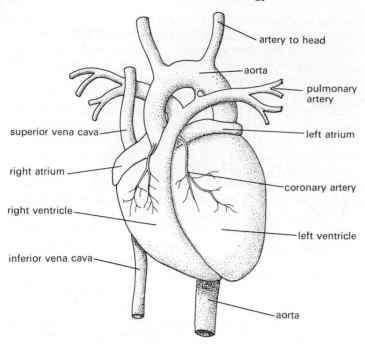

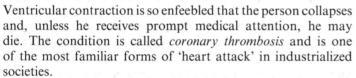

Fig. 11.13   External view of mammalian heart
(pulmonary veins not shown)

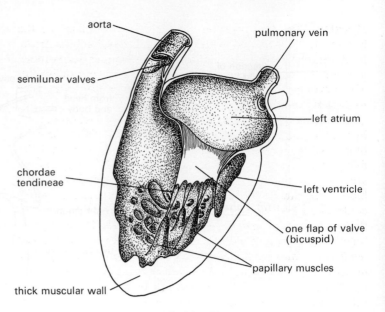

Fig. 11.14   Left side of heart cut open

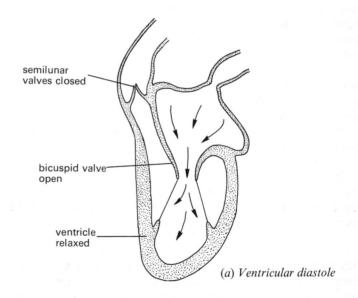

(a) *Ventricular diastole*

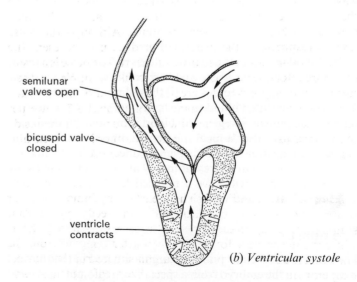

(b) *Ventricular systole*

Fig. 11.15   Heart action (left side only)

Ventricular contraction is so enfeebled that the person collapses and, unless he receives prompt medical attention, he may die. The condition is called *coronary thrombosis* and is one of the most familiar forms of 'heart attack' in industrialized societies.

Coronary heart disease is associated with a condition known as *atherosclerosis*. This is a deposition of fatty material in the lining of the arteries which results in a decrease in their diameter. Consequently the blood flow through the affected arteries is diminished and if a blood clot forms it may block the artery completely. The direct causes of atherosclerosis are not certainly known but since it is a disease prevalent in industrialized communities, the causes are sought in the habits and diets of modern society.

Apart from any inherited predisposition to the disease, the main contributory factors are thought to be high blood pressure, lack of exercise, a high level of fats in the blood plasma and smoking. The fats are deposited in the arterial lining probably as a result of outward filtration of blood fluids through the arterial walls. High blood pressure increases this filtration rate and smoking makes the arteries more permeable so that the rate of deposition of fat is increased. Obesity and nervous tension both increase blood pressure.

Large amounts of fat in the diet, particularly fats from animal sources, produce a high level of fat in the plasma and may increase the chances of its being deposited in the arterial lining. Exercise probably reduces the level of fats in the blood as well as improving the blood supply to the heart muscle (p. 123).

Although none of these conditions is universally accepted as a proven cause of coronary heart disease there is enough circumstantial evidence to indicate that obesity, fatty diet, heavy smoking, lack of exercise and nervous stress predispose the subject to a heart attack.

Heartbeat. (a) *Diastole* (Fig. 11.15*a*). In this phase of the heartbeat, the atria and ventricles are relaxed. Each ventricle, as it relaxes, returns to a shape that increases its volume, allowing blood from the atria and main veins to flow into it. The valves between the atria and ventricles offer no resistance to flow in that direction.

(b) *Systole* (Fig. 11.15*b*). Contraction of the atria precedes, by a fraction of a second, that of the ventricles. This expels the blood from the atria into the ventricles and completes the filling of the latter. The flaps of the bicuspid and tricuspid valves are brought together by the back eddies from this flow of blood. The muscle of the ventricles now undergoes a powerful contraction and increases the blood pressure within them so that the tricuspid and bicuspid valve flaps are forced tightly together and prevent blood returning to the atria. The *chordae tendineae*, which are pulled taut by contraction of the *papillary muscles* (Fig. 11.14), prevent the valves being turned inside out by the pressure. When the ventricular blood pressure exceeds that in the aorta and pulmonary artery, the *semilunar* valves (Fig. 11.11) are forced open and blood enters these arteries, whose walls are consequently distended by the sudden rise in pressure. While the ventricles are contracting, the atria have relaxed and are once more filling with blood. When the ventricles relax and the pressure in them falls, return of blood from the arteries is prevented by the closure of the semilunar valves.

The heart sounds (described as 'lubb-dupp'), which can be heard through the thorax, either directly or with a stethoscope are associated with the rhythmic muscular contractions and the closure of the valves. The precise causes of the sounds are a matter for controversy. However, variations in the sounds (heart 'murmurs') are used in the diagnosis of heart defects which result from the faulty operation of the valves.

During normal activity, the heart pumps about 5 litres of blood per minute while beating about 70 times. During exercise, the heartbeat may rise to more than 100 per minute and the output of blood to over 20 litres per minute.

The pulse. The sudden expansion of the main arteries as blood enters them from the contracting ventricles is transmitted as a pressure wave through all the principal arteries of the body. The wave can be felt as a pulse when an artery is near enough to the surface of the body, as in the wrist and neck. Such a region is known as a pressure point. Although the pulse rate corresponds to the heart beat it does not represent the arrival of blood just expelled from the heart. The pulse is a result of the elasticity of the arteries and is like a ripple on their surface travelling at about 7 metres per second from the heart. The blood in the arteries travels at only about 0.5 metre per second. A familiar pressure point is in the artery on the inside of the wrist just over the radius bone. Another is in the carotid artery on either side of the windpipe in the neck.

Control of the heartbeat. The heart's rhythmic muscular contraction is basically automatic and needs no nervous stimulation. If kept in the correct solution of salts, a frog's heart will continue to beat for some hours after removal from the body and the same is true of the mammalian heart at the correct temperature if an artificial circulation of nutrient and oxygenated liquid is maintained in the heart muscle. The

natural rhythm of the atrium (75 per minute) is, however, faster than that of the ventricle (40 per minute) but the muscle fibres of the heart are able to conduct electrical impulses and the contraction of one region of muscle acts as a stimulus for the rest of the heart so that the atrial rhythm is imposed also on the ventricle.

The 'pace-maker' is a zone of tissue at the junction of the vena cava and the right atrium. It is called the *sinu-atrial node*, and it initiates contraction of the atria which in turn stimulates the ventricles. In the mammalian heart, the atria are separated from the ventricles by a ring of non-conducting fibrous tissue, and the only direct muscular connection between atrium and ventricle is a band of modified muscle fibres, the *atrio-ventricular bundle*, which conducts the impulse from atria to ventricles. These muscle fibres spread over the internal surface of the ventricles and cause them to contract as a whole rather than from top to bottom, since a stimulatory wave would pass only slowly through the ordinary heart muscle.

The normal sequence of events is thus as follows: the sinu-atrial node initiates contraction; the impulse from the SA node is conducted rapidly over both atria which contract together; the impulse is blocked by the fibrous ring between atria and ventricles except at one point, the *atrio-ventricular node*, where it is picked up by the atrio-ventricular bundle and distributed to the ventricles which contract shortly after the atria. The SA and AV nodes receive motor nerve fibres from the autonomic nervous system (p. 151). Sensory fibres carry impulses to the brain from pressure receptors in the aorta and carotid artery (Fig. 11.16). The motor fibres of the parasympathetic system (p. 151) carry impulses which originate in the cardiac centre in the medulla of the brain (p. 154); they run in the *vagus nerve*. The nerve fibres carry a constant stream of impulses which tends to slow down the natural rhythm of the heart. Any increase in the stimulation of the vagus nerve slows the heart rate even more. The sympathetic fibres come from the spinal cord. Stimulation of the sympathetic fibres or an increase in the adrenaline content of the blood promotes a more rapid rhythm.

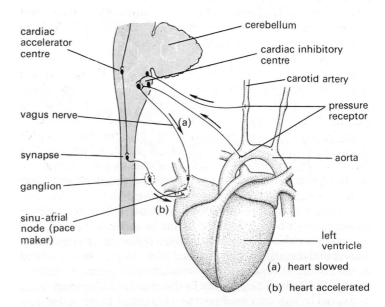

cardiac accelerator centre

cerebellum

cardiac inhibitory centre

carotid artery

pressure receptor

vagus nerve

(a)

synapse

aorta

ganglion

(b)

sinu-atrial node (pace maker)

left ventricle

(a) heart slowed

(b) heart accelerated

Fig. 11.16   Nervous control of heartbeat (the exact location of the accelerator centre is not known. It could be in the hypothalamus or in the medulla)

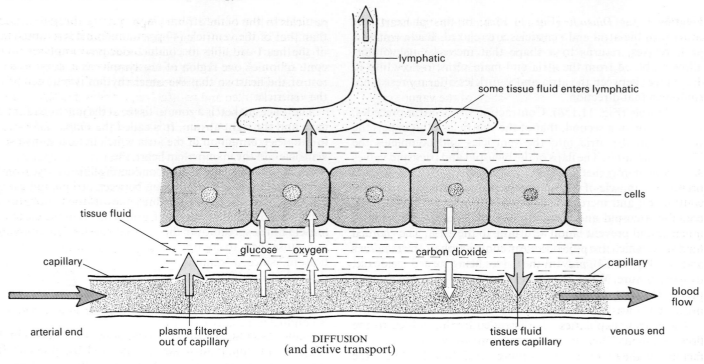

Fig. 11.17   Blood, tissue fluid and lymph

If the blood pressure in the aorta or carotid artery rises, impulses are sent from the pressure receptors through sensory fibres to the medulla. Here they make synaptic connections (p. 147) with other fibres and complete a reflex arc that causes impulses to be sent from the medulla to the heart, reducing the heart rate. By reducing the output of the heart, this reflex allows the blood pressure to fall to its normal level. Conversely, if the blood pressure in the arteries falls below a certain level, the pressure receptors send fewer impulses to the brain, stimulation of the heart by the parasympathetic motor fibres in the vagus nerve is diminished and the heart rate is permitted to increase. These reflexes tend to adjust the heart rate to meet the needs of the body and maintain the blood pressure.

At times of stress or excitement, the sympathetic nervous system sends motor impulses to the SA node and increases the heart rate and the volume of blood (stroke volume) it pumps out. Adrenaline from the adrenal gland acts directly on the SA node and has the same effect. The output of the heart is thus increased and the muscles are supplied with additional oxygen and glucose which will be required during vigorous activity.

Thyroxine, the hormone from the thyroid gland, also has an excitatory effect on the SA node.

Blood pressure. In order to overcome atmospheric pressure, which compresses the tissues, and the resistance in the fine capillaries to the flow of blood, the heart has to maintain a relatively high blood pressure in the circulatory system. The pressure of the blood varies with (a) the region of the circulatory system under consideration, e.g. an artery or vein, (b) the phase of the heartbeat (diastole or systole), and (c) the physiological state of the body. The pressure in the arteries during systole, as measured by a doctor's sphygmomanometer (an inflatable cuff placed round the arm and connected to a pressure gauge), is usually in the region of 120 millimetres of mercury and the pressure during diastole is about 75 mmHg, but both vary with

age and from one individual to another. The blood pressure is regulated by, amongst other methods, the reflex nervous mechanism of the heart, described above.

Regulation of blood distribution. The total volume of blood in the body is less than the volume of the blood vessels that contain it. At any one time, the complete or partial constriction of arterioles and capillaries in a particular region will reduce the volume of blood reaching that region and help to maintain a steady circulation and return of blood to the heart. After a meal, for example, more blood will flow to the alimentary canal as a result of dilation of the arterioles and capillaries in that region. This facilitates the rate of digestion and absorption but tends to reduce the supply of blood to the muscles.

The arteries and arterioles receive motor fibres from the sympathetic nervous system under the control of the *vasomotor* centre in the medulla. Normally, the steady flow of impulses in these fibres keeps the vessels in a state of partial contraction. More intense stimulation may close the arterioles in a particular region (for example the fingers, making them white and numb). Nearly all the nervous stimuli that normally cause pain or emotion influence the vasomotor centre, and may cause constriction of the arterioles. Sensory impulses producing the sensations of fear act through the vasomotor centre and sympathetic system, causing the blood vessels of the skin and abdominal organs to constrict, leading to pallor and the 'hollow' feeling in the abdomen.

The blood supply to the brain and muscles, however, is not reduced and in fact is likely to be increased as a result of the rise in arterial pressure when such a large proportion of the system is shut down. This increased supply of blood to the brain and muscles enables a person to react quickly and vigorously in an emergency.

Adrenaline has an effect similar to the constrictor effect of the nervous system. The localized vasodilator effect of histamine has been mentioned on p. 74.

Fainting occurs when the return of blood to the heart and consequently to the brain is reduced, for example as a result of prolonged standing or loss of blood. Emotional shock (fright, horror) may also induce fainting as a result of reduced heart output because of the stimulation of the parasympathetic nervous system and the depressor action of the vagus nerve.

Exchange between capillaries and cells

At the arterial end of the capillary bed (Fig. 11.10) blood pressure is high and forces plasma out through the thin capillary walls. The fluid so expelled has a composition similar to plasma, containing dissolved glucose, amino acids and salts, but has a much lower concentration of plasma proteins. This exuded fluid permeates the spaces between the cells of all living tissues and is called *tissue fluid*. From it the cells extract the glucose, oxygen, amino acids, etc. which they need for their living processes, and into it they excrete their carbon dioxide and nitrogenous waste.

The narrow capillaries offer considerable resistance to the flow of blood. This slows down the movement of blood, so facilitating the exchange of substances by diffusion between the plasma and tissue fluid (Fig. 11.17). The capillary resistance also results in a drop of pressure so that at the venous end of a capillary bed the blood pressure is less than that of the tissue fluid and much of the latter passes back into the capillaries.

The fact that the plasma contains more proteins than the tissue fluid gives the blood a low osmotic potential (p. 23) which tends to cause water to pass from the tissue fluid into the capillary. At the arterial end of the capillary network the blood pressure is greater than this osmotic pressure, so forcing water out, but at the venous end water from the tissue fluid enters the capillary by osmosis.

Lymphatic system

The capillaries are not the only route by which the tissue fluid returns to the circulation. Some of it returns via the lymphatic system. The proteins in the tissue fluid are unable to re-enter the capillaries but can drain into thin-walled vessels with blind ends which are found between the cells. These *lymphatics* join up to form larger vessels which eventually unite into two main ducts and empty their contents into the large veins entering the right atrium.

The fluid in the lymphatic vessels is called *lymph*. Its composition is similar to plasma but it contains less protein. It also contains a certain kind of white cell, the lymphocytes, which are made in the lymph nodes.

The larger of the two lymphatic ducts is the *thoracic duct* which collects lymph from the intestine and the lower half of the body (Fig. 11.19). The *lacteals* from the small intestine open into the lymphatic system. After a meal containing fats the lymph is a milky-white colour due to the emulsion of fat droplets absorbed in the lacteals.

On its way through the lymphatics to the ducts, the lymph passes through a number of *lymph nodes*, which appear as swellings on the lymphatics. The lymph nodes are particularly numerous in the armpit and groin. In each node there are spaces filled with a network of fibres, and attached to these fibres are white cells like macrophages which can trap and ingest foreign

particles in the lymph. The lymph passing through a node is thus filtered of any invading bacteria before it is returned to the circulation. In addition the lymph node produces lymphocytes, some of which are added to the lymph while some enter the blood capillaries which permeate the node. A septic wound may give rise to swollen and painful lymph nodes. Bacterial antigens from the wound reach the node via the lymphatics and stimulate them into activity. A poisoned finger, for example, may give rise to a swollen lymph node in the armpit. The tonsils and spleen act in a similar way to the lymph nodes.

The lymph flow takes place in one direction only, from the tissues to the heart, and there is no specialized pumping organ. The flow is caused partly by contraction of the lymphatics and the pressure of the lymph which accumulates in the tissues, but one of the most important factors in the movement of lymph is

Fig. 11.18   Deep lymphatic vessel cut open to show valves

muscular exercise. Some of the lymphatics have valves in them (Fig. 11.18) similar to those in veins and the pressure from contracting muscles around them compresses the lymphatic vessel and forces the lymph along in the direction determined by the valves.

If the accumulation of lymph exceeds the rate of removal in the lymphatics, a local swelling occurs. As a result of injury,

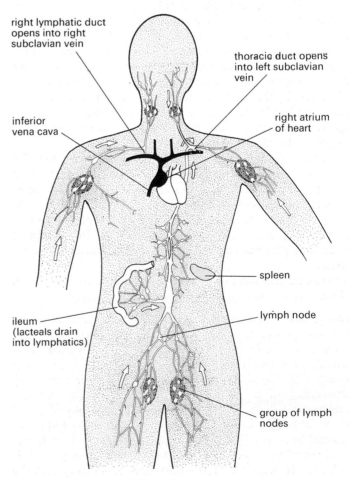

Fig. 11.19   Main drainage routes of lymphatic system

histamine is produced which makes the capillaries in the injured region more permeable, and the escape of fluid from them is too rapid for removal in the lymphatics.

Spleen. The spleen is a dark red, compact body about 12 cm long situated in the abdomen to the left of the stomach and protected by the lower ribs. It consists of spongy tissue enclosed in a fibrous capsule. The spleen produces lymphocytes and also contains other white cells, *macrophages*. The macrophages ingest and so remove from the blood worn-out red cells, cell fragments and any bacteria that may have entered the blood stream. The haemoglobin from the degenerating red cells is broken down and the iron-containing residue stored in the cells of the spleen and liver.

In diseases such as malaria (p. 224) in which many red cells are destroyed, the spleen in coping with the excess cell debris becomes considerably enlarged and projects beyond the rib cage and thus becomes more vulnerable to damage.

Practical Work

Experiment 1   **Blood smear**

To examine blood under the microscope, a thin film must be spread on a slide. The finger is pricked with a sterile lancet which must be used by one person only to avoid cross infection. One technique is to squeeze the first finger between the middle finger and thumb of the same hand and pierce the soft area at the tip. The capillaries are nearer the surface, however, on the side of the top joint of the finger, just to one side of the insertion of the fingernail.

The drop of blood is placed at one end of a clean, dry slide. A second slide is placed as shown in Fig. 11.20 so that the

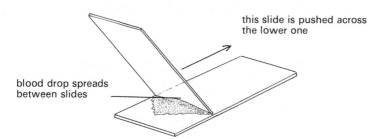

this slide is pushed across the lower one

blood drop spreads between slides

Fig. 11.20   Making a blood smear

blood drop spreads across the region of contact by capillary attraction. By pushing the top slide across the lower one, a thin film of blood is made which will show red cells quite clearly under the microscope.

To see white cells, it is best to study a prepared slide.

Experiment 2   **Valves in the veins**

If a light tourniquet is applied to the upper arm the veins in the forearm can be made to stand out. The lower end of one of these is blocked off near the wrist by pressing it with a finger. The blood can be expelled from the vein by running a finger with light pressure along its length towards the elbow. When this has

been done the vein will remain collapsed up to a certain point; above this the vein will fill up and swell once more. The boundary between the filled and collapsed regions indicates the position of a non-return valve.

Experiment 3   **Effect of gravity on circulation**

Allow the left arm to hang straight down at the side of the body. Open and clench the hand repeatedly between once and twice a second. It should be possible to continue these movements for 3 or 4 minutes or up to 500 times without feeling acute discomfort. After a period of rest, hold the arm straight up and repeat the exercises. After about one minute, or 100 closures, the movement becomes almost impossible. One reason for this is the reduced blood supply resulting from the retarding effect of gravity on the circulation. It is interesting to speculate on which particular aspect of circulation, i.e. oxygen transport, waste removal, etc., is responsible for fatigue.

Experiment 4   **Capillaries**

These are best seen in the web of a frog's foot or a tadpole's tail where the red cells can be seen streaming through the narrow vessels.

Our own capillaries can be seen by soaking the back of the top joint of a finger in a clearing agent such as cedarwood oil and examining, by reflected light under a microscope, the area below the nail cuticle. Capillary loops can usually be seen even with a good hand lens.

Experiment 5   **Pulse rate**

The swelling of the arteries as a result of the surge of pressure from the heart can be felt in certain places and gives an indication of the rate of the heart's contractions. The pulse in the wrist is the most usual region for this. Count the number of pulsations over a period of 30 seconds and make a note of it. Then take some form of exercise, e.g. standing on and getting off a stool once in two seconds for about half a minute, and take the pulse rate again. Find out how long it takes to return to its original rate.

Questions

1   Although the walls of the left ventricle are thicker than those of the right ventricle, the volume of the ventricles is the same. Why is this necessary?
2   State in detail the course taken by (a) a glucose molecule and a fat molecule from the time they are ready for absorption in the ileum, and (b) a molecule of oxygen absorbed in the lungs, to the time when all three reach a muscle cell in the leg.
3   Why is a person whose heart valves are damaged by disease unable to participate in active sport?
4   A system for transporting substances in solution might as well be filled with water. What advantages has the blood circulatory system over a water circulatory system?
5   What is the advantage to an animal having capillaries which are (i) very narrow, (ii) repeatedly branched and (iii) very thin-walled?
6   Study the paragraphs on homeostasis on pp. 72 and 92. Explain why transport by the blood of nitrogenous waste, carbon dioxide and heat can be considered to have a homeostatic function.
7   Explain what would happen in the following transfusions: (a) blood from an AB donor to an A recipient; (b) blood from an O donor to an AB recipient; (c) blood from a B donor to an AB recipient; and (d) blood from an O, Rh + donor to an A, Rh − recipient.

12
Breathing

The various processes carried out by the body, such as movement, growth and reproduction, all require the expenditure of energy. In animals this energy can be obtained only from the food they eat. Before the energy can be used by the cells of the body it must be transferred from the chemicals of the food. This process of energy transfer is called tissue respiration (p. 27) and involves the use of oxygen and the production of carbon dioxide.

Oxygen enters an animal's body from the air or water surrounding it. In the less complex animals the oxygen is absorbed over the entire exposed surface of the body, but in higher animals there are special respiratory areas such as lungs or gills. Excess carbon dioxide is usually eliminated from the same area.

In the respiratory organ, oxygen dissolves in the blood and is carried to all living parts of the body where it is used in tissue respiration. An efficient respiratory organ has a large surface area, a dense capillary network or similar blood supply, a very thin epithelium separating the air or water from the blood vessels and, in land-dwelling animals, a layer of moisture over the absorbing surface. All vertebrates and many invertebrates have some muscular mechanism whereby they can ventilate the breathing organs, i.e. exchange the water or air in contact with their lungs or gills. In man and other mammals, the respiratory organs are lungs.

Lungs

The lungs of man are enclosed in the thorax (Fig. 12.1). They have a spongy elastic texture and can be expanded or compressed by changes of pressure in the thorax in such a way that air is repeatedly taken in and expelled. They communicate with the atmosphere through the windpipe or *trachea* which opens through the *glottis* to the *pharynx* (Fig. 10.3). In the lungs a gaseous exchange takes place; some of the atmospheric oxygen is absorbed and carbon dioxide from the blood is released into the lung cavities.

Lung structure. The trachea divides into two *bronchi* which enter the lungs and divide into smaller branches (Fig. 12.2). These divide further into *bronchioles* which terminate in a mass of little thin-walled, pouch-like air sacs or *alveoli* (Figs. 12.3 and 12.5).

(a) *Air passages*. Incomplete rings of cartilage keep the trachea and bronchi open and prevent their closing up when the

pressure inside them falls during inspiration. The cartilaginous rings also stop the trachea from kinking when the head is turned or inclined. The lining of the air passages is covered with numerous *cilia*. These are minute cytoplasmic filaments that constantly flick to and fro. Mucus is secreted by glandular cells also in the lining. Dust particles and bacteria which are carried in with the air during inspiration become trapped in the mucus film and are swept away in it at about 0.4 mm per second by the movements of the cilia up to the larynx and into the pharynx where they are swallowed. The ciliary beat is independent of the nervous system but can be arrested by tobacco smoke.

The epiglottis and other structures at the top of the trachea prevent food and drink from entering the air passages, particularly during swallowing (p. 62). Choking and coughing are reflex actions that tend to remove any foreign particles which accidentally enter the trachea or bronchi.

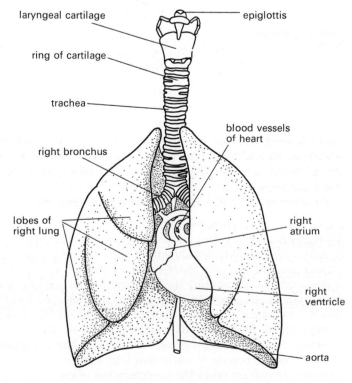

Fig. 12.1   Appearance of the lungs

(b) *Alveoli*. The alveoli have elastic walls consisting internally of a single layer of cells, or epithelium, 1 micron (1/1 000 mm) thick, and around each alveolus is a dense network of capillaries (Fig. 12.4) supplied with deoxygenated blood from the body pumped from the right ventricle through the pulmonary artery. There are about 700 million alveoli in a man's lungs with a total absorbing surface of about 50–80 square metres depending on the degree of expansion. There is evidence of minute cross-connections from one alveolus to another.

(c) *Pleural membranes* (Fig. 12.10). The pleural membrane is the lining that covers the outside of the lungs and the inside of the thorax. It produces pleural fluid which lubricates the surfaces in the regions of contact between lungs and thorax. As a result they can slide freely over one another with very little friction during the breathing movements.

Gaseous exchange

The lining of the alveoli is covered with a thin film of moisture (Fig. 12.6). The oxygen concentration of the blood in the lung capillaries is lower than in the alveolus, and oxygen in the air space dissolves in the film of moisture and diffuses through the epithelium, the capillary wall, the plasma and into a red cell, where it combines with the haemoglobin (p. 70). The capillaries reunite and eventually form the pulmonary vein which returns the oxygenated blood to the left atrium of the heart.

The carbon dioxide concentration in the alveoli is lower than in the lung capillaries. The enzyme *carbonic anhydrase* in the red cells breaks down the hydrogencarbonate (bicarbonate) salts and liberates carbon dioxide. This diffuses into the alveoli and is eventually expelled. About 250 cm³ oxygen is absorbed every minute and about the same volume of carbon dioxide expelled (Experiments 1 and 2).

Composition of inspired and expired air

	Percentage volume	
	Inspired	*Expired*
Oxygen	20.95	16.4
Carbon dioxide	0.04	4.1
Nitrogen	79.01	79.5
Water vapour	varies	saturated

There is no actual change in the volume of nitrogen in inspired and expired air but since the volume of carbon dioxide given out is slightly less than that of the oxygen absorbed, the percentage of nitrogen changes. The water vapour content of the air varies with the temperature and other atmospheric conditions but exhaled air is usually saturated with water vapour.

Diffusion gradient. A steep diffusion gradient (p. 22) of oxygen is maintained by (i) replenishment of air in the air passages by ventilation, (ii) the very short distance between the alveolar lining and the blood, (iii) the combination of oxygen with haemoglobin, so removing oxygen from solution, and (iv) the blood flow which constantly replaces oxygenated blood with deoxygenated blood. Similar factors work in the reverse direction for the diffusion of carbon dioxide. The conversion of hydrogencarbonate (bicarbonate) to carbon dioxide by carbonic anhydrase raises the concentration of carbon dioxide in the blood above that in the alveoli.

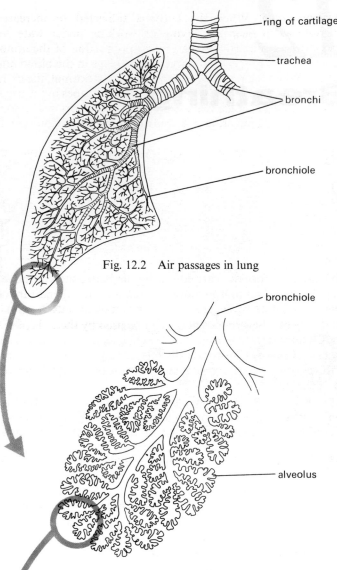

Fig. 12.2   Air passages in lung

Fig. 12.3   Air passages terminate in alveoli

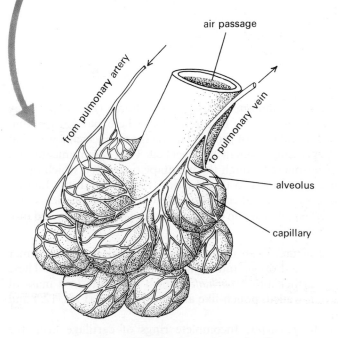

Fig. 12.4   Blood supply to alveoli

Decompression. When the body is subjected to increased pressure as in deep sea diving or working under water in pressurized compartments, a greater proportion of the atmospheric gases, oxygen and nitrogen, dissolves in the blood and tissues. When the pressure returns quickly to normal, the extra oxygen can be taken up by the red cells and escapes in the usual way through the lungs. There is no special transport mechanism for nitrogen, however, and it tends to come out of solution as

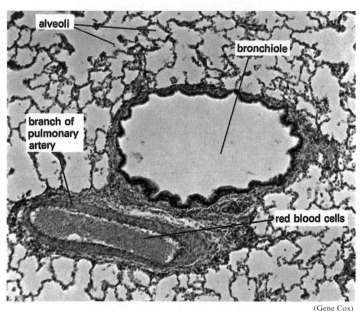

(Gene Cox)

Fig. 12.5   Appearance of lung tissue under the microscope (section) (× 100)

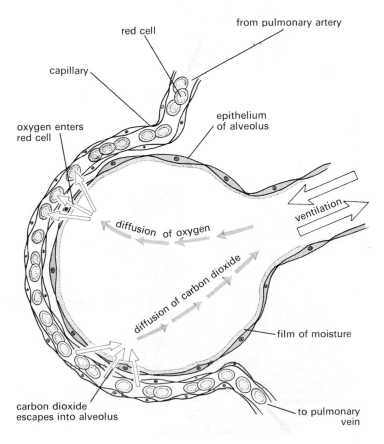

Fig. 12.6   Gaseous exchange in the alveolus

small bubbles of gas in any part of the body, rupturing tissues, blocking blood vessels and giving rise to acute pain at the joints. This decompression sickness or 'bends' is relieved by returning the victim to high pressure and reducing it very slowly, thus allowing the excess dissolved nitrogen to escape naturally from the lungs.

Smoking and health

There is little doubt today that smoking is injurious to health. There is overwhelming statistical and experimental evidence to associate smoking with the incidence of lung cancer and coronary heart attacks (p. 79). Quite apart from early death from these two diseases, heavy smokers suffer persistent coughs which damage the lungs, and an increased susceptibility to bronchitis and pneumonia. The absorptive surface of the lungs is so much decreased by the damage, that the individual becomes 'short of breath', i.e. the least exertion causes excessive panting and discomfort. The immediate effect of tobacco smoke on the lungs and respiratory passages is to inhibit the ciliary action that removes mucus and dust particles. Consequently, the fluid remains and accumulates in the lungs making them liable to infection. In the long term, certain of the chemicals in tobacco smoke may induce a cancer in the lung tissue which destroys the lung. (*See also* pp. 191 and 268.)

Carbon monoxide in tobacco smoke combines with the haemoglobin in the blood to form a compound called *carboxyhaemoglobin*. The raised level of carboxyhaemoglobin in the blood of smokers increases the permeability of their blood vessels and this in turn leads to a higher rate of deposition of fats in the artery linings and an increased risk of coronary heart attack (p. 79).

Giving up smoking progressively reduces the health hazards described above so that ten years after giving up, the death rates for ex-smokers are the same as for non-smokers.

The nose

The ciliated epithelium and film of mucus that line the nasal passages help to trap dust and bacteria. The air is also warmed to about 30 °C, depending on the external temperature, and moistened before it enters the lungs. In addition there are in the upper part of the nasal cavity sensory organs which respond to chemicals in the air and confer a sense of smell.

Inflammation of the mucous membrane lining the nasal cavity results in its swelling and blocking the free flow of air. This nasal congestion, characteristic of the common cold, can also give rise to discomfort in a warm but badly ventilated room.

Voice

The vocal cords are two folds protruding from the lining of the larynx. They contain ligaments which are controlled by muscles. When air is passed over them in a certain way, they vibrate and produce sounds. The controlling muscles alter the tension in the cords and the distance between them, and in this way they vary the pitch and quality of the sounds produced. Precise co-ordination of the breathing movements, vocal cords, lips, tongue and jaws is required to produce articulate speech. The ability to make sounds is inborn, but speech has to be learned.

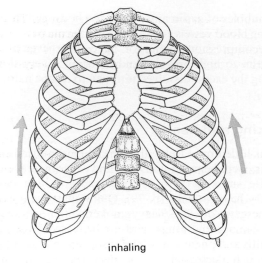

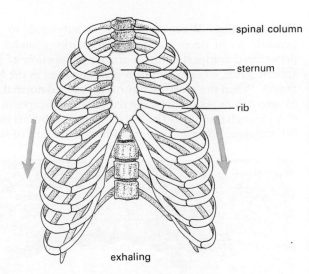

(a) *Inspiration. Ribs swing up and increase volume of thorax*

(b) *Expiration. Ribs swing down and reduce volume of thorax*

Fig. 12.7   Movement of rib cage during breathing

(a) *External intercostals*

(b) *Internal intercostals*

Fig. 12.8   Side view of rib cage to show intercostal muscles

Fig. 12.9   Model to show action of external intercostals

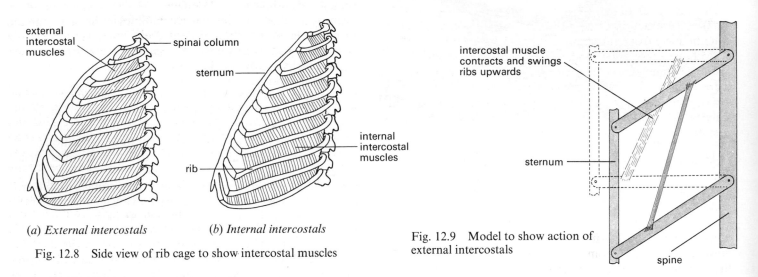

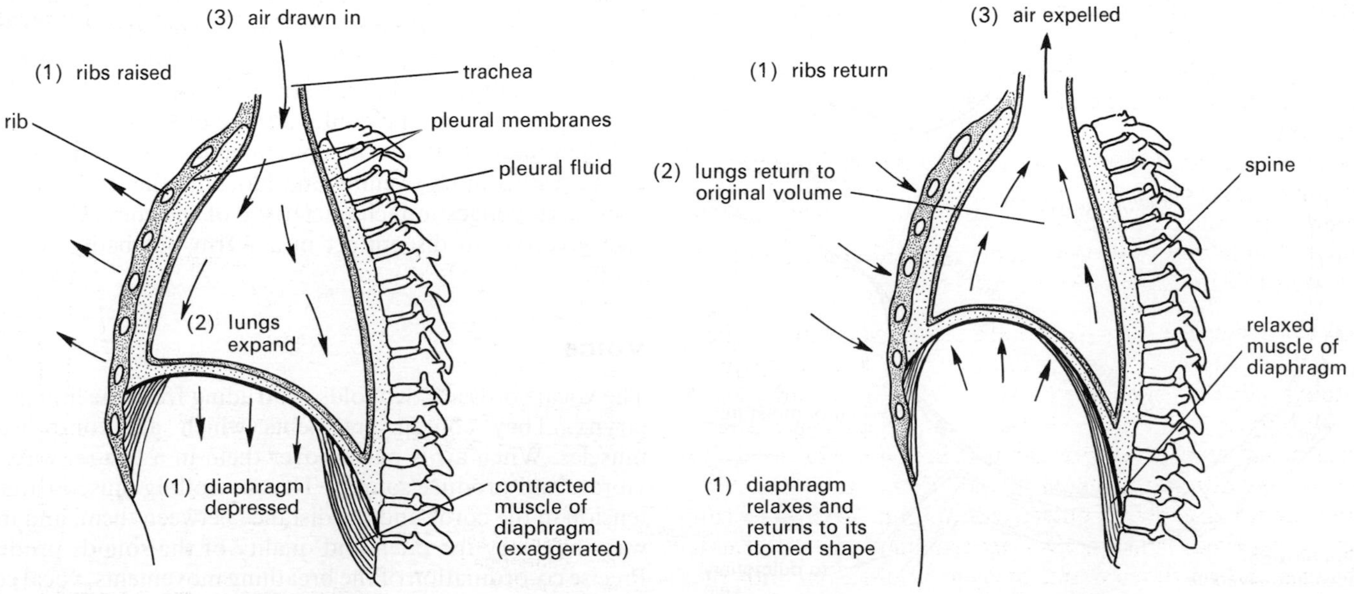

(a) *Inspiration*

(b) *Expiration*

Fig. 12.10   Sections through thorax to show mechanism of breathing

Ventilation of the lungs

The exchange of air in the lungs is brought about by movements of the thorax that alter its volume. The thorax is an airtight cavity enclosed by the ribs at the sides and the diaphragm below. The *diaphragm* is a muscular sheet of tissue extending across the body cavity between the thorax and abdomen. When relaxed it is a flattened dome shape extending upwards into the thoracic cavity with the liver and stomach immediately below it. The muscle fibres of the diaphragm radiate from a central sheet of fibrous tissue to the margins of the lower thorax, where they are attached to the ribs and body wall. Any change in the thorax volume is followed exactly by the lungs which are elastic, and too thin to oppose the movements.

Inspiration. During inspiration, the volume of the thorax is increased by two movements:

(a) the muscles round the edge of the diaphragm contract and pull it downwards rather like a piston (Fig. 12.10a). This movement incidentally has the effect of pushing the abdominal organs down so making the abdomen swell slightly.

(b) The lower ribs are raised upwards and outwards (Fig. 12.7) by contraction of the *external intercostal muscles* which run obliquely from one rib to the next (Figs. 12.8 and 12.9).

Both these movements increase the volume of the thorax and also the volume of the lungs which follow the movements. The increase in volume lowers the air pressure in the lungs so that atmospheric pressure forces air into them through the nose and trachea. The cartilaginous rings of the trachea keep it open despite the fall of pressure inside it.

Expiration. *Passive.* When we are resting or asleep, breathing out results mainly from the elasticity of the lungs squeezing out air as the diaphragm and rib muscles relax. Pressure from the abdominal organs, displaced when the diaphragm was pulled down, helps to push the diaphragm back to its relaxed position. *Forced breathing.* During exercise, the internal intercostal muscles (Fig. 12.8b) come into play. Their contraction pulls the ribs downwards and inwards, compressing the lungs and forcing air out.

More controlled expiration is achieved by contraction of the abdominal muscles, e.g. when singing or playing a wind instrument. Coughing is the result of forced expiration against intermittent closure of the glottis, producing explosive bursts of expiration.

Expiration, either forced or passive, expels from the lungs air containing less oxygen and more carbon dioxide and water vapour than when it entered.

In quiet breathing, the movements of the diaphragm alone are responsible for the ventilation of the lungs.

Lung capacity (Fig. 12.11). The total capacity of the lungs when fully inflated is about 5.5 litres in an adult man, but during quiet breathing only 0.5 litre of air is exchanged. This is called *tidal air*. During activity, the thoracic movements are more extensive, and deep inspiration can take in another 2 litres while vigorous expiration can expel an additional 1.3 litres. The thorax cannot collapse completely so that one litre of air can never be expelled. This *residual air* remaining in the alveoli exchanges carbon dioxide and oxygen by diffusion with the tidal air that sweeps into the bronchioles and air passages.

The 140 cm³ air which enters the trachea and bronchi at the end of each inspiration is expelled without having exchanged any oxygen or carbon dioxide with the lungs. This region is called the *dead space. Vital capacity* is the maximum volume of air which can be exchanged, i.e. the volume of air which can be expelled by forcible expiration after deep inspiration (Experiment 3).

Control of breathing rate

The rhythmical breathing movements are usually carried out quite unconsciously from 12 to 20 times a minute in an adult. A respiratory centre in the medulla of the brain is thought to control these reflex breathing movements, and experiments with the detached brains of animals have shown a pattern of nervous activity which corresponds to the breathing rhythm. These results suggest that the respiratory rhythm is initiated by the brain, but it is known also that the rate can be influenced by reflex action, the chemical composition of the blood and conscious control.

Variations in breathing rate are most obvious as a result of increased physical activity, e.g. running. The rate and depth of breathing both increase so that a greater volume of air is exchanged in a given time. This has the effect of supplying the active muscles with the extra oxygen they need for their increased tissue respiration, and removes the carbon dioxide which builds up at the same time.

(a) **Chemical effect.** (i) *Carbon dioxide.* Relatively small increases in the carbon dioxide content of the blood stimulate the respiratory centre and produce at first deeper breathing and then more rapid breathing. If the breath is held, the carbon dioxide concentration builds up in the blood and stimulates the respiratory centre until one is compelled to start breathing again. Forced breathing, i.e. deep inspiration and expiration for a minute, enables the breath to be held for much longer, probably as a result of the 'flushing out' of carbon dioxide from the blood so that a longer time is needed for the carbon dioxide concentration to build up to 'bursting point' in the medulla. Since the blood is normally 95 per cent saturated with oxygen it seems unlikely that forced breathing achieves its effects by increasing the oxygen content of the blood, and experiments show that breath-holding is very little improved by breathing oxygen beforehand.

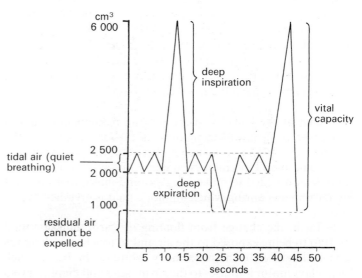

Fig. 12.11 Volumes of air exchanged in the lungs

The carbon dioxide concentration of the blood is most likely to rise during vigorous activity because of the rapid breakdown of carbohydrates in the muscles to provide energy. The accelerated breathing rate which results from the rise in carbon dioxide concentration helps to expel this gas as fast as it accumulates and also to maintain the amount of oxygen in the blood, so meeting the demands of increased tissue respiration during exercise. The increased rate of breathing which prevents the carbon dioxide rising above a certain level is an example of homeostasis (p. 92).

(ii) *Oxygen*. The oxygen concentration in the air can fall from 21 to 13 per cent with very little effect on the rate of breathing. In the carotid arteries are bodies containing chemical receptors which are stimulated by gross oxygen shortage and send sensory impulses to the medulla resulting in an increased breathing rate.

Flying at high altitudes may lead to oxygen shortage and is dangerous because the pilot is usually unaware of the deficiency since the carotid sensory receptors respond only to a severe fall in oxygen tension. The oxygen deficiency in the blood leads to errors of judgement, hilarity or aggressiveness similar to drunkenness. Pressurized cabins or supplementary oxygen prevent this condition.

(b) **Nervous reflex.** There is some evidence of the presence in the lungs of nerve endings which, when stretched, fire off sensory impulses to the medulla via the vagus nerve. These impulses appear to bring inspiration to a halt and allow the lungs to return to their relaxed volume. The motor nerves for breathing pass from the respiratory centre to the diaphragm and intercostal muscles via the spinal cord in the neck and thoracic region.

(c) **Conscious control.** 'Higher', conscious, parts of the brain can influence the respiratory centre and bring the breathing rhythm under conscious control, as in speaking and singing or simply holding the breath.

Acclimatization. Mountain sickness results from oxygen deficiency at high altitudes. The body becomes acclimatized by the kidneys' adjusting the acidity of the blood and, over a longer period, by the production of more red cells by the bone marrow.

Other factors affecting breathing. Swallowing momentarily interrupts breathing. Stimulation of the lining of the larynx by a food particle produces a reflex contraction of the abdominal muscles leading to coughing. Air is expelled violently from the lungs so dislodging and removing the foreign body. Sneezing is a comparable reflex action induced by the stimulation of the mucous membrane in the nasal cavity.

Breathing of the embryo. Although the mammalian embryo derives its oxygen and eliminates its carbon dioxide via the placenta (p. 105) by exchanges between the maternal and embryonic blood circulatory systems, the human foetus has been demonstrated to make irregular and intermittent breathing movements during which amniotic fluid is breathed in and out.

At birth, the change from floating in warm amniotic fluid (p. 106) to being exposed to the air and to touch stimuli, sends sensory impulses to the respiratory centre of the brain which then relays motor impulses to the rib muscles and diaphragm so that breathing begins.

Artificial 'respiration'

The effect of severe electric shock or partial drowning can result in the arrest of the breathing movements. If the brain cells are deprived of oxygen for more than a few minutes they are permanently damaged and full recovery of the victim becomes unlikely. Breathing can be restarted by a method of artificial 'respiration' or resuscitation which should be applied without delay. The most effective method is for the rescuer to apply his mouth to the victim's mouth and blow air into the lungs. The victim's head must be tilted well back to open the air passages and the nostrils pinched shut with the fingers. In this way the victim's lungs are inflated about 20 times per minute until his breathing begins once more. Although the air breathed into the unconscious person's lungs is 'used' air, it will contain about 16 per cent oxygen which is sufficient to nearly saturate the blood in the alveoli (*see also* p. 318).

Injury to the spinal cord in the neck may cause paralysis of the breathing muscles. In this case, ventilation is maintained by pumping oxygen or air into the lungs through an opening made in the windpipe.

Practical Work

Experiment 1   **Composition of exhaled air (1)**

Two large test-tubes (boiling tubes) are prepared as shown in Fig. 12.12 and 10 cm³ lime water placed in each. Check that the rubber tubing is connected to the long glass tube in one case and to the short glass tube in the other. Place both pieces of rubber tubing in the mouth and breathe in and out through the tubing for about 30 seconds. Air breathed out will bubble through the lime water in tube B, whereas air breathed in will pass through tube A.

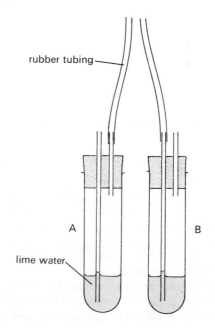

rubber tubing

A   B

lime water

Fig. 12.12   Testing exhaled air for carbon dioxide

Result. The lime water in tube B will go milky while that in A will remain clear.

Interpretation. Since it is carbon dioxide that turns lime water milky, the results suggest that exhaled air contains more carbon dioxide than inhaled air.

Experiment 2 **Composition of exhaled air (2)**

Prepare a large glass jar with two lids, one holding a bent wire with a candle stub as shown in Fig. 12.13*b* (a gas jar and deflagrating spoon will do if these are available). Light the candle, lower it into the jar and count the number of seconds it stays alight. Collect exhaled air in the jar as follows:

(a) lie the jar on its side in a basin of water and place a rubber tube inside it (Fig. 12.13*a*);

(b) stand the jar upside down in the basin so that it remains full of water;

(c) blow down the tube to fill the jar with exhaled air;

(d) remove the rubber tube and put the lid on the jar still upside down and under water.

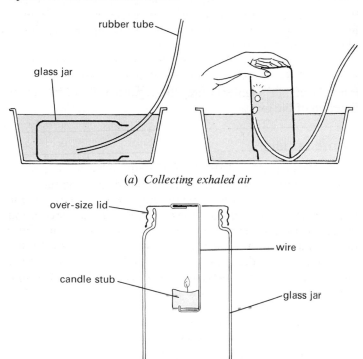

(a) *Collecting exhaled air*

(b) *Comparing the amounts of oxygen*

Fig. 12.13   Testing air for its oxygen content

Remove the jar from the basin and test the air in it with a lighted candle as before.

Collect and test a further sample of exhaled air, but this time expel nearly all the air from the lungs before blowing down the rubber tube into the jar.

Result. The candle will burn for a shorter time in exhaled air than in atmospheric air, and for a shorter time still in the air from the deeper regions of the lungs.

Interpretation. The candle needs oxygen in order to burn. The less oxygen there is present, the sooner will the flame go out. Exhaled air therefore contains less oxygen than inhaled air.

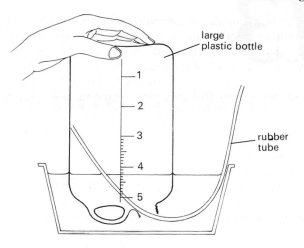

Fig. 12.14   Measuring lung capacity

Experiment 3 **Lung capacity**

Calibrate a large (plastic) bottle up to five litres by filling it with water one litre at a time and marking the levels. Invert the bottle full of water in a basin of water, remove the stopper under water and insert a rubber tube through the neck. Take a deep breath and exhale through the tube so that the exhaled air collects in the bottle, displacing the water. The level of water left in the bottle will give a measure of the lung capacity (*vital capacity*) (Fig. 12.14).

Push the rubber tube half-way up inside the bottle so that its end is clear of the water and blow out any water remaining in the tube. Support the bottle with your hand and breathe in and out through the tube as normally as possible. The bottle will rise and fall in the basin with each breath giving some idea of the volume of air exchanged during quiet breathing (*tidal air*).

Questions

1 Outline the events that take place in the course of vigorous exercise leading to a change in the rate and depth of breathing both during and after the activity. (*See also* Chapter 5, pp. 27 and 28.)

2 The lungs and ileum are adapted for absorption. Point out the features they have in common which facilitate absorption.

3

	inhaled air	*exhaled air*	*alveolar air*
oxygen (%)	21	16	14
carbon dioxide (%)	0.03	4	5.5

The table above gives the approximate percentage volume composition of air inhaled, exhaled or retained in the lungs. Explain how these differences in composition are brought about by events in the lungs.

4 An artificial pneumothorax is a method of resting an infected lung. Air is injected into the pleural cavity and the lung collapses. After a few months, the air is absorbed and the lung works normally again. Try to explain why the introduction of air into the pleural cavity should cause the lung to stop working and say why it is possible for a person with a collapsed lung to lead a normal life.

13

Excretion and the Kidneys

Excretion means getting rid of unwanted substances from the body. These substances, called *excretory products*, may be (a) by-products from the chemical reactions in cells, (b) substances taken in with the diet in greater quantities than the body needs, (c) poisonous substances taken in with food or produced by bacterial activity in the intestine, (d) drugs or (e) chemicals, such as hormones (p. 156), produced by the body but not needed once they have done their job.

Accumulation of any of these substances could interfere directly with the normal functioning of cells or alter the concentration of tissue fluid on which the cells depend (see 'Homeostasis' below).

The process of excretion removes these substances as soon as their concentration in the blood rises above a certain level, called the *threshold* level.

Excretory products. The waste-products of cell metabolism (p. 30) are carbon dioxide, water and nitrogenous compounds. Carbon dioxide and water are produced in the course of tissue respiration (p. 27).

Nitrogenous compounds result from the breakdown of proteins and amino acids. Deamination in the liver, removes the $—NH_2$ from excess amino acids (p. 69). The ammonium ion $(NH_4—)$ which results from this reaction is highly toxic and is converted at once by the liver cells into a less harmful compound called *urea* $(NH_2)_2CO$. Bacteria in the colon also produce some ammonia and this is converted to urea by the liver in the same way. The nitrogenous compound, *uric acid*, is also produced but in much smaller quantities and a very small amount of ammonia remains in the blood.

Water and salts taken in with the food are usually in excess of the body's needs and are removed by excretion.

Hormones are changed by the body into less active substances and excreted in the urine. Drugs and toxic substances such as alcohol may be excreted unchanged or altered by chemical reactions in the body before being excreted. Tests on urine are used to estimate the level of hormones or drugs in the blood.

Faeces (p. 66) are not usually included as excretory products because they consist largely of bacteria, dead cells and undigested food. However, the colour of the faeces is derived from the bile pigments, bilirubin and biliverdin (p. 71). These are produced when haemoglobin is broken down and are therefore excretory products.

Excretory organs. In man, the excretory organs are the kidneys and the lungs. The lungs eliminate excess carbon dioxide from the body and in this sense may be considered excretory. Water also is lost from the lungs, and water, sodium chloride, a little

urea and lactic acid are lost from the skin in the sweat. It can hardly be claimed that these losses are excretory since they do not take place in response to changes in the chemical composition of the body fluids. Loss of water vapour from the lungs is inevitable with every breath expired and sweat production is a response to increase in temperature. The removal of bile pigments may be regarded as an excretory function of the liver.

Homeostasis

Although most cells of the body are adaptable to changes in their environment, their delicately balanced chemical reactions work best within narrow limits of temperature, acidity, etc. A fall in temperature will slow down all the chemical reactions in the cell; a drop in pH may inhibit some enzyme systems; a rise in the concentration of solutes may withdraw water from the cells by osmosis. Changes such as these are kept within very narrow limits by various regulatory processes in the body, and the maintenance of the composition of the internal medium is called *homeostasis*.

The internal medium of most animals is the tissue fluid (p. 83) which is in contact with all the living cells of the body. The composition of the tissue fluid depends on the composition of the blood from which it is derived and, therefore, the homeostatic mechanisms of many animals act by adjusting the composition of the blood.

The skin helps to regulate blood temperature (p. 98), the liver adjusts its glucose concentration (p. 69), the lungs keep the carbon dioxide concentration down to a certain level (p. 90), and the kidneys control its composition in three principal ways: they (a) eliminate all soluble nitrogenous waste compounds above a certain concentration, (b) remove excess water, (c) expel salts above a certain concentration, and (d) excrete excess hydrogen ions to maintain its pH. These activities are both excretory, in that they remove the unwanted waste products of metabolism, and osmoregulatory (p. 23), in that they keep the osmotic potential of the blood more or less constant.

Kidney structure and function

The two kidneys are compact, oval structures with an indentation on the side nearest the mid-line of the body. They are red-brown, enclosed in a transparent membrane and attached to the back of the abdominal cavity (Fig. 13.1). The *renal artery* branching from the aorta brings oxygenated blood to them, and the *renal vein* takes deoxygenated blood away to the vena cava. A tube, the *ureter*, runs from each kidney to the base of the bladder in the lower part of the abdomen.

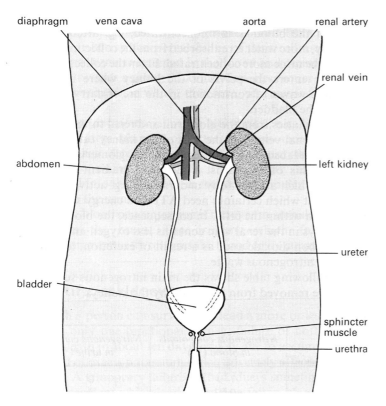

Fig. 13.1   Position of kidneys in the body

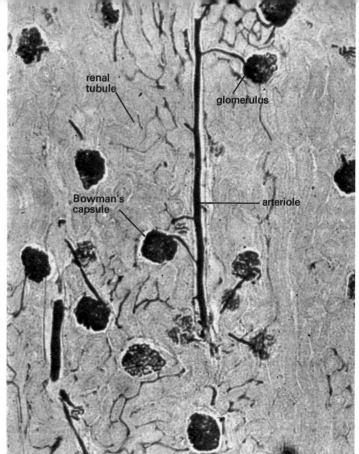

(Brian Bracegirdle/Biophoto Associates)

Fig. 13.3   Section through kidney cortex (× 80)

The kidney tissue consists of many capillaries and tiny tubes called renal tubules, held together with connective tissue. A section through the kidney shows an outer darker region or *cortex*, and a lighter inner zone, the *medulla*. Where the ureter leaves the kidney is a space called the *pelvis*, and projecting into this are cones or pyramids of kidney tissue.

Detailed structure. The renal artery divides up into a great many arterioles and capillaries, mostly in the cortex (Fig. 13.2). Each arteriole leads to a *glomerulus* which is a capillary repeatedly

branching and coiled into a small knot of vessels (Fig. 13.3). The glomerulus is almost completely surrounded by a cup-shaped structure called a *Bowman's capsule* which leads to a coiled renal tubule. This tubule, after a series of coils and loops, joins other tubules and passes through the medulla to open into the pelvis at the apex of a pyramid (Fig. 13.4). The glomerulus, Bowman's capsule and associated renal tubule are called a *nephron*.

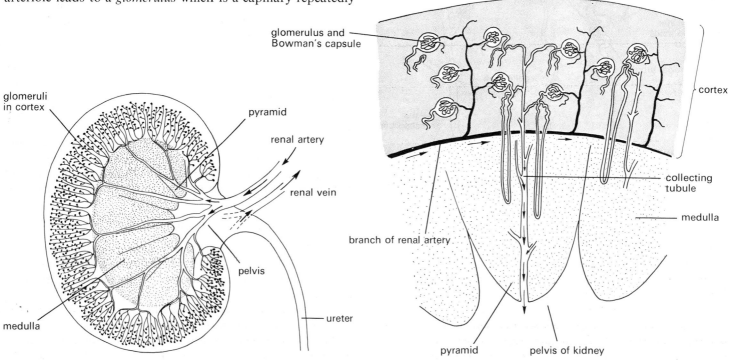

Fig. 13.2   Vertical section through kidney

Fig. 13.4   Section through cortex and medulla

93

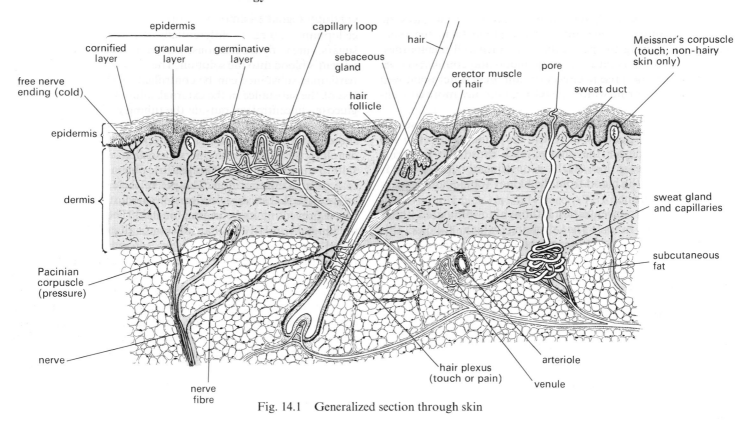

Fig. 14.1 Generalized section through skin

Skin structure

The skin consists of two main layers, an outer *epidermis* and an inner *dermis*. The details of skin structure vary, however, with the region of the body under consideration but the following description applies to most regions (*see* Fig. 14.1).

Epidermis

(a) **Germinative layer.** This is a continuous layer of cells which can divide actively and so produce new epidermis. In this layer also occurs the pigment *melanin* which determines the skin colour and absorbs the ultraviolet radiation. Infoldings of the germinative layer produce the sweat glands, the sebaceous glands, hair follicles and fingernails. Since the epidermis contains no blood vessels, the germinative cells must derive their food and oxygen by diffusion or active transport from the capillary networks in the dermis.

(b) **Granular layer.** This layer consists of living cells recently produced from the germinative layer. As these cells are gradually pushed to the outside by the accumulation of new cells beneath them, deposits of a substance called *keratin* are formed in them and they lose their nuclei, become flat and finally wear away or fall off.

(c) **Cornified layer.** These flat, dead cells full of keratin granules constitute the cornified layer which makes a tough, bacteria-resisting, waterproof coat, so forming a barrier between man and his environment. The cells of this region are continually worn away and replaced from beneath. In non-hairy skin, i.e. the palms of the hands and soles of the feet, the cornified layer may be very thick (Fig. 14.2). Non-hairy skin also has a thick epidermis with a regular pattern of ridges (fingerprints), and a relatively thin dermis with numerous sweat glands but few or

no hair follicles. Hairy skin has a thick dermis and a thin epidermis with numerous hair follicles (Fig. 14.3) and covers the whole body except the palms and soles.

Dermis

The dermis is a layer of connective tissue with relatively few cells but many collagen fibres and a small number of elastic fibres in it. The collagen fibres give the skin its tensile strength and its tough, slightly elastic properties. These fibres are probably made by cells called *fibroblasts* whose appearance varies according to age but are usually apparent only as nuclei. Fibroblasts are important in wound healing. Apart from fibroblasts, the only other cells generally distributed in the dermis appear to be *macrophages*, cells similar to white blood cells which can destroy foreign particles and bacteria by engulfing them and can move about to some extent.

In the dermis are also blood capillaries, nerve endings and lymphatic vessels. Sweat glands, hair follicles and sebaceous glands, although they appear in the dermis, are in fact produced by the epidermis.

Capillaries. The capillaries supply the skin with the necessary food and oxygen and remove its waste products. The sweat glands and hair follicles have a network of capillaries supplying them. The capillaries beneath the epidermis not only nourish it but play an important part in temperature control (p. 98).

Sweat glands. The sweat gland is a coiled tube consisting of secretory cells which absorb fluid from the surrounding cells and capillaries and pass it into the tube, through which it reaches the skin surface. The composition of sweat varies but it consists mainly of water with some salts dissolved in it, notably sodium chloride (0.3 per cent), and small quantities of urea (0.03 per cent) and lactic acid (0.07 per cent).

In hot conditions accompanied by hard work, a man may lose up to 10 litres of sweat, containing 30 g sodium chloride per day. Both the liquid and the salt must be replaced or the circulatory system may fail. If water alone is taken to replace the lost sweat, the salt and water balance of the blood and tissues is upset leading to heat cramp, sometimes called 'miner's cramp', i.e. cramping pains in the muscles most frequently used.

Hair follicle. This is a deep pit of granular and germinative layers whose cells multiply and build up a cylindrical hair inside the follicle. The cells of the hair become impregnated with keratin and they die. The constant adding of new cells to the base of the hair causes it to grow. The hairs on the head continue to grow for up to four years after which the hair falls out and a new period of growth begins. The growth of a body hair may be for only a few months. The colour of the hair depends on the amount, if any, of the melanin incorporated in the cells. The curliness of hair depends on the shape of its cross-section; a circular cross-section is characteristic of straight hair, an elliptical section produces wavy hair and a flattened or kidney-shaped section produces woolly hair. Contraction of the hair muscle pulls the hair more upright and at the same time compresses the sebaceous gland forcing a little of its secretion, sebum, on to the skin.

Sebaceous glands. The sebaceous glands open into the top of the hair follicles and produce an oily secretion, *sebum*. The function of the sebum is not clear. It has generally been thought to keep the skin supple and waterproof and to inhibit bacterial growth, but there is not much evidence for this.

Fingernails. These consist of a modified granular layer produced by a groove lined with germinative cells. They grow at the rate of 0.5–1.2 mm per week.

Subcutaneous fat. The layers beneath the dermis contain numerous fat cells in which fat is stored. Apart from being a reserve of food, this layer, by virtue of its insulating properties, has the effect of reducing heat losses.

Damage and repair

Burns. Superficial burns which affect only cornified or granular layers are soon repaired by the normal division of the germinative cells. Deeper burns which destroy the germinative layer heal more slowly as the new germinative layer has to grow out from the undamaged hair follicles to cover the exposed area.

If the burns are so deep that the hair follicles are destroyed, new epidermis cannot grow although scar tissue will be formed by the dermis. In this case skin grafting is used. Even superficial burns, if they occur over a wide area, can be dangerous if not given expert attention. This is partly due to the increased chances of bacterial invasion through the unprotected dermis, and partly a result of the leakage of blood plasma which may affect the concentration and volume of the blood causing impairment of the circulation.

Repair. (a) *Natural.* When the skin is damaged, as in a cut, the formation of a blood clot (p. 73) reduces both the escape of blood and the entry of germs. The capillaries of the damaged region dilate and become more permeable allowing plasma, its proteins and its white cells to escape into the tissues giving rise to the familiar reddening and swelling. The white cells of the blood and the macrophages of the dermis combat any bacteria which have gained access, aided by antibodies which may be present in the blood plasma (*see* p. 74). Bacteria which escape these defences may enter the lymphatics and on passing through a lymph node may be ingested by the phagocytic cells contained there. Beneath the scab, macrophages and fibroblasts move in, the former digesting the blood clot and cell debris, the latter laying down new collagen fibres. Nerve endings and capillaries invade the new dermis so formed. The epidermis is restored by multiplication and migration of cells from the margins of the undamaged germinative layer at the rate of about 0.5 mm per day until the wound is closed and the scab is ready to fall off. With deep wounds, healing takes place more or less as described, with the difference that far more collagen fibres are laid down making a tough, rather inflexible and insensitive scar.

Fig. 14.2   Section through non-hairy skin (the sweat ducts are contorted in passing through the cornified layer) (× 80)

(Brian Bracegirdle/Biophoto Associates)

Fig. 14.3   Section through hairy skin (× 30)

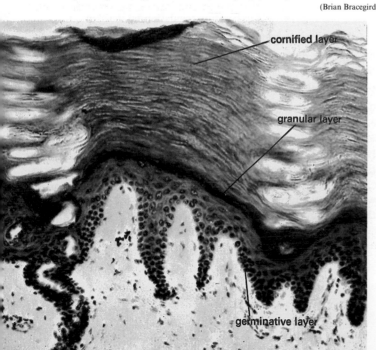

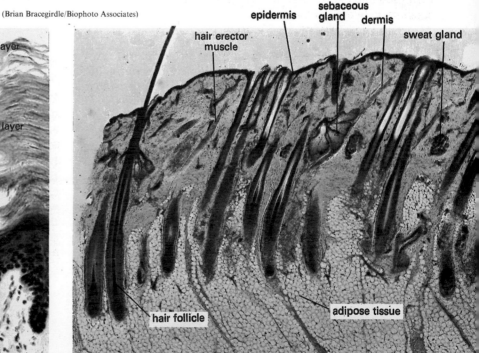

(b) *Skin grafting*. Extensive wounds and burns in which the hair follicles are destroyed can be repaired by skin grafts. Small slices of skin, deep enough to include the germinative layer and some dermis, are taken from a donor area such as the abdomen, arm or thigh and placed over the region to be covered. After a few days, the capillaries of the receiving area have invaded and made connections with the capillaries in the donated skin and in a few weeks the graft has become a part of the damaged area. The donor area heals normally by over-growth of the germinative layer from the hair follicles or, if a deeper layer of skin has been removed, healing is promoted by sewing the edges of the wound together. Grafted skin cannot be transferred from one person to another with success; the graft must come from the injured person's own body.

A transplant of skin from anyone other than an identical twin is rejected. Blood vessels fail to grow into it, it dies and falls away. The body reacts to transplants of skin and most other organs or tissues as it would to bacteria, incorrect red cells and foreign proteins. The transplant is attacked by phagocytes and antibodies and eventually destroyed.

The transplanting of kidneys, hearts, livers, etc. depends on using drugs to suppress this immune reaction of the body and this, in turn, leaves the patient very susceptible to attack by disease organisms.

Temperature control

Fish, amphibia, reptiles and all the invertebrate animals are *poikilothermic*. That is to say, their body temperature when at rest is the same as or only a few degrees above that of their surroundings and varies accordingly. This makes them very dependent on external temperature changes. For example, in cold conditions their low body temperature slows down most chemical changes and reduces the organism to a state of inactivity. Insects can be immobilized by a sudden fall in temperature.

Homoiothermic or constant temperature animals are more independent of their surroundings because their internal body temperature is kept constant and does not alter with fluctuations in external temperature. Man, by both voluntary and involuntary means, is able to maintain a constant internal temperature even in extremes of heat or cold. Maybe for this reason humans have a wider distribution over the Earth's surface than any other vertebrate, being able to maintain a stable temperature in arctic and tropical conditions (*see* Fig. 1.6, p. 5).

Heat loss and gain

Many of the chemical changes in living protoplasm release heat energy. Chemical changes in glands and in contracting muscle produce a great deal of heat; in fact, over a period of 24 hours, 95 per cent of the energy transferred in chemical reactions in the body appears as heat. The circulatory system distributes this heat round the body. If no heat were lost from the body, the temperature of even a totally inactive person would rise by about 1 °C per hour. In fact, the body loses heat from its surface to the atmosphere mainly by convection and radiation, though in tropical conditions heat may be gained by these means. Fluid is constantly diffusing through the skin and evaporating into the air so removing heat from the body. This is not the same as sweating, as the sweat glands are not in constant activity. An outer layer of fur, feathers or clothing reduces these heat losses. Heat lost with expired air cannot be reduced.

With an inactive subject in an environmental temperature of about 29–30 °C, the rates of heat loss and gain are about the same and the body temperature remains between 35.8 °C and 37.3 °C. The temperature of the body is not the same in all regions but is usually measured in the mouth, armpit or rectum and taken as an indication of the blood temperature. The mouth temperature is the most reliable indicator of changes in the temperature of the blood. Body temperature also varies during the day with activity and from one person to the next. The specific marking of 36.9 °C (98.4 °F) on a clinical thermometer represents an average temperature of healthy individuals but since their temperatures may range from 35.9 to 37.8 °C (96.6–100 °F), readings of 1 °C (2 °F) above or below this average do not indicate ill-health. When heat loss and gain do not balance, regulatory mechanisms in the body under the control of the brain come into action and compensate for overheating or overcooling.

Overheating

This may be brought about by a high environmental temperature, vigorous activity, disease, by absorbing radiation from the sun, or by many other external causes. If the temperature of the blood reaching the hypothalamus of the brain (p. 155) is a fraction of a degree higher than 'normal', nerve impulses are sent to the skin producing two marked effects.

1   **Vasodilation.** Certain of the arterioles beneath the epidermis dilate (get wider). Consequently more blood flows near the surface, losing heat through the epidermis into the air by convection and radiation (Fig. 14.4a). In white-skinned people,

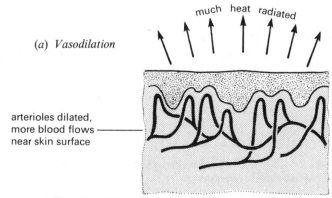

(a) *Vasodilation*

much heat radiated

arterioles dilated, more blood flows near skin surface

Fig. 14.4   Temperature regulation by the skin

vasodilation may cause obvious flushing (reddening) of the skin on account of the increased blood just beneath the epidermis. With an environmental temperature of 34 °C, about 12 per cent of the blood pumped out by the heart may pass to the skin.

2   **Sweating.** Vasodilation and vasoconstriction are effective in controlling the body temperature when external conditions vary between 25 °C and 29 °C. Above this range sweating begins, and below 25 °C shivering sets in, depending on other factors such as clothing and activity. Sweating begins when the external temperature rises above 29–31 °C, depending on the weight of clothing, or when for any reason the blood reaching the hypothalamus is 0.5–1.0 °C above 'normal'. Nerve impulses starting mainly in the hypothalamus and conveyed by the autonomic nervous system (p. 151) stimulate the sweat glands

into activity. Fluid from the blood is filtered into the glands and passes through their ducts so that a layer of moisture is produced on the skin surface. As the sweat evaporates it takes its latent heat from the body and so reduces the body temperature. Evaporation from the forehead, upper lip, neck, chest and trunk is responsible for temperature regulation. Sweating from the palms, soles and armpits seems more dependent on emotional stress than on body temperature.

Vigorous activity in direct sunlight or in a hot, humid climate may cause a rise in body temperature which impairs the working of the hypothalamus. Temperature control is lost, sweating stops and the body temperature rises to over 41 °C. This degree of overheating and the dehydration and salt imbalance which accompany it give rise to a variety of potentially lethal conditions, variously named as *heat stroke*, *heat exhaustion*, or *hyperpyrexia* depending on the physiological effects and symptoms. The effects are popularly called 'sunstroke', but any internal or external conditions which prevent adequate heat loss can be the cause, not just sunshine.

Overcooling

If the body begins to lose more heat than it is generating, the following compensatory changes may take place:

1 **Decrease in sweat production,** thus minimizing heat lost by evaporation.

2 **Vasoconstriction.** Constriction, or closing, of the arterioles that supply the surface capillaries in the skin reduces the volume of blood flowing near the surface and hence diminishes heat losses (Fig. 14.4*b*). Vasoconstriction makes a white-skinned person look pale or blue.

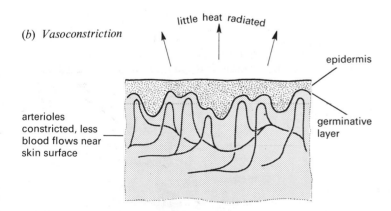

(*b*) *Vasoconstriction*

little heat radiated

epidermis

germinative layer

arterioles constricted, less blood flows near skin surface

3 **Shivering.** This reflex action operates when the body temperature begins to drop. It is a spasmodic contraction of the muscles. These contractions produce heat which helps to raise the body temperature.

Furry mammals and birds can fluff out their fur or feathers by contraction of the erector muscles which are attached to them in the skin. This increases the layer of trapped air round the skin and so improves insulation by reducing convection and conduction. In man, a similar contraction of the muscles of the hairs produces only 'goose-pimples' but adding layers of clothing increases the thickness of the insulating air layer round the skin.

Voluntary temperature control

By constructing houses with efficient heating and ventilation, by adding or removing clothing, by taking exercise or cold showers, man exerts conscious, voluntary control over his rate of heat loss or gain. Vigorous exercise can produce a rise of from 1 to 4 °C.

Comfort. The sensations of feeling hot or cold result from the stimulation of receptors in the skin. The body temperature remains virtually constant even though the brain is registering heat loss or gain through the skin. A person *feels* cold when the skin temperature receptors respond to a loss of heat to the atmosphere and the skin may become cold as a result of vasoconstriction, but the temperature of the blood does not alter.

A comfortable environment is one which results in a skin temperature of about 33 °C. The humidity of the environment is an important factor in temperature regulation and comfort. If the air is saturated with water vapour little or no sweat will evaporate from the body and this important means of temperature control is ineffective. A hot, humid environment may lead to overheating of the body and sensations of discomfort.

Hypothermia

In some conditions where heat loss exceeds heat production, the body is unable to compensate rapidly enough and the deep body temperature begins to fall. This is a dangerous condition known as *hypothermia*. It may occur in elderly people spending long hours inactive in cold buildings. It is also the main cause of death due to 'exposure' in cold, wet conditions. The affected person becomes tired and weak and loses consciousness. Unless the victim's temperature is raised carefully, hypothermia leads in a few hours to heart failure.

In certain heart operations the patient's body is deliberately cooled to 27 °C. At this temperature the rate of metabolism is slowed by 30 to 40 per cent and the oxidation of glucose ceases. In these conditions the demand for oxygen by the tissues is much reduced and the circulation can be interrupted for 10 to 15 minutes without damaging the nerve cells of the brain which are particularly affected by lack of oxygen. The patient is given a light anaesthetic and an injection to suppress the shivering reflex; the body is cooled by ice packs or by rubber tubing carrying chilled brine. The low temperature is sufficient to ensure the maintenance of unconsciousness.

Questions

1 Why is it more accurate to describe fish as 'variable-temperatured' animals rather than 'cold-blooded'?
2 When a dog is hot, it hangs its tongue out and pants. Why should this have a cooling effect?
3 Why do you think we experience more discomfort in hot humid weather than we do in hot dry weather?
4 You may 'feel hot' after exercise of 'feel cold' without your overcoat and yet your body temperature is not likely to differ by more than 0.5 °C on the two occasions. Explain this apparent contradiction (*See also* p. 130.)
5 Draw up a balance sheet showing all the possible ways in which the human body can gain and lose heat.
6 In what ways does the skin contribute to the homeostasis of the body? (*See* p. 92.)

15
Reproduction

Sexual reproduction involves the fusing together of two cells called *gametes*. One of these reproductive cells comes from a male and the other from a female. The fusion of two gametes is called *fertilization* and the resulting composite cell is called a *zygote*. The most important aspect of fertilization is the fusion of the nuclei of these male and female gametes, since these nuclei contain the genetic information that determines the inherited characteristics of the individual that develops from the zygote.

Fertilization then is the fusion of the nuclei of male and female gametes to form a zygote from which a new individual can develop. In animals the male gametes are *sperms* which are produced in the reproductive organs called *testes*. The female gamete is an *ovum* which is produced in reproductive organs called *ovaries*.

In mammals, sperms are placed in the vagina of the female and the eggs are subsequently fertilized internally. They are not laid after fertilization but are retained in the female's body while they develop to quite an advanced stage. They are born more or less fully formed and are protected by their parents and fed on a secretion of milk from the mammary glands until they can feed independently.

Sexual reproduction in man

Female reproductive organs (Figs. 15.1–2). The *ovaries* are two oval bodies each about 35–40 mm long, lying in the lower part of the abdomen. They are attached by a membrane to the uterus and supplied with blood vessels. Close to each ovary is the expanded funnel-shaped opening of an oviduct, also known as the *Fallopian tube*. This is the tube down which the ova pass after they are released from the ovary.

Both oviducts open into a wider tube, the *uterus* or *womb*, lower down in the abdomen. When there is no embryo developing in it, the uterus is only about 80 mm long with walls 10 mm thick consisting of unstriated (involuntary) muscle and a glandular lining. The uterus communicates with the outside through a muscular tube, the *vagina*. The *cervix* is a ring of muscle closing the lower end of the uterus where it joins the vagina. There is normally only a very small aperture joining these two organs at this point. Two folds of skin, the *labia majora* and the *labia minora*, enclose the *vulva* at the entrance to the vagina. The urethra from the bladder opens into the vulva just in front of the vagina.

Male reproductive organs (Figs. 15.3 and 15.4). The two testes lie outside the abdominal cavity in man, in a special sac called

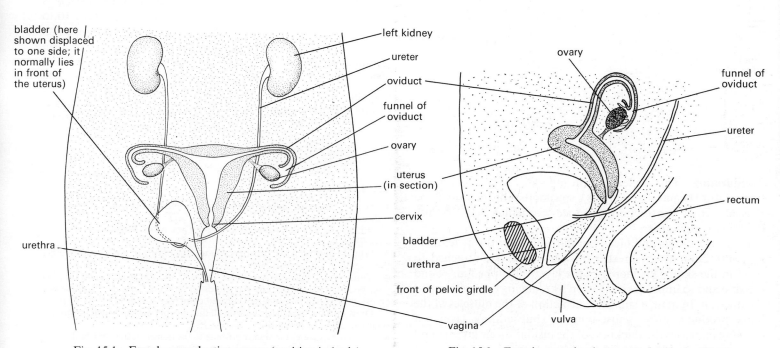

Fig. 15.1 Female reproductive organs (position in body)

Fig. 15.2 Female reproductive organs (vertical section)

the *scrotum*. This enables the testes to remain at a temperature about 2 °C lower than the rest of the body, a condition which favours sperm production. Each testis consists of a large number of sperm-producing tubules. Between these tubules are interstitial cells which produce a male hormone, *testosterone*, and possibly other substances. The sperm-producing or *seminiferous* tubules join to form sperm ducts leading to the *epididymis*, a coiled tube about 6 metres long on the outside of the testis. The epididymis in turn leads into a muscular *sperm duct* (Fig. 15.5). The two sperm ducts, one from each testis, open into the top of the urethra just after it leaves the bladder. Surrounding the urethra at this point are the *prostate gland* and, further down, *Cowper's gland*. At different times the urethra conducts either urine or sperms, and extends through the penis which consists of connective tissue with numerous small blood spaces in it.

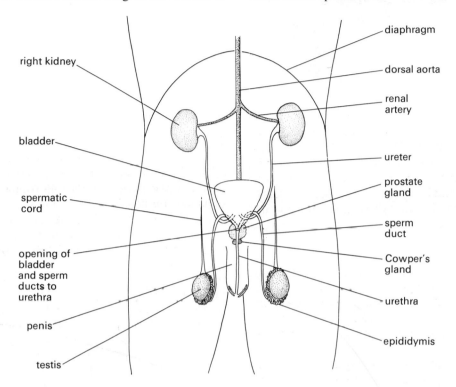

Fig. 15.3   Male reproductive organs (position in body)

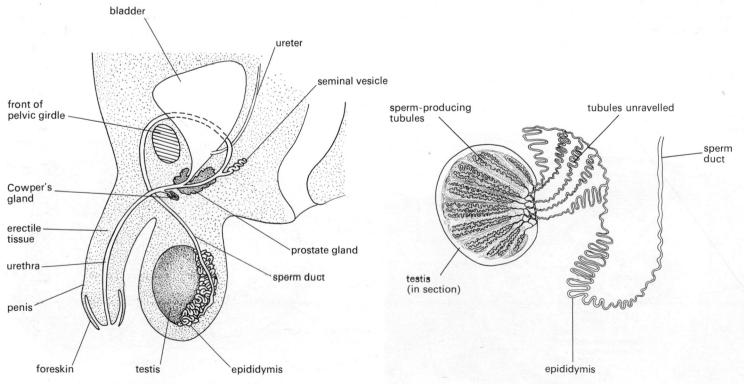

Fig. 15.4   Male reproductive organs (vertical section)

Fig. 15.5   Testis and sperm duct

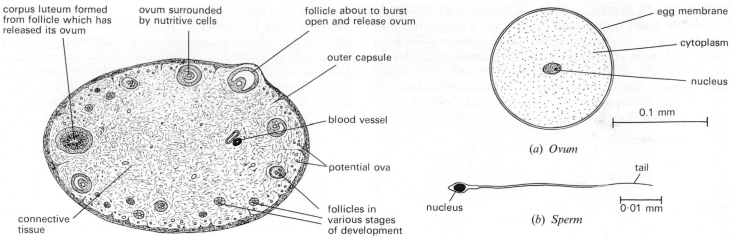

Fig. 15.6   Section through ovary

Fig. 15.8   The gametes

(a) Ovum

(b) Sperm

Ovulation. The ovary consists of connective tissues, blood vessels and potential egg cells (Fig. 15.6). It is thought that 70 000 potential egg cells are already present at birth and are not, like sperms, formed continuously during adult life. Of these potential egg cells only about 500 will ever become mature ova. The ovaries also secrete hormones called *oestrogens* which control the secondary sexual characteristics (p. 107) and initiate the thickening of the lining of the uterus which occurs each month.

Between the ages of about eleven and fourteen years, the ovaries become active and begin to produce mature eggs. The beginning of this period of life is called *puberty* and is characterized by a series of physiological and psychological changes (*see* p. 107). In the ovary, some of the ova start to mature and the cells around them divide rapidly. A fluid-filled cavity is eventually produced, which partly encircles the ovum and its coating of follicle cells. This structure is now called a *Graafian follicle*. When the Graafian follicle is ripe it is about 10–20 mm in diameter and projects as a clear vesicle from the surface of the ovary (Fig. 15.7). Finally it bursts and releases the ovum into the funnel of the oviduct whose ciliated cells waft it

into the tube. At this stage the ovum (Fig. 15.8a) is a spherical mass of protoplasm about 0.13 mm in diameter, with a central nucleus and with some of the cells from the follicle still adhering to the outside (Fig. 15.11a). An ovum is produced from one or other of the ovaries every four weeks and it may spend several hours travelling down the oviduct to the uterus. The oviduct is lined with ciliated cells whose cilia help to propel the ovum towards the uterus.

Sperm production (Fig. 15.9). The lining of the seminiferous tubules making up the testis consists of actively dividing cells which give rise ultimately to sperms (Fig. 15.8b). A sperm is a nucleus surrounded by a little cytoplasm which extends into a long tail (Fig. 15.10). The sperms are quite immobile when first produced and pass into the epididymis where they are stored. During copulation, muscular contractions of the epididymis, sperm duct and accessory muscles force the accumulated sperms through the urethra. Secretions from the prostate gland and seminal vesicles provide a fluid medium for the sperms and add nutrients and enzymes. In a short time, the sperms become motile, propelled by lashing movements of their tails. Sperms not ejaculated in this way are broken down and their products reabsorbed in the sperm ducts at the same rate as they are produced in the testis.

Fig. 15.7   Mature Graafian follicle (× 40)

(Brian Bracegirdle/Biophoto Associates)

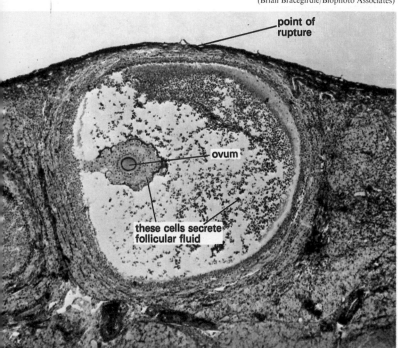

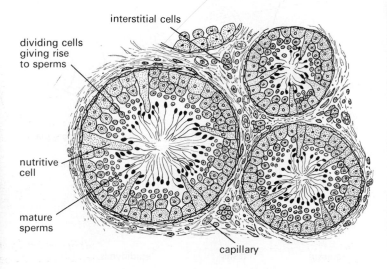

Fig. 15.9   Microscopic appearance of testis in section

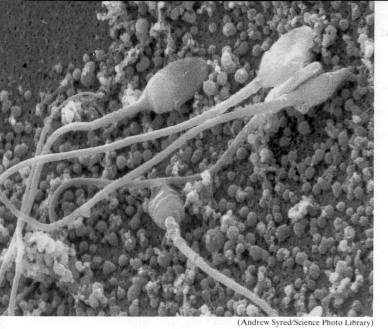

(Andrew Syred/Science Photo Library)

Fig. 15.10   Human sperms × 24,000

Fertilization. Fertilization occurs internally when the sperms meet the ovum as it passes down the oviduct. They are introduced into the female by the penis which is placed in the vagina. To facilitate this action the penis becomes erect and firm, largely as a result of the arterioles dilating and so allowing blood to enter the blood spaces more rapidly than it escapes, which increases the turgidity of the tissues round the urethra. Similarly, the lining of the vulva and vagina secrete mucus which makes entry of the penis easier. Contact with the walls of the vagina stimulates the sensory endings in the penis and sets off a reflex action that results in the *semen* (consisting of 60 per cent fluid from the seminal vesicles, 20 per cent prostatic secretion, and the accumulated sperms) being ejaculated into the vagina. This event is the climax of copulation. The sperms

deposited in the vagina swim through the cervix into the uterus. From here they travel to the oviduct, possibly by their own swimming movements, though this is not established for certain. If there is an ovum in the oviduct, the sperms seem able to penetrate between the follicle cells still adhering to it and one of the sperms sticks to and subsequently penetrates the zona pellucida and cell membrane. The zona pellucida then becomes impenetrable to any other sperms and the nucleus of the one successful sperm fuses with the ovum nucleus (Fig. 15.11).

Although a single ejaculation may release more than a hundred million sperms, only one will actually fertilize the ovum. Experiments with animals suggest that each sperm produces a minute quantity of enzyme, which disperses the remaining follicle cells adhering to the ovum, so facilitating fertilization. Whether this is an important process in man is not clear.

In the ovary, the cells of the follicle from which the ovum was released continue to divide and grow for a while, filling the follicle with a solid mass of tissue called the *corpus luteum*. The corpus luteum produces a hormone, *progesterone*, which acts on the uterus causing its lining to thicken and its vascular supplies to increase in preparation for the implantation and nourishment of the embryo. If fertilization does not occur, the corpus luteum degenerates. If fertilization does occur, the corpus luteum persists and continues to produce progesterone for another three months.

The period during which a released ovum can be fertilized is thought, by some authorities, to be as little as eight hours and by others to be up to forty-eight hours. Results of artificial insemination suggest that sperms can fertilize an ovum only up

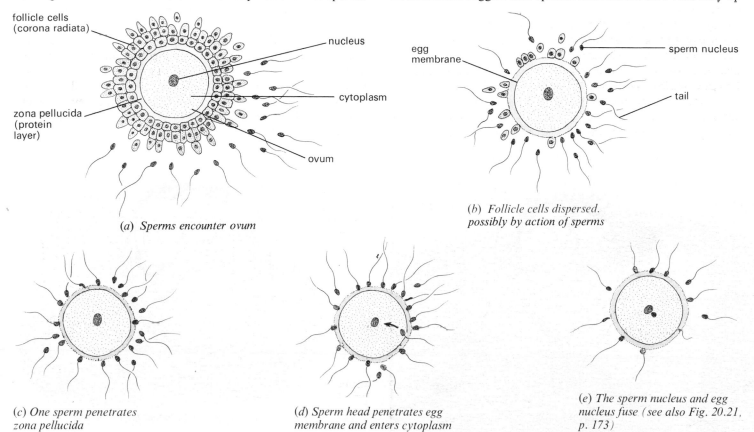

follicle cells (corona radiata)

nucleus

cytoplasm

zona pellucida (protein layer)

ovum

(a) *Sperms encounter ovum*

egg membrane

sperm nucleus

tail

(b) *Follicle cells dispersed, possibly by action of sperms*

(c) *One sperm penetrates zona pellucida*

(d) *Sperm head penetrates egg membrane and enters cytoplasm*

(e) *The sperm nucleus and egg nucleus fuse (see also Fig. 20.21, p. 173)*

Fig. 15.11   Fertilization of human ovum (the diagrams show what is thought to happen but the events are not known for certain)

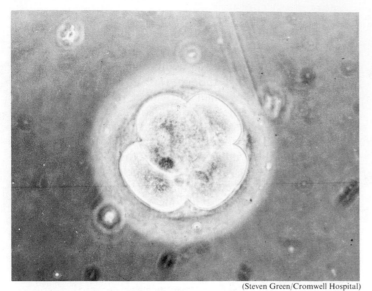

(Steven Green/Cromwell Hospital)

Fig. 15.12   Human embryo just after the 4-cell stage (× 500). The cells are enclosed in the zona pellucida

to twenty-four hours after arriving in the female tracts even though they may live for about three days. Thus there is a relatively short period in which fertilization can occur. Since the times of ovulation cannot be determined easily, it is not always possible to ensure that copulation takes place within a period of twenty-four hours before or thirty-two hours after ovulation, if children are wanted, or outside this period if they are not.

Pregnancy and development. The zygote undergoes rapid cell division (Fig. 15.12) as it passes down the oviduct and into the uterus. At first it floats freely, absorbing nutriment from the uterine secretions and then, four to seven days after its release from the ovary, it adheres to and sinks into the uterine lining. This is called *implantation* and may take place at various points in the uterus, but frequently it is high up on the posterior surface. As the embryo begins to form, finger-like processes grow from it into the uterine lining, digesting the epithelium and then absorbing nourishment. These villi do not form part of the embryo but, later, contribute to a special organ called the *placenta* which supplies the embryo with both food and oxygen (Fig. 15.13).

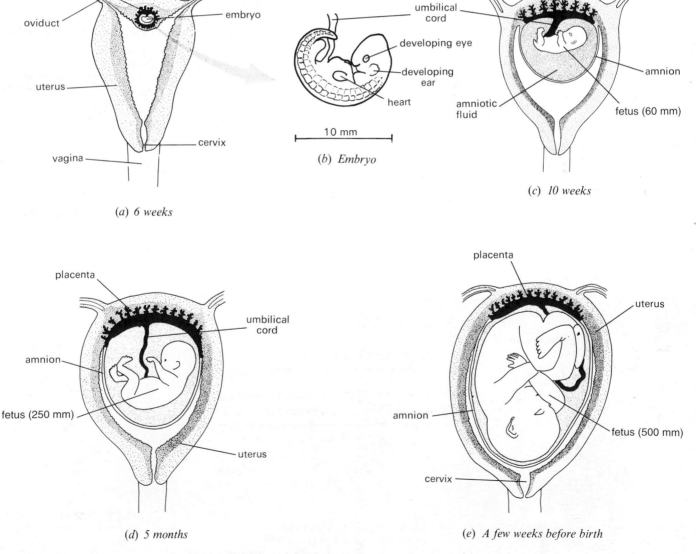

(a) 6 weeks

(b) Embryo

10 mm

(c) 10 weeks

(d) 5 months

(e) A few weeks before birth

Fig. 15.13   Growth and development in uterus (not to scale)

The uterus, which at first has a volume of only 2 to 5 cm³, extends with the growth of the embryo to 5 000–7 000 cm³, enlarging the abdomen and displacing the other organs to some extent. The uterus lining at the same time develops a greater supply of blood vessels and becomes increasingly muscular.

The embryo's cells divide to form tissues; the tissues swell, roll, extend and so form organs of the body. By the end of five weeks the heart and a circulatory system have formed, the heart is beating and circulating the blood but the embryo is still only about 10 mm long. After two months, the limbs and the main organ systems have been laid down and the embryo is now refered to as a fetus.

Although the fetus depends for its food and oxygen on the mother's blood, its circulatory system is never directly connected with the maternal blood vessels. If it were, the adult blood pressure would burst the delicate capillaries forming in the fetus and many substances in the mother's circulation would poison it.

The placenta. The placenta becomes a large disc of tissue, adhering closely to the uterine lining (Fig. 15.14). From the fetal part of the placenta, villi protrude into the uterine lining which has thickened and in which capillaries have broken down to form more extensive blood spaces. The membranes separating the fetal capillaries from the maternal blood spaces are very thin so that dissolved substances can pass across in both directions (Fig. 15.15). In the placenta, oxygen, glucose, amino acids and salts in the mother's blood pass from the uterine blood spaces into the capillaries of the fetus, while carbon dioxide and urea from the fetus pass across in the opposite direction (Fig. 15.16). The differences in concentration on either side of the placenta are sufficient to account for the diffusion of oxygen and carbon dioxide, but do not entirely explain the quantities of glucose transferred, and the method by which fats cross the placenta is not yet known. Amino acids pass from uterine to fetal blood against a diffusion gradient, presumably by a form of active transport. Vitamins and some maternal antibodies also reach the fetus.

In addition, the membrane in the placenta which separates fetal and maternal blood prevents most harmful substances from reaching the fetus. However, drugs in the mother's blood stream, carbon monoxide from cigarette smoke, and some viruses, e.g. the rubella (German measles) virus (*see* p. 197) do cross the placenta. Some fetal red blood cells sometimes escape across the placenta into the mother's circulation (*see* p. 76).

A pregnant woman will absorb more iron than usual from her diet and pass it to the fetus. The demands of the latter for calcium and phosphorus may be so great that if the mother's diet is deficient in available forms of these elements they will be taken from her own bones.

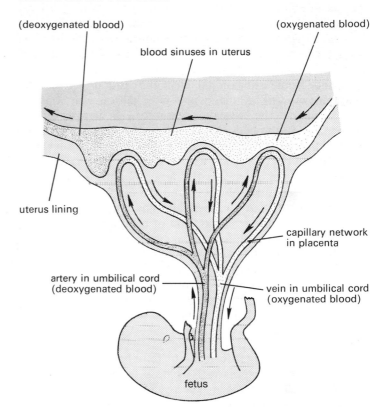

Fig. 15.15 Diagram of relationship between blood supply of embryo, placenta and uterus

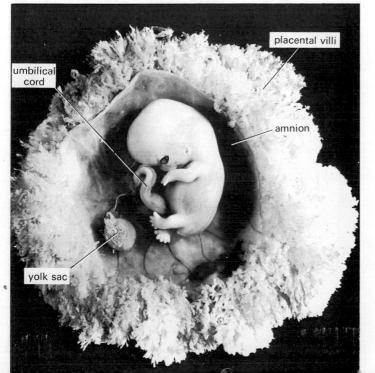

Fig. 15.14 Human fetus, 7 weeks (× 1.5)

(Prof. W. J. Hamilton)

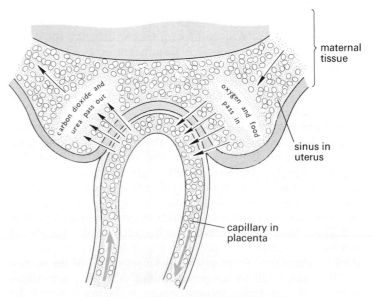

Fig. 15.16 Diagram to show exchange of oxygen and food from uterus to placenta

The placenta, as well as exchanging food, waste products and respiratory gases, also produces hormones similar to oestrogen and progesterone that prevent menstruation and ensure continued pregnancy. The capillaries in the placenta are supplied with blood from two arteries in the *umbilical cord*. The blood returns to the fetus in a single vein (Fig. 15.18).

The fetus is surrounded by a 'water sac', the *amnion*, whose fluid buoys it up while the delicate organs and tissues are growing. It also spreads pressure evenly around the fetus, minimizing damage from movements of the mother's body.

The fetus swallows the amniotic fluid and also takes it in and out of its lungs with shallow, irregular 'breathing' movements. It also excretes fetal urine into the amniotic fluid but does not defaecate. After about five months' growth, the fetus moves its limbs quite vigorously inside the amniotic cavity and uterus.

The time from fertilization to birth, the *gestation period*, lasts about nine months in humans, but over half the weight of the fetus is gained in the last six to eight weeks. In fifty per cent of pregnancies, birth occurs within seven days either side of the 280 days from the last menstrual period. The mechanism which starts *labour* is not yet known. It seems, however, that the distension of the uterus in the latter months of gestation restricts the blood supply in it, and the output of hormones which retain the fetus is reduced. The corpus luteum has degenerated and the concentration of progesterone falls to a level at which the uterus is more readily stimulated to contract. It is also suspected that a hormone produced by the pituitary body has some influence on the beginning of labour.

A baby born after only seven months of gestation has a good chance of survival if given special care and extra oxygen. If born earlier than this, the baby's chances of survival are slight. The average birth weight of a baby is about 3 kg.

Birth. A few weeks before birth, the fetus has come to lie head downwards in the uterus with its head just above the cervix (Fig. 15.17). The uterus makes rhythmic contractions even when empty but before birth these contractions become more frequent and more regular. This is the onset of labour, and the interval between the contractions becomes shorter and the contractions themselves more forcible. The opening of the cervix gradually dilates enough to let the child's head pass through and the uterine contractions are then reinforced by voluntary contractions of the abdominal muscles. The amnion breaks at some stage during labour and the fluid escapes through the vagina. Finally, the muscular contractions of the uterus and vagina expel the child, head first, through the dilated cervix and vagina.

In a few minutes, the arteries in the umbilical cord constrict and the latter can be tied and cut. About 10 to 20 minutes later the placenta breaks away from the uterus and is expelled separately as the 'afterbirth'. The uterus contracts to its normal size soon after the birth, but bleeding from the placental area continues for some days and the raw area is very liable to infection, so commonsense precautions must be taken such as refraining from sexual intercourse for a few weeks. Before the nature of infection was properly understood, many mothers died of 'childbirth' fever as a result of infections of the uterus.

The sudden fall in temperature and other new sensations from the skin and muscles experienced by the newly born baby stimulate it to take its first breath, usually accompanied by crying, when amniotic fluid is expelled from its lungs. In a few days, the remains of the umbilical cord attached to the baby's abdomen shrivel and fall away, leaving in the abdominal wall a scar called the *navel*.

For the first three to four days there is a loss in weight but once feeding is properly established the weight increases rapidly. The child does not grow uniformly, however, since the proportions at birth are quite different from those at maturity. For example, the relative increase in the legs is × 5, arms × 4, trunk × 3, but the head only × 2.

Fetal circulation (Fig. 15.18). The circulatory system of a fetus shows marked differences from that of an adult as a result of the connection to the placenta and the non-functioning of the lungs. Much of the blood flowing down the fetal aorta passes via two *umbilical arteries* to the placenta, where its composition changes as indicated on p. 105. The oxygenated blood returns from the placenta to the fetus through the *umbilical vein*, passing through the liver or bypassing the latter in the *ductus venosus*, to reach the vena cava. Only a small fraction of the blood from the inferior vena cava reaches the right ventricle; most of it passes through the *foramen ovale* between the two atria to enter the left atrium (Fig. 15.19). The deoxygenated blood from the head enters the right atrium through the superior vena cava and passes into the right ventricle in the usual way. From the right ventricle the blood is pumped into the pulmonary artery, but the *ductus arteriosus* which in the fetus connects the pulmonary artery to the aorta, allows blood to bypass the lungs and enter the aorta. At birth, the blood supply to the placenta is cut off by the constriction of the umbilical arteries and the blood pressure rising in the left side of the heart closes the valve in the foramen ovale which seals up within a week. The ductus arteriosus begins to close within a few hours so that all the blood from the right ventricle passes through the lungs.

Fig. 15.17   Model of human fetus just before birth

(Reproduced with permission from the Birth Atlas published by Maternity Centre Association, New York)

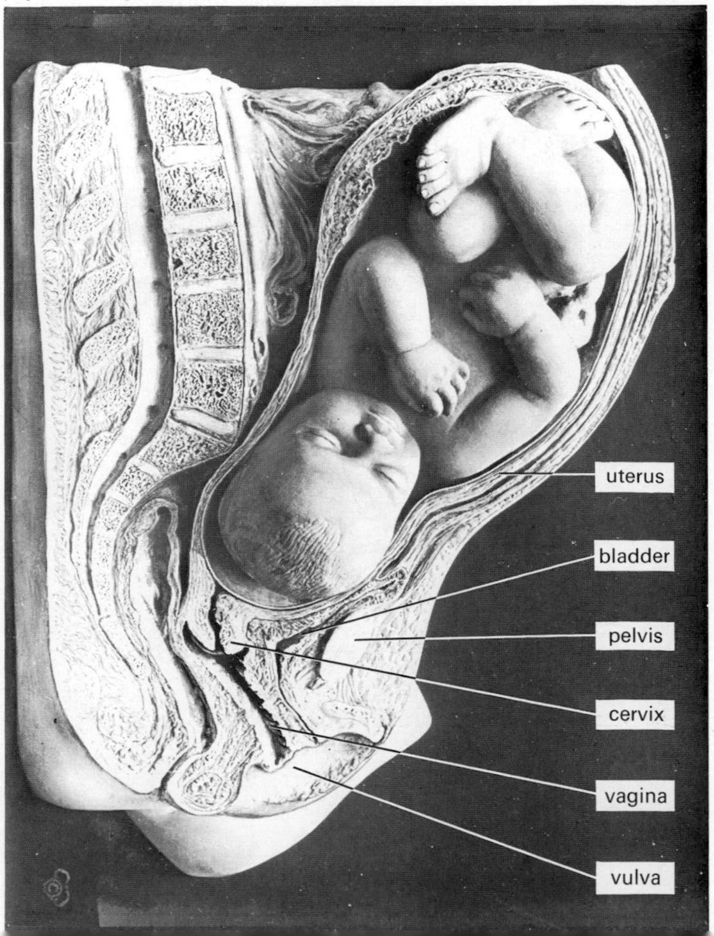

uterus

bladder

pelvis

cervix

vagina

vulva

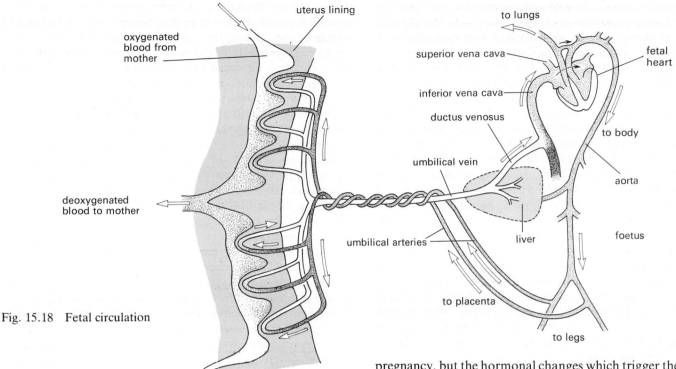

Fig. 15.18 Fetal circulation

- uterus lining
- oxygenated blood from mother
- deoxygenated blood to mother
- placenta
- to lungs
- superior vena cava
- inferior vena cava
- ductus venosus
- umbilical vein
- umbilical arteries
- liver
- fetal heart
- to body
- aorta
- foetus
- to placenta
- to legs

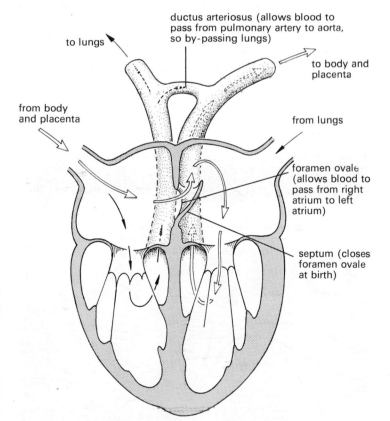

- ductus arteriosus (allows blood to pass from pulmonary artery to aorta, so by-passing lungs)
- to lungs
- from body and placenta
- to body and placenta
- from lungs
- foramen ovale (allows blood to pass from right atrium to left atrium)
- septum (closes foramen ovale at birth)

Fig. 15.19 Circulation in the fetal heart

Feeding and parental care. About twenty-four hours after birth, the baby starts to suckle at the breast. During pregnancy the mammary glands, or breasts, enlarge as a result of an increase in the number of milk-secreting cells. No milk is secreted during pregnancy, but the hormonal changes which trigger the birth of the baby also activate the secretory cells of the breasts. The breasts are stimulated to release milk by the first sucklings. The continued production of milk is under hormonal control but the volume produced is related to the quantity taken by the child during suckling.

Milk contains nearly all the food, vitamins and salts that babies need for their energy requirements and tissue building, but there is no iron present for the manufacture of haemoglobin. All the iron needed for the first weeks or months is stored in the body of the fetus during gestation. The mother's milk supply increases with the demands of the baby, up to one litre per day. It is gradually supplemented and eventually replaced entirely by solid food, a process known as *weaning*.

Cow's milk is not wholly suitable for human babies. It has more protein, sodium and phosphorus and less sugar, vitamin A and vitamin C than human milk. Manufacturers modify the components of dried cows' milk to resemble human milk more closely and this makes it more acceptable if the mother cannot breast-feed her baby.

Cows' milk and proprietary dried milk both lack human antibodies, whereas the mother's milk contains antibodies to any diseases from which she has recovered. It also carries white cells which produce antibodies or ingest bacteria. These antibodies are an important defence against infection at a time when the baby's own immune responses are not fully developed.

Breast-feeding provides milk free from bacteria whereas bottle-feeding carries the risk of introducing the bacteria of intestinal diseases. Breast-feeding also offers emotional and psychological benefits to both mother and baby.

In man, parental care follows the basic pattern common to all mammals but is conscious and planned. The period of dependence of the young on the parents is very prolonged compared with other mammals and involves not only food, warmth and shelter but also education, training and concern for the psychological and moral welfare of the offspring.

Puberty and secondary sexual characteristics. In addition to producing gametes, the ovaries and testes make hormones. At puberty these hormones are released into the blood stream, and as they circulate round the body they give rise to the physical and mental changes which we associate with masculinity or femininity. Although the sex chromosomes X and Y (*see* p. 172) are known to determine sex, the means by which they do so has not been discovered. The embryonic reproductive bodies do not become recognizably testes or ovaries until after the first seven weeks of development. A working hypothesis is that the XY constitution determines whether testes or ovaries develop from embryonic tissue and that the remaining organs depend on the production of specific hormones from the testis or ovary. It is known that a hormonal imbalance can upset the genetically pre-determined sex, but it is unlikely that the explanation is as simple as this. Until the fetus is three months old, the sex cannot be determined by examination of the external genitalia.

Puberty is the beginning of sexual maturity. In girls it is marked by the onset of menstruation which indicates the start of ovulation. Generally, menstruation starts between twelve and fourteen years but it may begin at any time between nine and eighteen years. The breasts may have started to enlarge prior to puberty, largely due to deposition of fat. Other changes are the increase in the size of the uterus and vulva, the growth of hair in the pubic region and the armpits (axillae) and the widening of the pelvic girdle.

Comparable changes take place a little later in boys. There is an increase in growth rate, hair grows in the pubic region, in the axillae and on the face. Enlargement of the larynx causes the voice to 'break', i.e. become deeper. The penis, scrotum and prostate gland enlarge and sperms are produced by the seminiferous tubules in the testes.

The external genitalia and the perceptible changes which take place at puberty are called the *secondary sexual characteristics*.
Menstruation. Of the five hundred or so ova produced in the life of a woman.   more than about twelve are likely, even in

theory, to be fertilized and form embryos; in practice the number is usually far less than this. Nevertheless, at the time of release of each ovum, the uterus lining becomes thicker with additional layers of cells; the ovum sinks into it if fertilized. The blood supply is increased at the same time. If the ovum is not fertilized production of progesterone by the corpus luteum decreases and the new uterine lining disintegrates (Fig. 15.20).

The unwanted cells and 50 to 250 cm³ blood (which does not clot) are lost through the cervix and vagina. This *menstruation*, as it is called, occurs twelve to fourteen days after the ovum is released, about once in four weeks. The interval between menstrual periods varies with individuals and from one period to the next. After fertilization, menstruation ceases and this is one of the first indications of pregnancy.

Menopause. Between the ages of about forty-two and fifty-two a woman loses her ability to reproduce. Ovulation ceases and so does menstruation. The uterus, breasts and genitalia decrease in size and sexual desire declines.

Twins. (a) *Identical twins*. Sometimes a developing embryo divides into two groups of cells at an early stage in cell division and the two parts each develop into a normal embryo, though often they share the same placenta. Such 'one-egg' twins are the same sex and are identical in nearly every physical respect, although differences in position and blood supply in the uterus may cause them to be different initially in weight and vigour.

(b) *Fraternal twins*. If two ova are released from the ovary and fertilized simultaneously, twins will result. These twins may be of different sexes and are not necessarily any more alike than other brothers and sisters of the same family.

About one in every eighty-eight births results in twins of which one-third are identical twins. Triplets and other multiple births may result from simultaneous fertilization of several ova or the separation of a single zygote into four or more cell masses, one of which subsequently fails to develop.

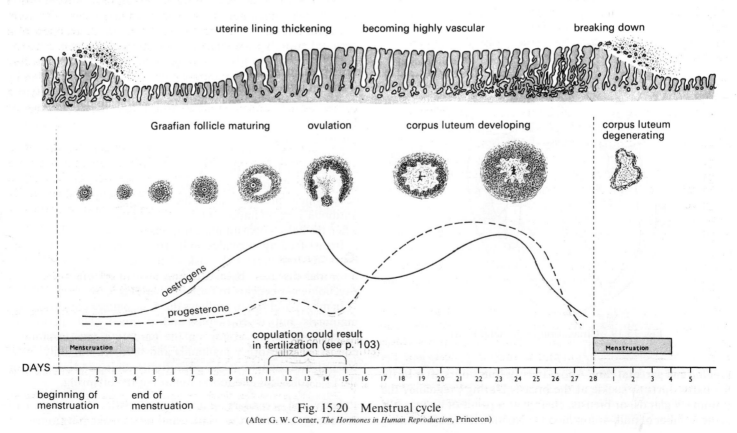

Fig. 15.20   Menstrual cycle
(After G. W. Corner, *The Hormones in Human Reproduction*, Princeton)

Family planning and fertility

As little as four weeks after giving birth, it is possible, though unlikely, that a woman may conceive again. Breast feeding may reduce the chances of conception. Nevertheless, it would be possible to have children at about one-year intervals from puberty to menopause. Most people do not want, or cannot afford to have, as many children as this. All human communities, therefore, practise some form of birth control to space out births and limit the size of the family. For most couples, particularly in countries where food and income are inadequate, it is better to use resources for a small number of well-nourished children than to raise a large family of underfed and sickly infants.

(a) **Natural methods of family planning.** If it were possible to know exactly when ovulation occurred, intercourse could be avoided for 3–4 days before, and one day after ovulation (*see* p. 103). At the moment, however, there is no simple, reliable way to recognize ovulation, although it usually occurs 12–16 days before the onset of the next menstrual period. By keeping careful records of the intervals between menstrual periods, it is possible to calculate a potentially fertile period of about 10 days in mid-cycle when sexual intercourse should be avoided if children are not wanted. On its own this method is not very reliable but there are some physiological clues which help to make it more accurate. During or very soon after ovulation, a woman's temperature rises about 0.5 °C. It is reasonable to assume that after 3 days of the elevated temperature a woman will be infertile. Another clue comes from the type of mucus secreted by the cervix. As the time for ovulation approaches, the mucus becomes more fluid. Women can learn to detect these changes and so calculate their fertile period.

By combining the 'calendar', 'temperature' and 'mucus' methods it is possible to achieve about 80 per cent 'success', i.e. only 20 per cent unplanned pregnancies. Advocates of the method may claim higher rates of success and, of course, it is a very helpful way of finding the fertile period for couples who want to conceive.

(b) **Artificial methods.** The main methods used are:
The sheath or condom. A thin rubber sheath is placed on the erect penis before sexual intercourse. The sheath traps the sperms and prevents them from reaching the uterus.
The diaphragm. A thin rubber disc, placed in the vagina before intercourse, covers the cervix and stops sperms entering the uterus. Condoms and diaphragms, used in conjunction with chemicals that immobilize sperms, are about 95 per cent effective.
Intra-uterine device (IUD). A small plastic strip bent into a loop or coil is inserted and retained in the uterus. Whether it interferes with fertilization or implantation is not certain. It is about 98 per cent effective but there is a small risk of developing uterine infections, particularly if sexual relationships are promiscuous.
The contraceptive pill. The pill contains chemicals which have the same effect on the body as the hormones oestrogen and progesterone. When mixed in suitable proportions, these hormones suppress ovulation and so prevent conception. The pills need to be taken each day for the twenty-one days between menstrual periods. There are many varieties of contraceptive pill in which the relative proportions of oestrogen- and progesterone-like chemicals vary. They are 99 per cent effective but long-term use of some types is thought to slightly increase the risk of cancer of the breast and cervix, though this is controversial at the moment.

Sterilization. Conception can also be prevented by simple surgical means. In the male, the sperm ducts are tied or cut (*vasectomy*); in the female, the oviducts are tied. Vasectomy is the more common operation, but either method prevents ova and sperms from meeting, and apart from this, sex life is unaffected. The operation in most cases is irreversible.

World population. With the increasing application of medical knowledge, fewer people are dying from infectious diseases. The birth rate, however, has not fallen off in proportion, with the result that the world population is doubling every fifty years or less. In the developing countries with limited natural resources, this leads to shortages of food and living space. In the industrialized countries, the population increase has contributed to the pollution of the environment as a result of waste disposal and intensive methods of food production.

There is clearly a physical limit to the number of people who can live on the Earth, though authorities differ in their estimates of this number. Therefore it seems essential, at the very least, to educate people (a) into accepting the need to limit their families and (b) in methods of achieving this. The alternative to voluntary population control is control by famine, disease and disaster.

Infertility. This condition may result from various causes, from blockage of the oviduct to failure to produce adequate amounts of living sperms. Some of these causes may be corrected by simple surgery or drugs.

Miscarriage and abortion. Miscarriage, or spontaneous abortion, are terms applied to the birth of a fetus at a stage too early to survive, e.g. after only three or four months' gestation. Treatment with the appropriate hormones may prevent this in women who have had a number of miscarriages. The term abortion is usually applied to the deliberate destruction of a fetus by drugs, its removal from the uterus by surgery or to the induction of its birth at a stage when it could not possibly survive. Abortion is a common practice as a measure of family limitation in some countries and it is also carried out when a continued pregnancy could endanger the mother's health or life.

Caesarian section. In some women, the gap between the sacral vertebrae and the front of the pelvic girdle is too small to permit the passage of the fetus at birth. In such cases, at the normal time for birth, the abdomen and uterus are opened surgically and the baby removed. Two or three children in succession may be born in this way.

Questions

1 In what ways does a zygote differ from any other cell in the body?
2 What is the advantage to the embryo of the early development of its heart and circulatory system?
3 What differences are there in the numbers, structure and activity of the male and female gametes in man?
4 How does the production of ova differ from the production of sperms?
5 List the changes in the composition of the maternal blood that are likely to occur when it passes through the placenta.
6 Explain why there are only a few days in each menstrual cycle when fertilization is likely to occur.
7 What changes in diet should be made by a woman once she becomes pregnant? Suggest reasons for these changes.
8 What are the advantages of breast-feeding over bottle-feeding?

16

The Skeleton, Muscles and Movement

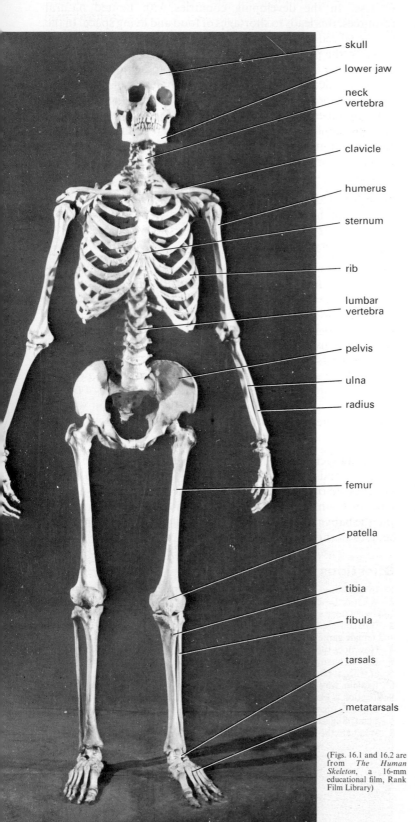

skull
lower jaw
neck vertebra
clavicle
humerus
sternum
rib
lumbar vertebra
pelvis
ulna
radius
femur
patella
tibia
fibula
tarsals
metatarsals

(Figs. 16.1 and 16.2 are from *The Human Skeleton*, a 16-mm educational film, Rank Film Library)

Skeletal tissues are hard substances formed by living cells. Usually they contain non-living mineral matter such as calcium salts. The structures made of such non-living material can nevertheless grow and change as a result of the activities of living cells which dissolve away and replace the hard materials.

Functions of the skeleton

The functions can be grouped conveniently under the headings of support, protection, movement and locomotion, and muscle attachment.

Support. In land-dwelling vertebrates a rigid skeletal support raises the body from the ground and allows rapid movement; it suspends some of the vital organs, prevents them from crushing each other, and maintains the shape of the body despite vigorous muscular activity (Fig. 16.1).

Protection. Certain delicate and important organs of the body are protected by a casing of bone. The brain is enclosed in the skull, the spinal cord in the vertebral column, while the heart, lungs, and, to some extent, the liver and spleen are surrounded by a cage of ribs between the sternum and spine (Fig. 16.2). The organs are thus protected from distortion resulting from pressure, and injury resulting from impact. The rib cage plays a positive part in the breathing mechanism (p. 89) in addition to protecting the organs in the thorax.

Movement. Many bones of the skeleton act as levers. When muscles pull on these levers they produce movements, such as the chewing action of the jaws, the breathing movements of the ribs and the flexing of the arms. Locomotion is the result of the co-ordinated action of muscles on the limb bones and is discussed more fully on p. 120. Movements of the skeleton require a system of joints and muscle attachments.

Muscle attachment. To produce effective movement of any part of the skeleton, the muscles must be attached securely to it at each end. One end of the muscle must be attached to the part of the skeleton to be moved while the other end is anchored to a part of the skeleton to be held stationary with respect to the moving part. This is discussed more fully on p. 118.

Fig. 16.1   Skeleton of man

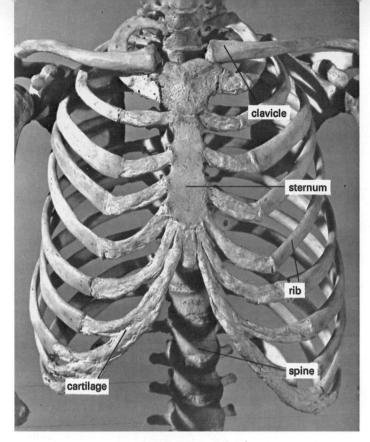

Fig. 16.2   Skeleton of the thorax

Cartilage, bone and muscle

Cartilage, typically, is a semitransparent, firm, elastic but somewhat gelatinous material found in the ear pinna, the epiglottis, the trachea and on the surface of bones where they form a joint. In mammals the skeleton of the embryo is first laid down as cartilage and replaced by bone before birth.

The gelatinous material or *matrix* of cartilage is secreted by cells called *chondrocytes*, which come eventually to lie in cavities within the matrix (Fig. 16.3 and 16.5a). The chondrocytes can divide and produce new cells which secrete

(Brian Bracegirdle/Biophoto Associates)

more matrix and so bring about the growth of cartilage. Such growth, in humans, takes place in the embryo when the skeleton is first formed in cartilage, and later, only at the junctions between the head and shaft of bones that are still growing in length. Capillaries do not penetrate the matrix and in consequence the chondrocytes must receive their food and oxygen by diffusion through the matrix.

Fibres of two kinds, tough, inextensible *collagen* fibres and stretchy, elastic fibres, are dispersed to varying degrees in the cartilaginous matrix. Since, for the most part, they run in different directions to each other, they prevent the cartilage splitting under pressure.

Elastic fibres are abundant in the ear pinna, while collagen fibres are predominant in the fibro-cartilage of a tendon insertion (Fig. 16.3b and p. 118); here, the fibres are lined up in the direction of the pull, thus transmitting the full force of muscular contraction to the bone.

Bone formation and structure. Most of the bones of the human embryo are formed in cartilage which is then replaced by bone, leaving cartilage only on the joint surface. The early cartilaginous model first of all undergoes calcification by precipitation of calcium phosphate in the matrix. Next, since the calcified matrix is impermeable to dissolved food and oxygen, the chondrocytes die and the calcified cartilage starts to disintegrate. Finally, cells called *osteoblasts* invade the space so formed (Fig. 16.4) and lay down bone on what is left of the calcified cartilage, using calcium and phosphate ions supplied by the blood. This is called *ossification*. At the same time, capillaries penetrate the cavities of the ossifying cartilage bringing food, oxygen and the necessary calcium and phosphate compounds with which the osteoblasts make bone. Eventually each osteoblast surrounds itself with bone and at this stage is called an *osteocyte*.

The process of ossification is more complicated than suggested above and the final product is by no means simply a homogeneous mineral deposit of calcium phosphate. The bone is penetrated by collagen fibres which give far more tensile strength to bone than a simple mineral could possess. If a

Fig. 16.3   Cartilage

(a) *left Hyaline cartilage (×50), e.g. from the tracheal rings. There are relatively few fibres in the matrix*

(b) *below Fibrous cartilage (×90), e.g. from a tendon insertion. The collagen fibres predominate*
(Brian Bracegirdle/Biophoto Associates)

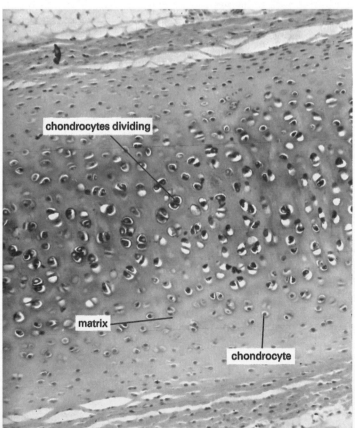

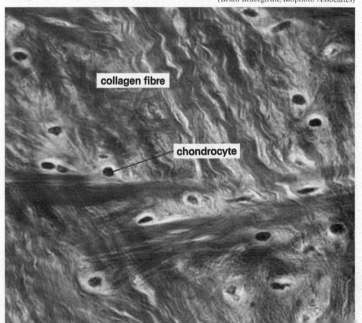

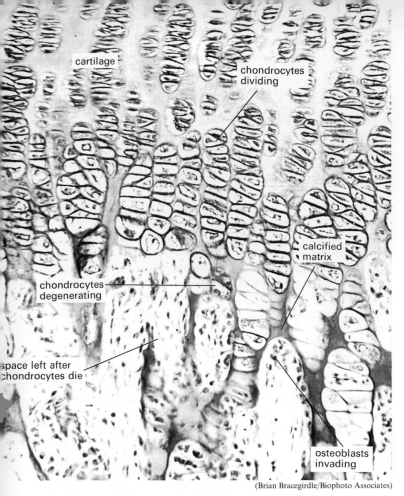

cartilage

chondrocytes dividing

calcified matrix

chondrocytes degenerating

space left after chondrocytes die

osteoblasts invading

(Brian Bracegirdle/Biophoto Associates)

Fig. 16.4   Zone of ossification (× 300)

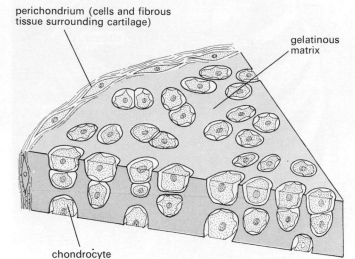

perichondrium (cells and fibrous tissue surrounding cartilage)

gelatinous matrix

chondrocyte

(a) Cartilage

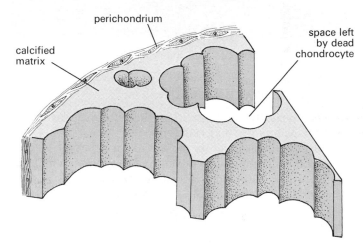

perichondrium

calcified matrix

space left by dead chondrocyte

(b) The chondrocytes die and the matrix is eroded

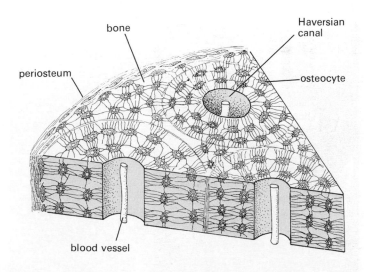

bone

Haversian canal

periosteum

osteocyte

blood vessel

(c) The spaces are invaded by osteoblasts which lay down bone in concentric cylinders

Fig. 16.5   Conversion of cartilage to bone (ossification)

bone is placed in dilute acid the calcium salts are dissolved out, leaving the collagen fibres and other connective tissues intact. The bone retains its shape but (while it is moist) it is quite flexible (Experiment 1, p. 123). Ossification is depicted diagrammatically in Fig. 16.5.

Certain bones, such as those roofing the skull, are not modelled first in cartilage but produced by ossification directly in the dermis of the skin. In the skull this gives rise to a series of plates which grow at the margins until they meet and join at the sutures to form a continuous skeletal system (Fig. 16.27).

Periosteum. Surrounding each bone is a tough, fibrous sheath. Its outer regions consist principally of collagen fibres, while its inner surface contains osteoblasts whose activities can produce growth in thickness of the bone. The periosteum is continuous across joints where it forms a capsule whose inner surface secretes synovial fluid which lubricates the surfaces of the joint (see p. 115).

Centres of ossification and growth of bone. The conversion of cartilage to bone in a limb begins first in the centre of the shaft and, shortly after, at the two ends which will form the heads or *epiphyses* of the bone. As ossification proceeds from these three centres (Fig. 16.6) the newly formed bone of the shaft approaches that of the epiphyses until only a narrow band of cartilage, the *metaphysis*, is left. It is in this region that the bone continues to grow in length. In the metaphysis, the chondrocytes divide and make the cartilaginous disc thicker. This new cartilage in its turn is ossified from the side nearest the shaft and so extends the shaft (Figs. 16.6*d* and 16.7). Only when the skeleton has attained its full proportions do the shaft and epiphyses fuse together (Fig. 16.8).

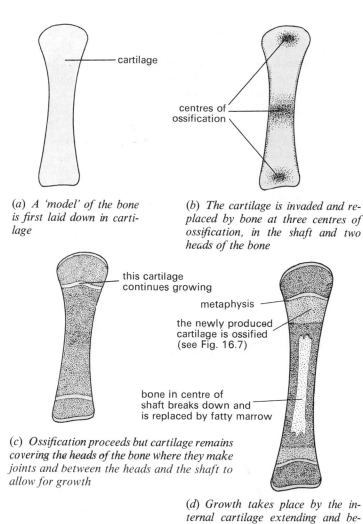

cartilage

(a) A 'model' of the bone is first laid down in cartilage

centres of ossification

(b) The cartilage is invaded and replaced by bone at three centres of ossification, in the shaft and two heads of the bone

this cartilage continues growing

metaphysis

the newly produced cartilage is ossified (see Fig. 16.7)

bone in centre of shaft breaks down and is replaced by fatty marrow

(c) Ossification proceeds but cartilage remains covering the heads of the bone where they make joints and between the heads and the shaft to allow for growth

(d) Growth takes place by the internal cartilage extending and becoming ossified

Fig. 16.6   The growth of a long bone

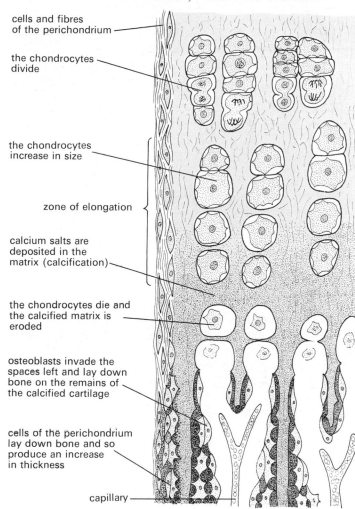

cells and fibres of the perichondrium

the chondrocytes divide

the chondrocytes increase in size

zone of elongation

calcium salts are deposited in the matrix (calcification)

the chondrocytes die and the calcified matrix is eroded

osteoblasts invade the spaces left and lay down bone on the remains of the calcified cartilage

cells of the perichondrium lay down bone and so produce an increase in thickness

capillary

Fig. 16.7   Growth and ossification of a long bone

(b) Adult (epiphyses fused)

(a) Six years old (epiphyses unfused)

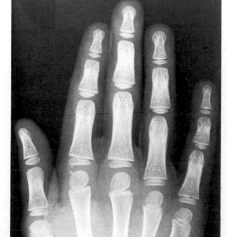

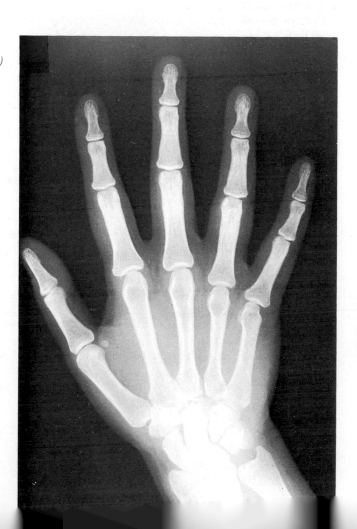

Fig. 16.8   X-ray photograph of the hand

(St Bartholomew's Hospital, Department of Medical Illustration)

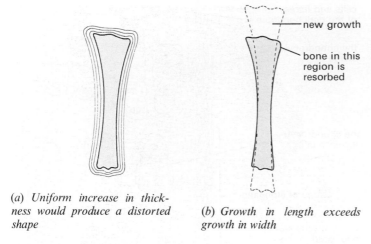

(a) *Uniform increase in thickness would produce a distorted shape*

(b) *Growth in length exceeds growth in width*

Fig. 16.9   Pattern of growth in long bones

It is clear from Fig. 16.9 that increase in length from the metaphysis and increase in thickness from the periosteum do not account for the necessary change in the shape of the bone as it grows, particularly the change in the expanded ends of the shaft. During growth, parts of the bone are added to as described above, and parts are dissolved and resorbed. There is no universal agreement on the precise mechanism by which resorption takes place. It may be that parts of the bone are dissolved away by tissue fluid or that special cells, the *osteoclasts*, erode away the areas of bone.

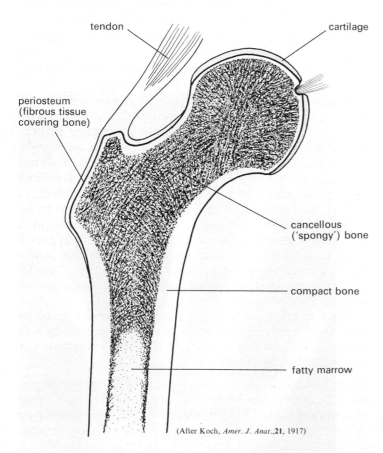

(After Koch, *Amer. J. Anat.*, **21**, 1917)

Fig. 16.10   Section through head of femur

Bone resorption also takes place in the centre of the shaft, leaving a cylindrical space occupied by marrow. The finally formed bone may be of two kinds, *compact* or *cancellous*. Compact bone is formed on the outside of the shafts and epiphyses, while cancellous bone is in the centre of the epiphyses and the centre of most bones other than the long limb bones, e.g. in the ribs and vertebrae. Cancellous bone appears as a network of thin, bony columns, *trabeculae*, interspersed with the tissues that form the blood cells (Fig. 16.10).

In its final form, bone consists of living osteocytes occupying spaces, *lacunae*, in the ossified matrix which they have produced (Fig. 16.11). Withdrawal of fine cytoplasmic processes which

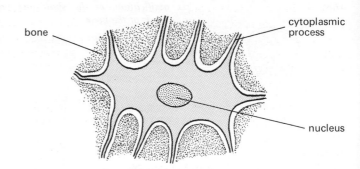

Fig. 16.11   Single osteocyte

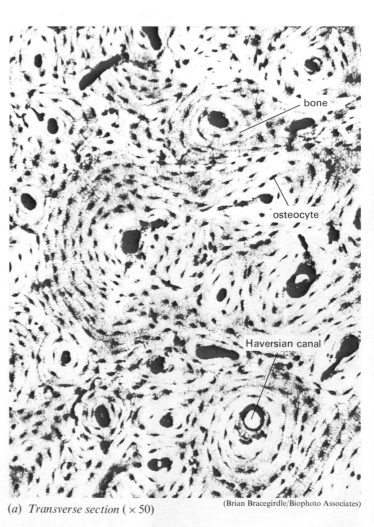

(a) *Transverse section* ($\times 50$)

(Brian Bracegirdle/Biophoto Associates)

Fig. 16.12   Photomicrographs of sections through bone tissue (the

once united the osteocytes leaves a delicate network of channels through which tissue fluid can penetrate, bringing nutrients, salts and oxygen to the osteocytes and removing their excretory products. The osteocytes are so arranged that they form cylinders of bone, not more than six cells thick (Figs. 16.12 and 16.5c); through these cylinders run capillaries which extend into the bone from the periosteum, having been incorporated into the bone tissue as it was laid down by the periosteum. The presence of this capillary network allows the bone to continue growth, repair fractures and effect a continual turnover of bone substance. Injection of radioactive phosphorus into experimental animals shows that phosphate in the blood is incorporated into existing bones within a few hours even though they have ceased growing. After about three weeks, the radioactive phosphorus has almost entirely disappeared from the bone. These results indicate that bone tissue is not an inert solid but in a state of constant change, its minerals being continually removed and replaced.

Rickets. If either calcium, vitamin D, or both are deficient in the diet, bone growth continues but ossification is inadequate. That is, the osteoblasts lay down a fibrous matrix that is not hardened by deposition of calcium salts. Such bones, if subjected to normal stresses of muscular action or body weight, distort easily leading eventually to crippling deformities (p. 35). Absence of

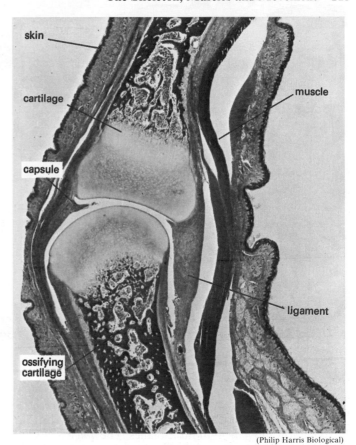

(Philip Harris Biological)

Fig. 16.13   Finger joint of human foetus

vitamin C also prevents normal bone formation in the shafts of bones, which are subsequently more liable to fracture.

Fractures. That bone can so readily repair fractures is evidence of its living properties and adaptability to new situations and stresses. When a limb bone is fractured, blood leaking from the torn ends of capillaries forms a clot which occupies the damaged area. Macrophages invade the clot and ingest red cells and cell debris. Shortly after, capillaries penetrate the clot. Osteoblasts in the periosteum round the margins of the fracture begin to reproduce rapidly both outside the shaft and in the marrow cavity. These osteoblasts start to lay down bone on the shaft and also migrate towards the blood clot where they lay down cartilage, forming a thickened layer or *callus* over the break, so uniting the bones. After this, ossification takes place in the normal way, with calcification, resorption and deposition of bone. The bone at the broken ends has died, probably as a result of the cessation of the normal blood supply. This dead bone is resorbed and replaced as the callus forms and, finally, the callus itself is resculptured by resorption to conform to the normal shape of the limb.

The time taken for healing depends on the diameter of the bone, its blood supply and the age of the individual. A rib heals quickly, a femur slowly, particularly the neck of the femur with its restricted blood supply.

Joints. Where two bones meet a joint is formed. Sometimes, as in the sutures between the bones of the skull, no movement is permitted (Fig. 16.27); in others, e.g. the vertebrae of the spine, only a very limited movement can occur; while the most familiar joints, called *synovial joints* (Fig. 16.13), allow a considerable degree of movement. The *ball and socket joints* of

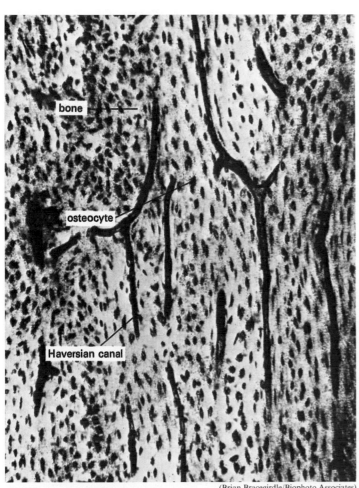

(*b*) *Longitudinal section* (× 20)

(Brian Bracegirdle/Biophoto Associates)

canals are blocked with dust produced while cutting the section)

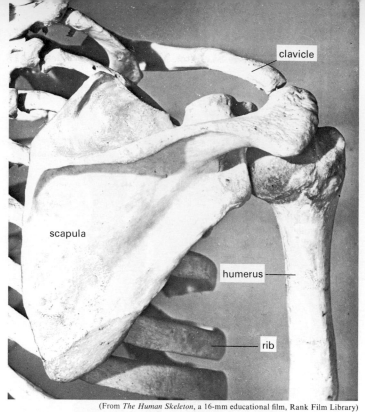

Fig. 16.14   The shoulder joint

of fibrous tissue, continuous with the periosteum, surrounds the joint, secreting and retaining the synovial fluid.

Between the vertebrae there are *cartilaginous joints* which allow a small degree of movement so that the vertebral column, as a whole, is flexible. The vertebrae are bound together by ligaments but separated by *intervertebral discs* (Fig. 16.19). These consist of a semi-liquid centre enclosed in a tough, fibrous, cartilaginous capsule.

Girdles. To produce movement of the body as a whole, the upwards or backwards thrust of the limbs against the ground must be imparted to the body. The force is transmitted through a skeletal structure called a girdle which is attached to the spinal column. The *pelvic* girdle (Fig. 16.20) is rigidly fused to the base of the spine so that in walking or jumping the force of the leg thrust is transmitted to the spine, which is the central support of the whole body. By this means also the weight of the body is supported when at rest. The shoulder blades which form part of the *pectoral* girdle (Fig. 16.14) are not fused to the spine but bound by muscles to the back of the thorax. This is not so effective a means of imparting a force to the body, but this function is not so important with the arms as it is with the legs and the free movement of the shoulders allows greater mobility of the arms. The shoulder in man is more mobile than in most mammals and the clavicle serves to limit these movements.

the humerus and scapula at the shoulder (Figs. 16.14 and 16.16*a*) and the femur and pelvis at the hip (Figs. 16.15 and 16.16*b*) allow movement in three planes. The *hinge joints* of the ulna and humerus at the elbow (Fig. 16.17), and the femur and tibia at the knee (Fig. 16.18) allow movement in only one plane.

The surfaces of the heads of bones, which move over each other, are covered with cartilage which is slippery and smooth. This, together with the liquid in the joint, called synovial fluid, allows friction-free movement. The relevant bones of the joint are held together by strong *ligaments* (Figs. 16.16*b* and 16.18*b*) which prevent dislocation during normal movement. A *capsule*

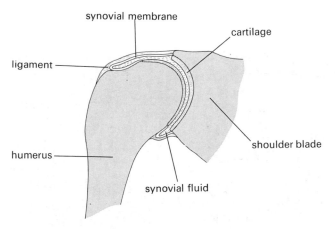

(*a*) *Shoulder (in section)*

Fig. 16.15   The hip joint

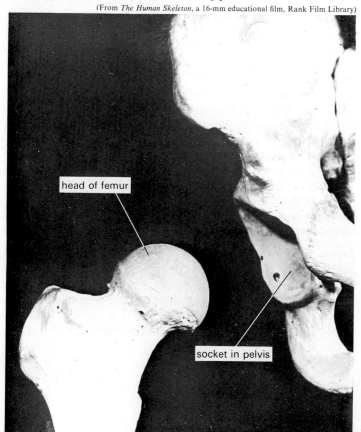

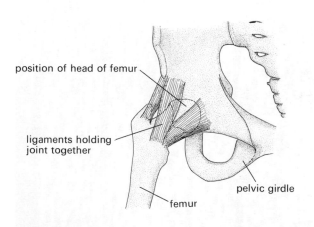

(*b*) *Hip, external view (muscles removed)*

Fig. 16.16   Ball and socket joints

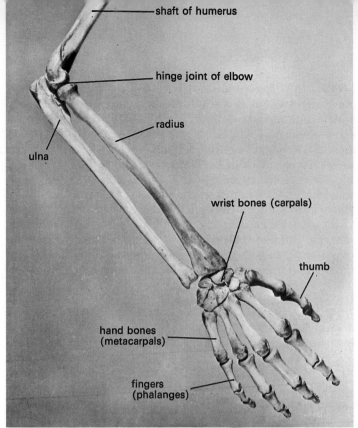

shaft of humerus

hinge joint of elbow

radius

ulna

wrist bones (carpals)

thumb

hand bones (metacarpals)

fingers (phalanges)

(From *The Human Skeleton*, a 16-mm educational film, Rank Film Library)

Fig. 16.17 Skeleton of the forearm

joins skull

7 neck vertebrae

12 thoracic vertebrae

5 lumbar vertebrae

5 sacral vertebrae

fused to pelvic girdle

tail vertebrae

(a) *Regions of the spine*

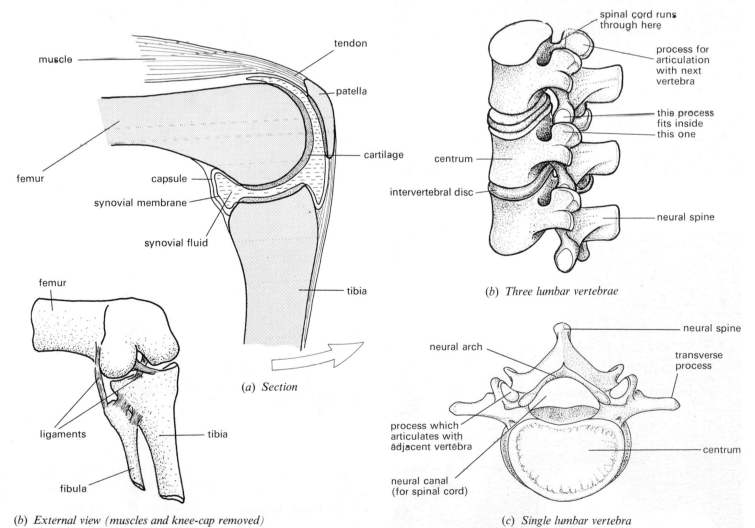

muscle

tendon

patella

cartilage

femur

capsule

synovial membrane

synovial fluid

tibia

(a) *Section*

femur

ligaments

tibia

fibula

(b) *External view (muscles and knee-cap removed)*

Fig. 16.18 Hinge joint of the knee

spinal cord runs through here

process for articulation with next vertebra

this process fits inside this one

centrum

intervertebral disc

neural spine

(b) *Three lumbar vertebrae*

neural spine

neural arch

transverse process

process which articulates with adjacent vertebra

neural canal (for spinal cord)

centrum

(c) *Single lumbar vertebra*

Fig. 16.19 The spinal column

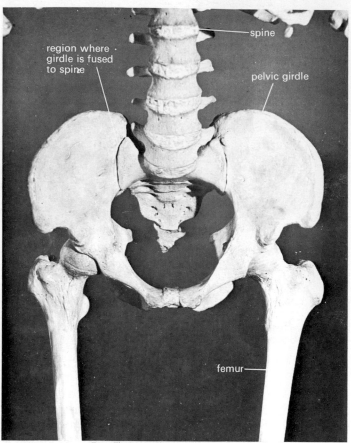

region where girdle is fused to spine

spine

pelvic girdle

femur

(From *The Human Skeleton*, a 16-mm educational film, Rank Film Library)

Fig. 16.20 The pelvic girdle

Muscle. (a) *Voluntary muscle.* Contractions of voluntary muscle are brought about as a result of nerve impulses reaching the muscle (although the contraction itself may often be described as involuntary if it results from a simple reflex). Voluntary muscle consists of elongated fibres up to 40 mm long and 10–40 microns in diameter. The fibres are grouped into bundles and are enclosed in sheaths of connective tissue (Figs. 16.21 and 16.22). At the ends of the limb muscles, the connective tissue sheaths are drawn out to form *tendons* which are attached to the periosteum covering the bone. At a tendon insertion collagen fibres of the periosteum are embedded in the bone substance giving a very firm anchorage. Muscle fibres if stimulated by a nervous impulse will contract to about two thirds or one half of their resting length. This makes the muscle as a whole shorter and thicker and, according to its attachment at each end, it pulls on a bone and so produces movement.

Muscles that produce movement usually act across joints in such a way that the bones are worked as levers with a low mechanical advantage. Fig. 16.23 makes this clear. The muscles can only contract a short distance, but since they are attached near a joint the movement at the end of a limb is greatly magnified. The biceps muscle of the arm may contract only 80 or 90 mm but the hand will move about 60 cm (Fig. 16.24).

Muscles can only contract and relax, they cannot elongate. A muscle that has contracted has to be pulled back to its resting length when it relaxes. Consequently, most muscles are in pairs, one of them producing movement in one direction and

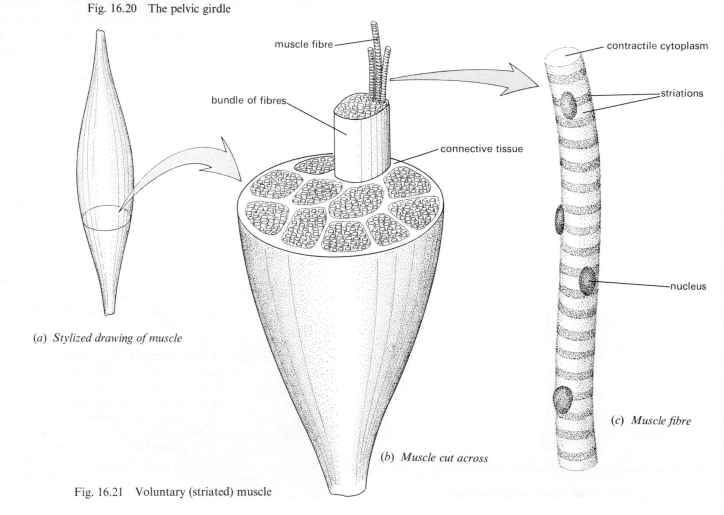

muscle fibre

bundle of fibres

connective tissue

contractile cytoplasm

striations

nucleus

(a) *Stylized drawing of muscle*

(b) *Muscle cut across*

(c) *Muscle fibre*

Fig. 16.21 Voluntary (striated) muscle

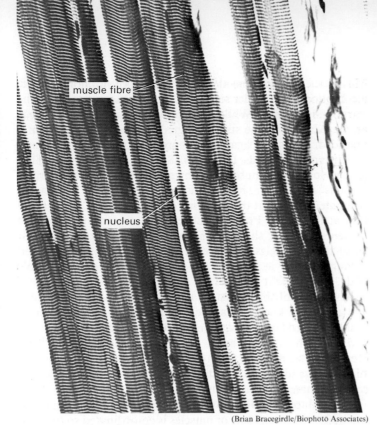

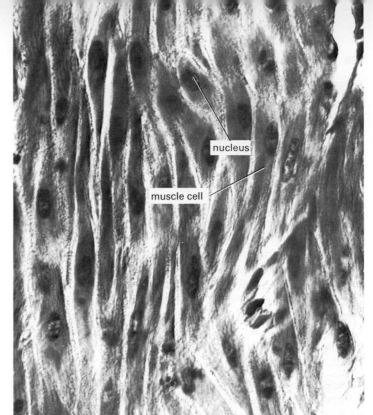

(Brian Bracegirdle/Biophoto Associates)

Fig. 16.22 Photomicrograph of striated muscle (× 500)

(Brian Bracegirdle/Biophoto Associates)

Fig. 16.25 Photomicrograph of unstriated muscle (× 800)

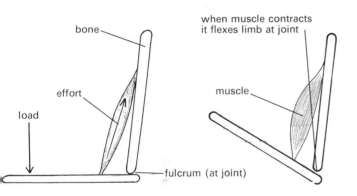

Fig. 16.23 Lever action of limb

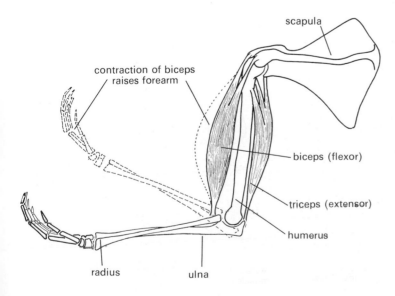

Fig. 16.24 Antagonistic muscles of the forearm

the other producing the opposite movement. When one partner contracts, it will extend the other if it is in a relaxed state. Such pairs of muscles are called *antagonistic*: in the case of a limb they may be called *flexor* and *extensor* muscles, according to whether they bend (flex) or straighten (extend) the limb; they may be called *adductors* and *abductors* if they pull a limb towards or away from the body respectively. Often, the antagonistic muscles are in groups; e.g. both the *brachialis* and the *biceps* muscles flex the arm at the elbow and antagonize the *triceps*, but the biceps does so only when the palm is facing upwards. The biceps serves, therefore, both to rotate the lower arm and to flex the elbow. Combinations of other pairs of muscles acting at the joints may produce rotary or circular movements.

Of such pairs or groups of antagonistic muscles, one is usually much stronger than the other. The biceps which flexes the arm is larger and more powerful than the triceps which extends it; the *gastrocnemius* and *soleus* of the calf muscles which extend the foot at the ankle are stronger than the *peroneus* muscles which flex it.

Locomotion is brought about by the co-ordinated movement of limbs, by sets of antagonistic muscles contracting and relaxing alternately. When the body is at rest, a proportion of the fibres in the antagonistic muscles remain in a state of contraction producing a *muscular tone* which holds the body in position.

(b) *Involuntary muscle.* The muscle in the walls of the arteries, arterioles and the alimentary canal has a structure different from that of voluntary muscle. It consists of elongated cells rather than fibres (Figs. 16.25, and 2.7c, p. 10) and is arranged in layers rather than in discrete muscles. The distribution of cells in circular and longitudinal muscle layers, however, allows an antagonistic action. Contraction of cells running in a circular manner round a cylindrical organ causes a reduction in diameter, while shortening of the longitudinally arranged cells brings about a reduction in length and, in some cases, an increase in diameter.

119

Involuntary muscle produces rather slow contractions which may be under the control of the autonomic nervous system (p. 151) as in vasoconstriction (p. 99), or stimulated in other ways to produce rhythmic contractions, as in peristalsis (p. 62). The bladder, ureters, arteries, arterioles and alimentary canal all contain involuntary muscle. Contraction of this muscle cannot be brought about by the conscious will.

Localized areas of circular muscle form sphincters such as those at the exit of the stomach, rectum or bladder. The latter two are usually under conscious control.

(c) *Heart muscle*. This tissue is different from both voluntary and involuntary muscle. Its rhythmic contraction can take place without nervous stimulation although the heartbeat is under the control of the central nervous system. All muscle fibres can conduct electrical impulses to some extent, but in the heart muscle this property is of particular importance since the majority of cells have no direct contact with nerves. The impulse that causes the muscle cells to contract in unison is conducted by the cells themselves.

Muscle attachment. The skeletal muscles must be attached to the limb bones at one end in order to produce movement but, in addition, they must have a rigid attachment at the other end so that only one part of the limb moves when the muscle contracts. Sometimes the 'stationary' end is attached to the upper part of the limb; e.g. the extensor muscles which extend the foot are attached to the top of the tibia or lower end of the femur. The muscles that move the femur, however, are attached to the pelvic girdle. Bones frequently have projections or ridges where muscles are attached (Figs. 16.26 and 16.27).

Locomotion

When a quadruped walks, its four limbs move in a co-ordinated sequence, each one in turn thrusting backwards on the ground, so propelling the animal forwards. Usually, one fore-limb and the opposite hind-limb are moved forwards while the body is supported on its other two limbs. Man has to maintain his upright posture on one leg while the other leg is swinging forwards during walking. Consequently, the action of the leg muscles during walking is rather complicated to describe and it will be clearer to consider how muscles and bones effect locomotion in a simpler situation, in this case, a runner at the start of a sprint race. While waiting for the starting gun, the legs are flexed and, on the starting signal, one of them is violently extended, thrusting against the ground to push the runner upwards and forwards. In Fig. 16.28 a simplified diagram of the leg and some of its muscles is shown. When muscle A contracts it pulls the femur backwards. Contraction of muscle B straightens the leg at the knee and contraction of C extends the foot at the ankle. Friction between the ground and the foot prevents the foot slipping backwards so that contraction of these three muscles produces a thrust which is transmitted through the pelvic girdle to the spine and thence to the whole body, propelling it forwards.

At the end of this thrusting leap, muscles A, B and C relax and allow the knee and ankle to be flexed and the femur to swing forwards by contraction of antagonistic muscles to A, B and C, of which only b and c are shown in the diagram. There are, of course, many more muscles involved than are shown in Fig. 16.28 and their co-ordinated action is more subtle and complex than can be described here.

To produce effective movement, it is essential that the contraction of the many sets of muscles is co-ordinated so that, for example, antagonistic muscles do not contract simultaneously. Contributing to this co-ordination there is a system of stretch receptors (p. 131) in the muscles; these fire nervous impulses to the spinal cord when the muscle is being stretched. Such internal sensory organs or *proprioceptors*, linked to the nervous system, feed back information to the brain about the position of the limbs and enable a pattern of muscular activity to be computed by the brain, so producing effective movement.

Fatigue. After repeated contractions over a prolonged period, a set of muscles may become fatigued. That is, they feel uncomfortable and will not go on working so fast as before. The cause of muscle fatigue is not known. It could be that the store of glycogen in the muscle is run down or the blood circulation cannot deliver oxygen and glucose fast enough or remove the

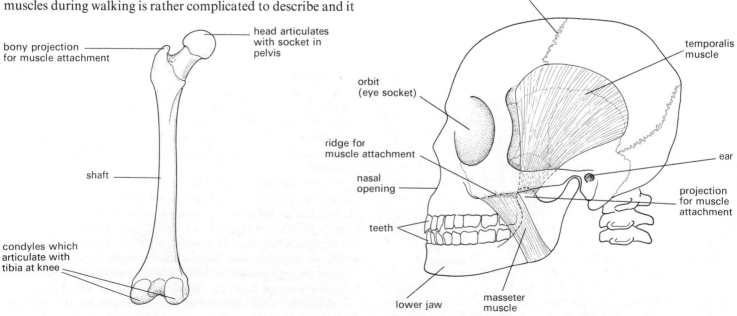

Fig. 16.26   The femur

Fig. 16.27   The skull and chewing muscles

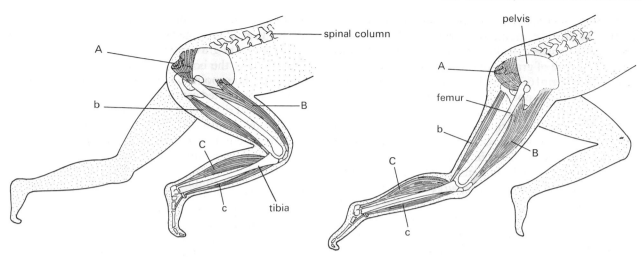

Fig. 16.28   Action of leg muscles in producing movement (the action could be the start of a sprint; only a few of the muscle systems have been drawn)

waste products, e.g. lactic acid (p. 28) fast enough. It is known that the level of lactic acid in the blood of athletes remains lower than in an untrained person during exercise, but this does not prove that lactic acid causes muscle fatigue.

Similarly, it is not possible to identify a single known cause of the general fatigue (exhaustion) of the whole body after hard work or exercise. Your rate of activity may be limited by the ability of your heart and blood system to deliver oxygen and glucose to the tissues, or to remove carbon dioxide and lactic acid. The nerves which supply the muscles may work less well or the brain may stop sending out impulses efficiently.

Your psychological state also influences the sensations of fatigue. The losers of a race usually feel far more fatigued than the winners.

Cramp is an involuntary and painful contraction of one or more muscles (muscular spasm), usually in a limb. One form is caused by loss of salt as a result of prolonged sweating. Other cramps occur after prolonged exertion or for no apparent reason.

Posture

A jointed skeleton alone cannot maintain posture, upright or otherwise, and if a man were to relax all his muscles, he would buckle at the knees and ankles, his spine would sag and he would fall to the ground. The upright posture is maintained mainly by the simultaneous contraction of appropriate sets of antagonistic muscles.

Fig. 16.29 shows some of the antagonistic muscles of the leg which contribute to walking and standing. It can be seen that contraction of the *tibialis anterior* would flex the ankle (i.e. tend to raise the foot) whereas contraction of the *soleus* and *gastrocnemius* (not shown) would extend it. When both sets of muscles contract or, rather, maintain themselves in a state of tension (*isometric contraction*), the tibia is held at about 90° to the foot. Similarly the controlled tension in the *rectus femoris* and the *biceps femoris* keeps the leg straight at the knee, while the *gluteus maximus* and the rectus femoris, acting together, prevent the pelvic girdle from rotating forward or backward. Thus, since the spine is fused to the pelvis, the trunk is kept upright.

If this were the whole story, a great deal of muscular energy would have to be expended just to keep the body stationary and upright but in fact the energy expenditure in maintaining posture is greatly reduced by (a) using only part of each muscle, (b) exploiting the ligaments and (c) balance.

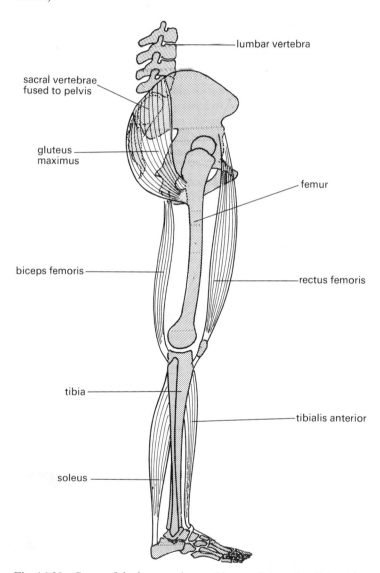

Fig. 16.29   Some of the leg muscles used in standing and walking (the many other muscle systems of the leg have been omitted for clarity)

(a) The postural tension of a muscle is maintained by only a small number of its fibres and this fraction is constantly changing with the result that fatigue is minimal.

flame, heating it strongly for 2 minutes. At first the bone will char and then glow red as the organic material burns away, but it will retain its shape. Allow the bone to cool and then try crushing the heated end against the bench with the end of a pencil.

Result. Both bones retain their shape after treatment, but the bone whose mineral component had been dissolved in acid is rubbery and flexible because only the organic, fibrous connective tissue is left. The bone that had this fibrous tissue burned away is still hard but very brittle and easily fragmented.

Interpretation. The combination in bone of mineral salts and organic fibres produces a hard, strong and resilient structure.

Questions

1   Construct a diagram similar to Fig. 16.18*a* to show a section through the elbow joint, using Fig. 16.17 for guidance. Show the attachment of the biceps and triceps tendons and state where you would expect to find the principal ligaments.
2   Write a list of the functions of cartilage in the human body (a) during growth and (b) at maturity.
3   Distinguish between ligaments and tendons with respect to their position and their function.
4   What is the principal action of (a) your calf muscle, (b) the muscle in the front of your thigh, and (c) the muscles in your forearm? If you don't already know the answers, try making the muscles contract and feel where the tendons are pulling.
5   Unlike most mammals, man stands upright on his hind-legs. What difference do you think this has made to his skeleton and musculature?

17
Teeth

Teeth are produced from the skin where it covers the upper and lower jaws. The crowns of the teeth break through the skin into the mouth cavity while their roots are enclosed in the bony sockets of the jaw. Human teeth begin their formation in the sixth week of embryonic development and start to erupt, i.e. emerge through the gum, at about 6 months after birth.

The first set of twenty *deciduous* or *milk teeth* eventually fall out and are replaced by thirty-two permanent teeth, between the ages of six and eighteen.

Structure

The generalized structure of human teeth is best shown in longitudinal section as in Figs. 17.1 and 17.2.

Enamel. This is a hard, brittle, non-living layer up to 2.5 mm thick, 96 per cent of which consists of a mineral comprising chiefly calcium and phosphate arranged in microscopic six-sided prisms. The enamel covers the crown of the tooth, forming a hard biting surface.

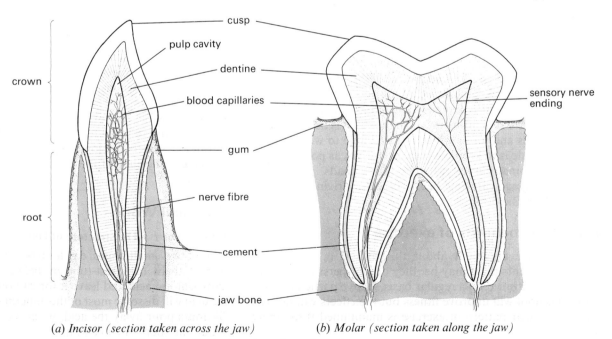

(a) Incisor (section taken across the jaw)   *(b) Molar (section taken along the jaw)*

Fig. 17.1   Vertical sections through teeth

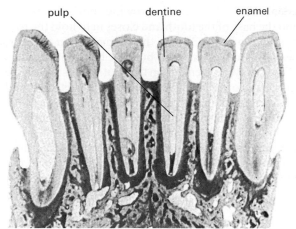

pulp   dentine   enamel

(G.B.I. Laboratories Ltd)

Fig. 17.2   Section through teeth and jaw of cat

Shape and function of teeth

The shapes of the teeth vary according to the part of the jaw in which they occur. In carnivorous mammals, such as the dog, the shapes of the front, side and back teeth differ greatly. The different teeth are specialized for holding the prey, cutting off flesh and cracking bones.

Human teeth show less extreme specialization (Fig. 17.4). The four chisel-like incisors in the front of each jaw pass each other and take bites out of solid food. Next to the incisors are two canine teeth, sharp and pointed but little longer than the incisors and much reduced compared with carnivorous mammals or apes. They seem to have a function similar to the incisors. Canines and incisors have single roots. Next come two premolars on each side, with two cusps (blunt points) and usually single roots. The three molar teeth on each side have

Dentine. Dentine is similar to bone in its structure. It is hard but not so brittle as enamel, having a higher proportion of fibrous material. It also has branching strands of living cytoplasm penetrating it. The cells, *odontoblasts*, from which these cytoplasmic strands extend, are able to add more dentine to the inside of the tooth (Fig. 17.3).

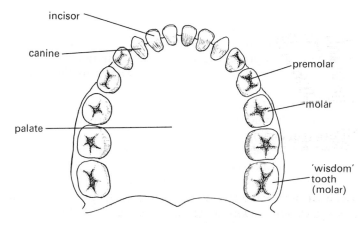

incisor

canine

premolar

molar

palate

'wisdom'
tooth
(molar)

Fig. 17.4   Arrangement of teeth in man's upper jaw

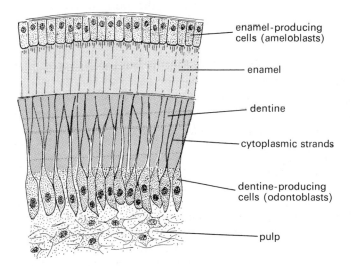

enamel-producing
cells (ameloblasts)

enamel

dentine

cytoplasmic strands

dentine-producing
cells (odontoblasts)

pulp

Fig. 17.3   Section through enamel and dentine of unerupted tooth

four cusps and in the upper jaw usually three roots, in the lower jaw two roots. The upper and lower molars and premolars meet, the cusps in one set fitting into the depressions of the other, so crushing the food (Fig. 17.5)

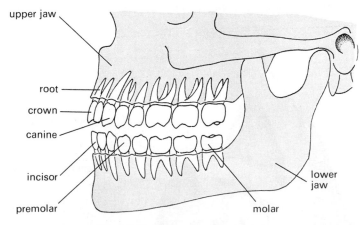

upper jaw

root

crown

canine

incisor

premolar

lower
jaw

molar

Fig. 17.5   Human jaws and teeth

Pulp. In the centre of the tooth is soft connective tissue called pulp. The pulp contains blood capillaries and sensory nerve endings which may penetrate the dentine. Oxygen and food brought by the blood enable the tooth to live and grow. The nerve endings respond mainly to changes in temperature, but produce only the sensation of pain.

Root. The root is not set rigidly in the jaw bone but is slung by collagen fibres. This method of suspension may prevent the crushing of the blood vessels and nerve fibres entering the tooth, when biting forces are applied.

Cement is a thin layer of bone-like material covering the dentine at the root of the tooth. The fibres that hold the tooth in the jaw are embedded in the cement at one end and in the jaw bone at the other.

The deciduous teeth (milk teeth) consist of 4 incisors, 2 canines and 4 molars in each jaw. As the permanent teeth develop, the roots of the deciduous teeth are resorbed (Figs. 17.6 and 17.7e) so that the teeth eventually fall out quite easily. The position of the deciduous teeth seems to determine the final arrangement of the permanent teeth so the former need to be allowed to complete their natural course of development.

Where the deciduous teeth have to be extracted as a result of decay, the permanent teeth may grow with unsatisfactory spacing. The first permanent teeth, i.e. the first molars, erupt at about six years, the second at about twelve and the last or 'wisdom' teeth usually after the seventeenth year or sometimes not at all. The deciduous teeth are usually all replaced between six and twelve years.

Development

During the development of the embryo, groups of cells in the epithelium of the mouth, overlying the jawbone rudiments, become especially active and give rise to tooth buds. In these, the epithelium thickens and grows into the tissues beneath (Fig. 17.7*a*). This ingrowth takes on an inverted cup-like form, the inner surface of which secretes enamel and is called the *enamel organ*. The cone of tissue enclosed by the cup is called the *dental papilla* and its outer layers produce dentine. The very elongated cells in the enamel organ become calcified by deposit of calcium phosphate. This process results in the formation of rods of enamel which are laid down at right angles to the tooth's surface and then fuse (Fig. 17.7*b* and *c*).

The outermost cells, the odontoblasts, of the dental papilla form dentine in much the same way as bone is produced (p. 112). A matrix containing collagen fibres forms between the cells and this matrix is then calcified, leaving strands of cytoplasm running from the odontoblasts through fine tubes in the calcified matrix. New dentine is added to the inner surface while the odontoblasts retreat deeper into the pulp. Even before the deciduous tooth erupts, the rudiment of the permanent tooth is forming alongside it (Fig. 17.7*c*).

Experiments in which whole tooth buds, the isolated enamel organ, or the dental papilla have been grafted into different parts of an animal show that the shape of the tooth is determined by the pattern made by the enamel organ as it grows inwards from the epidermis.

Fig. 17.6   Photomicrograph of milk and permanent tooth of kitten

(Gene Cox)

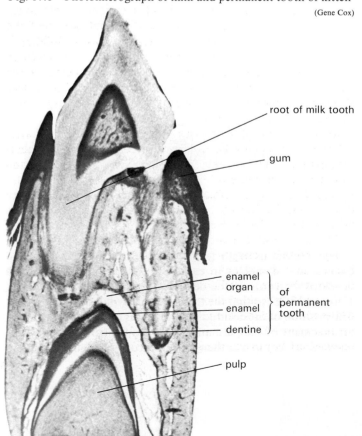

root of milk tooth

gum

enamel organ
enamel
dentine
} of permanent tooth

pulp

Eruption. While the tooth is growing, bone forms in the jaw around the base of the tooth and closes in to hold the root firmly in the socket of bone. Where the enamel organ covers the root it does not produce enamel, but just before it disintegrates it puts down a layer of cement on the root. In this region the *dental sac*, a sheath of connective tissue which has surrounded the developing tooth, enamel organ, dental papilla, etc., forms the collagen fibres which suspend the root in the socket. It is not clear what brings about eruption of the tooth but the outcome is that the crown pushes through the gum. The ruptured top of the enamel organ fuses with the epithelium of the gum forming a seal around the base of the tooth where it emerges from the gum.

Dental caries, plaque and periodontal disease

Caries is the technical name for tooth decay; *periodontal disease* is an infection of the gums and tooth sockets. Caries results in the erosion of the enamel and dentine of the teeth, leading to the formation of cavities. The absence of enamel and the thinner layer of dentine in these cavities results in the tooth being far more sensitive to temperature changes. If the cavities are untreated, the pulp may be invaded by pathogenic bacteria which ultimately cause a painful abscess at the root. Sometimes this can only be cured by extracting the tooth. If bacteria gain access to the circulatory system, they may cause septicaemia (blood poisoning).

Causes of caries. The principal theory put forward to explain how cavities arise suggests that the bacterial enzymes act on sugar and starch in the mouth and produce acid, mainly lactic acid, as a by-product. This acid dissolves the calcium salts in the teeth, so forming cavities (Fig. 17.8).

Plaque. Within 20 minutes of cleaning the teeth, a coating of saliva containing mucus and other substances forms over the teeth. At first, this layer contains few bacteria, but eventually it is colonized by micro-organisms, even if no food is taken in. These bacteria and their products form a layer of *plaque* over the teeth. The composition of plaque is 70 per cent bacteria and 30 per cent organic substances which are either produced by the bacteria or deposited from material in the mouth. This organic material helps the bacteria to adhere more securely to the tooth surface.

In this bacterial plaque are produced the enzymes, acids and toxins which might contribute to caries and gum disease. The population of micro-organisms changes as the plaque grows thicker, but perhaps two of the most important bacteria are *Lactobacillus acidophilus*, which produces lactic acid, and *Streptococcus mutans*. When the plaque has colonized most of the exposed tooth surface, it begins to spread downwards between the tooth and the gum and so causes gum inflammation or *gingivitis*.

If the bacterial plaque is not removed, it becomes mineralized by the incorporation of the phosphate and carbonate salts of calcium and magnesium. This mineralized plaque, called *calculus* or 'tartar', cannot be removed by normal teeth cleaning but has to be dispersed by scaling at the dentist's.

Refined sugar, sucrose. The intake of refined sugar has long been associated with dental caries and it has now been shown

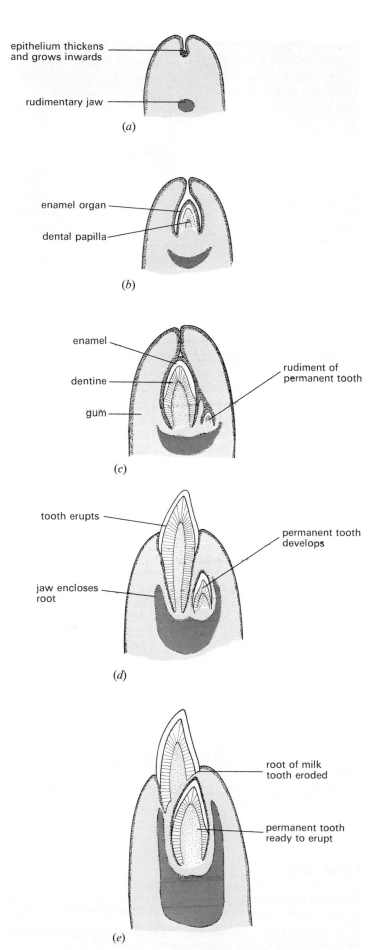

epithelium thickens
and grows inwards

rudimentary jaw

(a)

enamel organ

dental papilla

(b)

enamel

dentine

gum

rudiment of
permanent tooth

(c)

tooth erupts

permanent tooth
develops

jaw encloses
root

(d)

root of milk
tooth eroded

permanent tooth
ready to erupt

(e)

Fig. 17.7 Sections through lower jaw to show development and eruption of tooth

experimentally that sucrose increases the ability of *Streptococcus mutans* to colonize tooth surfaces by forming sticky polysaccharides which stick to the smooth surfaces of the teeth.

Periodontal disease. There is good evidence to connect the degree of plaque formation on teeth and the development of gum disease. The bacterial plaque spreads down the tooth and into the crevice at the junction of tooth and gum leading to inflammation of the gum (gingivitis). Gingivitis may not result in easily recognizable symptoms but leads at first to bleeding of the gums, bad breath and gum recession in which the gum retreats from the base of the tooth, exposing the cement. 'Getting long in the tooth' is a common expression for gum recession. However it is not a natural consequence of ageing but a result of chronic gingivitis. If treatment or preventive measures for gingivitis are ignored, the gums continue to recede and the fibres holding the teeth in their sockets break down. As this condition progresses, the jaw bone itself is resorbed, the teeth become loose and will fall out or have to be extracted. More teeth are lost as a result of periodontal disease than from caries.

Prevention of caries and periodontal disease

Prevention of caries. Caries results from acids attacking the enamel of teeth. The two lines of prevention are therefore to reduce the acids as far as possible and to make the enamel more resistant to them. Acids can be reduced by cutting down the intake of sugar and the enamel can be strengthened by fluoride.

The acids which cause cavities are produced by the action of bacteria on sugar. No amount of brushing will remove all the bacteria from the grooves and fissures on the surface of the teeth; so, the intake of sugar must be reduced if caries is to be avoided. Sugars occurring naturally in food (apart from honey) do not appear to contribute to caries. The main cause of caries is refined sugar (sucrose), i.e. the white (or brown) crystalline material in the sugar bowl or used in a wide variety of foods such as sweets, biscuits, jam, ice-cream, soft drinks, ketchup and many other processed foods.

It is not the concentration of sugar which matters but its constant presence in the mouth. It is better, for example, to eat several sweets at one time rather than keep introducing them into the mouth over a long period. The long-lasting lollipop is very harmful and the application of sugary substances to babies' 'dummies' causes early tooth decay. It is best to avoid sweets and sweetened food as much as possible and to replace these with sucrose-free foods, such as fruit and nuts, particularly between meals.

Brushing the teeth or rinsing the mouth after meals or after eating sweets has not been shown to reduce caries. The acids form within minutes, so brushing is too late to be effective and cannot reach the bacteria in the crevices and fissures of the teeth. Brushing, however, is effective against periodontal disease (see below), and if a fluoride-containing toothpaste is used, it does help to protect the teeth against caries.

At one time, it was thought that eating apples, raw carrot, celery and other crisp vegetable matter helped clean the teeth but it is now agreed that these materials have little effect except by replacing sweets and biscuits with less damaging food.

Fluoride ions, either taken in with food and drink or applied to the tooth surface, are taken up by the enamel and increase its resistance to decay, particularly in young people. The commonest way of acquiring fluoride is via toothpaste but it can be obtained in tablets or drops. Some regions add fluoride to the drinking water (see below).

There is a wide variation between individuals' susceptibility to caries. Some people may be able to eat sweets and neglect dental hygiene with relative impunity while others, more conscientious, may still suffer from caries. However, nearly everyone is susceptible to gum disease and this is reason enough for good oral hygiene.

Prevention of periodontal disease. Removal of plaque helps prevent gingivitis, and cutting down the intake of refined sugar helps to prevent plaque forming. Sugar provides the food needed by the mouth bacteria and allows them to make a particularly sticky polysaccharide which adheres to the surface of the teeth. However, even in the absence of sugar, some plaque will form and must be removed by brushing if gum disease is to be avoided.

Although plaque starts to form soon after brushing, it takes 24–36 hours to reach a level which leads to gingivitis. Consequently, effective brushing of the teeth once a day should be sufficient to prevent gum disease. However, it is essential that the brushing should remove all the plaque, particularly from the vulnerable crevices between gum and teeth.

It does not seem particularly valuable to brush the teeth after meals. Perhaps the best time is before going to bed because, during sleep, salivary flow is reduced and does not wash away the toxic by-products of plaque. Any fluoride in the toothpaste also lingers in the mouth and helps repair and harden enamel which has been attacked by acid during the day.

No particular method of brushing the teeth has been shown to be especially effective but brushing should take 3–4 minutes, firmly but not roughly executed and should reach all parts of the teeth. The brush should be angled so that the bristles enter the crevice between teeth and gums.

An effective toothbrush should have a short, straight head with fine nylon bristles in close tufts and be designated medium or soft. Toothpastes make brushing more pleasant and the mild abrasives they contain may help to remove plaque. Additives such as chlorhexidine have proved effective, in some cases, in reducing gingivitis, but not caries. Added fluorides, however, do increase resistance to decay.

No toothbrush will remove plaque from between the teeth and so brushing should be supplemented by drawing a waxed thread called 'dental floss', backwards and forwards between the teeth as directed by the dentist.

One way of checking whether the plaque has been removed is to spread a disclosing agent on the teeth after brushing. The disclosing agent is a harmless dye which, after rinsing the mouth, colours the patches of plaque not removed by the toothbrush. A dentist will advise on the most suitable disclosing agent to use.

Dental treatment. Regular visits to a dentist will enable him to remove calculus and detect cavities and gum disorders at an early stage. Decayed areas of teeth can be cleaned out and replaced with various materials, some metals, some plastics, and so save the tooth from further damage.

Fluoridation. Fluoride ions are a fairly common constituent of drinking water occurring naturally in concentrations of up to five parts per million (ppm) or more. It has been shown that in areas where water naturally contained fluoride ions, the incidence of decay was up to 60 per cent less than in areas

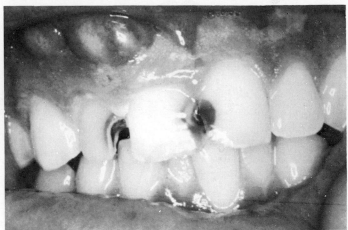
(Dr M. Hobdell/AHRTAG)

Fig. 17.8   Three of the upper incisors have cavities caused by decay

containing little or no fluoride. Experiments were conducted in some American towns by adding fluoride to drinking water in concentrations of 1 ppm, and the populations of these towns were compared with control areas with little fluoride in the water. A similar reduction in decay was found in the children with no evidence of undesirable side effects. Concentrations of 2 ppm or more, however, tend to cause some degree of mottling. Experimental fluoridation has been carried out in the U.S.A. for over 20 years and in Britain for over 15 years with encouraging results. The way in which fluoride acts is not fully understood, but it is known to be taken up by bones or teeth from the blood, though 95 per cent of it is excreted in the urine; and it is principally effective when the permanent teeth are still developing. In adults, fluoride is taken up into acid-damaged enamel and speeds its repair.

Opposition to fluoridation has arisen mainly on the grounds that it is a measure forced on all people, giving them no choice in the matter. Biologically it may seem a rational adjustment of our environment to meet optimum demands. Teeth seem to need a supply of fluoride just as they need calcium and phosphate, and the best way, it is claimed, of obtaining this supply in continuous small doses is via the drinking water.

Certainly, any adjustment of our environment which affects the health of millions of people needs very thorough consideration and though the case for fluoridation seems to have received careful study, there are still wide differences of opinion about the interpretation of the evidence and about the desirability of interfering with the water supplies to achieve medical benefits as distinct from merely making it safe to drink. Opponents of fluoridation point out that fluoride tablets are available to anyone who wants to fluoridate their own drinking water.

Questions

1   An eight-year-old boy pulls out a loose incisor and is alarmed because (a) it has no root and he thinks he has broken it off, and (b) he fears he will have a gap in his front teeth. How would you use your biological knowledge to reassure him?

2   In an erupted tooth, the dentine may continue to increase in thickness but the enamel cannot. Explain the difference.

3   Why does brushing the teeth not help to prevent decay and yet is of value in preventing gum disease?

18

The Sensory System

The sensory system makes an animal aware, though not necessarily in the sense of 'conscious', of conditions and changes both outside and inside its body. In the simpler animals, only very general stimuli such as light or darkness, heat or cold can be perceived by the sensory system. In the higher animals, including man, detailed information about the surroundings, such as distance, size and colours of objects, can be gained as a result of the specialization of the sensory organs and the elaboration of the nervous system.

General sensory system. This includes sense organs, most of which occur in the skin and are fairly evenly distributed; hence any part of the skin is sensitive to touch, heat, cold and pressure, any of which stimuli may also produce the sensation of pain. Examples of such sense organs are indicated in Fig. 18.1.

It must be emphasized that, in general, a particular sense organ can respond to only one kind of stimulus. That is to say, a sense organ sensitive to touch will not be affected by the stimulus of heat; an organ sensitive to chemicals will not respond to pressure. It is not yet certain just how specific some of the dermal sense organs are in their responses, for the sensory endings that produce the sensation of pain can be activated by a variety of stimuli, such as pressure, heat and cold. Moreover, the extent to which the various sensory endings can be distinguished by their structural appearance in sections is not yet determined. There are three basic types, (a) the *free-ending* type, in which the terminal branches of a sensory nerve cell penetrate the outer tissues of the dermis and possibly into the epidermis, (b) the *hair plexus*, a network of fibres surrounding a hair follicle, and (c) the *encapsulated* type, where a branched or coiled nerve ending is enclosed in a capsule of varying complexity, e.g. *Pacinian* and *Meissner's corpuscles* (Fig. 18.2). There may, however, be physiological differences between these encapsulated endings despite their apparently similar structure. For example, in the pinna of the ear only free endings and hair plexuses can be distinguished, and yet heat, cold, touch and pressure can all be detected there. It may be that one type of ending responds to light pressure, others to temperature changes, etc. It is possible, also, that their position in the dermis may determine the kind of stimulus to which they respond, e.g. the Pacinian corpuscle is not likely to be stimulated by a light touch but may be activated by more pronounced pressure on the skin.

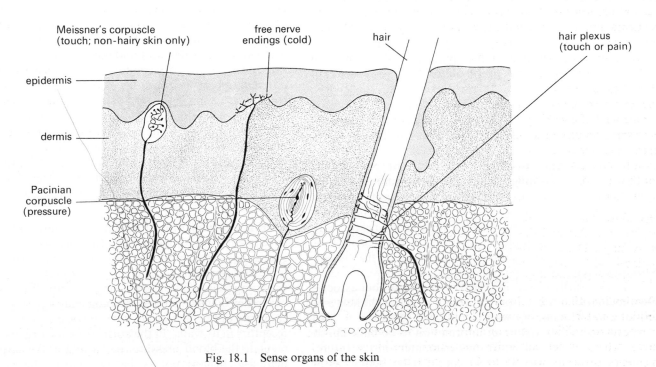

Fig. 18.1 Sense organs of the skin

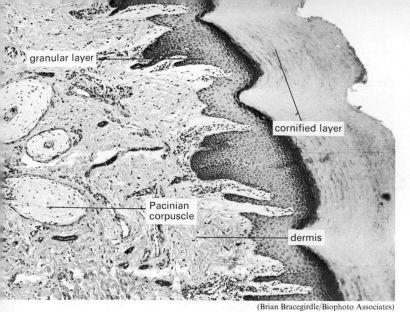

granular layer

cornified layer

Pacinian corpuscle

dermis

(Brian Bracegirdle/Biophoto Associates)

Fig. 18.2   Pacinian corpuscles in human skin (× 60)

Touch. The encapsulated end organs in the dermis respond to light pressure on the skin, though it is likely that some at least are stimulated by the distortion of the skin rather than by the direct pressure. In the hair-covered areas of skin, the sense of touch also depends on the free nerve endings wrapped round the hair follicles. Any movement of the hair follicle stimulates the nerve ending, the hair acting as a lever.

Heat and cold. Whatever the nature of the receptors that respond to temperature, experiments suggest that there is a difference in either physiology or distribution between those responding to heat and those responding to cold. A cold 'spot' (*see* below) will not respond to heat, or if it does it will produce the sensation of cold. It also appears that the response is made to a change of temperature rather than a steady state, i.e. once the skin ceases to gain or lose heat at a certain rate the receptor ceases to fire off impulses; it is said to be adapted to the new situation (Experiment 4).

Pain. It seems that any kind of stimulus that affects the free nerve endings described above can produce the sensation of pain. However, a pin may elicit a sensation of touch on one area of skin but pain on another. It is unlikely that excessive stimulation of the encapsulated endings can cause pain since, if the sensory fibres passing in the spinal cord from the free endings are severed, pain is prevented while all the other skin sensations are present.

The origin of a painful stimulus can be accurately located even though the stimulated area of skin may contain a number of overlapping free endings whose impulses reach the brain by more than one route.

Although we tend to regard sensations of pain as inconvenient and alarming, they have important biological advantages. By making us respond quickly and automatically by reflex action they cause us to remove the affected part from danger. Our response when touching something unexpectedly hot affords a good example. If there were no appreciation of harmful stimuli, untold damage might occur to the tissues. Withdrawal reflexes (p. 148) could be just as effective, however, without necessarily producing the feeling of pain, but association of pain with other sensory information such as the heat and light of a naked flame allows a conditioned reflex (*see* p. 150) to be established, or a store of information to be built up, so reducing the chances of hazardous encounters in the future. Where pain occurs without producing a reflex action, as in toothache, it serves as a warning that all is not well in that region and advice or treatment should be sought.

As might be expected, however, the significance of pain is not so straightforward as has been suggested. There is plenty of experimental evidence and everyday experience to indicate that our psychological state determines the intensity of pain we feel from any one stimulus. A person wholly preoccupied with some activity may be quite unaware of injuries received until his attention is called to them. In fact, it appears that we tend to experience the degree of pain which we expect to feel, or have learned to feel. Analgesic drugs, hypnosis, self-imposed mental states and many other effects can influence our experience of pain.

Distribution of sensory endings. Certain regions of the skin have a greater concentration of a particular type of sense organ. In general, the hairy regions of the skin have mainly hair follicle plexuses while in hairless regions, i.e. lips and fingertips, encapsulated end organs predominate. The fingertips have a large number of touch organs, making them particularly sensitive to touch. The front of the upper arm is sensitive to heat and cold probably as a result of the large numbers of those sense organs present. Some areas of skin have relatively few sense organs and can be pricked or burned in certain places without any sensation being felt. In the different regions of the body there is also a difference in the sensitivity to any one type of stimulus. In some areas, a light stimulus will produce an appreciable nerve impulse while in other areas a stronger stimulus is needed. In this respect, the tongue, nose and lips are most sensitive to touch, i.e. respond to weak stimuli, while the shin, sole of the foot and back of the forearm are the least sensitive (Experiments 1 and 2). The arms can detect changes of temperature of just over 0.2 °C while the fingers need a change of 0.5–1.0 °C to produce a sensation. In a newborn baby, the heat-sensitive endings are not fully functional, making the baby very vulnerable to injuries from heat, e.g. hot bath-water.

When the skin is explored with fine bristles or finely pointed hot or cold instruments it appears that areas sensitive to a particular stimulus are arranged in discrete 'spots', i.e. small areas of skin which will detect (say) touch, while the area next to it will not. Similar heat and cold spots can be mapped out. The spots do not correspond to single receptors since many different receptors must be affected by even the smallest stimulus. In some cases, the areas mapped are not even consistent from one day to the next. The density of the spots, however, varies with the sensitivity of the region tested. The physiological condition of the skin also affects the distribution of these spots, e.g. there is a more uniform sensitivity to heat and cold in an inflamed area; the spots have merged or 'vanished', perhaps as a result of the vasodilation that causes the reddening of the skin.

Internal receptors. Many internal organs and connective tissues have free nerve endings in them. In some cases they respond to stretching; for example, the alimentary canal if distended may produce a sensation of pain which is not produced by pinching or cutting. The inflation of the lungs is controlled to some extent by a reflex in which stretch receptors play a part (p. 89) and the emptying of the bladder (p. 94) is probably initiated by stretch receptors in its walls. In the aorta are sense organs that help to regulate the blood pressure (p. 82) and in the carotid arteries there are chemoreceptors that respond to changes in oxygen concentration. In certain regions of the brain, too, there is sensory apparatus that responds to changes in the osmotic potential (p. 155), carbon dioxide concentration (p. 89) and temperature (p. 98) of the blood.

The response of these receptors to changes in the internal environment enables corrections to be made by the body, and so keeps the blood and body fluids at optimum conditions for the vital chemical reactions of life to take place rapidly and predictably.

Proprioceptors. Proprioceptors are internal sense organs which occur most frequently in muscles. For the most part they are sensory endings embedded in a specialized group of muscle fibres called *muscle spindles* (Fig. 18.3). The spindles fire off

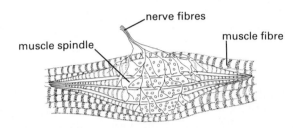

Fig. 18.3   Stretch receptor in muscle

impulses when stretched either by the extension of the muscle around them or by contraction of their own fibres. The muscle spindle, by feeding back sensory impulses into the spinal cord affects the motor impulses which are bringing about contraction, and so controls the movement of limbs. After a period of learning, the sensory information about the tension of muscles that reaches the brain from the proprioceptors enables an individual to know (consciously or otherwise) the precise orientation of his limbs at any moment, an essential requirement for co-ordinated activity. One can put food into the mouth, for example, without having to watch the hand all the time.

Information to the brain

Stimulation and conduction of impulses. The sense organs are connected to the brain or spinal cord by nerve fibres. When the sense organ receives an appropriate stimulus it sets off a burst of electrical impulses in the nerve fibre supplying it. The impulses travel along the nerve fibre to the spinal cord and on to the brain, sometimes producing an automatic or reflex action, or recording an impression by which the person feels the nature of the stimulus and where it was applied.

Sense organs of one kind, and those in a definite area, are connected principally but not exclusively with one particular region of the brain. It is the region of the brain to which the impulse comes that gives rise to the knowledge about the nature of the stimulus and where it was received. For example, if the region of the brain receiving impulses from pain nerve endings were eliminated or its activity suppressed by drugs, no amount of stimulation of the pain sensory endings would produce any sensation of pain at all, even though impulses were being fired off and travelling as far as the brain. On the other hand, if a region of the brain dealing with impulses from sense organs in

the leg is stimulated by any means, e.g. a weak electric current applied directly to the appropriate part of the brain, the sensations so produced seem to have come from the leg. People who have had an arm or leg amputated often feel as if the limb were still there and may experience real pain from this 'phantom limb'. These sensations must arise in the brain.

Another important consideration is the fact that the impulses transmitted along the nerve fibres are all exactly the same in quality. It is not the sensations that are carried but simply a surge of electrical energy, which is the same whether it is a heat organ or touch ending that sets off the impulse. Only in the brain can the stimulus be identified, according to the region of the brain where the impulse enters. For example, if the nerves from the arm and leg were changed over just before they entered the brain, stubbing one's toe would produce a sensation of pain in the arm or hand (Experiment 3).

Intensity of sensation. A strong stimulus usually produces a more pronounced sensation than a weak stimulus. This is probably the result of (a) a more rapid sequence of impulses fired off from the receptor, (b) the stimulation of a greater number of sensory organs in the area, and (c) the stimulation of a number of sensory organs that do not respond at all unless the stimulation is intense. Vigorous stimulation does not have any effect on the quality or intensity of the nerve impulses travelling in the sensory fibres, but increases the total number of impulses reaching the brain (Fig. 19.4, p. 147).

Interpretation. It will be appreciated that an enormous amount of sensory information is being constantly fed in to the nervous system. All this sensory data enables the organism to adjust itself to changes taking place internally and externally, so that in the first instance it maintains a constant and optimum internal environment for its metabolism, and in the second case can behave in an appropriate manner for the survival of itself and its species. Much of the sensory information never reaches conscious levels, e.g. changes in blood pressure or carbon dioxide content; the adjustments are made entirely on a reflex basis. Those sensory impulses which do reach consciousness produce sensations.

It seems that the sense organs on the whole respond to changes in conditions. As soon as conditions settle down, the sensations and probably the impulses cease or return to a low level. For example, we are unaware of the tactile stimulus of our own clothing once we have dressed; we become accustomed to unfamiliar smells after a time though a newcomer will notice them at once.

We are able to locate the source of skin sensation fairly accurately. Touch is precisely located, heat and cold with less precision. The ability to locate the source of stimulation depends on memorizing the muscular movements needed, on previous occasions, to reach it.

We often tend to attribute our source of information to one particular sense organ, though in most cases we are interpreting the sum effect of a whole range of sensory data. Texture, for example, is appreciated by sense of touch, temperature (i.e. loss or gain of heat from the fingers when touching) and even the sound which is heard when running the fingers over an object. If one of these sources of information is lacking, our judgement is likely to be faulty. Information from the pressure receptors of the buttocks is of importance to a pilot controlling an aircraft or a motorist controlling his car, although we tend to assume that the eyes and semicircular canals are the sole sources of information about position and movement.

The special senses

Sight, hearing and balance, smell and taste are called the special senses. The relevant sense organs each consist of a great concentration of cells that are sensitive to one kind of stimulus. These sensory cells may be associated with structures that direct the stimulus on to the sensory region.

Taste. The sense of taste is conferred by groups of sensory cells that are stimulated by chemicals. The receptors are grouped into about 9 000 taste buds (Fig. 18.4) containing both sensory

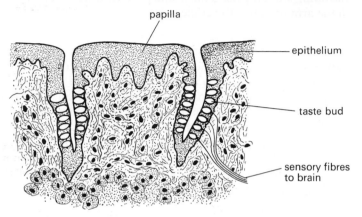

(a) Section through tongue epithelium

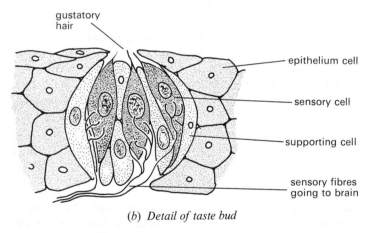

(b) Detail of taste bud

Fig. 18.4 Taste receptors in tongue

cells and supporting cells embedded in the epithelium round the base of the papillae on the tongue. Taste buds are also found in the soft palate, the epiglottis, the opening of the oesophagus and, at least in children, the cheek lining.

Only chemicals soluble in water can affect the taste buds, by dissolving in the moisture of the mouth. Their solutions enter the pores of the taste buds and stimulate the sensory cells, probably affecting the cytoplasmic filaments (*gustatory hairs*) projecting from their outer end. Although all the taste buds appear to be identical in structure, they are not equally sensitive to the chemicals described as sweet, sour, salt and bitter (the only tastes we can discriminate). Most taste buds appear to respond to all four classes of chemical but to differing extents.

For example, one taste bud may make its maximum response to salts, a lesser response to sour, and weak responses to sweet or bitter substances. The different types of taste bud are unevenly distributed over the tongue so that some parts of the tongue are more sensitive than others to a particular chemical.

The means by which the sensory cells are activated by chemicals is not known, but when the cells are stimulated by the appropriate solution, impulses are fired off in nerve fibres that pass to the medulla of the brain. Combinations of chemicals will affect the different taste buds simultaneously and so, according to the type and proportions of taste buds stimulated, a fairly wide range of discrimination is possible, over and above the four basic tastes. Lemonade, for example, will stimulate sweet and sour receptors simultaneously and its taste will be judged accordingly.

The fact that only four tastes can be distinguished indicates the limited importance of the sense of taste compared with the other senses. Few behaviour patterns are likely to be initiated by taste stimuli, which seem primarily to serve for discrimination of acceptable and undesirable food.

The wide variety of flavours attributed to food result from the simultaneous stimulation of taste buds of the tongue and olfactory (smell) organs in the nasal cavity, the latter organs being affected by the vapours emanating from food in the mouth. When the nasal cavity is congested, as with a heavy cold, the sense of flavour is lost although taste is unaffected, and the sufferer may realize the extent to which flavour depends on smell. The limited discriminatory power of taste can also be demonstrated by placing solutions or pieces of food on the extended tongue while the nostrils are pinched and chewing is forbidden. In these circumstances it is more difficult to distinguish between the taste of apple, turnip and onion, whose flavour and texture usually make them easy to recognize. It is also observed that, unlike most sensory organs, there is a long interval between application of the stimulus and appreciation of the sensation.

The tongue is also sensitive to touch, heat, cold and pain, and much of our information about food and our reactions to it

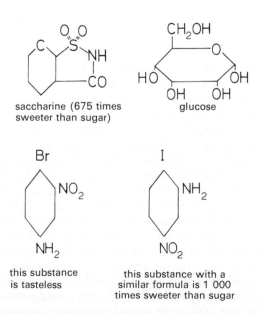

saccharine (675 times sweeter than sugar)

glucose

this substance is tasteless

this substance with a similar formula is 1 000 times sweeter than sugar

Fig. 18.5 Taste and chemical composition (it is difficult to see any connection between the formula of a substance and its taste)

depends on the texture as well as the taste of the food. Hot food is more easily tasted than cold and, in addition, has more flavour.

The relationship between a particular chemical and its taste has not been elucidated, although it is true that acid solutions are sour-tasting and that chlorides, sulphates and nitrates of certain metals, particularly sodium, are salty. Different substances that taste sweet, however, often seem to bear no relationship to each other in their chemical structure (Fig. 18.5).

Smell. The olfactory organs, sensitive to smell, are two small patches of epithelium located in the upper regions of the nasal cavity in narrow crevices on each side. The several million receptors are not grouped into buds as are the taste organs, but distributed more or less evenly between the supporting cells. The sensitive cells (Fig. 18.6) are derived from the cell bodies of

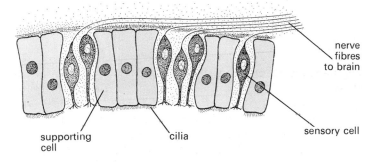

Fig. 18.6   Sensory epithelium of smell receptor

bipolar nerve cells (p. 146), a short fibre (dendron) projecting outwards and terminating in sensory filaments in the nasal epithelium and a longer fibre (axon) running into the forebrain. Tubular nasal glands secrete a fluid which moistens and flushes the surface of the sensory epithelium. During ordinary breathing, the main stream of air passes below the olfactory organs and only swirls or eddies of air reach them. In sniffing, however, the air is directed upwards on to the sensitive surfaces.

The olfactory cells, like those of taste buds, respond to chemical stimulation, though the chemicals must be volatile (i.e. easily vaporized) at ordinary temperatures, and in most cases soluble in fat solvents, to produce any sensation of smell. These two conditions apply mainly to organic compounds, though inorganic gases can be detected, for example chlorine, ammonia, hydrogen sulphide. Although it is often considered that our sense of smell is poor compared with that of other mammals, we can nevertheless recognize an enormous number of chemicals in extremely low concentrations; with one chemical at least, it is estimated that a concentration of one molecule per 5 000 molecules of air can be detected. In general, about 3 000 times as much material is needed to elicit a sense of taste as is necessary to give a sensation of smell. Far too many different odours can be distinguished for it to be likely that there is one kind of receptor for each type of substance, and it is suggested that discrimination depends on a small number of different types of receptor being stimulated in varying proportions.

The sense of smell is readily fatigued, so that after a few minutes' exposure to a new smell it is no longer perceived, although other observers on arrival will detect it at once. The fatigue may be due to the sensory cells failing to set off nervous impulses, or to the suppression of impulses arriving in the central nervous system from the olfactory area. The impulses pass via complicated nerve plexuses to the forebrain and cerebral cortex, and in vertebrates as a whole may be responsible for initiating complex behaviour patterns related to food seeking, mating or marking out territory.

Sight. The eyes are the organs of sight. They are spherical organs housed in deep depressions of the skull, called orbits. They are attached to the wall of the orbit by six muscles which can also move the eyeball (Fig. 18.7). The structure is best seen in a horizontal section, as shown in Fig. 18.8.

Hearing and balance. *See* p. 139.

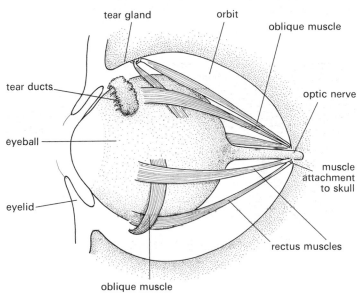

Fig. 18.7   Muscles of left eye (side view)

Structure and functions of parts of the eye

Eyelids. These can cover and so protect the eye. Closing the eyelids can be a voluntary or a reflex action to protect the cornea from damage. Regular blinking serves to distribute fluid over the surface of the eye and prevent drying. Modified sebaceous glands (*see* p. 97) secrete an oil at the margins of the eyelids.

Conjunctiva. This is a thin epithelium which lines the inside of the eyelids and the front of the sclera, and is continuous with the epithelium of the cornea.

Tear gland. The tear gland is housed in the upper part of the orbit above the eyeball, and its ducts open under the top eyelid (Fig. 18.7). It secretes a solution of sodium chloride and sodium hydrogencarbonate (bicarbonate) that keeps the exposed surfaces of the conjunctiva and cornea moist and washes away dust and other particles. An enzyme, *lysozyme*, is present in tear fluid, and has a destructive action on bacteria. Tear fluid normally evaporates from the eye as fast as it is secreted but some may be drained into the nasal cavity through the *lachrymal duct* which leads from the corner of the eye nearest the nose.

Eye muscles. These muscles are attached to the sclera at one end and to the wall of the orbit at the other (Fig. 18.7). Their

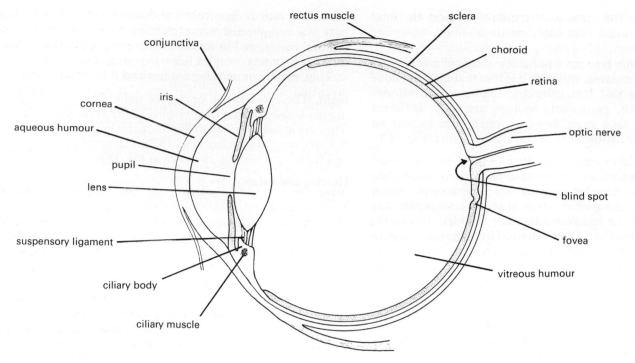

Fig. 18.8   Horizontal section through left eye

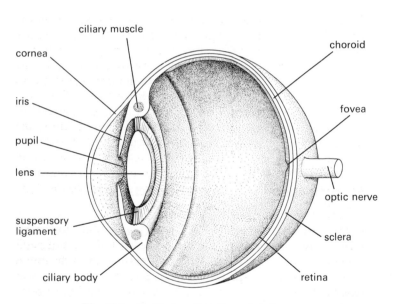

Fig. 18.9   Vertical section through left eye

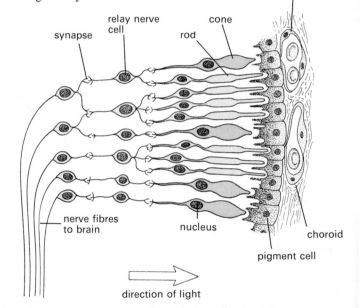

Fig. 18.10   Structure of the retina

co-ordinated contractions can make the eye move from side to side and up and down. The muscles are co-ordinated by the brain so that they move both eyes together in the same direction, and by reflex action can fixate on moving objects, i.e. follow the path of a moving object without conscious effort on the part of the observer.

Sclera. The sclera is a tough, non-elastic fibrous coating on the outside of the eyeball. It opposes the outward force of the eye fluids (*humours*) and maintains the shape of the eyeball.

Cornea. This is the transparent disc in the front part of the sclera. Light passing through the curved surface of the cornea is refracted and the rays begin to converge. There are free nerve endings in the cornea which, if stimulated, cause reflex blinking, tear secretion and produce sensations of pain.

Choroid. The choroid is a layer of tissue lining the inside of the sclera. It contains a network of blood vessels supplying food and oxygen to the eye and particularly to the retinal cells. It is also deeply pigmented, the black pigment reducing the reflection of light within the eye.

Aqueous and vitreous humours. The aqueous humour occupies the anterior chamber of the eye between the cornea and the lens. It is a watery solution, having a composition similar to blood plasma with dissolved salts and glucose, though there is more sodium and chloride and less urea than in plasma. Aqueous humour contains dissolved oxygen and supplies the lens and cornea, which contain no blood vessels, with nutrients and oxygen. The humour is secreted by the ciliary body and probably drains into a circular canal which runs around the junction of sclera and cornea, and which itself empties into the venous system. The pressure of the aqueous humour, some 25–30 mm of mercury, maintains the shape of the cornea and keeps the lens stretched against its natural elasticity.

The jelly-like vitreous humour is a protein gel filling the space between the retina and lens. It helps to maintain the shape of the eye. Both fluids play a part in refracting the light which enters the eye.

The lens. The lens continues the refraction of light which begins at the cornea and produces an image on the retina. It is made of transparent, ribbon-like fibres and living epithelial cells arranged in concentric layers like the scales of an onion. The material in the centre of the lens has a greater refractive power than that at the edges, and this property enables it to produce a sharper image than could a simple glass lens. The lens is held in position by the fibres of the suspensory ligament which radiate from its edge and attach it to the ciliary body (Fig. 18.9). The shape of the lens can be altered by the contraction or relaxation of the muscles in the ciliary body.

Ciliary body. This is the thickened edge of the choroid in the region around the lens. It contains blood vessels and muscle fibres, some of which run in a circular direction, that is, parallel to the outer edge of the lens. These muscles control the curvature of the lens during accommodation (*see* below). The ciliary body also secretes aqueous humour.

Iris. The iris consists of an opaque disc of tissue. It is continuous at its outer edges with the choroid. In the centre is a hole, the pupil, through which passes the light that will produce an image on the retina. The contraction or relaxation of opposing sets of circular and radial involuntary (unstriated) muscle fibres in the iris increases or decreases the size of the pupil, so controlling the amount of light entering the eye. The iris contains blood vessels and sometimes a pigment layer that determines what is usually called the 'colour' of the eyes. Blue eyes have no special pigment, the colour being produced by a combination of the black inner surface of the iris, the blood capillaries and the white outer layers.

Retina. The retina is a layer of cells which respond to light. There are two kinds of light sensitive cell called, according to their shape, *rods* and *cones* (Fig. 18.10). There are about 7 million cones and 12 million rods, the cones having a density of about 6 000 per mm² and the rods 150 000 per mm² over the general surface of the retina. Only the cones can discriminate coloured light but they need a stronger stimulus than the rods. It follows that colour discrimination is only possible in high light intensities. The rods and cones are connected to the brain by nerve fibres though, since the optic nerve contains only about 800 000 fibres, many of the receptor cells must feed their impulses into the same fibre.

As a result of the development of the retina from an outgrowth of the brain, the nerve fibres from the rods and cones lie on the inner surface of the retina. Thus light entering the eye has to pass through or between these fibres before it can stimulate the receptors.

Some detail about the mechanism of the rods but little about the cones is known. The rods contain a pigment, *visual purple* or *rhodopsin*. This is formed from two chemicals, *opsin* and *retinal*, the latter being derived from vitamin A. It is known that light bleaches this pigment, and the present theory supposes that bleaching is a result of rhodopsin splitting into opsin and retinal. This chemical decomposition in some way starts a nerve impulse in the nerve fibre supplying the rod. The rhodopsin is then resynthesized, involving an expenditure of energy by respiration.

The rods are extremely sensitive and it is thought that one quantum of light (the smallest unit of energy) will decompose one molecule of rhodopsin and that, in some parts of the retina, from 4 to 14 quanta falling on an area containing 500 rods will just succeed in producing a sensation of vision on most occasions. When a dark night sky is just visible, every rod receptor is receiving on average one quantum per minute.

Blind spot. The nerve fibres from the rods and cones pass across the front of the retina and all leave the eye at one point to form the optic nerve, which passes through the skull to the brain. At the point where the optic nerve passes through the retina there are no light receptors, so leaving a non-sensitive disc, the blind spot, about 1.5 mm across. If part of an image falls in this region, no impression is recorded in the brain (Experiment 6). We are not normally aware of this 'blank' in our vision because (a) it never coincides with an image on which we are concentrating, (b) it is compensated by the use of two eyes scanning the same field, and (c) the eyes are constantly making small movements so that the image is not projected on to the same part of the retina for more than a fraction of a second.

Fovea. The fovea is a small depression, 500 μm across, in the centre of the retina. It contains no rods but the cone density is about 12 500 per mm². There are no overlying capillaries, and the great concentration of light-sensitive cells makes it the region of the retina where greatest discrimination is possible. In addition, each cone is connected to a separate nerve fibre and, although cross-linkages occur later, a detailed pattern of impulses corresponding to the retinal image can be sent to the brain. When an observer concentrates on an object or part of an object, its image is thrown on to the fovea. Only in this region is there detailed appreciation of form and colour. Since the fovea contains only cones, it is not very sensitive to low light intensities.

Control of light intensity. The circular muscle fibres in the iris form a kind of sphincter. When they contract, the size of pupil is reduced and less light is admitted. Contraction of the radial fibres widens the pupil, so admitting more light. This is a reflex action set off by changes in the light intensity. In poor light the pupils are wide open; in bright light the pupils are contracted. In this way the retina is protected from damage by light of high intensity, and in poor light the wider aperture of the pupil helps to increase the brightness of the image. The eyes are linked by nervous paths so that each makes the same adjustment, no matter which eye is stimulated, and the adjustment takes about five seconds to complete.

Seeing in the dark. No creature can see in total darkness, but the sensitivity in low light intensities varies from one animal to the next. The sensitivity of the eye is increased in poor light intensity partly by widening the pupil, but more important are changes that take place in the retina. After half an hour in the dark the eye will respond to light intensities 10 000 times less bright than when light-adapted. This may be the result of the greater concentration of rhodopsin in the rods during darkness.

The fovea contains only cones, which respond only to high light intensities. Thus, concentrating on a poorly lit object, to focus its image on the fovea, is a less effective way of studying it than by looking to one side of it. This action throws the image on to a part of the retina containing many more rods than cones, the rods being more easily stimulated. Less detail can be seen but a more distinct outline is appreciated.

Image formation and vision. Light from an external object enters the eye. The curved surfaces of the cornea, the lens and the humours refract the light and focus it so that 'points' of light from the object are represented as corresponding points of light on the retina. The image thrown on to the retina is real, upside-down and smaller than the object (Figs. 18.11 and 18.12). Although the image is inverted (Experiment 5) the brain forms an upright impression of the object or, at least, the observer interprets correctly the position of external objects with respect to himself. Experiments have been conducted in which a person wears glasses that produce an upright image on the retina. Although this at first gives the impression that external objects are upside down, this sensation wears off and the experimenter gets used to interpreting the sensory information in the same way as before, reporting vision to be normal. When the spectacles are removed and the retinal image restored to its usual inverted position, the visual field is, at first, reported as upside-down.

The light-sensitive cells of the retina are stimulated by the light falling on them, and impulses are fired off in the nerve fibres which pass along the optic nerve to the brain where, as a result, an impression is formed of the shape, size and colour of the object. The nervous impulses travelling in the optic nerve reach a relay centre in the brain, after which some pass to the *oculomotor centre* in the midbrain. Here are initiated the automatic adjustments to the eyes, e.g. co-ordination of the eye muscles for following a moving object with the eyes, accommodation, pupil size and reflex blinking. Further fibres carry impulses to the cerebral cortex (p. 155) where the nerve impulses are interpreted and related to the object producing them.

The accuracy of the impression in the brain of the image depends on how numerous and how closely packed are the light-receiving cells of the retina, since each one can only record the presence or absence of a point of light and, in the case of cones, its colour. If there were only ten such cells, the image of a house projected on to five of them would record an impression of its size, the fact that it was differently coloured at the top and bottom and a vague representation of its shape, but no detail of windows, doors or fabric.

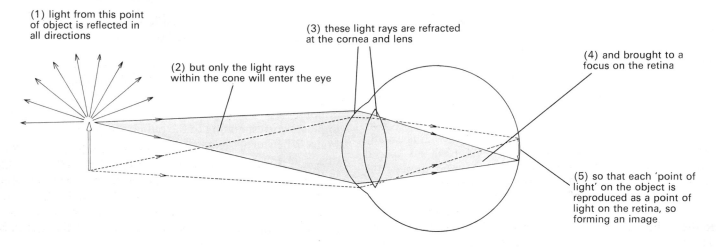

Fig. 18.11   Image formation on the retina (shown graphically)

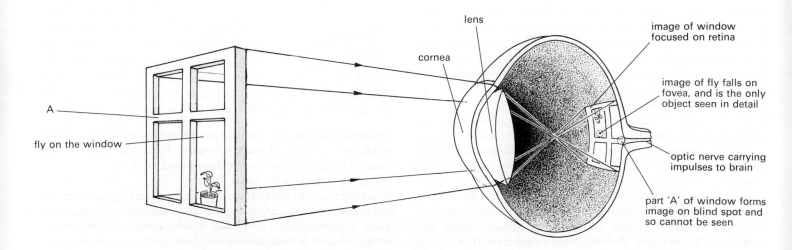

Fig. 18.12   Image formation in the eye

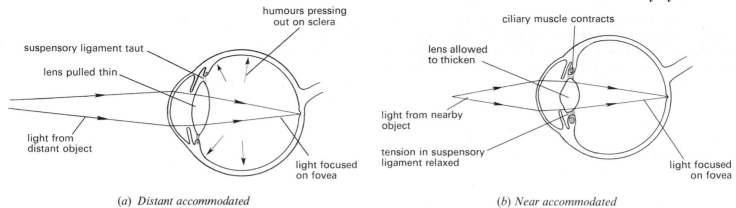

Fig. 18.13 Mechanism of accommodation

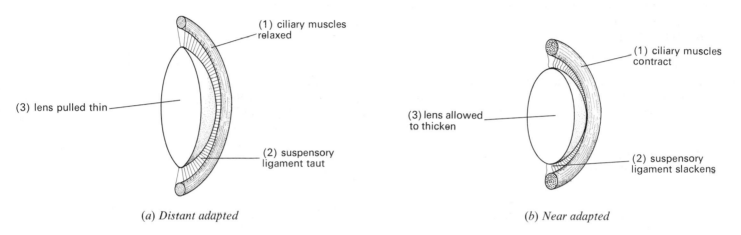

Fig. 18.14 Action of the ciliary muscle

Accommodation is the adjustments made in the eye to focus on near or distant objects. With a rigid lens of definite focal length and at a fixed distance from a screen, it is possible to obtain a sharply focused image of an object only if it is at a certain distance from the lens. The focal length of the lens in the eye can be altered by making it thicker or thinner. In this way, light from objects from about 25 cm distant to the limits of visibility can be brought to a focus. This ability of the eye to alter its focal length is called accommodation.

The majority of the refraction that brings about image formation takes place at the surface of the cornea, with the lens contributing only about one quarter of the total refraction. The lens is therefore mainly concerned with accommodation. A person with a cataract (opaque lens) can have the lens removed and, with the aid of spectacles, still focus a clear image on the retina.

The lens is surrounded by an elastic capsule and tends to change its shape, becoming thicker in the centre, but the aqueous humour pushing outwards on the cornea maintains a tension in the suspensory ligament which stretches the lens into a thinner shape. Thus, when the eye is at rest, the lens is thin, has a long focal length and is adapted for seeing distant objects (Fig. 18.13a). When a nearby object is to be observed, the ciliary muscles running round the ciliary body contract and so reduce its diameter. The ciliary body holds the suspensory ligaments, pulling on the margins of the lens, so any reduction in the diameter of the ciliary body reduces the tension in the suspensory ligaments and allows the lens to become thicker

(Fig. 18.14). A thicker lens has a shorter focal length, and light from a close object can be brought to a focus (Fig. 18.13b). Relaxation of the ciliary muscles allows the fluid pressure acting on the cornea and sclera to pull the lens back to its thin shape. When the eye is focused on a near object, the pupil is also contracted. Since this allows only the centre of the lens to be used, it sharpens the image.

Colour vision. There are three kinds of cone in the retina. All three respond to more than one colour but each is particularly sensitive to either blue, green or yellow light. Blue light with a wavelength of 430 nm (nanometres) stimulates the blue-sensitive cones most strongly and the green-sensitive cones very slightly. Green light at 536 nm stimulates the green-sensitive cones more than the other two. Yellow light at 565 nm has the maximum effect on the yellow-sensitive cones but also stimulates the green-sensitive cones quite strongly. Red light at 650 nm stimulates the yellow-sensitive cones more than the green-sensitive ones. This difference is interpreted by the brain as 'redness'. All three types, stimulated in the right proportions would give the sensation of white.

Colour blindness. About 8 per cent of men and 0.4 per cent of women suffer from one or other forms of colour blindness. Only rarely is this the total inability to distinguish colours from shades of grey with equal light intensity. More often it is the failure to discriminate between red, brown and green. (See p. 180 for an account of the inheritance of colour blindness.)

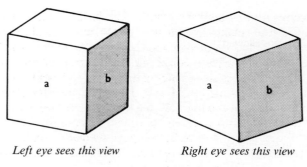

Left eye sees this view   *Right eye sees this view*

Fig. 18.15   Cube as seen by left and right eyes

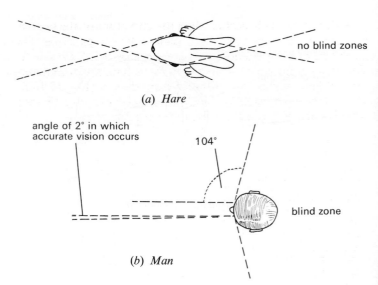

(a) Hare

angle of 2° in which
accurate vision occurs

104°

no blind zones

blind zone

(b) Man

Fig. 18.16   Angle of vision

Stereoscopic vision. Each eye forms its own image of an object under observation, so that two sets of impulses are sent to the brain. Normally the brain correlates these so that we gain a single impression of the object. Since each eye 'sees' a slightly different aspect of the same object (Fig. 18.15) the combination of these two images produces the impression of solidity based on the three-dimensional properties of the object.

If the eyes are not aligned normally or if the centres of the brain dealing with sight impressions are dulled, for example with alcohol, the two sensory impressions from the eyes are not properly correlated and we 'see double' (Experiment 8).

Judgement of distance. Our ability to judge distance probably depends on a number of factors; the apparent size of the object, overlapping of objects, parallax effects (see below). When we concentrate on close objects, our eye muscles must contract to rotate the eyeballs slightly inwards, and stretch receptors in these muscles may send impulses to the brain so providing information about distance, at least for very close objects. The stereoscopic vision described above probably helps to judge distance. It is more difficult to estimate distance using only one eye, though other information such as the relative size of an object and its apparent movement against the background (parallax) also contribute to our judgement of distance, and these factors do not depend on binocular vision.

Field of vision. If the eyes and head are held stationary, a man can see objects within an angle of about 200° from his face. The shape of the face and nose will influence this field in different individuals. Only objects that subtend an angle of about 2° at the eye will form an image on the fovea and so be observed in accurate detail (Fig. 18.16). This is a considerably narrower range of accurate vision than most people imagine, and means, for example, that only about one letter at a time in any word on this page can be studied in detail.

Eye defects. The causes and corrections of long and short sight are explained diagrammatically in Fig. 18.17. As one gets older, the lens loses some of its elasticity and hence its power to accommodate. The near point (shortest distance from the eye where vision is still perfect) for a 20-year-old is about 250 mm but it is more like 500 mm for a 40-year-old. This reduced accommodation is called *presbyopia* and can be corrected by using converging lenses for reading and close work.

(a) Long sight

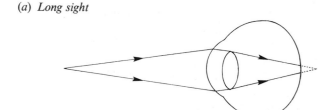

Long sight is caused by small or 'short' eyeballs. Light from a close object would be brought to a focus behind the retina, so the image on the retina is blurred

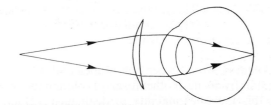

Long sight can be corrected by wearing converging lenses

(b) Short sight

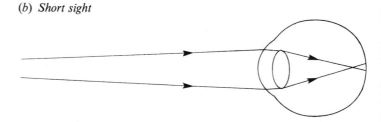

Short sight is usually caused by large or elongated eyeballs. Light from a distant object is focused in front of the retina, so the image on the retina is blurred

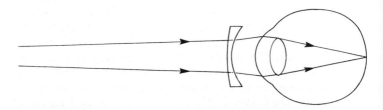

Short sight can be corrected by wearing diverging lenses

Fig. 18.17   Long and short sight

Astigmatism. This occurs when the curvature of the cornea and lens is not uniform in all planes. If, for example, the curvature from top to bottom is greater than that from side to side, it is not possible for the eye to focus both horizontal and vertical lines at the same distance from the eye (Fig. 18.18a). This defect is corrected by using cylindrical lenses (Fig. 18.18b) which effectively refract the light in one plane only.

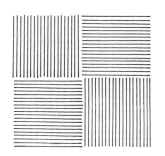 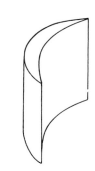

(a) *The astigmatic eye cannot focus the vertical and the horizontal lines at the same time*

(b) *Example of a cylindrical lens*

Fig. 18.18   Astigmatism

The ear (Fig. 18.19)

The ear contains receptors sensitive to sound vibrations in the air between frequencies of 20 Hz and 25 000 Hz, the precise range varying with age. (The unit of frequency is the hertz, symbol Hz. A frequency of 10 cycles per second is called 10 Hz.)

Outer ear. This is a tube opening on the side of the head and leading inwards to the ear drum. Its lining of skin contains sebaceous and *ceruminous glands.* The latter, which may be modified sweat glands, secrete wax. At the outer end of this tube there is an extension of skin and cartilage, the pinna, which in some mammals helps to concentrate and direct the vibrations into the ear and assists in judging the direction from which the sound came. Its function, if any, in man is not very clear. A membrane of skin and fine collagen fibres is stretched across the innermost end of the outer ear, closing it off completely. This is called the *ear drum.* A muscle, the *tensor tympani,* running from the ear drum to the wall of the middle ear maintains a tension in the ear drum by pulling it inwards.

Middle ear. The middle ear is an air-filled cavity in the skull. It communicates with the back of the nasal cavity (nasopharynx) through a narrow tube, the Eustachian tube. Three small bones, or *ossicles,* in the middle ear link the ear drum to a small opening in the skull, the *oval window,* which leads to the inner ear.

Inner ear. The inner ear is filled with a fluid, *perilymph,* and contains a coiled tube, the *cochlea,* with sensory endings in it. It is here that the sound vibrations are converted to nervous impulses.

Eustachian tube (Figs. 10.3 and 18.19). Air pressure in the middle ear is usually the same as atmospheric pressure. If changes take place in the pressure outside the ear drum, for example when gaining height rapidly in an aircraft, the pressure is equalized by the opening of the Eustachian tube to admit more air to or release air from the middle ear. Normally the Eustachian tubes are closed by a muscle and are opened only in swallowing or yawning, when a 'popping' sound may be heard in the ears. Violent nose-blowing may sometimes force air or mucus up the Eustachian tubes into the middle ear, resulting in temporary deafness or earache.

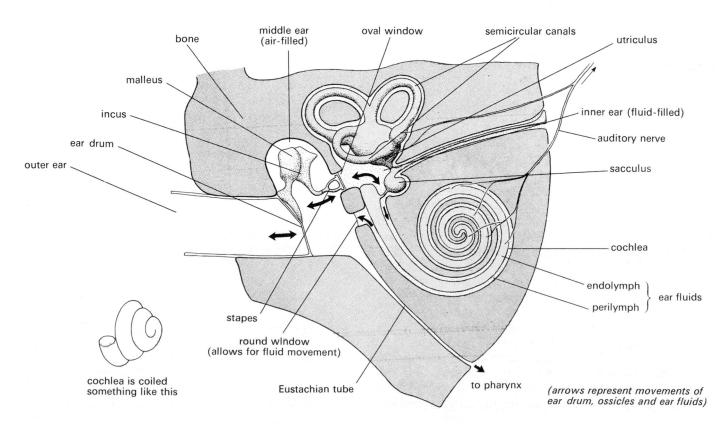

Fig. 18.19   Schematic diagram of the ear

Hearing. The vibrations in the air that constitute sound waves enter the outer ear and set the ear drum into vibration. The tiny displacements of the ear drum are transmitted through the three ossicles, which act as levers (Fig. 18.20), and cause the innermost of them, the *stapes*, to vibrate against the oval window. The greater area of the ear drum, 85 mm², compared with the oval window, 3.2 mm², and the leverage of the ossicles, cause an increase of about 22 times in the force of vibrations that reach the inner ear. The stapedial muscle tends to pull the stapes away from the oval window and, by means of a reflex action, damps down any violent oscillations of the ear ossicles produced by loud noises.

The oscillations of the stapes, acting like a miniature piston, set the fluids of the inner ear and cochlea into vibration. A membrane containing sensory endings runs the length of the cochlea, and when these endings are stimulated by vibrations,

perilymph of the inner ear these are conducted to the perilymph of the scala vestibuli; they are transmitted across the inner tube, or *cochlear duct* as it is termed, at right angles to its long axis and so reach the scala tympani and round window. The floor of the cochlear duct consists of the *basilar membrane* which contains transverse fibres. The roof of the cochlear duct is membranous and thin. Resting on the basilar membrane is a single layer of sensory cells, about 30 000 of them, with fine cytoplasmic hairs on their free, upper edge. These hairs are embedded in a gelatinous ribbon, the *tectorial membrane*. The hair cells and their supporting tissues are called the *organ of Corti*.

The distortion of the cochlear duct which results from pressure differences in the upper and lower compartments of the cochlea are shown diagrammatically in Fig. 18.22, and can be visualized as causing the tectorial membrane to give a tug on the hair cells. This is most probably the stimulus that induces the

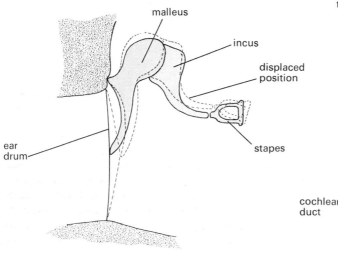

Fig. 18.20   Movement of the ear ossicles in transmitting sound

they fire off impulses which travel in the auditory nerve to the brain. When these nervous impulses reach the brain they are interpreted as sound, the quality, pitch and loudness of the sound being determined by the way in which the vibrations affect the cochlea.

Cochlea. The cochlea can be visualized as a tube of triangular cross-section enclosed within a cylindrical tube (Fig. 18.21) so that the outer tube is divided into upper and lower compartments by the inner tube. (*Note:* the terms 'upper' and 'lower' refer to the diagrams in this book, rather than to the real anatomical position.) The upper compartment or *scala vestibuli* opens to the perilymph in contact with the oval window, while the lower compartment or *scala tympani* communicates with the round window. Since fluids are not compressible, any change in pressure in the upper compartment will be transmitted at once to the lower compartment across the inner tube. The changes in pressure in the lower compartment will be relieved by the membrane of the round window bulging in or out. The upper and lower compartments do communicate at the tip of the cochlea by a small hole, the *helicotrema*, but this is too small to permit rapid movement of fluid during vibrations. The inner tube is distorted by alterations in the pressures of perilymph in the upper and lower compartments.

When the stapedial movements induce vibrations in the

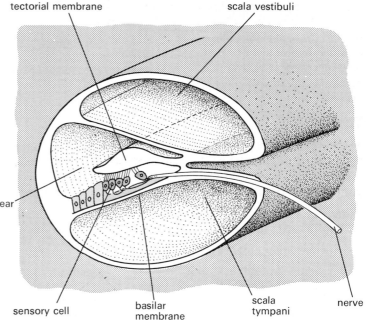

Fig. 18.21   Section through cochlea

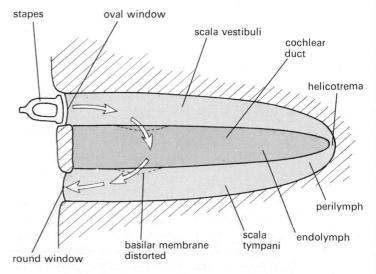

Fig. 18.22   Simplified diagram of cochlea in longitudinal section to show effect of sound vibration

hair cells to fire off a nervous impulse in the nerve fibres connected to them.

Determination of pitch. It was once thought that the transverse fibres or 'auditory strings' of the basilar membrane were responsible by virtue of their differing length and tension for an appreciation of pitch, i.e. high or low notes. There are about 24 000 of these strings, ranging in length from about 60 μm near the base of the cochlea to 500 μm at the tip, giving about $2\frac{1}{2}$ strings per tone. If a group of strings were able to resonate (vibrate) in response to only one particular frequency, then only the sensitive hair cells resting on these fibres would be stimulated by a single note of that pitch, i.e. one part of the cochlea would be stimulated by a high note and another part by a low note. The brain, by detecting which section of the cochlea had been stimulated, would be able to determine the relative pitch of the note.

Something like this is known to happen, but the role of the basilar fibres is not clear since they do not appear to be under tension. Experiments show that for sounds of low pitch (frequency up to 60 hertz or cycles per second), the upper end of the basilar membrane vibrates as a whole. Above 60 Hz, however, the region of maximum response of the cochlea becomes more localized, shifting down the cochlea as the frequency rises. Between 300 and 2 000 Hz the localized region of response moves even more rapidly down the cochlea, but above 2 000 Hz observation becomes too difficult. It has also been shown that the region of cochlea responding at one time to a certain note produces a pattern of electrical activity in a particular zone of the cerebral cortex. It is on these lines that an understanding of the ability to distinguish pitch seems to be unfolding.

Sensitivity. The sensitivity of the ear is least with sounds of low frequency, e.g. 100 Hz. This may have the advantage that the low-frequency sounds of the body, e.g. muscle contractions, are not constantly picked up and transmitted. Since the movements of the ear drum are very small at some frequencies it seems that the maximum sensitivity of the ear is very high, but it is known to fall off steadily with increasing age. Children can usually detect sounds of 30 000 Hz, and this ability decreases by about 80 Hz every six months after reaching maturity. The brain is remarkable in its ability to single out sounds that a person wants to hear and suppressing or ignoring a large proportion of all others.

Sense of direction of sound. When the sound from a single source is perceived by both ears, the sound will be heard more loudly in one ear than the other, and also very slightly earlier. The fact that the two ears are stimulated to different extents enables the animal to estimate the direction from which the sound came (Experiment 9). Most mammals can also move their ear pinnae to a favourable position for receiving the sound, and so obtain a more accurate bearing. A source of sound that is equidistant from both ears is difficult to locate, since it can be below eye-level, directly above or behind the head and still stimulate both ears equally (Experiment 9). A dog can locate the position of a sound in one of thirty-two positions all around it, while man is accurate in the perception of only one out of eight possible sources; i.e. the dog can distinguish sounds only 11° apart while man can rarely distinguish between sounds less than 45° apart (Fig. 18.23).

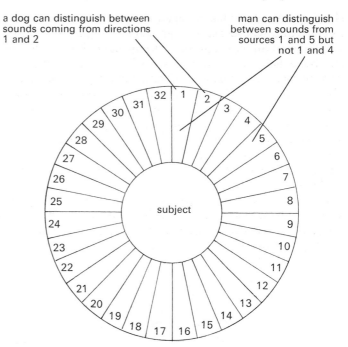

a dog can distinguish between sounds coming from directions 1 and 2

man can distinguish between sounds from sources 1 and 5 but not 1 and 4

Fig. 18.23 Discrimination of sources of sound

Speech. A large proportion of the sounds we produce during speech, particularly those sounds of low frequency, are transmitted through the bones of the skull to the inner ear and give us the impression of qualities of vibrancy and depth in our voices which are not credited to us by our listeners, who hear only the airborne sounds. Tape-recordings of one's voice are usually disappointingly flat and toneless for this reason.

The complicated co-ordination of lips, tongue and larynx during speech depends on accurate feedback of information via the ears, so that delicate adjustments can be made according to the sounds we hear ourselves making. For example, in singing a note, the first sound produced may need to be adjusted to the correct pitch as a result of the sensory feedback which indicates it to be above or below the pitch of the note required. It follows that totally deaf people will not learn naturally to speak in the usual way, because they simply cannot hear the sounds they are making and so learn to make the necessary corrections. Other sensory information, such as vibrations detected by the finger-tips, has to be used instead. Such deaf people are occasionally described as 'dumb', suggesting incorrectly that they lack the equipment for producing sounds.

Deafness. Deafness can be caused by long exposure to a high level of noise, drugs, ear infections or deposition of bone in the oval window. In the latter case, the membrane of the oval window is invaded by bone, so fusing the stapes to the skull and preventing the transmission of sound via the ear drum and ossicles. Such deaf people can still hear sounds transmitted through the skull, e.g. a ticking watch pressed to the temple, and the condition can be helped up to a point by amplification of sounds by a 'deaf aid' or by surgery in which the stapes is freed or removed altogether and replaced by a plastic strut.

Damage to the cochlea and auditory nerve, which can be caused by loud noises, produces incurable deafness. Infections of the middle ear sometimes produce fluids which burst the ear drum in escaping. Similar breaks or perforations can be produced by violent changes of pressure. Small perforations usually heal, but the thickening produced by scar tissue may impair the hearing.

We tend to be rather impatient with deaf people compared with our sympathetic attitude to the blind. This behaviour is quite irrational; deaf people need as much patience and understanding as the blind, perhaps even more, since their primary means of communicating with other people is cut off.

Balance

The semicircular canals, utriculus and sacculus are organs used in maintaining balance and posture (Fig. 18.24), the former responding mainly to changes in direction of movement, the two latter to changes in posture. The principle on which they all work involves the displacement during movement of gelatinous plates, which pull on sensitive, cytoplasmic hairs so initiating nerve impulses.

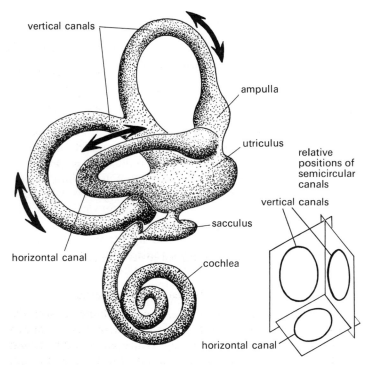

Fig. 18.24   Semicircular canals (arrows show the direction of rotation that has maximum stimulatory effect on each canal)

Utriculus and sacculus. The utriculus (Fig. 18.25) is a sac filled with *endolymph*. On its floor is an area of sensory hair cells with their supporting cells; this is called the *sensory macula*. The hairs of the sensory cells are embedded in a gelatinous plate, containing chalky granules called *otoliths* which increase its density. The gelatinous plate is not quite horizontal and so pulls on the hairs of the sensory macula causing a steady stream of impulses to reach the brain. Any change of posture will tend to displace the gelatinous plate and it will exert more or less pull on the hair cells, causing more or less rapid pulses to be fired off in the nerve.

As a result of these pulses, reflex actions occur in which the tone of the body muscles is adjusted to maintain the body in a stable position. For example, if a moving vehicle suddenly stops, standing passengers are thrown forward; the reflex

contraction of the muscles of the legs will tend to correct for this temporary upset of posture.

In the sacculus, the macula and gelatinous plates are more or less vertical, but possibly do not function as organs of balance in mammals, though it is thought that they play some part in hearing at low frequencies.

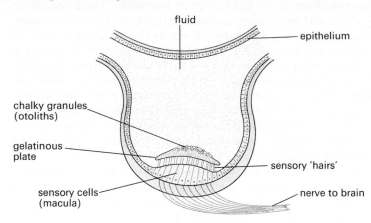

Fig. 18.25   Section through utriculus

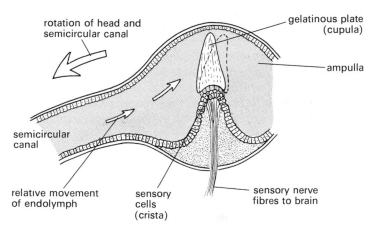

(*a*) *Section through ampulla*

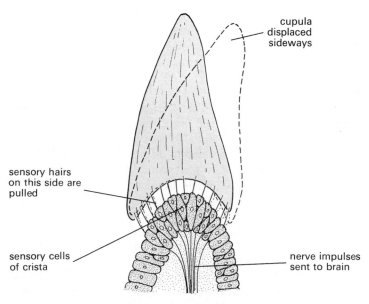

(*b*) *Detail of crista and cupula*

Fig. 18.26   Sensory mechanism of ampulla

Semicircular canals. The membranous ducts of the semicircular canals are each about 15 mm long and enclosed in corresponding but wider ducts in the bone of the skull. The space between the canals and the bone contains perilymph, while the canals are filled with endolymph. In the swelling or *ampulla* at the end of each canal is a raised mound of sensory and supporting cells, the *crista*. The tufts of hairs from the sensory cells of the crista are embedded in a gelatinous mass called a *cupula* which extends most of the way across the ampulla (Fig. 18.26a). A rotation of the semicircular canal in its own plane causes the endolymph to exert more pressure on one side of the cupula than the other. The fluid tends to lag behind the movement of its enclosing canals and, although the canals are probably too narrow for it actually to flow, a difference in pressure as little as 0.05 mmHg (in fish) is sufficient to produce a movement of the cupula.

The cupula on being displaced to one side pulls on the hair cells (Fig. 18.26b) and alters the pattern of nervous discharge originating from them. Rotation in one direction causes a greater frequency of discharge, while rotation in the other direction reduces the frequency of the discharge pattern. Each semicircular canal responds most strongly to rotations in its own plane.

The reflex actions resulting from the stimulus of the semicircular canals involve movements of the eyes and muscular adjustments to keep the body suitably orientated during movement.

Much of our information about posture and movement is derived from our eyes and muscle stretch receptors, as well as from the semicircular canals and utriculus. Any conflict in this information may produce dizziness. Motion sickness is often caused by constant rotational movements about a horizontal axis, e.g. swinging round corners in a car on a winding road.

Practical Work

SKIN: Experiment 1   **Spatial discrimination**

A simple apparatus such as is shown in Fig. 18.27 can be used. The distance apart of the points is measured, and starting at about 5 cm apart, the experimenter touches the two points simultaneously on parts of the skin of a 'volunteer' whose eyes are closed. In the more sensitive regions of the skin, the two points are felt as separate stimuli. Elsewhere, e.g. the back of the hand or the neck, a single stimulus is felt, perhaps because there are fewer touch endings. By reducing the distance between the points from time to time, the degree of sensitivity of different regions of the skin can be mapped out.

It is best to vary the stimulus, using sometimes one point and sometimes two so that the subject does not know in advance which is to be used.

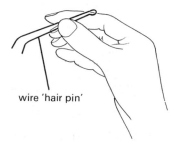

wire 'hair pin'

Fig. 18.27   Apparatus for applying double or single stimulus

Experiment 2   **Sensitivity to touch**

A patch of skin on the back of the wrist is marked with regular dots, using a rubber stamp such as is illustrated in Fig. 18.28. A similar pattern is stamped on a piece of paper so that the results can be recorded. Held by a pair of forceps or stuck to a wooden handle, a bristle such as a horse hair (Fig. 18.28) is pressed on the skin at each point marked by a dot, with enough force just to bend the bristle. The subject, who must not watch the experiment, states when he can feel the stimulus. A third person, with a duplicate set of marks on paper, indicates the positive or negative result of each stimulus, which can finally be expressed as a percentage. Using the same technique on different parts of the skin the relative concentration of touch organs can be estimated, though the sensitive 'spots' do not correspond to single nerve-endings.

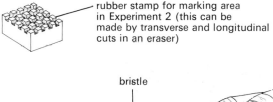

rubber stamp for marking area in Experiment 2 (this can be made by transverse and longitudinal cuts in an eraser)

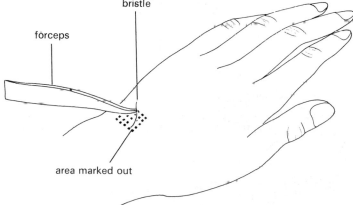

bristle

forceps

area marked out

Fig. 18.28   Testing sensitivity to touch

Experiment 3   **Location of the source of stimulation**

If a dried pea or marble is rolled about on a table between the crossed tips of the first and second fingers, an impression of two solid objects is received in the brain. The eyes should be closed while doing this so that only sensations of touch are received. Normally these regions of the fingers are stimulated simultaneously only by two separate objects, and it is this impression that has been learned by the brain.

Experiment 4   **Sensitivity to temperature**

Obtain three jars or beakers of about the same size and fill one with cold water (10–15 °C), one with hot water (40–45 °C) and the third with warm water (about 25 °C). Place the first finger of the left hand in the cold water and the first finger of the right hand in the hot water and leave both fingers immersed for at least one minute. After this time, remove the fingers from the hot and cold water and dip them *alternately* in and out of the warm water. The finger which has been in cold water will give the sensation of warmth while the other finger will register cold.

The mechanism of temperature sensitivity is not well understood but the result of this experiment suggests that the receptors respond not so much to the actual temperature as to the *change* of temperature. The cold finger registers an increase while the warm finger registers a decrease in temperature.

EYES: Experiment 5 **Inversion of the image**

If the apparatus shown in Fig. 18.29 is held close to the eye and the pin observed by looking through the pin-hole, an upright silhouette of the pin's head is seen. The apparatus is now reversed so that the pin is nearer to the eye, and moved until the pin-head can be seen against the outline of the pin-hole. In this case, an upright and enlarged shadow is cast on to the retina, and the brain makes the usual correction so that the impression gained is of the pin-head upside down.

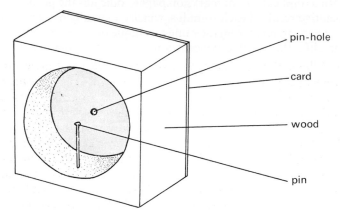

Fig. 18.29　Apparatus for showing inversion of the image

Experiment 6 **The blind spot**

Hold the book about 60 cm away. Close the left eye and concentrate on the cross with the right eye. Slowly bring the book closer to the face. When the image of the dot falls on the blind spot it will seem to disappear.

Experiment 7 **To find which eye is used more**

A pencil is held at arm's length in line with a distant object. First one eye is closed and opened and then the other. With one the pencil will seem to jump sideways. If the pencil 'jumps' when the right eye is closed, it means that the right eye was dominant in lining up the pencil.

Experiment 8 **The double image**

If a nearby object is observed, and a finger pressed against the lower lid of one eye so as to displace the eye-ball slightly, an impression of two separate images will result, one formed in each eye.

EARS: Experiment 9 **Location of sound**

A ticking clock is held, in turn, above, behind, and in front of a blindfolded subject, so that it is always equidistant from both ears. The subject is asked to indicate its position. These results are then compared with the number of successes scored when the clock is held in similar positions but to the sides of the subject.

TONGUE: Experiment 10 **Sensitivity to taste**

Solutions of sucrose (sweet), sodium chloride (salt), citric acid (sour) and quinine (bitter) are prepared. The subject puts out his tongue and the experimenter places a drop of one of the solutions at one point of the tongue using a glass rod or pipette. Without withdrawing the tongue, the subject tries to identify the taste. The various solutions are applied in turn to all parts of the tongue, washing the glass rod or pipette between each application. In this way it may be possible to determine (a) which regions of the tongue are most sensitive to a particular group of chemicals, and (b) the minimum concentration needed to produce a sensation of taste.

See *Experimental Work in Biology 8. Human senses* for other experiments (*see* p. 320).

Questions

1 Most animals have a distinct head end and tail end. Why do you think the main sensory organs are confined to the head end?

2 Chemicals such as sugar and saccharine both taste sweet and yet they are chemically quite different. Middle C on the piano has a frequency of 264 Hz whereas D has a frequency of 297 Hz. The difference is small and yet the two notes are easily distinguished.

　What are the properties of the sense organs concerned which make for poor discrimination of chemicals and precise discrimination of sounds?

3 In what functional way does a sensory cell in the retina differ from a sensory cell in the cochlea?

4 An eye defect known as 'cataract' results in the lens becoming opaque. To relieve the condition, the lens can be removed completely. Make a diagram to show how an eye without a lens could, with the aid of spectacles, form an image on the retina. What disadvantage would result from such an operation?

5 In poor light, an object can be seen more clearly in silhouette by looking to one side of it than by looking at it directly. Explain this phenomenon.

6 A person whose ear ossicles are ineffective can often hear a ticking watch pressed against his head better than he can if it is held close to his ear. Explain this effect.

7 On board a ship which is pitching and rolling it is fairly easy to maintain an upright posture when standing still. When walking, however, it is very difficult to maintain balance. Suggest reasons why this should be so.

19
Co-ordination

The various physiological processes in living animals have been described so far as if they were quite separate functions of the body, the total result of which produces a living organism. In fact, although this is true in a limited sense, all these processes are very closely linked and dependent on each other. The digestion of food, for example, would be of little value without a blood stream to absorb and distribute the products; release of energy in a contracting muscle would quickly cease if the lungs failed to supply oxygen via the circulatory system.

The working together of these systems is no haphazard process. The timing and location of one set of activities is closely related to the others. Some examples may give a clearer idea of this. During locomotion, while the muscles that pull the leg forward are contracting, the antagonistic muscles are relaxed without the walker's having to think consciously about it. During exercise, when the muscles need to lose excess carbon dioxide and obtain more glucose and oxygen, the breathing rate is automatically increased and the heart beats faster, so sending a greater volume of oxygenated blood to the muscles. When eating a meal, the position of the food is recorded by the eyes and as a result of this information the arms are moved to the right place to take it up, not by trial and error but with precision and accuracy. As the food is raised to the mouth, the latter opens to receive it at just the right moment. Chewing movements commence and saliva is secreted. At the moment of swallowing, many actions happen at the same time, as described on p. 62. In the stomach, the gastric glands begin to secrete enzymes which will digest the food when it arrives.

In the sequences described above, many bodily functions come into action at just the right moment, with the result that no unnecessary movements are made and enzymes are not wasted by being secreted when no food is present. The linking together in time and space of these and other activities is called co-ordination. Without co-ordination, the bodily activities would be thrown into chaos and disorder. Food might pass undigested through the alimentary canal for lack of enzyme secretion, even assuming it negotiated the hazard of the windpipe in the absence of unco-ordinated swallowing; extensor and flexor muscles of limbs might contract simultaneously instead of alternately; a runner would collapse after a few yards through lack of an increased blood supply to his muscles.

The activities of the organs and systems of the body are not only closely related to each other and the overall pattern of activity within the body, but to changes outside the body. In response to environmental changes the mammal makes adjustments that tend to maintain its internal conditions, e.g. its temperature, its posture, and the composition of its body fluids, and also reacts by patterns of behaviour that favour its survival, e.g. it moves towards food and away from danger. Our elaborate sense organs receive stimuli from the outside world and convert these into nerve impulses which are transmitted to the brain. As a result, these impulses may cause alterations in the pattern of activity of the organs in the body.

Co-ordination is brought about by the nervous system and the endocrine system. The former is a series of conducting tissues running to all parts of the body (Fig. 19.1), while the latter comprises a number of glands in the body which produce chemicals that are circulated in the bloodstream and stimulate certain organs when they reach them.

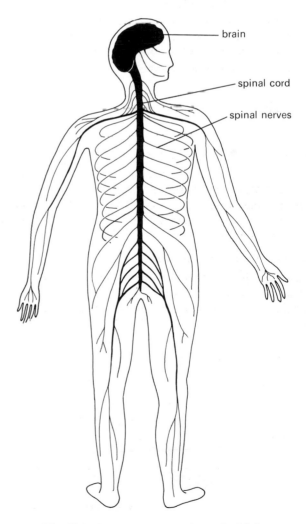

Fig. 19.1   Nervous system of man ($\times 1/14$)

brain

spinal cord

spinal nerves

The nervous system

Nerve cells. The basic units of the nervous system are nerve cells or *neurones*. Although they vary greatly in structure they may be visualized generally as consisting of a cell body, i.e. a central nucleus in a cytoplasmic mass, from which branch numerous filaments or *dendrites* (Fig. 19.2). Often one of these filaments is very long, and is called an *axon* if it conducts impulses away from the cell body and a *dendron* if the impulses are carried towards the cell body; in more general terms it is simply called a nerve fibre. In mammals the cell bodies are mostly confined to the brain or spinal cord (Fig. 19.3) while many fibres extend the whole distance to the organ being supplied. The fibre, consequently, is sometimes very long; e.g. from the spinal cord to the foot the fibre could be over a metre long, though only 1–20 μm in diameter. The fibre contains cytoplasm with few organelles, and in many cases the fibre is itself enclosed in a sheath of fatty material, the *myelin sheath*, which is formed by distinct cells, *Schwann cells*, wrapped several times around the fibre. Between the Schwann cells, which extend 0.3–1.5 mm along the fibre, are short zones of exposed axon called *nodes of Ranvier*. The myelin sheath has an important insulating effect,

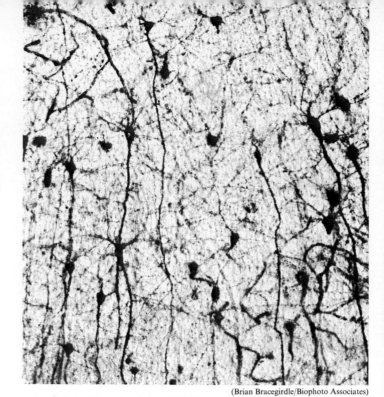

(Brian Bracegirdle/Biophoto Associates)

Fig. 19.3   Multipolar neurones in brain cortex ($\times$ 350)

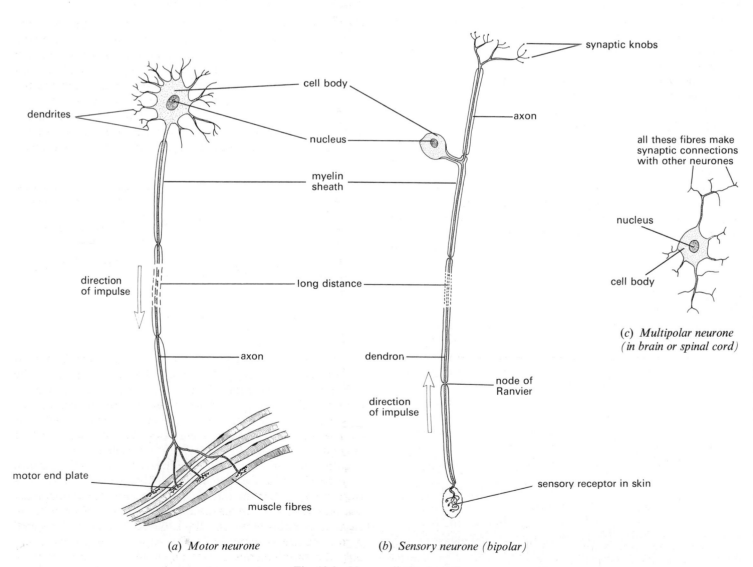

(a) *Motor neurone*

(b) *Sensory neurone (bipolar)*

(c) *Multipolar neurone (in brain or spinal cord)*

Fig. 19.2   Nerve cells (neurones)

causing the impulse to travel rapidly from one node to the next, greatly increasing the speed of conduction.

Nerve impulses. The nature of the conduction along a nerve fibre is not comparable with electrical conduction along a wire. In the latter there is a flow of electrons along the wire, the rate of flow depending on the voltage difference between the ends of the wire. In the nerve fibre a difference in potential of about 90 millivolts is built up between the inside and outside of the fibre, the inside being negative. It is thought that a stimulus applied to a sense organ causes a temporary reversal of this potential difference, so that the fibre is discharged along its length. The rate at which this wave of discharge travels varies between 1 metre and 120 metres per second, according to the diameter of the fibre and the presence or absence of the thick myelin sheath formed by the Schwann cells. Once the nerve impulse has passed down the fibre, the potential difference has to be restored again, probably by means of energy released in the respiration of the nerve cell. A good blood supply bringing fresh supplies of glucose and oxygen is essential to normal nervous conduction.

The nerve cell is ready for conduction again in a period of time ranging from 1/500 to 1/1 000 of a second. This is such a short interval that even a very brief stimulus will give rise to a series of electrical discharges at rates usually between 20 and 100 per second. A stimulus below a certain strength, i.e. below the *threshold of response*, will produce no impulse, but once the threshold is exceeded the impulse so initiated will travel to the end of the nerve fibre without diminishing in intensity. In fact, the intensity of the impulse depends only on the potential difference between the inside and outside of the fibre and bears no relationship to the intensity of the stimulus. A strong stimulus will, however, produce a more rapid burst of impulses than a weak stimulus (Fig. 19.4) and in this way the central nervous system can distinguish between stimuli of varying intensity.

Although impulses may be initiated by heat, light, touch, etc. in the respective sense organs, there is no difference in the quality of the nervous impulse travelling in the fibres. If one could 'tune in' to the optic nerve from the eye or the auditory nerve from the ear, one would detect only rapid bursts of

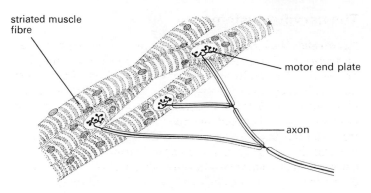

Fig. 19.5   Motor nerve ending (one nerve fibre may branch to supply 100 muscle fibres; in an eye muscle, however, each muscle fibre has its own nerve fibre giving very precise control)

electrical activity. It is the response of the region of the brain at which the impulses arrive that enables us to distinguish the type of stimulus. If the auditory and optic nerves could be transposed so that impulses from the eye reached the part of the brain normally served by the ear and vice versa, a light shone in the eye would give the sensation of noise and a shout in the ear would cause the subject to see flashes of light.

Motor endings and synapses. A motor nerve fibre, i.e. one carrying an impulse from the central nervous system to a muscle or gland, branches repeatedly in the muscle it is supplying, each branch terminating in a *motor end plate* (Figs. 19.5 and 19.7) on a single muscle fibre. When a burst of impulses reaches the motor end plate they cause the muscle fibres to contract.

Conduction of an electrical impulse from one neurone to the next takes place across *synapses*. Branches from the axon of one neurone make contact with the dendrites or cell body of the next

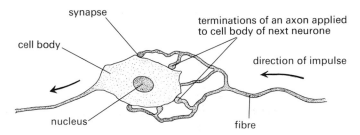

Fig. 19.6   Schematic diagram of synapses

neurone in line. In most cases, there are minute gaps (20 nm), at these points of contact. The arrival of a nerve impulse at the termination of the axon branch causes the release of a small quantity of a chemical (a *neuro-transmitter substance*) into the gap. This chemical stimulates the next neurone to generate its own nerve impulse. Several impulses may be needed before enough chemical is released to fire the next neurone.

Although synapses form a temporary barrier to the passage of a nerve impulse, they do enable one fibre to make connections with many other neurones and produce correspondingly complex reactions. Synapses also act as 'on-off' switches in computing appropriate reactions. On some cells of the spinal cord there are as many as 2 000 synapses. It is probable that only the simultaneous arrival of impulses at many of these synapses will set off an impulse in the nerve cell.

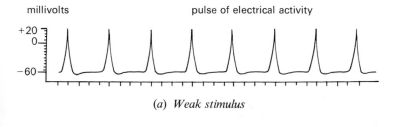

(a) Weak stimulus

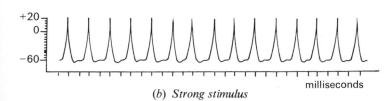

(b) Strong stimulus

Fig. 19.4   Pattern of electrical activity in a single nerve fibre as measured by sensitive recording apparatus. (*Note*—there is no difference in amplitude (strength) of the electrical discharge; only the frequency changes)

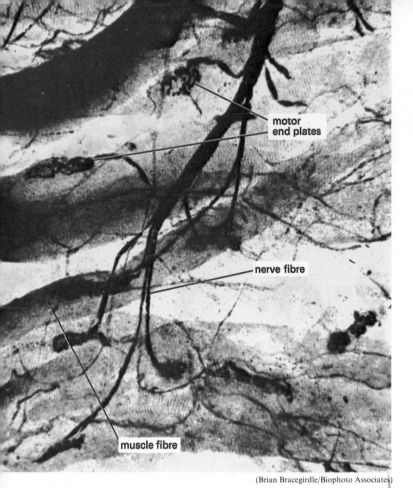

Fig. 19.7   Photomicrograph of motor end plates ($\times$ 500)

(Brian Bracegirdle/Biophoto Associates)

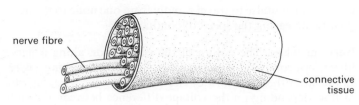

(a) Nerve fibres grouped together to form a nerve

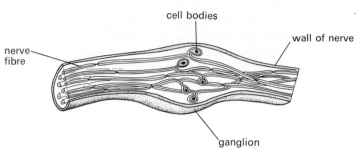

(b) When cell bodies occur outside the brain and spinal cord they produce a bulge (ganglion)

Fig. 19.8   Structure of a nerve

The nervous system. The thousands of nerve cells are organized into a nervous system. The cell bodies are grouped largely into the *central nervous system*, i.e. the brain and spinal cord, and the fibres are arranged in bundles with connective tissue to form nerves (Fig. 19.8) which are clearly visible during dissection. The *cranial nerves* leave the brain through holes in the skull, while the *spinal nerves* leave the spinal cord between adjacent vertebrae and usually carry both sensory and motor fibres. From between the vertebrae of the neck emerge nerves which supply the diaphragm and the skin and muscle of the neck and arms. The nerves from the thoracic region of the spinal cord supply the skin and muscles of the thorax, while those from the lumbar and sacral regions go to the legs and also supply the skin and muscle of the abdomen.

Each spinal nerve contains both sensory and motor fibres carrying impulses towards and away from the spinal cord, and also fibres of different diameter. Narrow fibres conduct more slowly than wider ones, with the result that impulses generated in one region will travel at different rates and arrive in the central nervous system at different times.

Reflex. One of the patterns of co-ordination that is simplest to explain in terms of nervous conduction is the reflex. A reflex action is a quick, automatic response to a particular stimulus. The response does not require conscious control and in some cases cannot be influenced by the conscious will. A *spinal reflex* is one that need not involve the brain for its successful completion. Coughing, sneezing and blinking can all be reflex actions. The path traversed by the nerve impulse during a reflex action is called a *reflex arc*, and such a path (Fig. 19.9) is described below for the withdrawal reflex that results from inadvertently touching something hot.

The sensory endings in the skin of the hand are stimulated by the heat, and set off a stream of nerve impulses which travel in the sensory fibres running up the arm in a nerve to the spinal cord in the neck. For simplicity, only one of these fibres will be considered. In the spinal cord, the sensory fibre makes a synapse with a short *relay neurone*. The volley of nerve impulses crosses this synapse, passes along the relay neurone and traverses a further synaptic connection with the dendrites and cell body of a motor neurone (Fig. 19.10). The motor fibre carries the impulses, in the same spinal nerve, to the biceps muscle of the arm. The muscle contracts, flexes the elbow and withdraws

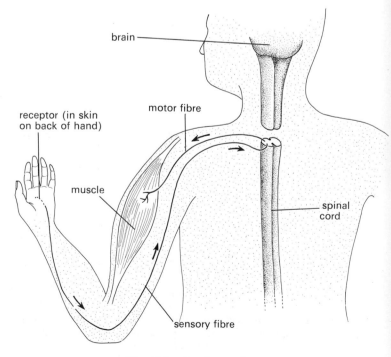

Fig. 19.9   A reflex pathway

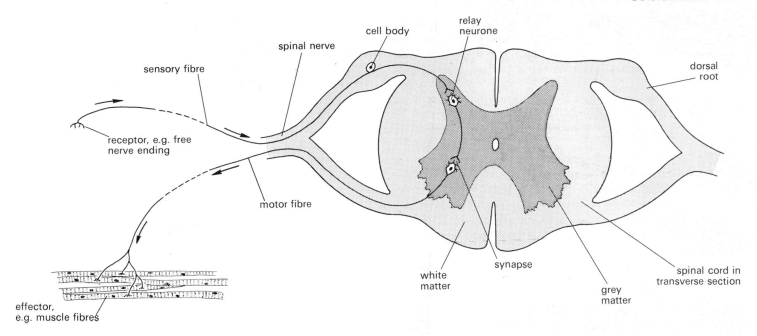

Fig. 19.10   One of the simplest connections for a reflex action

the hand from the hot object (Fig. 19.11). This prompt withdrawal prevents serious damage to the tissues of the fingers, and many human reflexes have this protective function.

Although, theoretically, the brain is not necessary for this reaction, the relay neurones will make synaptic connections with sensory fibres running to the brain, and impulses reaching the brain by this path will make the person aware of the heat of the stimulus and give rise to a sensation of pain. Motor impulses may consequently be sent from the brain and add to or modify the pattern of reflex response, e.g. cause a cry of pain. Moreover, the reflex can be suppressed by the brain. A hot object which, if touched unexpectedly, would produce a reflex withdrawal can nevertheless be deliberately touched or grasped.

The sensory data fed into the brain, namely the heat and pain arising from the stimulation of the skin and information about the object coming from the eyes, will be associated in the higher

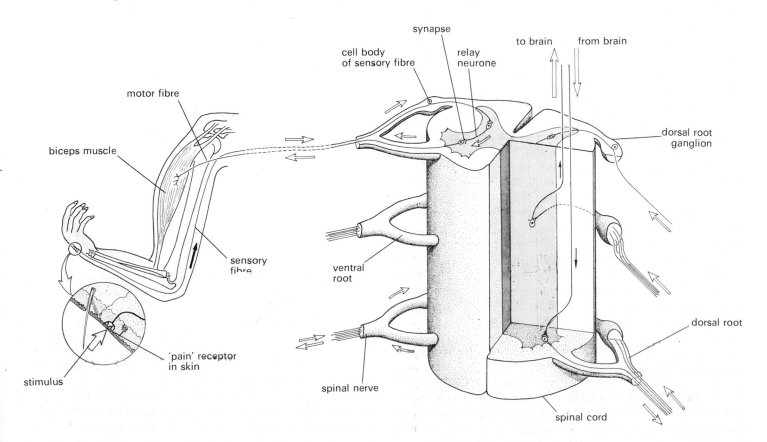

Fig. 19.11   The reflex arc (withdrawal reflex)

centres of the brain such as the cerebral cortex in a memory store, and so reduce the chances of the same hazardous situation arising again. When the same object is encountered in the future it will be recognized as likely to be hot and contact will be avoided.

If an injury severs the spinal cord some spinal reflexes are retained below the damaged area, but in a normal individual, all reflexes are influenced to some extent by the activities of the brain.

Reflex actions are rarely as simple as the description of the withdrawal reflex above would make it seem. They usually involve much more sensory data and a far more complex response by many sets of muscles. Coughing, for example, is a reflex initiated by the irritation of the trachea by a foreign particle and involves precise co-ordination of the contraction of the abdominal muscles and temporary closure of the glottis to produce the explosive bursts of exhalation which tend to dislodge the particle. In the withdrawal reflex described for the hand, further co-ordination is required to ensure that when the biceps muscle is suddenly contracted, its antagonistic muscle, the triceps, is relaxed. Synaptic connections (via the relay neurone) with the motor nerve cells supplying the triceps suppress nervous impulses which might normally pass to the triceps. In particular, they suppress the contraction of the triceps, which would normally follow from the stimulation of the stretch receptors, when the muscle is extended by the sudden contraction of the biceps. This, incidentally, is a good example of synaptic connections that inhibit rather than excite their connecting neurones.

Other examples of reflex actions are sneezing, in response to the irritation of the nasal mucous membrane; ejaculation of semen at the climax of sexual intercourse; secretion of saliva in response to taste stimuli on the tongue; contraction of the iris diaphragm in response to intense illumination; and the many responses under the control of the autonomic nervous system described on p. 151. In some cases the reflex pathways lie principally in the brain, e.g. iris diaphragm response and salivation, but the reflex arc involved is basically the same as in the spinal cord.

Conditioned reflex. The conditioned reflex is one way in which animals and perhaps man may learn. The Russian biologist, Pavlov, carried out many experiments on the conditioned reflex in dogs, and since one of them is something of a classic, it is described here. If a dog tastes or smells food, the salivary gland secretes saliva as a result of an inborn reflex, i.e. the dog does not have to learn this behaviour. If a bell is rung, no salivation occurs; but if a bell is rung at the same time as food is presented on a number of occasions, eventually the ringing of the bell alone will cause salivation. The appropriate stimulus of taste has been replaced or supplemented by an inappropriate stimulus of sound and the dog has been conditioned to respond to a new stimulus. Our guess is that, in some way, new nervous pathways have been established in the brain, perhaps by forming fresh synapses between neurones or facilitating the path of impulses across existing synapses. Fig. 19.12 shows this in pictorial form but our knowledge of what really happens is very scant. The conditioned reflex established in the laboratory cannot readily be extended to explain behaviour in the natural life of the animal. The dog in the experiment has to be trained to stand still throughout the experiment and all other sources of stimuli must be eliminated, whereas in its normal life the animal

would be receiving a wide variety of stimuli and selecting information from them while constantly responding to internal and external changes.

Establishment of conditioned reflexes may be the basis of training of animals, the conditioning being achieved sometimes

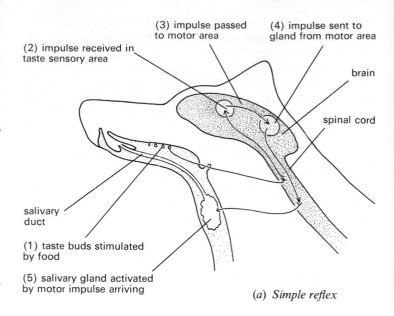

(a) Simple reflex

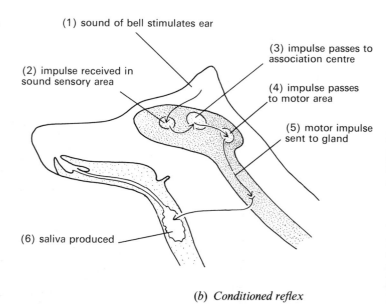

(b) Conditioned reflex

Fig. 19.12   Possible nervous pathway for conditioning

by reward and sometimes by punishment. To what extent conditioned reflexes as apparently simple as these play a part in human behaviour is difficult to say. We learn to walk, to swim, to ride bicycles and a great many other complex patterns of activity, which we carry out more or less efficiently without having to think consciously about what we are doing. If an experienced car driver rides as a passenger with another driver who approaches hazards at speed and pulls up rather late, the passenger may find his right foot pressing hard on the floor, as if to apply the foot-brake. You may have done similar things, e.g. switched off the light on leaving a room leaving the family in darkness, or wound up your watch automatically when taking it off in the middle of the day. These have all the appearances of conditioned reflexes and there are many similar examples, but it is misleading to think of learning as isolated conditioned reflexes.

Autonomic nervous system. The part of the nervous system that co-ordinates the internal and largely involuntary bodily activities such as digestion, vasoconstriction, heartbeat and blood pressure, is called the autonomic nervous system. The structure of its neurones and their system of connections to the organs is slightly different from that of the rest of the nervous system. The final nerve fibres supplying the organs do not have a thick myelin sheath and are thus sometimes called *unmedullated*. In addition, the cell bodies of these motor fibres do not lie in the spinal cord but are enclosed in ganglia outside it. The autonomic nervous system is further subdivided into the *sympathetic* and *parasympathetic* systems. In general, impulses in the sympathetic nervous system prepare an animal for activity, e.g. they speed up the heart and breathing rate and divert blood from the alimentary canal to the muscles, while the parasympathetic system is concerned with conservation of the animal's resources, as in feeding and sleeping. Most organs of the body receive fibres from both the sympathetic and para-

sympathetic nervous systems and impulses in these fibres produce opposite, i.e. antagonistic, effects. For example, impulses in the sympathetic fibres supplying the heart make it beat faster, whereas stimulation by the parasympathetic fibres slows it down. The final synapses of the motor fibres in the sympathetic system occur in the ganglia, but in the parasympathetic system the final synapses are in the organ being supplied. Fig. 19.13 shows a few of the organs that receive sympathetic and parasympathetic fibres and indicates the function of these fibres. It is not possible to generalize about the effects of the two sets of fibres and say that one always enhances while the other inhibits the organ's activity. For example, although the parasympathetic fibres reduce the number of heart contractions, they increase the rate of peristaltic contractions in the alimentary canal.

Structure and function of the central nervous system

The spinal cord. The spinal cord consists of a great number of nerve cells, both fibres and cell bodies, grouped into a cylindrical mass, running from the brain to the second lumbar vertebra. It is enclosed in two fibrous membranes, the *meninges*, and protected by the bone of the spinal column. From between

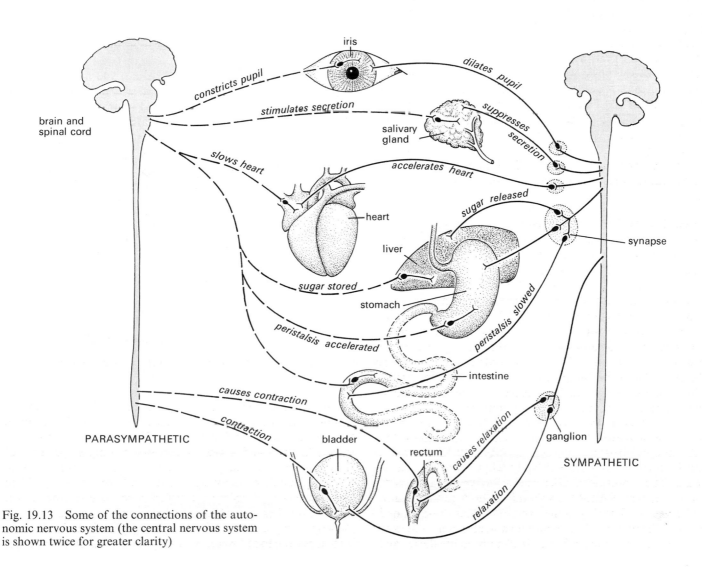

Fig. 19.13 Some of the connections of the autonomic nervous system (the central nervous system is shown twice for greater clarity)

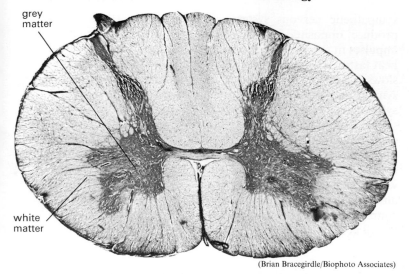

Fig. 19.14   Section through spinal cord (×8)

(Brian Bracegirdle/Biophoto Associates)

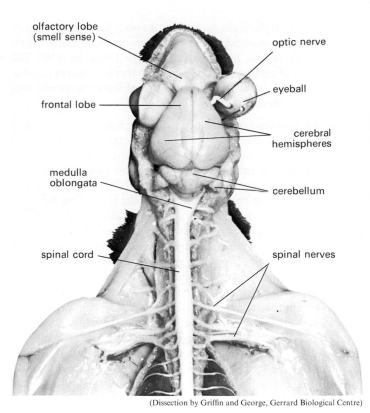

(Dissection by Griffin and George, Gerrard Biological Centre)

Fig. 19.15   Dissection of the brain and spinal cord of a rabbit (seen from above)

the vertebrae, spinal nerves emerge and run to all parts of the body. The fibres of these nerves may be concerned with spinal reflexes, or may carry sensory impulses to the brain or motor impulses from the brain to the muscles and other effector organs of the body. A central canal containing *cerebrospinal fluid* runs through the centre of the spinal cord.

The nerve cell bodies are grouped in the centre of the cord making a roughly H-shaped region of *grey matter* when seen in transverse section (Fig. 19.14). Outside this is the *white matter* consisting of nerve fibres running up and down the cord or passing out to the spinal nerves. The spinal cord is concerned with the spinal reflex actions described above, and the conduction of nervous impulses from one region of the spinal cord to another and to and from the brain (*see* Fig. 19.11). Most of the sensory information from the skin and muscles reaches the brain by way of the spinal cord, and all the 'commands' from the brain to the muscles are conveyed in motor fibres through the spinal cord and out into the spinal nerves.

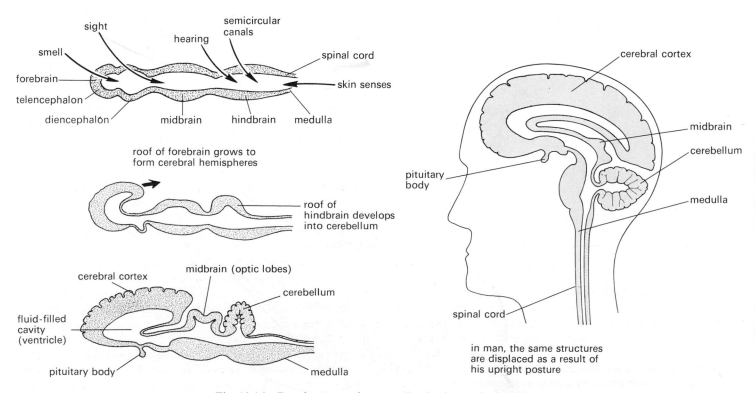

Fig. 19.16   Development of mammalian brain (vertical sections)

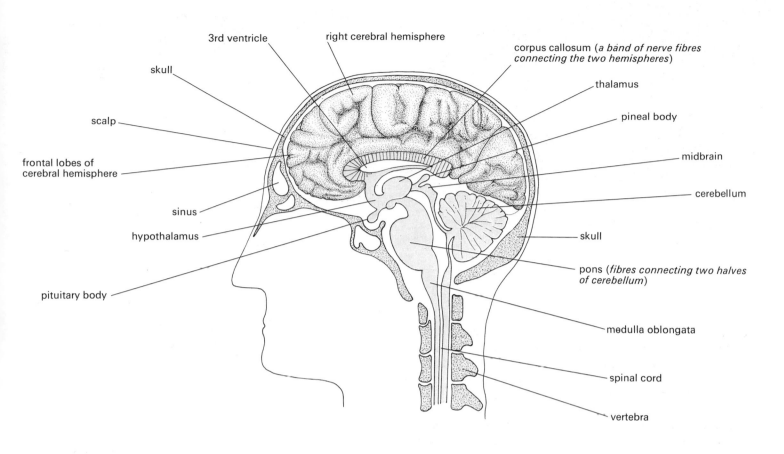

Fig. 19.17   Section through head to show brain

The brain. It seems likely that during evolution the increasing effectiveness and specialization of the sensory organs of the head, particularly the eyes, ears and nose, led to more and more sensory fibres entering the front part of the spinal cord. Consequently, this region has enlarged and developed to form the brain of the vertebrate animals. It consists of some 15 000 million neurones.

The brain is thus an enlarged, specialized, front region of the spinal cord (Fig. 19.15). Like the spinal cord it is basically cylindrical with a central canal containing cerebrospinal fluid. The nerve fibres constituting the white matter lie outside the central grey matter which consists of cell bodies. Twelve pairs of sensory and motor nerves enter and leave the brain, though not in the regular sequence seen in the spinal cord. The complicated structure of the brain is probably best understood by tracing in a simplified form its development in the embryo from the neural tube, the rudimentary central nervous system (Fig. 19.16). The anterior part of this tube enlarges in such a way that three regions or *vesicles* are distinguishable, the *fore-*, *mid-* and *hindbrain*. The enlarged regions of the cerebrospinal canal in these vesicles are called *ventricles* and when the brain is fully developed they may communicate by only narrow ducts. In the course of development two regions can be distinguished in the forebrain, the *telencephalon* in front and the *diencephalon* behind it.

The roof of the telencephalon grows out on each side to form the cerebral hemispheres which enlarge to such an extent that they come to overlie the whole of the rest of the brain

(Fig. 19.17). Outgrowths from the sides of the diencephalon form the *optic vesicles* which contribute to the retina of the eyes. Internally, the side walls of the diencephalon thicken to form the *thalami*, while a downgrowth from the ventral surface fuses with an upgrowth from the roof of the developing mouth and forms the *pituitary body*. The pituitary body retains its connection with the floor of the diencephalon or *hypothalamus*. The roof of the diencephalon, which remains thin and vascular, secretes cerebrospinal fluid into the third ventricle, and is called the *anterior choroid plexus*. The roof of the midbrain thickens to form the *optic lobes* while in the floor is a tract of fibres linking the thalami with the hindbrain. The anterior part of the roof of the hindbrain enlarges greatly to form the *cerebellum* while the posterior portion remains thin, vascular and non-nervous forming the *posterior choroid plexus*, which secretes cerebrospinal fluid into the fourth ventricle. The floor and sides of the hindbrain thicken to form the *medulla*.

In the primitive (and hypothetical) brain the primary vesicles merely accept sensory impulses and send out the appropriate motor responses, i.e. a reflex associated with a particular sense organ. The forebrain, for example, receives sensory information from the organs of smell, the midbrain from the eyes, and the hindbrain from the ears and semicircular canals. The structures such as the cerebral hemispheres, thalamus, optic lobes and cerebellum are not, however, principally concerned with one particular sense organ but are *association centres*, receiving information from a variety of sources and then

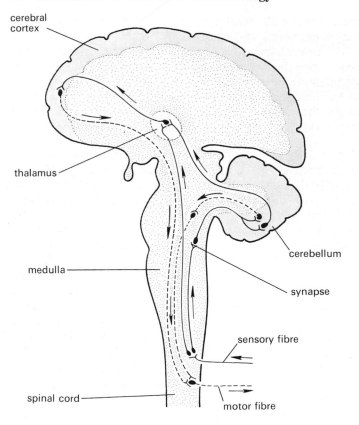

Fig. 19.18 Some of the main sensory and motor pathways

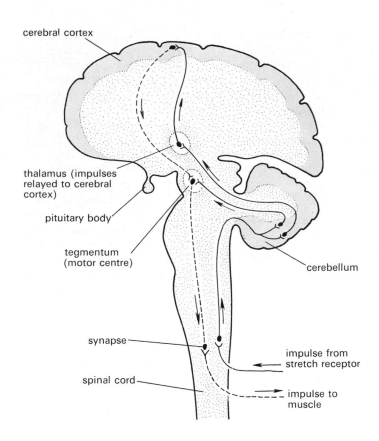

Fig. 19.19 Some of the nervous pathways involving the cerebellum and tegmentum

relaying impulses to other regions of the brain (Fig. 19.18) to produce a greater flexibility of behaviour.

If the nervous system were nothing more than a system of direct pathways from sensory organs to effectors, (a) the number of different possible responses would be very limited, and (b) they would always be the same in a given situation. An animal would respond to the smell of food by approaching it irrespective of whether it was hungry or not. A jackal would approach a carcass in response to its smell regardless of the sensory data from its eyes indicating that a lion was in possession. These are purely hypothetical illustrations, but a set of unalterable nerve circuits would produce a set of unalterable responses and sensory information from different sources could not effectively interact.

The jackal's brain will, in fact, receive sensory data from nose, eyes and stomach indicating, though not necessarily in the conscious sense, food, danger and hunger. These impulses are received in the primary sense centres of the fore-, mid- and hindbrain but are relayed to association centres where they may suppress or reinforce each other to produce a response with maximum survival value. Stored information from memory centres will also reach the association centres, and the outcome will not be automatic and unnecessary flight from the lion or uncontrollable and hazardous approach to the food, but a delaying action until the combination of stimuli produces a different behaviour pattern.

Knowledge of the function of different parts of the brain comes from a variety of sources; electrical stimulation of small areas of the exposed brain during operations for removal of brain tumours, for instance, may produce motor responses if a motor area is stimulated or a recognizable sensation if a sensory area is stimulated; detailed anatomical studies reveal the nervous connections between the sense organs, the spinal cord, effectors and association centres so that functions may be inferred from these connections; sensitive electrical measuring instruments can detect the bursts of electrical activity that occur in localized areas of the brain when parts of the body are moved or stimulated in some way; accidental damage (in man) and experimental brain lesions (in animals) indicate, to some extent, the function of the part of the brain removed according to the activities that the animal can no longer perform normally. All these methods have their defects; stimulation of only a small area of the brain at its surface is a quite abnormal state of affairs, and removal of one part of the brain interferes with so many nervous pathways that it is difficult to tell whether the new behaviour observed is a result of the region removed or the other centres affected by its removal.

Medulla. The medulla is largely concerned with the automatic adjustment of bodily functions, e.g. heart and breathing rates, blood pressure and temperature regulation, and is the primary sense centre for taste, hearing, balance and posture; i.e. the sensory fibres from the tongue, the cochlea, the semicircular canals, some touch receptors and some muscle spindles make their first synapse here. The medulla receives sensory information from internal organs by way of the *vagus nerve* and, after synaptic connections in the tissue of the medulla, motor impulses are fired off, also in the vagus, to the heart, diaphragm, etc. Rhythmical discharges also arise from it which influence the rhythmical changes in the body, e.g. the breathing rhythm.

Although most of these actions are under reflex control via the medulla, the latter is influenced by higher centres in the brain, particularly the hypothalamus. A situation involving excitement or emotion is likely to be interpreted first by the cerebral cortex which will then directly or indirectly influence the regulatory centres in the medulla, quickening the heartbeat, increasing the breathing rate, releasing more glucose into the blood stream, and so on. Similarly, most of the sensory impulses reaching the synapses in the medulla will be relayed to the higher brain centres. The impulses from the semicircular canals, for example, are passed on to the cerebellum and cerebral cortex. Relaying the information to the cerebral cortex enables the individual to become aware of what is happening as well as making reflex adjustments at the level of the medulla. A great many fibres pass through the medulla without making any synaptic connections on their way to or from the brain.

The cerebellum. The cerebellum receives, predominantly, sensory fibres from the stretch receptors of the muscles, semicircular canals, utriculus and sacculus. These fibres may proceed directly from the sense organ or the spinal cord to the cerebellum or via synapses in the medulla or other association areas. Outgoing fibres pass, not directly to effector organs, but to motor centres in the *tegmentum* (floor of the midbrain) and medulla or to the cerebral cortex via the thalamus (Fig. 19.19). This pattern of connections and the experimental results of removing parts of the cerebellum suggest that the cerebellum exercises control over posture, balance and, in particular, the finely co-ordinated patterns of muscular activity. Although the reflex responses may be co-ordinated in the cerebellum, the large tract of fibres that runs into it from the cerebral cortex indicates that the cerebellum probably translates the 'commands' from this higher centre into positive action by the muscles.

Unlike the spinal cord, there is a layer of nerve-cell bodies on the outside of the cerebellum, the *cerebellar cortex*, the surface of which is greatly increased by deep folds. In this cortex the incoming impulses in only a few fibres are enabled to make connections with a vast number of outgoing fibres via relay neurones and the densely branching dendrites of the *Purkinje cells* at the head of the outward-going fibres. In this way, the impulses are amplified and spread over a wide range of outgoing fibres so that a small, localized sensory impulse could produce a powerful and widespread response.

Without the cerebellum, movement would be jerky, violent and spasmodic instead of smooth and controlled, as if the cerebral cortex were commanding a pattern of activity but without the means of controlling its detailed execution. For instance, if one reaches for a glass of water, the initial act of extending the arm may be swift and inaccurately directed, but as the hand nears its objective the action slows down, the aim is more precise and finally, a slow, positive muscular action grips the glass. Without cerebellar control, the arm may overshoot the mark, the fingers close too late or too weakly. In this kind of action the feedback of information from the stretch receptors to the cerebellum is of great importance, indicating the position of the limb and the speed with which it is moving and, in conjunction with visual information, correcting any errors of judgement as fast as they arise.

The cerebellum probably damps down the oscillations that arise when the stretch receptors of antagonistic muscles are stimulated, e.g. when the biceps contracts, the stretch receptors of the triceps are extended and these initiate a reflex which would shorten the triceps and thus extend the biceps, whose stretch receptors would produce reflex contraction, and so on.

Hypothalamus. The hypothalamus is formed from a thickening of the walls in the lower region of the diencephalon. Part of the hypothalamus extends downwards to contribute to the pituitary body. It receives fibres from the cerebral cortex and sends impulses back to the cerebral hemispheres and to the tegmentum. The hypothalamus influences a great variety of bodily activities concerned with maintaining a constant internal environment and with finding and digesting food. Much of its motor activity is brought about by the secretion of hormones from the pituitary body. For example, it is in the hypothalamus that the osmoreceptors are situated. These detect changes in the concentration of the blood passing through the hypothalamus and cause the pituitary body to release more or less ADH which, in turn, makes the kidney reabsorb more or less water from the urine (p. 94). The temperature of the blood is also monitored in the hypothalamus, and impulses are sent out that result in vasoconstriction or vasodilation and sweating, according to whether the blood temperature is too low or too high. There are, however, other centres in the brain which play a part in temperature control.

The hypothalamus is the brain centre for the parasympathetic nervous system and accordingly plays a part in controlling blood pressure, heart rate and peristaltic movements in the alimentary canal.

Cerebral hemispheres. These enormous outgrowths of the roof of the forebrain have a cortex of nerve cells, deeply folded and grooved in such a way as to increase the total surface. In this cortex a tremendous number of possibilities arise for the interconnection of sensory, motor and relay fibres. Nearly all impulses from sense organs are relayed to the cerebral cortex which is thus in possession of all the relevant information about events in the environment and to some extent within the organism. It seems likely that in the course of learning, certain nervous pathways become established between the various sensory centres, so that complex patterns of nervous activity build up and leave a 'memory store'. From all the sensory input and memory store the cortex calculates the best course of action, i.e. that with the greatest immediate or long-term survival value, and sends out appropriate motor impulses either directly to the spinal cord or via the hypothalamus and tegmentum.

In a general sense, the pattern of behaviour for the whole animal is decided by the cerebral hemispheres, and the detailed execution is carried out by lower centres in the brain. Motor impulses start in the cortex, may be relayed to the tegmentum, translated by the cerebellum and pass down the spinal cord. Even at this level, the connections in the spinal cord will ultimately determine the pattern of movement by, for example, a limb. There are, however, fibres running from the motor areas of the cortex, which pass through the brain and down the spinal cord without making any synapses until they are about to leave the spinal cord.

Experiments in which the cortex is stimulated by fine electrodes, or in which electrical activity is detected in the cortex when a sense organ is stimulated, show that there is a certain localization of function in the cortex as indicated in Fig. 19.20. A particular region of the cochlea when stimulated would produce activity in a precise area of the cortex, and similarly a particular group of retinal cells is ultimately connected with a specific area of the cortex, but the interconnections between these areas are, in a way, more important than this localization.

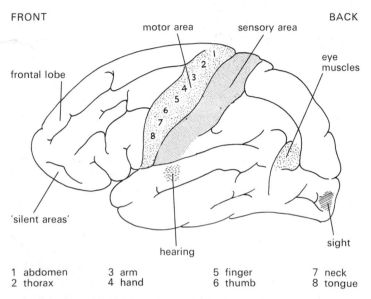

FRONT   BACK

Fig. 19.20   Localization of areas in the left cerebral hemisphere

| 1 abdomen | 3 arm | 5 finger | 7 neck |
| 2 thorax | 4 hand | 6 thumb | 8 tongue |

Consciousness is probably the outcome of activity in the cortex, but we are able to suppress from consciousness a great deal of the sensory information which pours constantly into the cerebral hemispheres and single out for attention only those aspects on which we want to concentrate. One can listen to a single conversation in the midst of a crowded room where everyone is talking.

A complex group of neurones called the *reticular formation* in the floor of the midbrain appears to be able to 'switch' the cortex on or off. During sleep, sensory information is still being relayed to the cortex but without reaching consciousness. On waking, the reticular formation arouses the cortex and the sensory input becomes meaningful sensations. It seems that the reticular formation can alert the cortex during wakefulness as well, to make a person 'sit up and take notice'.

The hypothalamus and medulla at the time of birth probably have built-in nerve circuits which maintain the heartbeat, blood pressure, etc. These are inborn, or inherited, patterns of nervous activity. The cerebral cortex may start life as a blank slate but with vast possibilities of incredibly complex connections. The cortex thus acquires its characteristic circuitry and activities as a result of learning during the lifetime of the individual.

Prefrontal lobes. The most anterior portion of the cerebral cortex (the 'silent area') does not produce any movement or sensation when stimulated experimentally and does not appear to be associated with any specific part of the body. Cases of severe anxiety neurosis are sometimes alleviated by severing the tracts of fibres running from these prefrontal lobes to the rest of the brain. From this and other evidence, it is suggested that this area has something to do with self-restraint, the inhibition of natural drives, perhaps even the 'conscience', which distinguish man from other mammals.

Functions of the brain. To sum up:
1   The brain receives impulses from all the sensory organs of the body.
2   As a result of these sensory impulses it sends off motor impulses to the glands and muscles causing them to function accordingly.
3   In its association centres it correlates the various stimuli from the different sense organs.
4   The association centres and motor areas co-ordinate bodily activities so that the mechanisms and chemical reactions of the body work efficiently together.
5   It stores information so that behaviour can be modified according to past experience.

The endocrine system

Co-ordination is also effected by chemicals called *hormones*, secreted from the *endocrine glands* (Fig. 19.21). Unlike the *exocrine glands*, such as the digestive glands or sweat glands, the endocrine glands have no ducts or openings. The chemicals they produce enter the blood stream as it passes through the glands and they are circulated all over the body. When the hormones reach particular parts of the body they cause certain changes to take place. Their effects are much slower and more general than nerve action and they control rather long-term changes such as rate of growth, rate of activity and sexual maturity. When they pass through the liver, the hormones are converted to relatively inactive compounds which are excreted, in due course, by the

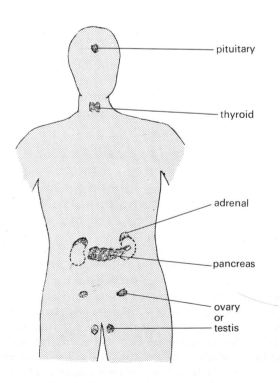

Fig. 19.21   Position of endocrine glands in the body

kidneys (hence the tests on urine for the hormonal products of pregnancy). The liver in this way limits the duration of a hormonal response which might otherwise persist indefinitely.

Thyroid. The thyroid gland is in the neck, in front of the windpipe. It produces a hormone, *thyroxine*, which controls the rate of growth and development. In adults thyroxine controls the rate of chemical activity, particularly respiration: too little tends to lead to overweight and sluggish activity; too much can cause thinness and over-activity. Deficiency of the thyroid in infancy causes a certain type of mental deficiency called *cretinism*, which can be cured in the early stages by administering thyroxine.

Parathyroid. The four parathyroid glands are small bodies attached to the back of the thyroid, two on each side. The parathyroid hormone raises the concentration of calcium in the blood by (a) causing it to be released from bone tissue, (b) promoting its reabsorption in the kidney tubules and (c) increasing its absorption in the ileum.

Adrenal. The adrenal or supra-renal glands are situated just above the kidneys. The outer layer of the adrenal body, the cortex, produces several hormones, including *cortisone*, one of whose functions is to accelerate the conversion of proteins to glucose (p. 67). Secretion by the adrenal cortex is stimulated by certain pituitary hormones.

The inner zone, the medulla, of the adrenal gland is stimulated by the nervous system and produces *adrenaline*. When the sense organs of the animal transmit to the brain impulses associated with danger or other situations needing vigorous action, motor impulses are relayed to the adrenal medulla which releases adrenaline into the blood. When this reaches the heart it quickens the heartbeat. In other regions it diverts blood from the alimentary canal and the skin to the muscles; it makes the pupils dilate and speeds up the rate of breathing and conversion of glycogen to glucose. All these changes would increase the animal's efficiency in a situation that might demand vigorous activity in running away or putting up a fight. In ourselves, they do the same but, together with the nervous system, they produce also the sensation of fear: thumping heart, hollow feeling in the stomach, pale face, etc. In humans, adrenaline may be secreted in many situations which promote anxiety or excitement, and not only in the face of danger.

The conditions that cause a release of extra adrenaline will already have stimulated the sympathetic nervous system (p. 151) and the sensations described above cannot be ascribed solely to one or the other, except in the sense that the sympathetic nervous system produces a rapid and immediate response while adrenaline secretion acts more slowly and over a longer period.

Pancreas. As well as containing cells that secrete digestive juices, the pancreas contains endocrine cells, which control the use of sugar in the body. The endocrine cells are of two kinds, α and β, and are grouped into patches of tissue called *Islets of Langerhans*. The α cells produce a hormone called *glucagon*; the β cells produce *insulin*. These two hormones, together with adrenaline, control the level of glucose in the blood. The relationships are complex and not fully worked out. In the simplest terms, glucagon promotes the conversion of glycogen to glucose when the blood sugar concentration falls; insulin is released in response to a raised level of glucose in the blood.

Insulin increases the rate of conversion of glucose to glycogen and storage in the liver. Adrenaline accelerates the conversion of glycogen to glucose in times of stress and excitement.

As described on p. 68, these changes serve to maintain a fairly steady concentration of glucose in the blood but all three hormones have other effects in the body. Glucagon also enhances the conversion of fat to fatty acids for transport in the blood. Insulin, as well as lowering the glucose level in the blood, also promotes the uptake of glucose by the body cells and increases protein synthesis in some cells.

The failure of the pancreas to produce sufficient insulin leads to one form of *diabetes*. The diabetic cannot effectively regulate the blood sugar level. It may rise to above 160 mg/100 cm^3 and so be excreted in the urine, or fall to below 40 mg/100 cm^3 leading eventually to convulsions and coma. This diabetic condition can be treated by regular injections of insulin and a controlled intake of carbohydrate.

Another form of diabetes, which occurs more commonly in older people, is not so much the result of insufficient insulin but the inability of the body cells to respond to it. This condition is treated by regulating the diet.

Reproductive organs. The ovary produces several hormones called *oestrogens* of which oestradiol and oestrone are the most potent. These oestrogens (a) control the development of the secondary sexual characteristics at puberty (*see* p. 108), (b) cause the lining of the uterus to thicken just before an ovum is released, and (c), in some mammals at least, oestradiol brings the female 'on heat', i.e. prepares her to accept the male. Progesterone, the hormone produced from the corpus luteum (p. 103) after ovulation, promotes the further thickening and vascularization of the uterus. Progesterone also prevents the uterus from contracting until the baby is due to be born.

Testosterone is the male sex hormone, produced by the testis. It promotes the development of the masculine secondary sexual characteristics.

Duodenum. The presence of food stimulates the lining of the duodenum to produce a hormone, *secretin*, which on reaching the pancreas via the blood stream, initiates the production of pancreatic enzymes. In this way, the enzymes are secreted only when food is present.

Pituitary. The pituitary gland is an outgrowth from the base of the forebrain (Figs. 19.16 and 19.17). It releases into the blood several different hormones. Some of them appear to have a direct effect on the organ systems of the body. For example, *antidiuretic hormone* (ADH) controls the amount of water reabsorbed into the blood by the kidneys (*see* p. 94). *Growth hormone* influences the growth of bone and other tissues. Growth, however, is affected by other endocrine glands as well, in particular the thyroid and pancreas, and the growth hormone may exert its influence through these glands rather than directly on the tissues.

In fact, the majority of the pituitary hormones do act upon and regulate the activity of the other endocrine glands to such an extent that the pituitary is sometimes called the 'master gland'. It is a pituitary hormone which, acting on the ovary, causes the Graafian follicle to develop and secrete its own hormone, oestrogen. Another pituitary hormone stimulates the thyroid gland to grow and to produce thyroxine, and a third acts on the cortex of the adrenal gland and promotes the production of cortisone (*see* p. 158).

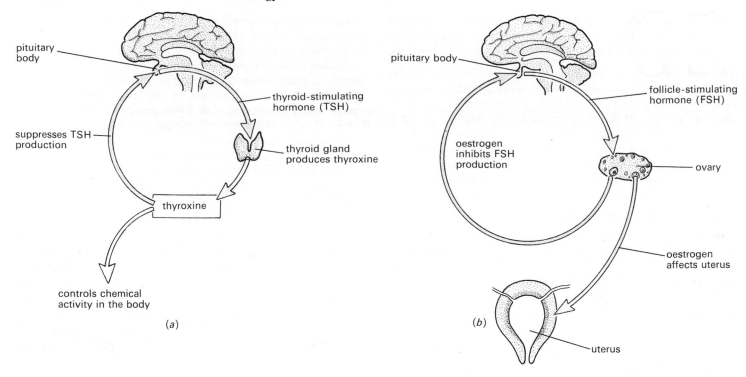

Fig. 19.22   Feedback

Homeostasis

The foregoing account shows how the hormones co-ordinate the organs of the body to meet various contingencies (e.g. adrenaline), to produce rhythmic patterns of activity (e.g. the sex hormones), and to maintain control over long-term processes such as the rate of growth (thyroid and pituitary). It can also be seen that they fulfil a homeostatic function in regulating the composition of the internal environment (*see* p. 92). If the blood sugar level rises, the pancreas is stimulated to secrete insulin which increases the amount of glucose removed from the blood and stored as glycogen in the muscles and liver. A fall in the blood sugar level suppresses the production of insulin from the pancreas. An increase in the osmotic potential of the blood results in the release of ADH from the pituitary gland and the consequent reabsorption of water from the kidney tubules.

Interaction and feedback

For effective control, two opposing systems are needed. A car needs an accelerator and brakes, a muscle must have its antagonistic partner (p. 119). The hormones too have antagonistic effects. Glucagon promotes the release of sugar into the blood while insulin has the opposite effect. A fine adjustment of the balance of such antagonistic hormones helps to maintain the controlled growth, development and activity of the organisms in constantly changing conditions.

The balance is maintained partly by the 'feedback' effect of hormones, i.e. a system whereby 'information' is 'fed back' to a source 'telling it' about events in the body and so enabling it to adjust its output accordingly. A pituitary hormone (TSH) stimulates the thyroid to produce thyroxine (Fig. 19.22*a*) but thyroxine production is kept in check by the fact that when

thyroxine reaches the pituitary via the circulation, production of thyroid-stimulating hormone is suppressed. The feedback of thyroxine to the pituitary regulates the output of the latter. The ovarian follicles are stimulated to produce oestrogen by follicle-stimulating hormone (FSH) from the pituitary, but when the oestrogen in the blood reaches a certain level it suppresses the secretion of FSH by the pituitary. A delay in the feedback effect leads to rhythmic changes. For example, it may take two weeks for the level of oestrogen in the blood to affect the pituitary, by which time the uterus lining has thickened and the ovum has been released from the follicle. The output of follicle-stimulating hormone is diminished as a result of increasing oestrogen and this in turn reduces the output of oestrogen from the ovary which, in the absence of fertilization and the development of the corpus luteum, leads to the breakdown of the uterine lining, characteristic of menstruation (Fig. 19.22*b*).

Questions

1 List the differences between control by hormones and control by the nervous system.
2 Trace by diagram or description the possible reflex arc involved in (a) sneezing, (b) blinking. Do not attempt to describe the effector systems in detail and treat the brain as simply an enlarged region of the spinal cord.
3 All nervous impulses, whether from the eyes, ears, tongue or skin are basically the same. This implies that the information reaching the brain is little more than a rapid series of electrical pulses of identical strength. How then is it possible for us to distinguish between light and sound, heat and touch?
4 It is possible to train a dog to seek food concealed behind one of several identical doors by flashing a light over the appropriate door. Suggest a nervous pathway by which this behaviour is established.
5 Study Fig. 19.13 and then suggest why stimulation of the sympathetic nervous system should lead to greater efficiency in a situation that demands rapid and vigorous action.

20
Chromosomes and Heredity

Most living organisms start their existence as a single cell, a *zygote* (fertilized egg, p. 100). This single cell divides into two cells, four, eight and so on to produce eventually the millions of cells which make up the new organism.

If the new cells all behaved in the same way, they would produce only a mass of structureless tissue. The cells, however, as they develop, become different from each other in structure and function. Therefore, in the processes which turn a zygote into an organism there must be forces directing the cells to determine that some become muscle, some skin, some bone or blood, and these cells must be directed into groups in the right order and in the right place to produce tissues, organs and ultimately the complete, integrated, co-ordinated organism.

Moreover, a zygote does not produce just any organism. It will produce one which resembles the parents from whom the zygote was derived. The zygotes of a human and a cat may look identical, a nucleus surrounded by a little cytoplasm, but they will develop in quite different ways. The cat zygote will produce a cat and not a man. A mouse zygote will produce a mouse and not a rat. The study of the mechanism by which the characteristics of the parents are handed on to the offspring is known as genetics.

The zygote of a bird develops into a chick inside the egg without any outside interference other than incubation. It follows therefore, that the 'instructions' for building a bird from a single-celled zygote must reside somewhere inside the zygote. What is more, the 'instructions' must be present in the two gametes which fuse to form the zygote. They could be in the cytoplasm, the nucleus or both.

When one examines the gametes of most animals, the egg usually has a relatively large volume of cytoplasm associated with its nucleus. The male gamete, on the other hand, consists of little more than a nucleus with a very thin layer of cytoplasm round it, and a tail. However, there is nothing to suggest that the male's contribution to the 'instructions' or *genotype* of the zygote is any less than the female's so it looks as if the bulk of the genotype for building a cat or a man resides in the nucleus.

The next question is, can the 'instructions' be seen or studied in some way? One would not expect to see printed directions but there might be structures to be seen which would give some idea about the nature of the 'instructions'. Thus, a study of the nucleus would seem to be the most profitable course, particularly at a time when the nucleus is dividing because the genetic information, the *genotype*, must be handed on intact and undiminished to each cell. For example, if the two cells resulting from the first division of a frog's zygote are separated, each cell can develop into a complete frog. This is true of cells even at the 4- and 8-celled stage. Thus the 'frog-building instructions' are intact and complete in each of these cells after cell division.

Thus, if one could observe a structure or structures, reproduced exactly in the nucleus and shared equally between the two nuclei at cell division, this might give a clue to the site of the genetic information. The next section, consequently, examines the events in the nucleus at cell division in some detail.

Cell division

In the early stages of growth and development of an organism, all the cells are actively dividing to produce new tissues and organs. Later, particularly when the cells become specialized, this power of division is lost and only a limited number of unspecialized cells retain the power of division, e.g. the cells of the germinative layer in the skin which produce new epidermis. In those cells that continue to divide, the sequence of events leading to cell division is basically the same. Firstly, the nucleus divides into two and then the whole cell divides, separating each

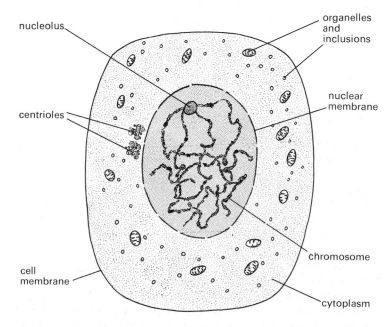

Fig. 20.1   Cell at early prophase

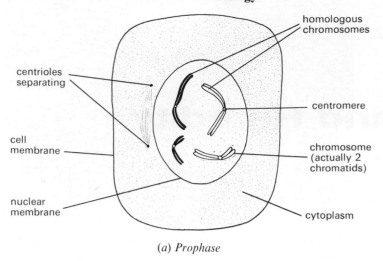

centrioles separating

cell membrane

nuclear membrane

homologous chromosomes

centromere

chromosome (actually 2 chromatids)

cytoplasm

(a) Prophase

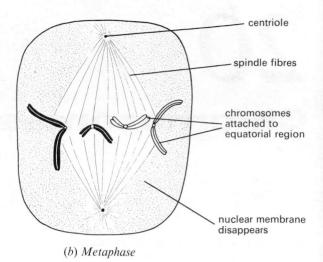

centriole

spindle fibres

chromosomes attached to equatorial region

nuclear membrane disappears

(b) Metaphase

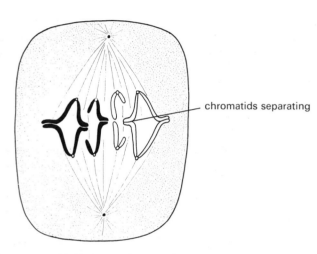

chromatids separating

(c) Early anaphase

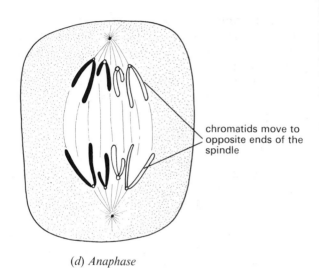

chromatids move to opposite ends of the spindle

(d) Anaphase

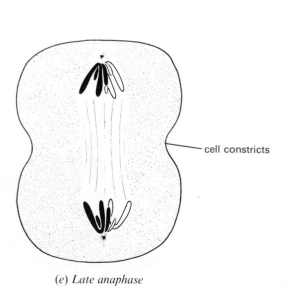

cell constricts

(e) Late anaphase

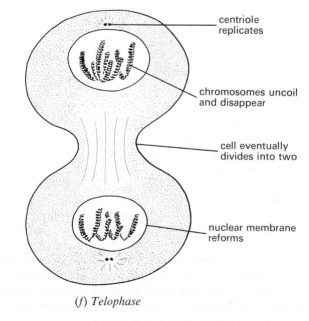

centriole replicates

chromosomes uncoil and disappear

cell eventually divides into two

nuclear membrane reforms

(f) Telophase

Fig. 20.2   Mitosis

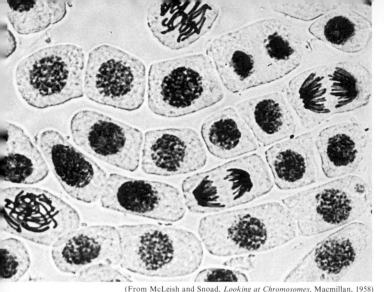

(From McLeish and Snoad, *Looking at Chromosomes*, Macmillan, 1958)

Fig. 20.3   Cell division in the root tip of a plant (× 500)

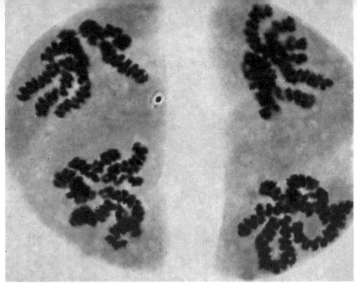

(A. H. Sparrow, Ph.D., Brookhaven National Laboratory, U.S.A.)

Fig. 20.4   Plant chromosomes at telophase of meiosis (p. 171) showing coiling (× 2 000)

nucleus in a unit of cytoplasm, so that two cells now exist where previously there was only one. Both cells may then enlarge to the size of the parent cell. Such cell division and enlargement give rise to growth.

The detailed sequence of events which takes place when the nucleus of a cell divides has been worked out over the last eighty years and is called *mitosis*.

Mitosis

Prior to division, the nucleus of the cell enlarges and in the nucleus there appear a definite number of fine, coiled, thread-like structures called *chromosomes* (Fig. 20.1). The behaviour of these chromosomes during cell division is usually described as a series of stages, prophase, metaphase, etc., though, in fact, the events occur in a smoothly continuous pattern and do not occupy equal periods of time (*see* Fig. 20.2).

1   **Prophase.** The chromosomes become more pronounced (that is, they react more readily to stains and chemical fixatives). They shorten and thicken (Fig. 20.9*a*), probably by coiling like a helical spring, but with the coils so close to each other that they are not visible at low magnifications (Figs. 20.4 and 20.5). The nuclear membrane dissolves, leaving the chromosomes suspended in the cytoplasm, and at the same time the one or more *nucleoli* disappear.

2   **Metaphase.** In the cells of animals and some of the simpler plants there is a pair of minute bodies called the *centrioles* which lie just outside the nucleus. By this stage they have separated and moved to opposite ends of the cell. From a region near each centriole there radiate cytoplasmic fibres which meet and join near the centre of the cell. This system of 'fibres' makes a web-like structure called the *spindle* (Fig. 20.2*b*) and the chromosomes become attached by their centromeres (Fig. 20.5) to the equatorial region of the spindle. By this time it is apparent that each chromosome consists of two parallel strands, called *chromatids* (Fig. 20.9*b*), joined in one particular region, the *centromere*. In forming two chromatids, the chromosome *replicates*; that is, it reproduces an exact copy of itself but the two identical chromatids remain in contact along their length. This replication has already occurred at prophase but is more evident during metaphase.

3   **Anaphase.** The two chromatids now separate at the centromere and begin to migrate in opposite directions towards either end of the spindle (Fig. 20.9*c*). Experiments show that the spindle fibres play some part in separating the chromatids. The appearance is that of the chromatids first repelling each other at the centromere and then being pulled entirely apart by the shortening spindle fibres, although such a mechanism has not yet been verified.

4   **Telophase.** The chromatids, now chromosomes, collect together at the opposite ends of the spindle (Fig. 20.9*d*) and become less distinct, probably by becoming uncoiled and therefore thinner. The one or more nucleoli reappear, and a nuclear membrane forms round each group of daughter chromosomes so that there are now two nuclei present in the cell. At this point in animal cells, the cytoplasm between the two nuclei constricts, and two cells are formed. Both may retain the ability to divide, or one or both may become specialized and lose their reproductive capacity.

From a study of mitosis it seems very likely that the chromosomes are the site of the genetic instructions, since they reproduce themselves when they form chromatids and the chromatids are shared equally between the cells by the events of mitosis. A further study of chromosomes provides more evidence that they carry genetic information.

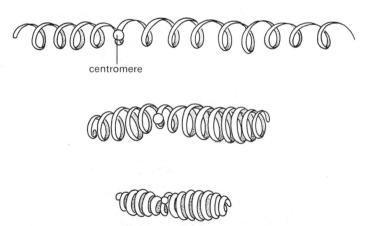

centromere

Fig. 20.5   Diagram showing how chromosomes appear to shorten and thicken during prophase

161

(a) Man (46)

(b) Kangaroo (12)

(c) Domestic fowl (36)

(d) Drosophila (8)

Fig. 20.6   Chromosomes of different species

(From C. C. Hurst, *The Mechanism of Creative Evolution*, Cambridge University Press, 1933)

(World Health)

Fig. 20.7   Studying human chromosomes. A member of a team of scientists under Professor Jerôme Lejeune at the Institut de Progénèse, Paris, identifies and prepares pictures of human chromosomes

Chromosomes

Chromosomes are so called because they take up certain basic stains very readily (*chromos* = colour, *soma* = body), but they can also be observed by phase contrast microscopy in the unstained nuclei of dividing, living cells. When the cell is not dividing, the chromosomes cannot be seen in the nucleus, even after staining. Nevertheless, it is thought that they persist as fine, invisible threads, isolated patches of which still respond to dyes and show up as flakes or granules of deeply staining material. The chromosomes consist of protein and a substance called *deoxyribonucleic acid* (DNA: *see* p. 166), but the exact relationship between these two components in forming the chromosome has not been fully worked out.

Counts of chromosomes show that there is a definite number in each cell of any one species of plant or animal: e.g. mouse, 40; crayfish, 20; rye, 14; fruit fly (*Drosophila*), 8; and man, 46 (*see* also Fig. 20.6). This confirms our expectation that the chromosomes determine the difference between one species and another. It can also be seen (Fig. 20.8) that the chromosomes exist in pairs, although not actually joined together, each pair

(World Health)

Fig. 20.8   Preparing a 'karyogram': the chromosome silhouettes from the photomicrograph on the left are cut out and arranged in order on the right-hand chart

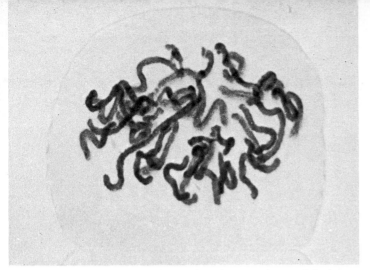

(a) Prophase; the chromosomes have become short and thick

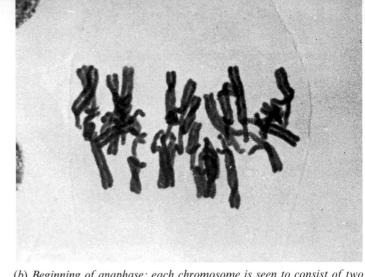

(b) Beginning of anaphase; each chromosome is seen to consist of two chromatids which are attached to the equator of the spindle and are beginning to separate

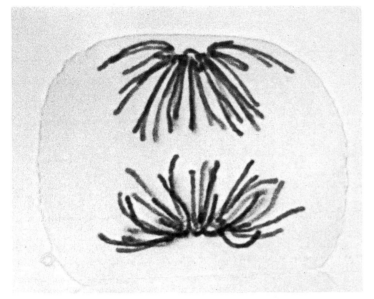

(c) Anaphase; the chromatids have completely separated. The spindle is not visible in this photograph

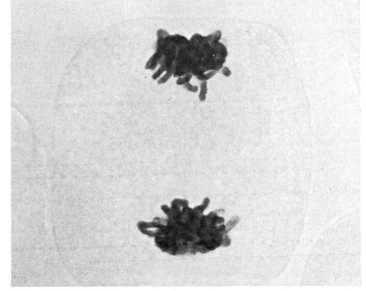

(d) Telophase; the chromosomes are becoming less distinct

Fig. 20.9   Stages in mitosis in a plant root cell ($\times 1\ 800$)

(From McLeish and Snoad, *Looking at Chromosomes*, Macmillan, 1958)

having a characteristic length and, during anaphase, a characteristic shape (Fig 20.9*c*) governed by the position of the centromere at which the chromatids are pulled apart; e.g. a V shape if the centromere is central, or a $\sqrt{\ }$ shape if it is close to one end. In other words, human cell nuclei contain 23 pairs of chromosomes, mouse cells 20 pairs and so on, one member of each pair having been derived from the male and one from the female parent. The members of each pair are called *homologous chromosomes*, and the total number of chromosomes in each cell is called the *diploid number*.

Although the constituent chemicals of the cytoplasm of a cell are constantly being broken down and rebuilt from fresh material, the chemicals of the chromosomes remain remarkably stable. Other investigations show that during cell division, no protoplasmic material is shared so exactly as that of the chromosomes in the nucleus. Such evidence points again to the chromosomes as the main source of the chemical information which determines that a cell should become like its parent cell, and that in their development, the cells of the organism will endow the animal or plant with all the characteristics of its

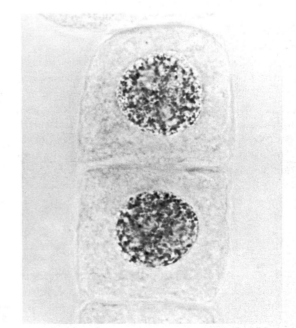

(e) The end of cell division; the daughter nuclei are separated by a new cell wall

163

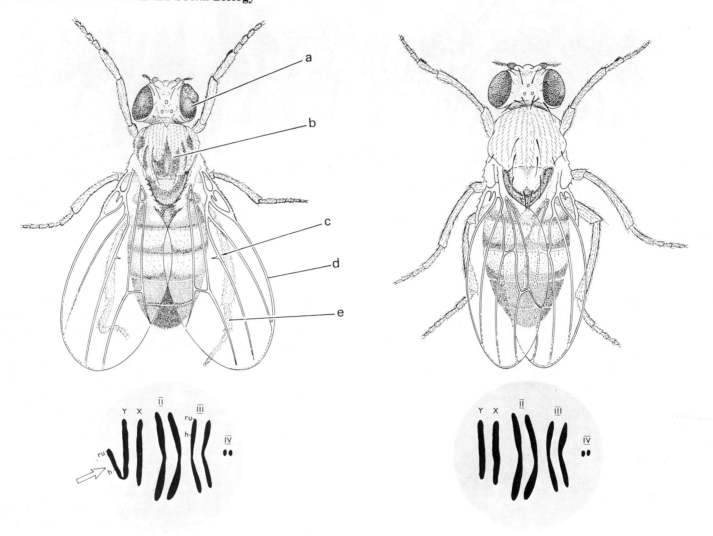

(After Muller, *Journal of Genetics*, 1930)

Fig. 20.10   Effects of a chromosome mutation in *Drosophila*. The cells of the fruit fly on the left have an extra segment of chromosome no. III which has become attached to the Y chromosome. As a result, this fly has (*a*) mis-shapen eyes, (*b*) dark patterned thorax, (*c*) imperfect cross veins, (*d*) broad wings, (*e*) incurved hind legs. The normal fly is shown on the right

species. Although the supporting evidence is very sketchily outlined here, the hereditary material must almost certainly lie on the chromosomes of the nucleus and be passed on to each daughter cell by the process of mitosis. In a similar way the nuclei of the sperm and egg carry a set of chromosomes from the male and female parent, and these chromosomes determine that the zygote grows and develops to an animal or plant of the same species as the parents, reproducing to quite a minute degree, individual characteristics of both parents.

The fruit fly, *Drosophila melanogaster*, which has four pairs of chromosomes in its nuclei is, for various reasons, a suitable subject for study. By breeding many hundreds of flies through many generations, geneticists have become very familiar with the detailed anatomy of the fly and the appearance of its chromosomes. It has been observed on many occasions that some unexpected change in the external appearance of a fly is associated with a change in the chromosome pattern (*see* Fig. 20.10). This chromosome aberration must take place at an early stage in the development of the fly for it to affect so many parts of the body, or it may have occurred in one of the gametes from which the zygote was formed.

Cells from the salivary glands of *Drosophila* and other flies have very large chromosomes called giant chromosomes, on which bands can be seen (Fig. 20.12). The size, shape and position of these bands is quite consistent and characteristic for any pair of chromosomes. If, due to some accident in replication, one or other of the bands is lost, there is a corresponding malformation in the adult fly (Fig. 20.11). Although the bands can be seen on only these rather unusual giant chromosomes, it is thought that they represent the site of genes or gene activity on all the chromosomes in the body.

Genes. A gene is a theoretical unit of inheritance, theoretical in the sense that the word was coined long before chromosome structure was investigated in detail or the DNA theory of inheritance put forward. The gene is one of the 'words' in the genetic 'instructions' (*see* p. 159). For example one gene will specify whether the cat is to have black fur or white fur. Another gene will determine whether the fur is long or short. Today the gene is thought to consist of a group of chemicals situated in the chromosome.

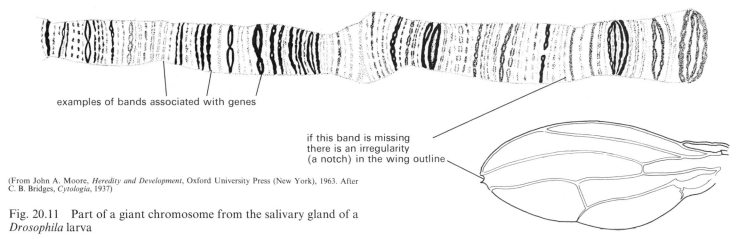

examples of bands associated with genes

if this band is missing
there is an irregularity
(a notch) in the wing outline

(From John A. Moore, *Heredity and Development*, Oxford University Press (New York), 1963. After C. B. Bridges, *Cytologia*, 1937)

Fig. 20.11   Part of a giant chromosome from the salivary gland of a *Drosophila* larva

Gene function

The picture that emerges from this and other evidence is that the genes which determine the characteristics of the organism are somehow arranged in line down the chromosome. These genes control the production of enzymes, which in turn determine what functions go on in a cell, and eventually in the organs and entire organism. If anything happens to a gene it will affect the organism, e.g. in mice there is a gene which determines that the coat will be coloured. If this gene is missing, the mouse will be without pigment; it will be white with pink eyes. In this case, as in many others, more than one gene will in fact play a part in determining the characteristic.

The number of genes in man is not known but it could be about 1 000 per chromosome. At mitosis, each chromosome, and therefore each of any of the genes it carries, is exactly reproduced.

Two problems arise from this account. Since every cell in the body carries an identical set of chromosomes and since cell structure and function are determined by the genes on the chromosomes, why is not every cell of the body identical? Furthermore, what possible part can a gene for brown eyes play when it is in the nucleus of a cell lining the stomach wall? Briefly, when we follow the development of a particular cell, it seems that the way in which one of its genes will affect the cell depends not only on the gene itself but also on the physiology of the cell, which in turn is related to its particular position in the body. For example, the chemical environment in a certain cell in the scalp allows the gene for black hair to operate in a particular way. Just what the same gene does in another part of the body is not certain; its action may simply be suppressed, but it is known that most genes have more than one effect, and the characteristic by which they are recognized is not necessarily their most important function; e.g. the genes responsible for producing colour in the scales of one kind of onion also confer a resistance to fungus disease because they determine the presence of certain chemicals which act as a fungicide. The colour, however, is the more obvious characteristic. The gene in *Drosophila* which produces the effect of diminutive wings also reduces the expectation of life to half that of normal flies. The wing characteristic is the more immediately obvious effect but the effect on life span may be far more important and damaging to the species. The idea that the expression of a gene depends to some extent on the physiology of the cell and the situation in which it finds itself is illustrated by the experimental work with certain amphibian embryos. If a piece of tissue, which would normally become skin, is taken from the abdomen and grafted into a region overlying the developing eye, the graft will be incorporated into the eye as a lens. It has the same chromosomes and genes, but its new position has altered its fate. This effect is by no means true of all animals and is certainly not the case in insects in which the fate of individual cells seems to be determined at a very early stage in development and is not affected by moving the cells to a new situation.

Fig. 20.12   Four giant chromosomes from a cell in the salivary gland of the midge larva, *Chironomus tentans*, showing transverse banding (magnification approx × 600)

(Courtesy of Professor Wolfgang Beerman, Max Planck Institute, Tübingen, from *Sci. Amer.*, April 1964)

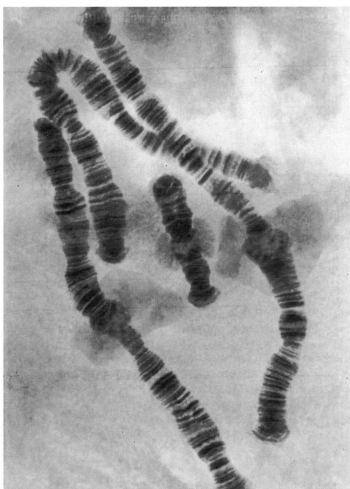

How genes work

The role of DNA. It was mentioned earlier (p. 162) that chromosomes consist of protein and a nucleic acid, deoxyribonucleic acid (DNA). Although the precise relationship between the DNA and the protein is not known for certain, the structure of DNA has been intensively studied. This chemical consists of long molecules coiled in a double helix. The strands of the helix are chains of sugars and phosphates, the sugar being a 5-carbon compound, *deoxyribose*. The two helices in a DNA strand are linked together by cross-bridges made by pairs of organic nitrogenous bases joined to the sugar molecules (Figs. 20.13 and 20.14). Although there are only four principal kinds of base in the DNA molecule, *adenine, cytosine, thymine* and *guanine*, it is thought that the sequence of these bases is the important factor in heredity, and that a gene may consist of a particular sequence of up to 1 000 base pairs in a DNA molecule.

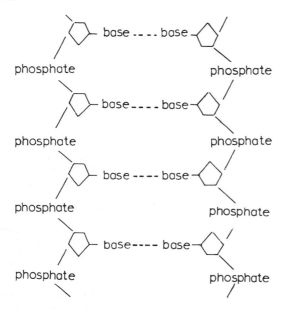

Fig. 20.13   Part of a DNA molecule
(○=*deoxyribose*)

The different sequence of bases *along the length* of the DNA molecule seems to act like a code, instructing the cell to make certain proteins, the order of bases indicating the sequence of amino acids to be joined up in order to make the protein. For example, the sequence CAA (cytosine–adenine–adenine) specifies the amino acid *valine*; three thymines in a row, TTT, specify *lysine*, while AAT specifies *leucine*. So the sequence of bases CAA–TTT–AAT would direct the cell to link up the amino acids valine–lysine–leucine to make the appropriate peptide, and ultimately proteins are formed (*see* p. 17).

The role of RNA. Cell proteins are not made by the nucleus but by the ribosomes in the cytoplasm. The coded information in the nuclear DNA must somehow be transposed to the cytoplasm. This transfer is carried out by a nucleic acid, *ribonucleic acid* (RNA), which differs slightly in composition from DNA. Each part of the DNA that is active in the chromosomes builds up a replica of itself in RNA. These lengths of RNA, called *messenger RNA*, become detached from the chromosomes, leave the nucleus through the nuclear pores and become attached to ribosomes (Fig. 20.15*a* and *b*).

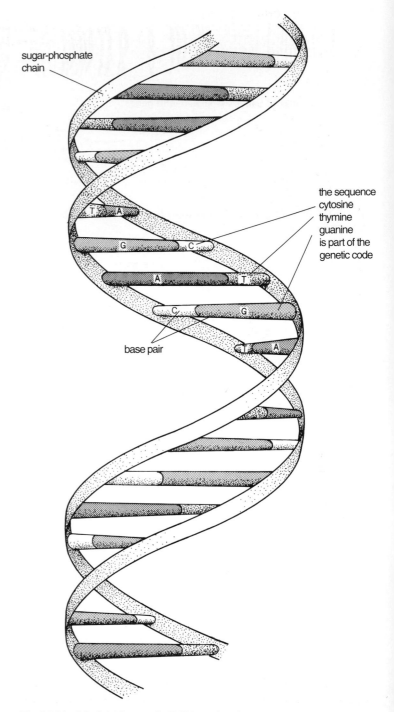

Fig. 20.14   Model of part of a DNA molecule

In the cytoplasm there are many amino acids and some of these become attached to a second kind of RNA called *transfer RNA*. Three of the bases on each molecule of transfer RNA correspond to one of the groups of three bases on the messenger RNA. The amino acid molecules attached to the transfer RNA will eventually encounter a ribosome to which is attached a strand of messenger RNA. If the section of messenger RNA attached to the ribosome has the three bases that correspond to those on the transfer RNA, the latter will combine temporarily with the messenger RNA. As the messenger RNA moves along

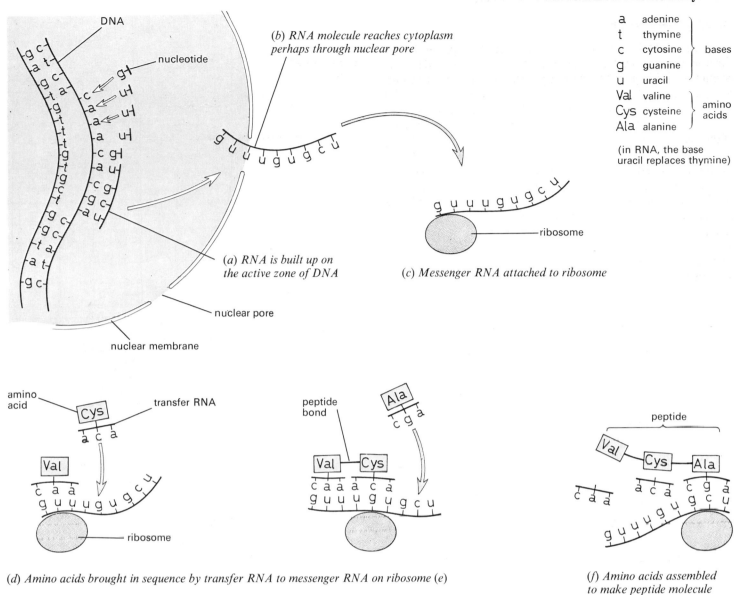

(a) RNA is built up on the active zone of DNA

(b) RNA molecule reaches cytoplasm perhaps through nuclear pore

(c) Messenger RNA attached to ribosome

a	adenine
t	thymine
c	cytosine
g	guanine
u	uracil

bases

Val	valine
Cys	cysteine
Ala	alanine

amino acids

(in RNA, the base uracil replaces thymine)

(d) Amino acids brought in sequence by transfer RNA to messenger RNA on ribosome (e)

(f) Amino acids assembled to make peptide molecule

Fig. 20.15 The role of DNA and RNA

the ribosome, the triplets of bases will be presented in turn, and each triplet will allow its corresponding transfer RNA to attach itself, bringing an amino acid with it. The amino acids are linked together by peptide bonds (*see* p. 17) in the correct sequence to make polypeptides and proteins (Fig. 20.15*d–f*).

Most of the proteins made are enzymes which direct the pattern of chemical activity in the cell. Thus DNA, by determining the kinds of enzyme formed in the cell, will control the cell's activities. This in turn will affect the nature of the cell, the organ of which it is a part and eventually the organism as a whole. A change in the sequence of bases in the DNA molecule will result in a different order of amino acids and, hence, a different and probably ineffective enzyme. This will usually act adversely on the metabolism of the cell.

A rat with coloured fur has a gene that controls the production of the enzyme *tyrosinase*. This enzyme converts *tyrosine*, a colourless amino acid, to *melanin*, a black pigment. An albino rat has no gene for tyrosinase production, and consequently no pigment is formed from the tyrosine in its body (*see* p. 169).

Normal humans have a gene that controls the production in the blood of an enzyme that accelerates the breakdown of a chemical, *alcapton*. Persons having no gene for the enzyme excrete unchanged alcapton in the urine which darkens on exposure to the air. This relatively harmless effect is associated with pigmentation in other parts of the body and, later, with arthritis. The condition is inherited as a recessive factor (p. 175). This is a rather peculiar example of the mechanism of inheritance, but if a gene controlled the production of an enzyme essential in a much earlier stage of a series of vital reactions, the absence of the gene could have devastating effects even to the extent of causing premature death. Conversely, since normal physiology is the result of hundreds of chemical changes catalysed by hundreds of enzymes, it is not surprising that characteristics such as intelligence, stature and activity come under the influence of many genes.

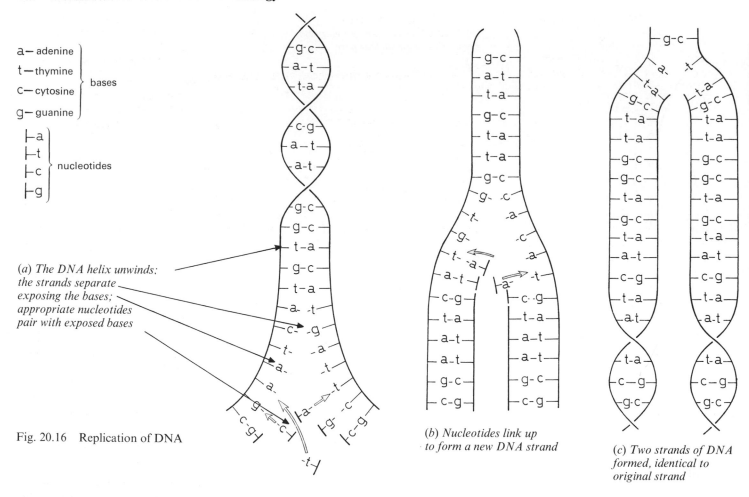

a — adenine ⎫
t — thymine ⎬ bases
c — cytosine ⎪
g — guanine ⎭

⊢a ⎫
⊢t ⎬ nucleotides
⊢c ⎪
⊢g ⎭

(a) The DNA helix unwinds; the strands separate exposing the bases; appropriate nucleotides pair with exposed bases

Fig. 20.16   Replication of DNA

(b) Nucleotides link up to form a new DNA strand

(c) Two strands of DNA formed, identical to original strand

Replication of genes. When a chromosome forms two chromatids prior to cell division, its constituent DNA also replicates. It is thought that the component strands of each double helix unwind, exposing the organic bases. In the nucleoplasm are the corresponding bases attached to sugar and phosphate molecules. This combination of organic base, sugar and phosphate is called a *nucleotide*. The bases on the appropriate nucleotides become attached to the exposed bases of the separated DNA strands, and so build up a new molecule of DNA with exactly the same sequence of bases as the original partner. In this way the genetic code is preserved intact to be passed on with the chromosomes to the new cell (Fig. 20.16).

Mutations

A mutation is a spontaneous change in a gene or a chromosome which may produce an alteration in the characteristic under its control. Fig. 20.10 shows a chromosome mutation. A fairly frequent form of mental deficiency known as Down's syndrome (mongolism) results from a chromosome mutation in which the ovum carries an extra chromosome, so that the child's cells contain 47 chromosomes instead of 46.

A mutation in a single cell may not be very important in heredity, but if the cell is a gamete mother cell, a gamete or a zygote, the entire organism arising from this cell may be affected. Since the mutant form of the gene is inherited in the usual way, the mutation will persist in subsequent generations.

On the whole, genes are stable structures because DNA is a stable chemical, but once in a hundred thousand replications or more a gene may mutate.

Gene mutations occur when a section of the DNA in a chromosome is not copied exactly at cell division. For example, part of the haemoglobin molecule consists of a sequence of eight amino acids:

valine—histidine—leucine—threonine—proline
—*glutamic acid*—glutamic acid—lysine

This order will have been determined by the sequence of the four bases on part of a DNA strand (*see* p. 167). In persons suffering from a disease called sickle cell anaemia one of the DNA bases has not been correctly paired prior to the cell division that formed the gamete from which the individual developed. As a result, the sixth amino acid directed into the haemoglobin fragment depicted above is *valine* instead of glutamic acid. This small difference so alters the properties of the haemoglobin that in low concentrations of oxygen it becomes relatively insoluble and forms rod-like particles which distort and eventually destroy the red blood cells. The production of the 'faulty' DNA is a mutation, and since this DNA will faithfully reproduce itself in all the cells of the body, including the sex cells, the mutated gene will be passed on to the offspring. This is an example of a harmful mutation which is heritable.

Harmful mutations. Consider the sequence of bio-chemical reactions which occur in cells of the liver (Fig. 20.17).

The letters 'a', 'b' and 'c' represent enzymes which control the reactions indicated by the arrows. Each enzyme is a protein and is built up from amino acids according to the pattern determined by the base sequence of DNA in a gene. Mutations

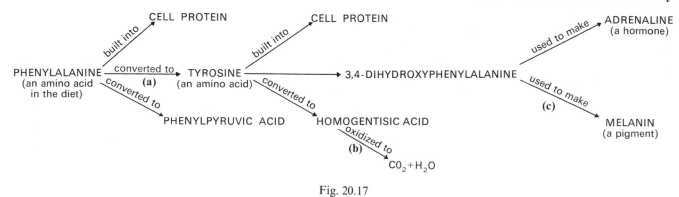

Fig. 20.17

may occur in the genes which will result in the failure to produce one or other of these enzymes.

If enzyme 'a' is lacking, tyrosine cannot be made from phenylalanine. Since tyrosine is available from the normal diet this does not matter, but blocking this pathway leads to excessive production of phenylpyruvic acid. This compound collects in the cerebrospinal fluid and is thought to cause the brain damage leading to mental retardation characteristic of the disease *phenylketonuria*.

If there is no enzyme 'b', homogentisic acid (alcapton) is not oxidized to carbon dioxide and water but excreted unchanged in the urine which darkens on exposure to air, a symptom of *alcaptonuria*. Alcapton also collects in the cartilage of the joints causing a form of arthritis.

In the absence of enzyme 'c', the production of melanin from 3,4-dihydroxyphenylalanine is not possible, in which case the person will be an albino, lacking pigment in his skin, hair, and eyes.

As mentioned on p. 162, the chromosomes in the nucleus occur in pairs, one chromosome of each pair having come from the male parent and the other from the female parent. It follows, therefore, that since they are located on the chromosomes, the genes also occur in pairs. If a person inherits a mutated gene from only one parent, it is likely that its effects will be masked by the corresponding normal gene from the other parent, and the harmful condition will not appear in the individual. The mutant gene is then said to be *recessive* to the normal gene, and the normal gene is *dominant* to the mutated form (*see also* p. 175). The clinical conditions described above will therefore appear only if an individual inherits two mutated recessive genes, one from each parent.

Most mutations that produce an observable effect seem to be harmful if not actually lethal. This is not surprising, since any change in a well- but delicately balanced organism is likely to upset its physiology. Most mutations, however, are recessive and so in the presence of a normal gene do not usually produce obvious symptoms. In humans there occurs a form of dwarfism known as *achondroplastic dwarfism*, in which the limb bones do not grow normally. This condition arises as a result of a dominant mutation having a frequency of about 1 in 20 000.

Although the majority of mutations are potentially harmful, there is a small number that are beneficial, either at the time when they occur or at a later date when the population carrying them encounters changing conditions. The beneficial mutations thus tend to be preserved in a population because they confer some advantage on the organisms. It is from such beneficial or neutral mutations that variations arise in populations and, in the course of evolution, give rise to new species.

Mutation in bacteria. Mutations can occur whenever DNA is reproducing itself, an event which takes place prior to cell division. Some bacteria can divide every twenty-five minutes, and it follows that there is a very high chance of mutations occurring per unit of time. One such mutation produces a resistance to specific drugs and occurs about once in every thousand million cell divisions. This seems very infrequent but a colony of ten bacteria could produce a population of that order in four or five hours, and it is likely that a thriving colony will contain several individuals resistant to, say, penicillin. In the 'normal' environment the mutation may have no advantage, but in the presence of penicillin all but the mutants will be destroyed. The progeny of the surviving mutants will inherit the resistance to penicillin and in this way a population of resistant bacteria can soon become established. For this reason, the widespread use of antibiotics is discouraged.

Mutation rate. One cannot predict when a gene is going to mutate but the frequency of its occurrence can be determined in some cases; for example in achondroplastic dwarfism it is possibly as high as one mutation in 20 000, compared with one in 100 000 for many genes. The rate of mutation is characteristic of particular genes in particular species, but the frequencies are such that in a human ejaculate of, say, 200 million sperms, there are likely to be a considerable number of nuclei bearing gene and chromosome mutations.

Exposure to radioactivity, X-rays and ultraviolet radiation is known to increase the rate of gene and chromosome mutation.

Radiation and mutations. The cause of mutation is not known, but exposure to X-radiation, gamma radiation, ultraviolet light, etc., is known to cause an increase in the mutation rate in experimental animals such as fruit flies and mice. The artificially induced mutations are the same as those that occur naturally, but the frequency with which they occur is greatly increased.

There is a fairly constant background of radiation on the Earth's surface as a result of cosmic rays. Individuals also receive radiation from X-rays used in medicine, television tubes and luminous watch dials. Workers in atomic power stations and other people handling radioactive materials in industry or research may receive additional radiation. The radioactive fall-out from atomic explosions has increased the background radiation.

It is of obvious importance to assess the effect of any increase in radiation on the health of individuals and, as a result of mutations in their reproductive cells, the health of their children.

There is, so far, insufficient information to determine the

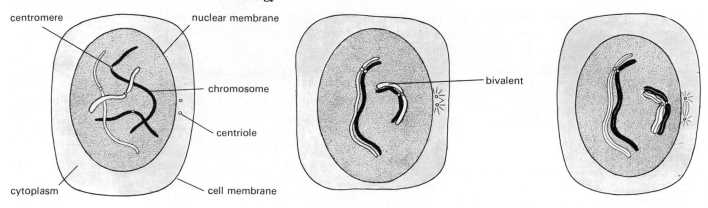

centromere   nuclear membrane

chromosome

centriole

cytoplasm   cell membrane

bivalent

PROPHASE
(*a*) *The diploid number of chromosomes appears*

(*b*) *Homologous chromosomes pair with each other, shorten and thicken*

(*c*) *Replication has occurred and the chromatids become visible*

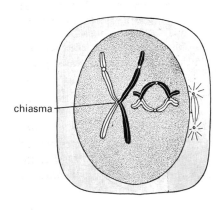

chiasma

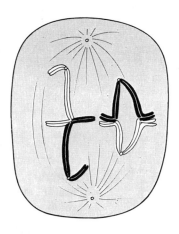

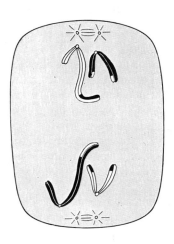

(*d*) *Homologous chromosomes move apart except at the chiasmata where chromatids have exchanged portions*

ANAPHASE
(*e*) *A spindle forms and homologous chromosomes move to opposite ends taking exchanged portions with them*

(*f*) *Homologous chromosomes separated but not enclosed in nuclear membranes*

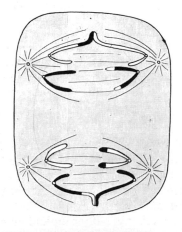

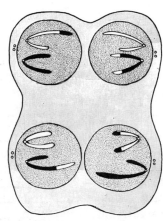

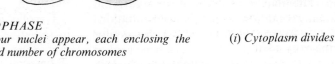

SECOND MEIOTIC DIVISION
(*g*) *Spindles form at right angles to the first one and the chromatids separate*

TELOPHASE
(*h*) *Four nuclei appear, each enclosing the haploid number of chromosomes*

(*i*) *Cytoplasm divides to form four gametes*

Fig. 20.18   Meiosis in a gamete-forming cell

correlation between the radiation dose and the mutation rate in man. The maximum safe dose in respect of direct effects on the individual such as leukaemia is still a matter of controversy. Nevertheless it is probably safe to say that any increase in the mutation rate is likely to have harmful effects on the population. Consequently, the exposure of individuals to the hazards of radiation is limited, though somewhat arbitrarily, by law.

Variation

The exact replication of chromosomes and genes and their equal distribution between cells at mitosis produces conformity; the organism breeds true to type, e.g. a sheep reproduces sheep and not goats or cows. Nevertheless, the offspring will differ in many respects from its brothers and sisters and from its parents. It is possible for two black mice to have some white babies as well as black ones. A child of blood group O could be born to a mother of blood group B and a father of group A.

The variations in question must arise in the first instance from neutral or beneficial mutations, but the appearance and distribution of the varieties in the population will depend on how the parental chromosomes and genes are rearranged in the offspring. For example, in a white mouse the absence of a gene for pigment is the result of a mutation, but the numbers of white mice appearing in a population will depend on the frequency with which a sperm carrying the mutated gene fertilizes an ovum with the same gene. The distribution of genes in the gametes is a direct result of the way in which the chromosomes separate at the cell division that leads to gamete formation. This sequence of chromosome separation is called meiosis.

Meiosis

Cells in the reproductive organs which are going to form gametes, either sperms or ova, undergo a series of mitotic divisions resulting, in the case of sperms, in a vast increase of numbers. The final divisions, however, which give rise to mature gametes are not mitotic. Instead of producing cells with 46 chromosomes in man, they form gametes with only 23 chromosomes. When, at fertilization, there is a fusion of the two gametes, the resulting zygote contains the diploid number of 46 chromosomes, and this number is present in all the cells of the offspring. The halving of the chromosome number which occurs at gamete formation, ultimately maintains the diploid number of chromosomes characteristic of the species. If gametes were produced by mitosis, a human egg and sperm would each contain 46 chromosomes and when they fused at fertilization would give rise to a zygote with 92 chromosomes. The gametes from the resulting organism would in turn give rise to offspring with 184 chromosomes and so on.

1 **Prophase.** In meiosis, the chromosomes appear in the nucleus (i.e. they appear when the cell is fixed and stained or observed by phase contrast microscopy) in much the same way as described for mitosis, but although it is reasonably certain that two chromatids are present in each chromosome, the chromosomes still appear to be single threads. Another difference from mitosis seen at this stage is the failure of the chromosomes to shorten by coiling.

In complete contrast to mitosis the homologous chromosomes now appear to *attract* each other and come to lie

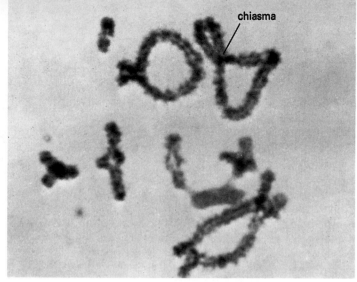

(From colour slide set, *Meiosis in Chorthippus brunneus*, published by Harris Biological Supplies Ltd)

Fig. 20.19 Meiosis in grasshopper testis (late prophase) ($\times 2\,000$)

alongside, so that all parts of the two chromosomes correspond exactly. The pairs of chromosomes so formed are called *bivalents*, e.g. a cell with a normal complement of six chromosomes would have, at this stage, three bivalents.

In this paired state the chromosomes shorten and thicken by coiling, and now each chromosome is seen to consist of two chromatids. As soon as this occurs, however, the pairs of chromatids seem to repel each other and move apart, except at certain regions called *chiasmata*. In these regions the chromatids appear to have broken and joined again but to a different chromatid. The significance of this exchange of sections of chromosomes or 'crossing over', is discussed on p. 179. All these changes occur during prophase while the nuclear membrane is still intact.

2 **Metaphase.** The nuclear membrane disappears and the bivalents approach the equatorial region of the spindle.

3 **Anaphase.** The paired chromatids of each bivalent now continue the separation that began in prophase and move to opposite ends of the spindle in a manner superficially similar to that of chromatids in mitosis. The outcome is that only half the total number of paired chromatids reaches either end of the spindle. Thus although there may originally have been six chromosomes in the nucleus, there are now only three paired chromatids at either end of the spindle.

4 **Second meiotic division.** A nuclear membrane does not usually form round the paired chromatids at this stage. Instead, two new spindles form at right angles to the first one and the chromatids of each pair separate and become chromosomes.

5 **Telophase.** The four groups of chromosomes are now enclosed in nuclear membranes so forming four nuclei, each containing half the diploid number of chromosomes (the *haploid* number). Finally, the cytoplasm divides to separate the nuclei, giving rise, in the case of males at least, to four gametes. In sperm formation (*spermatogenesis*) in most animals the four cells will develop 'tails' to become sperms (Fig. 20.20).

Since it gives rise to cells containing half the diploid number of chromosomes, meiosis is sometimes called the *reduction division*.

Formation of ova: oogenesis. During the formation of ova, the cytoplasm is not shared equally. After the first meiotic division, one of the daughter nuclei receives the bulk of the cytoplasm and the other nucleus is separated off with only a vestige of cytoplasm to form the first *polar body*, which, although it may undergo the next stage of its meiotic division, cannot function as an ovum and subsequently degenerates (Fig. 20.20).

In a similar way, the next meiotic division of the remaining egg nucleus produces a second polar body and a mature ovum. In many vertebrates, the first polar body is not formed until after the potential ovum is released from the ovary, and the second polar body is not formed until after the penetration of the sperm in fertilization. In man, when a sperm bearing 23 chromosomes fuses with a 23-chromosome ovum, a 46-chromosome zygote is formed.

New combinations of genes in the gametes. In mitosis the full complement of chromosomes derived from both parents is first doubled and then shared equally between daughter cells. The result of this replication is that the daughter cells receive identical genetic information. In meiosis, on the other hand, the genetic information on the chromosomes is not shared in exactly the same way between all the gametes. If homologous chromosomes were identical in their gene content, this variability in chromosome distribution would have no effect. Since, however, an individual's parents are bound to be genetically dissimilar in many respects, there will be many gametes with combinations of genes quite different from either of the individual's parents (*see* p. 179).

Fertilization (Fig. 20.21)

The cytoplasm of the sperm fuses with that of the ovum and the male nucleus passes into the ovum, coming to lie alongside the egg nucleus: the zygote is formed, but in many cases there is no fusion of nuclear material at this stage. Each nucleus simultaneously undergoes a mitosis with the axes of the spindles parallel to each other, but at the telophase stage the adjacent chromatids, originally from different parents, become enclosed in the same nuclear membrane, thus restoring the diploid number of chromosomes. These events are followed by the first cleavage, the zygote dividing into two cells. Subsequent mitotic division produces a multicellular organism with the diploid number of chromosomes in all its cells.

Determination of sex. In humans, one pair of the smallest chromosomes is known to determine sex. In the female, these two chromosomes are entirely homologous and are called the *X chromosomes*, while in the male, one is smaller and is called the *Y chromosome* (Fig. 20.6*b*). Femaleness normally results from the possession of two X chromosomes and maleness from possession of an X and a Y chromosome. At meiosis, the sex chromosomes are separated in the same way as the others (Fig. 20.22), so that all the female gametes will contain an X chromosome, but half the male gametes will contain an X and half will contain a Y chromosome. If a Y-bearing sperm fertilizes an ovum, the zygote will be XY and give rise to a boy. Fertilization of an ovum by an X-bearing sperm gives an XX zygote which develops to a girl. There should be an equal chance of X or Y sperm meeting an ovum, and therefore equal numbers of boy and girl babies should be born. In fact, slightly more boys than girls are born in most parts of the world. The reason for this is not clear, but it also happens that the mortality rate for boy babies and men is slightly higher than for girl babies and women, which tends to restore the balance.

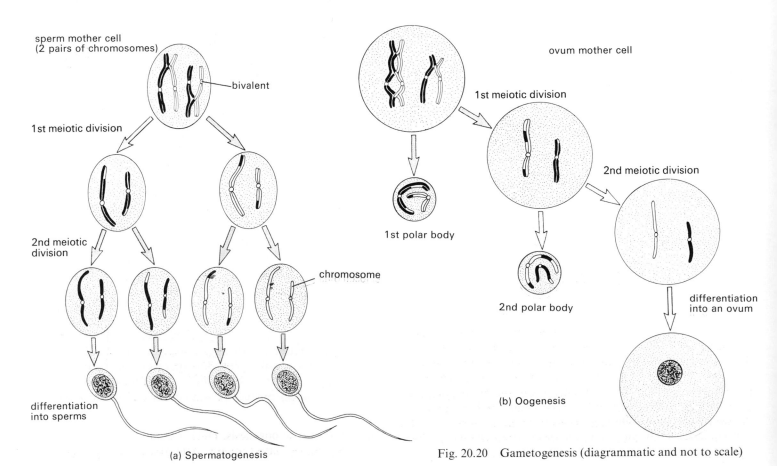

(a) Spermatogenesis

(b) Oogenesis

Fig. 20.20   Gametogenesis (diagrammatic and not to scale)

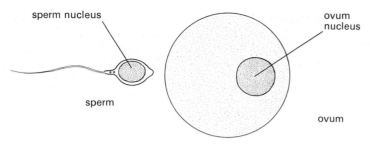

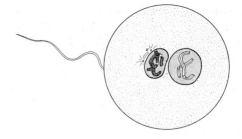

(a) Sperm meets ovum

(b) Sperm nucleus enters ovum and both nuclei undergo mitosis

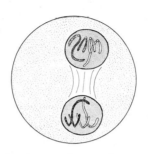

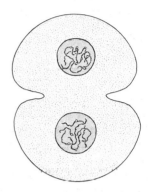

(c) The spindles are parallel but separate

(d) The chromatids at each end of the spindle are enclosed in a common nuclear membrane

(e) The zygote divides to form two cells

Fig. 20.21 Fertilization (polar bodies not shown)

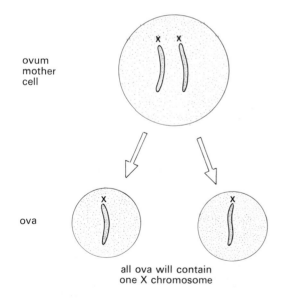

ovum mother cell

ova

all ova will contain one X chromosome

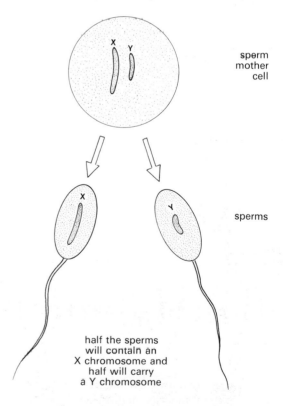

sperm mother cell

sperms

half the sperms will contain an X chromosome and half will carry a Y chromosome

Although the X and Y chromosomes determine sex, it does not necessarily follow that male and female characteristics are determined by genes found only on the sex chromosomes. In man, genes for male and female characters may be scattered fairly evenly throughout all the chromosomes, but the presence of the Y chromosome in an XY zygote may tip the balance in favour of maleness. Femaleness results from the absence of the Y chromosome but this is not the case in all the animals studied; e.g. it is true for the mouse but not for *Drosophila*.

Fig. 20.22 Determination of sex (diagrammatic only)

Practical Work

Experiment 1   **Squash preparation of chromosomes using acetic orcein**

Material. Allium cepa (onion) root tips. Support onions over beakers or jars of water. Keep the onions in darkness for several days until the roots growing into the water are 2–3 cm long. Cut off about 5 mm of the root tips, place them in a watch glass and

(a) cover them with 9 drops acetic orcein and 1 drop molar hydrochloric acid;

(b) heat the watch glass gently over a very small Bunsen flame till steam rises from the stain, but do not boil;

(c) leave the watch glass covered for at least five minutes;

(d) place one of the root tips on a clean slide, cover with 45 per cent acetic (ethanoic) acid and cut away all but the terminal 1 mm;

(e) cover this root tip with a clean cover-slip and make a squash preparation as described below.

Making the squash preparation. Squash the softened, stained root tips by lightly tapping on the cover-slip with a pencil: hold the pencil vertically and let it slip through the fingers to strike the cover-slip. The root tip will spread out as a pink mass on the slide; the cells will separate and the nuclei, many of them with chromosomes in various stages of mitosis (because the root tip is a region of rapid cell division) can be seen under the high power of the microscope ($\times 400$).

Preparation of reagents

(i) **Acetic orcein.** 2 g orcein; 100 cm^3 glacial acetic (ethanoic) acid. Dilute a small portion with an equal volume distilled water just before use.

(ii) **Molar hydrochloric acid** (i.e. one gram-molecule of HCl per litre). Make up 87.3 cm^3 of concentrated acid to 1 litre by adding distilled water.

Questions

1 Sometimes, at meiosis, the bivalent chromosomes fail to separate properly, with the result that the gamete so formed contains the diploid number of chromosomes. If such a diploid sperm were to fertilize a normal, monoploid ovum, (a) what effect would you expect this to have on the zygote and offspring, and (b) supposing the zygote grew into a normal individual, what might happen when the individual produced gametes by meiosis?

2 A horse and a donkey are related closely enough to be able to reproduce when mated together. The offspring from this mating is a mule and though healthy in all other respects is sterile. Suggest an explanation, to do with chromosomes and meiosis, for this phenomenon.

3 Revise the chemistry of proteins on pp. 17–18. Trace the steps involved in building up a tripeptide from alanine, glycine and serine starting from the appropriate segment of a DNA molecule. The DNA code for these amino acids is as follows: alanine, CGA; glycine, CCA; serine, TCA.

21
Heredity and Genetics

From its parents an individual inherits the characteristics of the species; e.g. man inherits highly developed cerebral hemispheres, vocal cords and the nervous co-ordination necessary for speech, a characteristic arrangement of the teeth and the ability to stand upright with all its attendant skeletal features. In addition, he inherits certain characteristics peculiar to his parents and not common to the species as a whole, e.g. hair and eye colour, blood group and facial appearance. The study of the method of inheritance of these 'characters' is called genetics.

In sexual reproduction a new individual is derived only from the gametes of its parents. The hereditary information must therefore be contained in the gametes. For many reasons, this information is thought to be present in the nucleus of the gamete and located on the chromosomes (*see* pp. 163–5).

Genes and inheritance

The term gene was originally applied to purely theoretical units or particles in the nucleus. These particles, in conjunction with the environment, were thought to determine the presence or absence of a particular characteristic. On p. 165 it was suggested that the genes may correspond to regions on the chromosomes and may consist of a large group of organic bases linked in a particular sequence in the chromosome.

In some cases, the presence of a single gene may determine the appearance of one characteristic, as in the eye colour of *Drosophila* (p. 164), but most human characteristics are controlled by more than one gene. This *multifactorial* inheritance, and the impossibility with humans of breeding experiments, make it difficult to collect and present simple, clear-cut genetical information about man. In addition, the readily observable human characteristics known to be under the control of single genes tend to be either severe abnormalities such as albinism or haemophilia, or they are rather trivial, e.g. the ability to roll the tongue or taste an obscure chemical. Consequently, in this chapter, the first example of certain types of heredity will be drawn from mice rather than men.

Single-factor inheritance. If a pure-breeding i.e. homozygous (*see* below) black mouse is mated with a pure-breeding brown mouse, the offspring will not be intermediate in colour, i.e. dark brown or some combination of brown and black, but will all be black. The gene for black fur is said to be *dominant* to that for brown fur, because although each of the baby mice, being the product of fusion of sperm and egg, must carry genes for both blackness and brownness, only that for blackness is expressed in the visible characteristics of the animal. The gene for brown fur is said to be *recessive*. The black babies are called the *first filial* or F_1 *generation*. If, when they are mature, these F_1 black mice are mated amongst themselves, their offspring, the F_2 *generation*, will include both black and brown mice, and if the total number for all the F_2 families are added up, the ratio of black to brown babies will be approximately 3 to 1. It must not be assumed, however, that if two black F_1 mice have four babies, three will be black and one brown. In a mating which produced, say, eight babies, it would not be at all unusual to find all black, or five black and three brown, etc. The ratio 3:1 appears only when large numbers of individuals are considered.

The appearance of brown fur in the second generation is proof of the fact that the F_1 black mice carried the recessive gene for brown fur, even though it did not find expression in their observable features.

In explanation, it will be assumed that a pure-breeding black mouse carries, on homologous chromosomes (*see* p. 163), a pair of genes controlling the production of black pigment. The genes are represented in subsequent diagrams (Figs. 21.1 and 21.2) by the letters BB, the capital letters signifying dominance. In the same position on the corresponding chromosomes in brown mice are carried the genes bb for brownness. The genes B and b are called *allelomorphic genes* or *alleles*.

The allelomorphic genes B and b influence the same characteristic, namely coat colour, but in different ways. *Two* genes, BB, Bb or bb, must be present, because the individual receives one chromosome from each parent. During the formation of gametes the process of meiosis (p. 171) will separate the homologous chromosomes, so that the gametes will contain only one allele from each pair. All the sperms from the pure-breeding black parent will carry the B allele and all the eggs from the brown parent will carry the b allele. When the gametes fuse, the zygotes will contain both B and b alleles, but since B is dominant to b, only the former allele is expressed, i.e. the offspring will all be black.

When, later on, these black F_1 mice produce gametes, the process of meiosis will separate the chromosomes carrying the B and b alleles (*see* Fig. 21.2) so that half the sperms of the male parent will carry B and half will carry b. Similarly, half the ova from the female will contain B and half b. At fertilization there are equal chances that a B-carrying sperm will fuse with either an egg carrying the B allele or an egg with the b allele, so producing either a BB or a Bb zygote. Similarly there are equal chances of a b-carrying sperm fusing with either a B- or a b-carrying ovum to give bB or bb zygotes.

This results in the theoretical expectation of finding, in every four F_2 offspring, one pure-breeding black mouse BB, one pure-breeding brown mouse bb, and two 'impure' black mice Bb.

The separation at meiosis of the alleles B and b into different gametes is called *segregation*. The pure-breeding black (BB) and brown (bb) mice are called *homozygous* for coat colour and the 'impure' black mice (Bb) are called *heterozygous*. The heterozygous mice will not breed true, i.e. if mated with each other their litters are likely to include some brown mice. The homozygous BB mice mated together can produce only black offspring and the bb homozygotes only brown offspring.

Genotype and phenotype. The BB mice and Bb mice will be indistinguishable in their appearance, i.e. they will both have black fur, and they are thus said to be the same *phenotypes*; in other words they are identical in appearance for a particular characteristic, in this case blackness. Their genetic constitutions, or *genotypes*, however, are different, namely BB and Bb. In short, the black phenotypes have different genotypes.

Multiple-factor inheritance. The 'one gene–one character' effects described above illustrate very clearly the Mendelian* principles of inheritance, but they are the exceptions rather than the rule. Rarely do single genes control one obvious trait. Colour in sweet peas, for example, is controlled by two pairs of alleles, CC and RR. Allele C controls the production of the colour base and allele R the enzyme which acts on it to make a colour. The recessive cc will produce no colour base and rr will have no enzyme. CCrr and ccRR combinations will thus be unable to produce coloured flowers. Many genes operate to produce coat colour in mice. In man, eight of the chemical changes involved in blood clotting are known to be under genetic control so that several genes are responsible for coagulation, and absence of any one of them may lead to a blood-clotting disease such as haemophilia (p. 180). A human characteristic such as skin pigmentation is controlled by multiple genes. The number of genes involved is not known, but it would have to account for a wide range of phenotypes.

---

* The term is derived from the name of an Austrian monk, Gregor Mendel, who in the 1850s first discovered the type of inheritance described here.

SINGLE-FACTOR INHERITANCE

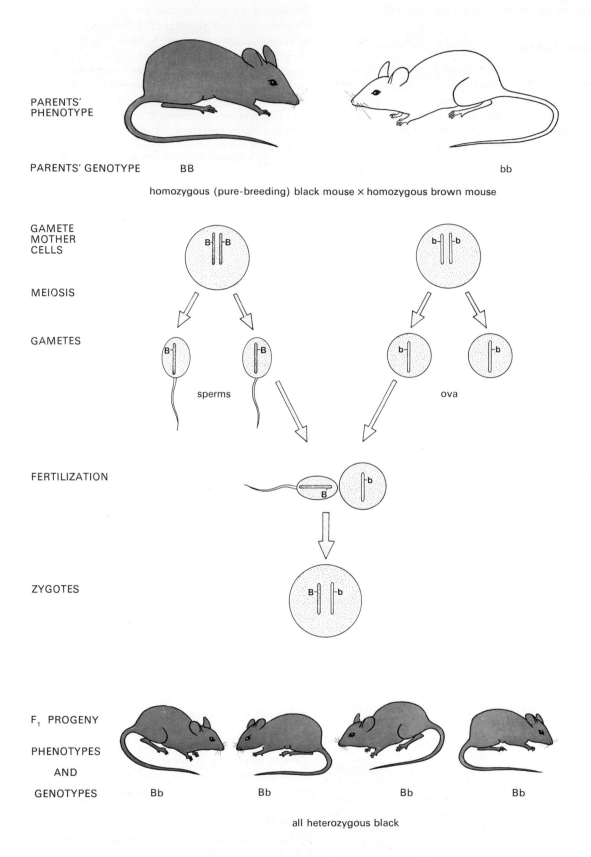

PARENTS'
PHENOTYPE

PARENTS' GENOTYPE   BB   bb

homozygous (pure-breeding) black mouse × homozygous brown mouse

GAMETE
MOTHER
CELLS

MEIOSIS

GAMETES

sperms   ova

FERTILIZATION

ZYGOTES

F₁ PROGENY

PHENOTYPES
AND
GENOTYPES   Bb   Bb   Bb   Bb

all heterozygous black

Fig. 21.1   Inheritance of a single factor for coat colour in mice (first cross)

SINGLE-FACTOR INHERITANCE

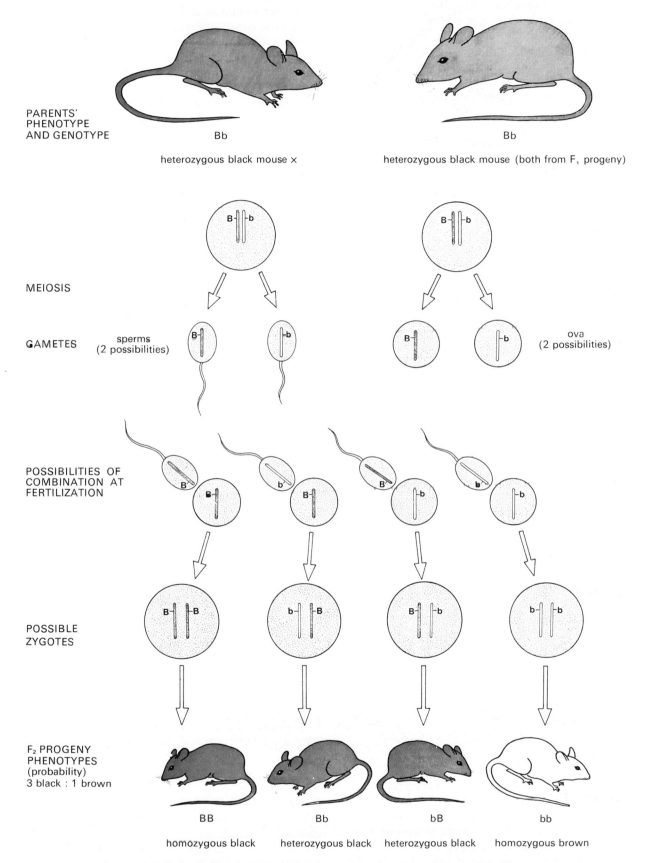

PARENTS'
PHENOTYPE
AND GENOTYPE

Bb

Bb

heterozygous black mouse ×

heterozygous black mouse (both from F₁ progeny)

MEIOSIS

GAMETES   sperms
(2 possibilities)

ova
(2 possibilities)

POSSIBILITIES OF
COMBINATION AT
FERTILIZATION

POSSIBLE
ZYGOTES

F₂ PROGENY
PHENOTYPES
(probability)
3 black : 1 brown

BB

Bb

bB

bb

homozygous black   heterozygous black   heterozygous black   homozygous brown

Fig. 21.2   Inheritance of a single factor for coat colour in mice (second cross)

Single factor inheritance in man

The familiar variations between individuals are controlled by multiple genes. A phenotypic character which depends on a single gene, usually manifests itself as a disorder such as albinism or phenylketonuria as described below. In many cases, eye-colour of Europeans seems as if it were controlled by a single pair of alleles determining pigmentation of the iris, although it is known to be a multiple gene effect. Brown eyes result from extra pigment being deposited in the iris. Brown-eyed people carry at least one dominant allele for the characteristic and blue-eyed people lack this allele. Truly blue-eyed people are therefore homozygous recessive (bb), while brown-eyed people could be homozygous (BB) or heterozygous (Bb). On this basis, it would be possible for two brown-eyed heterozygotes to have blue-eyed children. Blue-eyed homozygous parents, both lacking the dominant allele, would not expect to have any brown-eyed offspring. However, there is a wide variation in so-called 'blue' eyes and some blue-eyed phenotypes may carry non-dominant alleles for pigment.

Similarly in white-skinned races, red (ginger) hair is the result of an allele which is recessive to all other genes for hair colour. A red-haired person must be homozygous for the allele (nn) though both his parents could have black hair (Nn × Nn).

An example of single factor inheritance common to all races is the Rhesus factor in the red blood cells (see p. 76). In this case the Rh + allele is dominant to the Rh − allele. The inheritance of the other blood groups is discussed below.

When one gene only is responsible for an important physiological change, its absence or modification will have serious consequences. Therefore most known instances of single-factor inheritance in man are associated with rather freakish abnormalities. These are usually rare conditions, e.g. occurring once in 10 000 to 100 000 individuals, but there are a great number of different kinds of genetic abnormality.

Examples of known single-factor inheritance involving a dominant allele in humans are one form of night-blindness, one form of *brachydactyly* in which the fingers are abnormally short owing to the fusion of two phalanges, and achondroplastic dwarfism in which the limb bones fail to grow.

The achondroplastic dwarf will have the genotype Dd. This means that half his gametes will carry the dominant allele D for dwarfism, and if he marries a normal woman (dd) there is a chance of one in two that they will have an affected child.

	Affected man		Normal woman	
Genotype	Dd	×	dd	
Possible gametes	D	d	d	d
Possible zygotes	Dd	Dd	dd	dd
	affected		normal	

Similarly, there is a chance of one in four that two achondroplastic dwarfs Dd will have a normal child.

Some of the genetic disorders resulting from recessive alleles are albinism, phenylketonuria, alcaptonuria, sickle cell anaemia, red-green colour blindness and haemophilia. The last two are inherited as sex-linked genes and discussed under this heading on p. 180. Phenylketonuria and alcaptonuria are described on p. 169. Albinism is a condition which results from the absence of pigment from the skin, hair and iris. The affected person has white skin, very blond hair and pink irises to the eyes.

In experimental animals or plants, the type of inheritance and the genetic constitution can often be established by breeding together the brothers and sisters of the F_1 generation, or by back-crossing one of the F_1 individuals with the mother or father and producing numbers of offspring large enough to give results that have statistical significance. These methods are obviously not applicable to man and our knowledge of human genetics comes mainly from detailed analyses of the pedigrees of families, particularly those showing abnormal traits such as albinism, from statistical analysis of large numbers of individuals from different families for characteristics such as sex ratio, intelligence, susceptibility to disease, etc., and from individual studies of identical twins (see p. 181).

Incomplete dominance and co-dominance. Sickle cell anaemia results from the inheritance of a recessive allele (h) which affects the haemoglobin (p. 70). A child inheriting two recessive alleles (hh) from both parents will produce red cells of which one third are likely to take on a distorted 'sickle-like' shape in low oxygen concentrations. This distortion of the red cells makes them fragile and short-lived, leading to anaemia. The heterozygotes (Hh) can be normal healthy individuals but, even so, one per cent of their red cells are liable to distortion at low oxygen concentrations. The allele H for normal haemoglobin is thus not completely dominant over the recessive h allele.

The inheritance of the ABO blood groups in man includes an instance of co-dominance. According to whether their blood will mix without clotting during a transfusion, people are classified into four major blood groups A, B, AB and O (see p. 75). The blood group is controlled by three alleles, A, B and O, acting at the corresponding site on homologous chromosomes. A person will inherit two of these alleles, one from each parent. Allele O is recessive to both A and B, but A and B are co-dominant, i.e. if a person inherits allele A from one parent and allele B from the other, he will be group AB, neither allele being dominant to the other. It follows that group O people must have the genotype OO while group A persons could be AA or AO and group B individuals BB or BO. The following example shows the possible blood groups of children born to a group A man and a group B woman both of whom are heterozygous for these genes.

Phenotype	group A		group B	
Genotype	AO		BO	
Gametes	A and O		B and O	
F_1 *Genotype*	AB	AO	OB	OO
Phenotype	group AB	group A	group B	group O

New combinations of genes

It was stressed in Chapter 20 that at mitosis the genetic information is passed on complete and intact to both cells. At gamete formation, however, meiosis gives rise to variations. It is evident that although offspring resemble their parents, they are not identical to them or to their brothers or sisters. This variability is largely the result of new combinations of genes which were present in the two parents. For example, a recessive gene may be present in the genotype of both parents but not be expressed until it appears in one of the offspring. Parents with the genotype Bb for hair colour may both have black hair and yet any one of their children could have red hair (bb).

On p. 171 it was explained that as a result of meiosis, only one of each pair of chromosomes could be present in the gamete. Since the genes are on the chromosomes it follows that only one from each pair of alleles can be present in a gamete. Which one of the pair appears in any one gamete is a matter of chance, and there are 2^{23} possible different combinations of chromosomes in the human gamete. The separation of alleles at gamete formation is called *segregation* of genes.

A European family might have a mother with red curly hair, bbCc, and a father with straight black hair, Bbcc (the allele for curliness, C, is dominant over straight, c). Each parent would produce two types of gamete with respect to these alleles, bC or bc from the mother and Bc or bc from the father. Thus they could have children with curly red hair or straight black hair, like one or other of the parents, but there is an equal chance of their children having straight red hair or curly black hair, combinations which are not present in either parent (*see* Fig. 21.3).

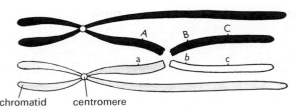

(a) *Prophase; homologous chromosomes have paired up (the diagram shows the chromatids breaking at the chiasma but it is not known if this is actually what happens)*

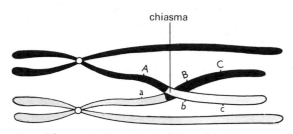

(b) *Prophase (the terminal portions of the adjacent chromatids have become attached to the opposite chromatid)*

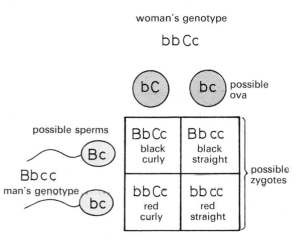

Fig. 21.3   New gene combinations

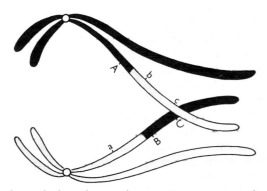

(c) *Metaphase; the homologous chromosomes seem to repel each other except at the chiasma*

Linkage and crossing over. During that early stage of meiosis when homologous chromosomes pair up, the maternal and paternal chromosomes exchange portions as mentioned on p. 171 and shown in Fig. 21.4. This leads to an even greater variability in the gene combinations of the gametes. In the absence of crossing over, the maternal genes A-B-C on the same chromosome would always appear together no matter how the chromosomes were assorted in meiosis. Similarly the paternal genes a-b-c would always remain together. For example, in *Drosophila*, since the genes for black body, purple eyes and vestigial wings occur on the same chromosome, one might expect that a black-bodied *Drosophila* would always have purple eyes and vestigial wings. Crossing over between chromatids, however, gives the possibility of breaking these *linkage groups* as they are called, so that new combinations, ABc, Abc, aBC, abC, aBc, AbC, could arise in the gametes, two of them being black body with red eyes and normal wings, or black body with purple eyes and normal wings.

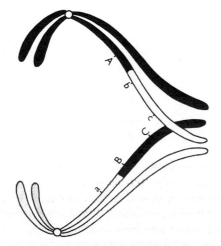

(d) *Anaphase; the chromosomes separate, but as a result of crossing over, the genes A, B, C and a, b, c on the 'inner' chromatids are rearranged*

Fig. 21.4   Crossing over

Sex linkage. Certain genes that occur on the X chromosome are more likely to affect a male than a female. The gene or genes for a certain form of colour blindness in man are carried on the X chromosome. Normal vision is dominant to colour blindness so that if a colour-blind woman, who must be homozygous for the character, marries a normal man, all their sons but none of their daughters will be colour blind. This can be explained by the fact that the Y chromosome is homologous with only a small section of the X chromosome and the non-homologous part of the X chromosome carries genes which are not represented on Y. It is assumed that the Y chromosome plays no part in the determination of colour vision. Fig. 21.5 shows how this type of sex linkage produces its effect.

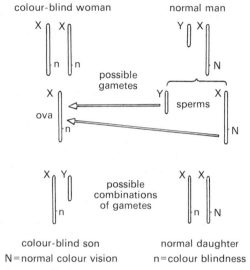

colour-blind son   normal daughter
N = normal colour vision   n = colour blindness

Fig. 21.5   Sex linkage, showing the possible distribution of X and Y chromosomes between the gametes and the chances of combination in the zygotes

The normal daughter is heterozygous for colour blindness and is therefore a 'carrier' for the recessive gene. If she marries a normal man the possible combinations of genes in the children are shown by:

Parents:	XN Xn		XN Y	
	carrier woman		normal man	
Gametes:	XN Xn		XN Y	
Possible combinations of gametes:	XN XN	Xn XN	XN Y	Xn Y
	normal girl	girl carrier	normal boy	colour-blind boy

The theoretical expectations are that all the girls will be normal but half of them will be carriers, half the boys will be normal and half of them colour blind. The types of children expected from the marriage between a woman carrier and a colour-blind man, or a normal woman and a colour-blind man can be worked out in a similar way.

Other X-linked factors are *haemophilia* and brown enamel on the teeth. Haemophilia causes a delay in the clotting time of the blood. Although there are two kinds of sex-linked haemophilia, at least three other clotting disorders are known which are controlled by genes not on the sex chromosomes.

Sexual characteristics such as bass voice, beard and muscular physique in males, mammary glands and wide pelvis in females, are not the result of sex-linked genes but the different expressions of the same genes present in both sexes. Both sexes carry genes controlling the growth of hair, mammary glands and penis, but in the physiological environment of maleness or femaleness they have different effects, with the result that, for example, the mammary glands in males are small and functionless; the penis in females is represented by only a small organ, the *clitoris*.

Only a few rare abnormalities are thought to be linked to the Y chromosome and even these are now open to doubt.

Discontinuous and continuous variation

The individuals within a species of plants or animals are alike in all major respects; indeed, it is these likenesses that determine that they belong to the same species. Nevertheless, even though an organism recognizably belongs to a particular species, it may differ in many minor respects from another individual of the same species. A mouse may be black, brown, white or other colours, the size of its ears and tail may vary; but despite these variations, it is still recognizably a mouse.

Discontinuous variation. The variations in coat colour are examples of discontinuous variation because there are no intermediates. If black and brown mice are bred together they will produce black or brown offspring. There are no intermediate colours and no problems arise in deciding in which colour category to place the individuals. It is not possible to arrange the mice in a continuous series of colours ranging from brown to black with almost imperceptible differences of colour between adjacent members of the series. The way sex is inherited is another example of discontinuous variation. With the exception of a small number of abnormalities, one is either male or female and there are no intermediates.

Discontinuous variation in humans is rather more difficult to illustrate. There are, for example, four major blood groups designated A, B, AB and O. Blood from different groups cannot be mixed without risking a clumping of the red cells. A person must be one or other of these four groups; he cannot, for example, be intermediate between group O and group A. In general terms, eye colour in white races appears to be inherited in a discontinuous manner; one has blue eyes or pigmented eyes, but there are some individuals who would be difficult to classify. Clear-cut examples of discontinuous variation occur among the more serious variants, e.g. one is either an achondroplastic dwarf or one is not; intermediates do not occur.

The features of discontinuous variation are clearly genetically determined; they cannot be altered during the lifetime of the individual. You cannot alter your eye colour by changing your diet. An achondroplastic dwarf cannot grow to full height by eating more food. An albino cannot acquire a darker skin colour by sunbathing. Moreover, the variations are likely to be under the control of a small number of genes. One dominant gene makes you an achondroplastic dwarf; the absence of one gene for making pigment causes albinism.

Continuous variation. When one tries to classify individuals according to height or weight rather than eye colour, the decisions become more difficult and the classes more arbitrary. There are not merely two classes of people, tall and short, but a whole range of intermediate sizes differing from each other by barely measurable distances. Categories can be invented for convenience, e.g. people from 1.4 to 1.6 m, 1.6 to 1.8 m, 1.8 to 2.0 m, but they do not represent discontinuous variations of 0.2 m between individuals.

There is no reason why continuous variations should not be genetically controlled but they are likely to be under the influence of several genes. For example, height might be influenced by twenty genes, each gene contributing a few centimetres to the stature. A person who inherited all twenty would be tall whereas a person with only five would be short. Although this example is purely hypothetical, it is known that height is at least partially genetically determined because tall parents have, on average, tall children and vice versa, but how many genes are involved is not known.

Continuous variations are also those most likely to be influenced by the environment. A person may inherit genes for tallness but if he is undernourished in his years of growth he will not grow as tall as he might if he had received adequate food. In fact, most continuous variations result from the interaction of the genotype with the environment. A person may grow fat if he eats too much food; he will lose weight if he goes on a diet. This seems to be an entirely environmental effect until one realizes that another person may eat just as much food and yet remain slim because of his different, inherited constitution. Whether one catches a disease or not would appear to be dependent on exposure to the disease germs, an exclusively environmental effect, and yet it is apparent that one may inherit susceptibility or resistance to a disease. If a person with inherited susceptibility to an infectious disease is never exposed to the infection, he will not develop the disease.

Heredity or environment?

It is possible to experiment with plants and animals to discover whether an observed variation is due primarily to the genetic constitution or to environmental differences. A species of plant growing in a valley may have larger leaves and taller stems than individuals of the same species growing on a mountain side. If the two varieties are collected, planted in the same situation and grown through one or two generations and still show the differences of leaves and stem size, one may assume that the differences are genetically controlled. If, however, the two varieties, after growing in the same environment, produce offspring which are indistinguishable, the original variations must have been due solely to the environmental differences.

Similar experiments with man are not feasible or desirable. Even when situations occur which resemble the experiment, such as the 'uniformity' of an institutional environment for orphaned children, the observations are always susceptible to more than one interpretation. Thus, there is usually a great deal of argument about very little evidence when people discuss whether our intelligence, for example, is predominantly due to the genes we inherit or the conditions of home and school in which we were brought up. One source of evidence in the 'nature v nurture' controversy is the study of identical twins.

Identical twins

Twins may be either identical or fraternal. If they are fraternal, they are the result of the simultaneous fertilization of two separate ova by two separate sperms. The resulting zygotes will thus contain sets of chromosomes very different from each other, as explained on p. 172, and the twins, although they develop simultaneously in the uterus, will not necessarily be any more alike than if they were brothers or sisters born at different times, e.g. they can differ in sex. Identical twins, on the other hand, result when a single fertilized ovum, usually after a period of cell division, separates into two distinct embryos. The two embryos will thus have in their cells identical sets of chromosomes, since they are derived by mitosis (p. 161) from a single zygote. The twins often share a placenta, although they may be enclosed in separate amnions. Such twins, having the same genotypes, usually resemble each other very closely and are invariably of the same sex, though variations in their position and blood supply while in the uterus may produce differences at birth.

Since the one-egg twins carry identical sets of genetic 'instructions', it can be argued that any differences between them are due, not to their genes, but to the effects of their environment. Identical twins, therefore, are a useful source of evidence for assessing the relative importance of heredity and environment. For example, the average difference in height of fifty pairs of identical twins reared together was only 1.7 cm while the average difference for the same number of non-identical twins was 4.4 cm. These and other points of comparison are given in the table below.

Average differences in selected physical characteristics between pairs of twins

Difference in:	50 pairs of identical twins reared together	50 pairs non-identical twins reared together	19 pairs of identical twins reared apart
Height (cm)	1.7	4.4	1.8
Weight (lb)	4.1	10.0	9.9
Head length (mm)	2.9	6.2	2.2
Head width (mm)	2.8	4.2	2.85

(From Freeman, Newman and Holzinger, *Twins: A study of heredity and environment*, Univ. Chicago Press, 1937)

(Barnaby's Picture Library)

Fig. 21.6   Identical twins

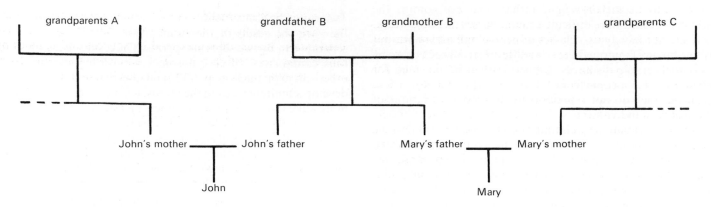

Fig. 21.7 Lineage of first cousins

Detailed histories of identical twins sometimes give impressive results. There is the case of girl twins, separated soon after birth; one was brought up on a farm and the other in a city, and both became affected with TB at the same age. Identical twin sisters were separated and adopted shortly after birth but both became schizophrenic within two months of each other in their sixteenth year. Such individual examples are of interest but cannot lead to any far-reaching conclusions about inheritance of human characteristics in general.

Applications of genetics to human problems

(a) **Screening.** Tests can be carried out to predict, with varying degrees of certainty, whether a person is carrying the genes for a heritable disorder. These tests can be made on the potential parents, on babies or even fetuses in the uterus.

A blood test can reveal a raised level of phenylalanine in a baby with two recessive alleles for phenylketonuria, PKU (p. 169) or a low level of thyroxine in a baby which has inherited congenital hypothyroidism (p. 157). Samples of amniotic fluid taken from a pregnant woman (amniocentesis) will contain cells shed from the fetus. These cells can be examined microscopically to see if they contain the extra chromosome indicative of Down's syndrome (p. 168).

Potential parents can also be screened to see if either or both are carrying recessive genes for a hereditary defect. Abnormal levels of sodium chloride in sweat may indicate (though not with 100% certainty) that the individual is a carrier (i.e. heterozygous) for the cystic fibrosis gene. Carriers of the blood diseases thalassaemia and sickle cell anaemia can be detected by abnormalities in the size or shape of some of their red cells.

In many instances, it is now possible to detect the presence of a harmful gene in a DNA sample from a person's cells. This test may be used as a first resort or to confirm a less reliable screening test, e.g. for cystic fibrosis.

Who should be screened? Potential parents with a family history of a genetic disease are obvious candidates for screening to see if they too carry the relevant genes.

In Britain, babies are routinely screened for PKU and congenital hypothyroidism. This has the obvious advantage that the conditions can be effectively treated by special diet or hormone supplements respectively, before irreversible damage is done.

In Cyprus, mass screening for thalassaemia reduced the incidence of the disease by 95 per cent in ten years.

However, the outcome is not always so positive. The heterozygotes for thalassaemia revealed by screening in Greece were regarded as unsuitable marriage partners. Similarly there are fears that widespread screening for more and more conditions might lead to undue anxiety and also discrimination by employers and insurance companies.

(b) **Counselling.** Screening should always be followed by genetic counselling. Counselling is possible, however, even without prior screening. If the family history is known to include affected individuals, it is possible to assess the *chances* of an individual inheriting the gene (see below).

If screening reveals that both parents are carriers of a faulty gene, they can be told that the chance of having an affected child is one in four (p. 175). They can then decide whether to try for a baby, and what to do if it proves to be affected. If amniocentesis reveals that the fetus carries a serious genetic disorder, the parents can decide whether or not to terminate the pregnancy.

You will, by now, appreciate that genetic screening introduces many ethical problems. It would be very undesirable to use it, for example, in trying to control the sex, IQ, physique or other normal attributes of an offspring.

(c) **Consanguinity.** The study of human genetics enables predictions to be made on the chances of the recombination of two harmful recessive genes in the children of marriages between first cousins.

Fig. 21.7 shows diagrammatically the theoretical pedigree of two cousins, John and Mary. Cousin John is assumed to be heterozygous for a comparatively rare recessive gene, Nn. (He would be known to be Nn for certain, only if one of his parents was nn.) John could have inherited this gene from his grandparents B and in this case there is also a 1 in 4 chance that Mary has inherited the gene. There is thus a chance of 1 in 8 ($\frac{1}{2} \times \frac{1}{4}$) that cousin Mary is also Nn, in which case the chance of an affected child arising from their marriage is 1 in 4. The overall chances of an affected child if John is Nn and marries Mary, are thus 1 in 32 ($\frac{1}{8} \times \frac{1}{4}$).

If the gene is fairly rare in the general population, e.g. occurs once in 100 individuals, the chances of John marrying an Nn person from the general population are 1 in 100. The overall chances of an affected child then are $1/100 \times 1/4$, i.e. 1 in 400.

Although these considerations would apply equally well to beneficial genes, cousin marriages are not usually encouraged and brother-sister marriages forbidden by law. This does not reduce the total number of homozygous recessives which occur in a population but does reduce the chances of their occurring in a particular family. Consanguinity is bound to occur sooner or later, otherwise we should need to have had an impossibly large number of ancestors.

Intelligence

Intelligence is a product of a person's genetic constitution and the effect of his environment; i.e. he may inherit from his

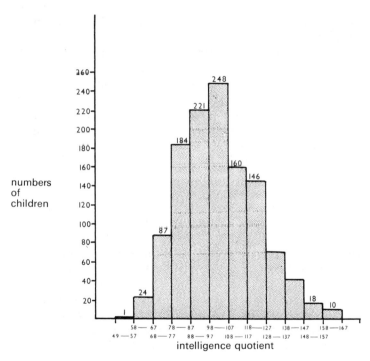

Fig. 21.8   Distribution of IQ rating in a random sample of 1 207 Scottish 11-year-old children

(From C. O. Carter, *Human Heredity*, Penguin Books, 1962)

parents the mental equipment for intelligent thought, but this will not produce his full potential of intelligent behaviour unless he is educated.

That part of intelligence which is genetically controlled, i.e. over half, is almost certainly influenced by a large number of genes and is not susceptible to simple analysis. When a graph or histogram is plotted to show the different numbers of individuals possessing a particular IQ value, the type of picture obtained is that shown in Fig. 21.8, often called a 'normality' curve or *curve of normal distribution*. Similar curves are also obtained for factors such as height and skin colour and can be explained on the basis of several genes influencing the characteristic. Instead of the straightforward presence or absence of a condition, such as albinism, there is a continuous variation with every grade of intermediate.

Family trees

A great deal of information about human genetics is obtained by studying family trees similar to the one in Fig. 21.7. By convention, males are represented by squares and females by circles. Parents are linked by a horizontal marriage line, from which descend vertical lines to indicate the children:

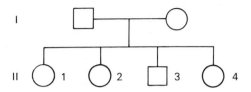

Fig. 21.9   Method of showing a family tree

Fig. 21.9 shows the pedigree of a couple who have 4 children, 2 girls, a boy and another girl in that order. The numerals I and II represent the two generations. The brothers and sisters in generation II are called *siblings* or *sibs*. When the girl sibling II 4 marries and has 3 children, the pedigree would become:

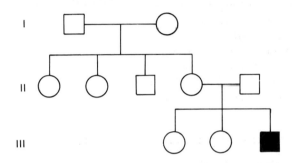

Fig. 21.10   Pedigree with one affected individual

Boy III 3 is represented with a filled square because he shows a genetic trait, e.g. albinism or haemophilia.

By studying pedigrees such as these, it is often possible to discover whether a trait is controlled by a dominant, recessive or sex-linked gene.

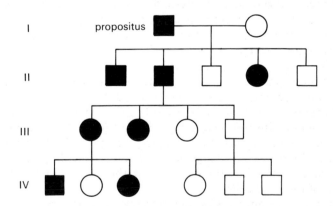

Fig. 21.11   Pedigree showing inheritance of a dominant trait

Fig. 21.11 shows a 4-generation pedigree for a dominant trait. The affected male (called the *propositus*) at the top of the family tree, must be heterozygous for the trait otherwise all his children would be affected. The spouses of II 2, III 1 and III 4 are not shown; it is assumed that they are unaffected (and therefore homozygous) because the condition is a rare one.

It might be argued that the gene causing the condition is recessive, the male propositus being homozygous for the gene and his wife heterozygous. (If she were homozygous normal, none of the offspring would be affected.) In this case, the expectation would be for half the children to be affected. However, more than 50 per cent of the offspring of affected individuals in successive generations are themselves affected, so this explanation seems less likely than the 'dominant' theory. This is particularly so if the condition is rare, because the unaffected spouses from the general population are quite likely to be homozygous normal. In this case none of their offspring would show the trait if the affected parent was homozygous recessive.

If an affected individual turned up in III 4's family, however, the recessive theory would have to be considered more seriously.

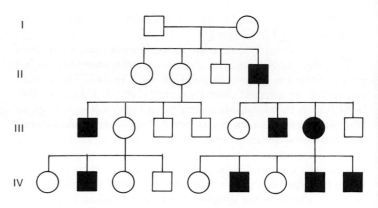

Fig. 21.13   Pedigree indicating sex-linkage

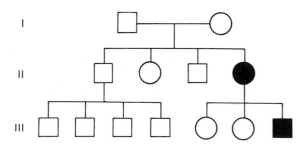

Fig. 21.12   Inheritance of a recessive trait

Fig. 21.12 shows the inheritance of a recessive trait. The birth of an affected child, II 4, to unaffected parents, means that both parents must be heterozygous for the condition and the relevant gene is recessive. The only other explanation is that II 4's condition is the result of a spontaneous mutation.

Generation III shows that the affected woman must have married a heterozygous man, since one of their children is also affected.

In Fig. 21.13 it is apparent that about 50 per cent of the males in the pedigree are affected. The females rarely show the condition. The fact that, in the family of the one affected woman, all the boys are affected is also significant. This is a pedigree characteristic of a recessive gene linked to the X-chromosome and absent from the Y-chromosome.

Questions

1 Two black guinea pigs are mated together on several occasions and the offspring are invariably black. However, when their black offspring are mated with white guinea pigs, half of the matings result in all black litters and the other half produce litters containing equal numbers of black and white babies.

From these results, deduce the genotypes of the parents and explain the results of the various matings, assuming that colour in this case is determined by a single pair of genes (alleles).

2 (a) What are the possible blood groups likely to be inherited by children born to a group A mother and group B father? Explain your reasoning.

(b) A woman of blood group A claims that a man of blood group AB is the father of her child. A blood test reveals that the child's blood group is O. Is it possible that the woman's claim is correct? Could the father have been a group B man? Explain your reasoning.

3 The first child born to a married couple with normal phenotypes is an albino. What are the chances that their second child will also be an albino? Explain your reasoning.

4 A woman whose blood group is O, Rh− marries a group AB, Rh+ man. What are the possible blood groups of their children? Explain your reasoning.

5 Individuals of a pure-breeding line of *Drosophila* are exposed to X-rays to induce mutations in their gametes. Most mutated genes are recessive to normal genes. How could one find out if mutations had occurred?

6 Two black rabbits thought to be homozygous for coat colour were normal and produced a litter which contained all black babies. The F_2, however, resulted in some white babies which meant that one of the grandparents was heterozygous for coat colour. How could you find out, by breeding experiments, which parent was heterozygous?

22
Population Statistics

Censuses

Population means the number of people living in an area. The area may be a town or a district, a country or the whole world. To discover the number of people in a selected area it is necessary to take a **census**, and the practice of recording the numbers of people in a country was probably started by the Romans. A census was being taken throughout the Roman Empire at the time when Christ was born, and the records of populations have been taken for particular reasons ever since that time. 'Bills of Mortality' were required by law in Britain after the outbreaks of bubonic plague in the sixteenth century and these originally gave information about the numbers of people who had died each week from the disease. In the following century more detailed information was called for which included reference to the trade or profession of the person who had died, together with a note of their sex and age.

By 1758 the Swedish government had instituted a regular census and in 1839 the General Register Office was established in London to collect information about births, marriages and deaths throughout Britain. This information, although useful, as we shall see, did not disclose the size of population and it was only the institution of a national census taken every ten years that revealed not only the size of the population of Britain but the way in which it changed with the passage of time.

The scientific study of population changes is termed demography, and the information or data on which the study is based is obtained by a census, and by the registration of vital statistics (literally, the statistics of life). A modern census involves a counting of people either where they are on the day appointed for the census—a *de facto* census—or else according to where they usually live—a *de jure* census. The information is obtained either by sending people (called enumerators or canvassers) into the community to do the counting and write down the statistics of sex, age, occupation and so on, or else by the householder method, in which the head of a house is provided with a form on which he or she has to fill in all the statistics on the appointed day. In Britain the householder method is used at ten-year intervals. By 1961 it was realized that the rate of change of the population was such that ten years was too long to wait to bring certain pieces of information up-to-date, and in 1966 for the first time a subsidiary census was taken in which statistics were collected from 10 per cent of the population.

Vital statistics

The registration of births, marriages and deaths provides a continuous flow of information. Birth statistics provide at least information about the numbers of live births and stillborn infants (babies born dead) and the sex of each baby. This information may be related to the health and marital status (that is, married or unmarried) of the mother. Marriage records indicate the age and sex of the persons married and whether they have been married before or not. Death certificates carry not only the name of the dead person and details of where he died but the cause of death and, if this was due to disease, how long the deceased had suffered from it. Where there is doubt about the cause of death, a post-mortem ('after death') examination is required to be carried out in most countries of the world and this examination must be done by qualified medical men. The World Health Organization has drawn up a classification of diseases, injuries and other causes of death with a view to standardizing the recording of death statistics throughout the world, but the classification suggested is not very detailed.

The information gained by the census and registration enables the *vital rates* to be calculated. These express the frequency of events such as birth or death in proportion to the total population.

The **crude birth rate** is the number of live births per 1 000 of the population per year. This is calculated from the total number of live births per year and the mid-year population, as follows:

$$\frac{\text{number of live births in a year}}{\text{population at the middle of that year}} \times 1\,000$$

e.g. 900 live births in a population of 30 000 gives a crude birth rate of

$$\frac{900 \times 1\,000}{30\,000} = 30.$$

This information is essential if a country is to prepare for the education of its children and provide the right number of school places. If long-term planning is to be effective, it is necessary also to know the *fertility rate*. This is calculated as follows:

$$\frac{\text{number of live births in a given year}}{\text{number of women aged 15 to 45 in that year}} \times 1\,000.$$

A population with a large proportion of young women in this

age group will have a higher fertility rate than one with a large number of elderly women, incapable of child-bearing.

The *specific fertility rate* gives a more detailed picture by showing the numbers of babies born to each 1 000 women in five-year age bands. Women in the 20- to 25-year-old age band are likely to have a higher fertility rate than those in the 40- to 45-year-old age band.

Fertility varies from one social group to another. Generally the better-educated members of a population have a lower fertility rate. This does not mean that they are biologically less able to produce children but that they feel there are reasons for limiting the size of the family. If the data obtained in a census include details of education, age and family size, then instead of talking of a 'general trend' it is possible to calculate a precise fertility rate for that group.

The *infant mortality rate* is a useful guide to the standards of hygiene and sanitation in a community. When the standard is poor, many young babies die, chiefly from intestinal diseases and respiratory infections. The infant mortality rate is calculated as follows:

$$\frac{\text{number of deaths in a given year of babies under 1 year old}}{\text{number of babies born in that year}} \times 1\ 000.$$

The infant mortality rate in Norway in 1970 was 12.8, in Britain 18.8; in Jamaica the rate was 30 and in Pakistan, 142. Merely to quote the rate can be misleading and does not indicate the fact that hurricanes and famine had very seriously disturbed life in Pakistan, resulting in outbreaks of diseases to which young babies are particularly vulnerable.

The information from death certificates is used firstly to calculate the *crude death rate*, which is:

$$\frac{\text{total number of deaths in a population}}{\text{total population at the middle of the year}} \times 1\ 000.$$

When compared with the crude birth rate, this figure shows whether a population is growing, stable or declining. Comparison of birth and death rates over a number of years gives the *rate of population change*.

The crude death rate is indeed crude in the sense that it gives no indication of whether people are dying in old age or in infancy, from disease or starvation or as the result of war. To study the first two of these factors, *age-specific mortality rates* have to be worked out. The method is similar to that used in calculating infant mortality, that is the number of deaths in a particular age group is related to the total number of persons in that age group.

When age-specific mortality rates are known, it is possible to construct life expectancy tables. Such tables were first used by people whose business was to sell life insurance. To do this it is necessary to calculate the customer's probability of dying before his next birthday and, from this, his *expectation of life.*

A *life table* starts with a population of 1 000 at birth. Knowing the age-specific mortality rate for the population concerned it is then possible to work out the numbers of people who are likely to survive at the end of each year, and therefore their expectation of life. The figures given below are part of a life table for males and females in Britain in the years 1866 and 1966, and represents how many more years the various age groups can expect to live.

Expectation of life at age	1866		1966	
	Males	Females	Males	Females
0	40.2	42.5	68.4	74.7
15	43.8	44.3	55.4	61.3
25	36.5	37.4	46.0	57.5
45	22.8	24.1	27.1	32.5
65	10.7	11.5	12.0	15.7

From this table it can be seen that over 100 years the expectation of life at the time of birth has risen dramatically whereas in middle age the increase in life expectancy is very much smaller. However, many more people are now surviving to middle and old age than in 1866.

The *increase* in the expectation of life at age 15 over the expectation at age 0 in both sexes in 1866 deserves comment. This reflects the high infant and child mortality rates of the time. Children who survived into adolescence, having avoided or recovered from such diseases as typhoid fever, tuberculosis, malaria and diphtheria, could indeed look forward to a better chance of surviving to middle age.

This change in life expectation is seen more clearly in the two graphs in Fig. 22.1. The graph or *population pyramid* for 1851 has a shape characteristic of a population with a high birth rate and a high infant mortality rate. Improvements in the standard of hygiene and sanitation in the intervening hundred years have reduced the infant and childhood mortality rates. The pyramid for 1961 shows a population that is ageing, with a death rate increasing rapidly above the age of 45. The reasons for this are discussed in Chapter 23. By superimposing the two pyramids, drawn to the same scale (Fig. 22.2), it can be seen that not only have the numbers of people in different age groups changed, but the total population, represented by the *area* of each pyramid, has increased. The population of every country in the world is believed to have increased in the last hundred

Deaths in the United Kingdom: analysis by age and sex

	All ages	Under 1 year	1–4	5–9	10–14	15–19	20–24	25–34	35–44	45–54	55–64	65–74	75–84	85 and over
Males														
1902	340 664	87 242	37 834	8 429	4 696	7 047	8 766	19 154	24 749	30 488	37 610	39 765	28 320	6 562
1972	342 605	8 393	1 460	1 024	801	1 779	2 092	3 661	7 629	25 184	64 379	109 448	85 535	21 220
Females														
1902	322 058	68 770	36 164	8 757	5 034	6 818	8 264	18 702	21 887	25 679	34 821	42 456	34 907	10 099
1972	331 333	6 198	1 238	665	416	770	878	2 107	5 267	15 897	36 260	77 261	113 076	71 300

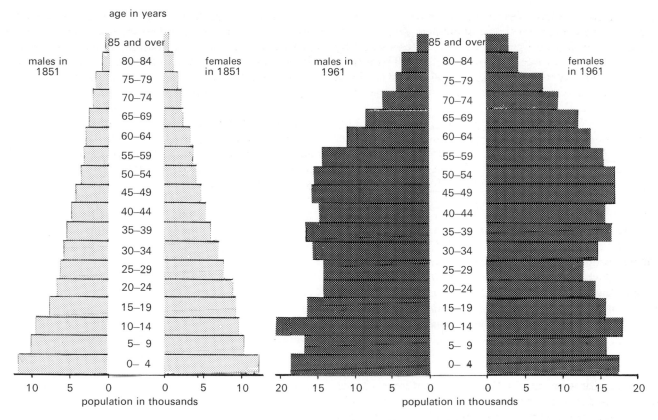

Fig. 22.1 Population pyramids showing the structure by age and sex of the population of England and Wales in 1851 and 1961

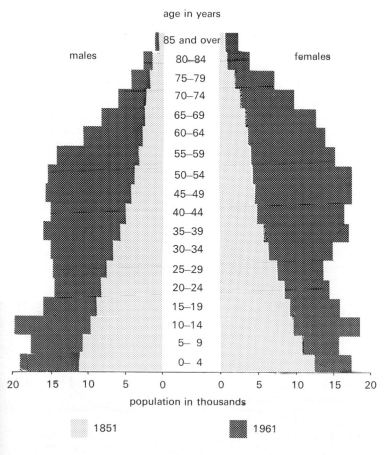

Fig. 22.2 Population pyramids superimposed

years and in some countries the *rate* of increase has been much greater than in Britain.

When the population of a country increases, more food has to be produced to keep the people nourished. More houses also have to be built, with increased water supply and facilities for sewage disposal. More schools and hospitals are required. If the increase in food production and in the provision of housing and other facilities does not take place at the same rate as the increase in population then the *standard of living* falls. The information about population change and about the composition by age of the population enables a thoughtful government to plan ahead and try to provide facilities at the time when they will be needed. Thus if an increase in the birth rate is recorded, not only will there be more mouths to feed but more primary school places will be needed in five to seven years' time. In eleven years, more secondary school places will be needed. Equally, if the birth rate is falling then predictions can be made about the reduction in the number of school places and the number of teachers required. Yet such predictions were either not made or not acted upon in Britain in the late 1970s and early 1980s when education authorities were faced with the problem of 'falling rolls'.

Farming methods have become more productive in many parts of the world but there seems to be no reason to believe that productivity will increase indefinitely.

In 1798 an English priest, the Reverend Thomas Malthus, wrote his *Essay on the Principle of Population*. His main theme was that a population would reproduce and increase in numbers by geometric progression, unless checked. Food supply for that population would only increase by arithmetic progression.

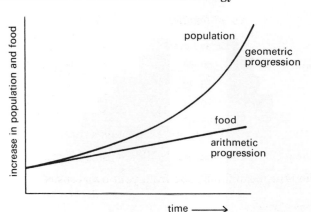

Fig. 22.3   A graphic comparison of geometric and arithmetic increase

Geometric progression (or increase) is illustrated by the series 1, 2, 4, 8, 16, in which the numbers are *multiplied* by two at each step, and by the series 1, 4, 16, 64, 256, where each successive number is multiplied by four. *Arithmetic progression* is illustrated by the series 1, 2, 3, 4, 5, where one unit is *added* to each number, and by the series 1, 4, 7, 10, 13, where three units are added at each step (*see* Fig. 22.3).

In simple terms, Malthus said that food supply does not increase as rapidly as population. In fact, improvements in the methods of raising crops and rearing animals for food have increased productivity much more dramatically than Malthus would have predicted, but the population of many countries of the world continues to increase at rates that outstrip food production and this leads to undernourishment at least and often to death by starvation.

Charles Darwin was made aware of the Malthusian principle before he started to formulate his theory of evolution. He was impressed by the fertility rate (which he called *reproductive capacity*) of all species of animal and plant and said in his book *The Origin of Species*, 'There is no exception to the rule that every organic being naturally increases at so high a rate that if not destroyed, the earth would soon be covered by the progeny of a single pair'. He also recognized that there are many 'checks to increase' which limit the size of populations. For human populations the checks have included disease and famine over which, until recent times, man had very little control. Even now his control is far from complete, but the degree to which the incidence of fatal diseases such as smallpox, cholera and tuberculosis has been reduced since 1955 has resulted in an enormous increase in population throughout the tropics. With more mouths to be fed, the threat of starvation has increased. Starving people are not as well able to work and increase food production as healthy, well-fed people. While food shortage is indeed a check to population increase in animal communities, it is not desirable in human communities. The alternative is to restrict the number of births by contraception (*see* Chapter 15, p. 109).

In an industrialized country such as Belgium the birth rate is controlled to such an extent that the population increases at the rate of only 0.4 per cent per year. It would take 175 years for the population of Belgium to double at this rate of increase. The average population growth rate for tropical South America is 3 per cent, and at this rate the population will double in only 24 years, thus doubling the demand for food and other resources.

The change in world population is shown in Fig. 22.4.

From about A.D. 1500, awareness of the importance of pure water, the need to remove sewage, and improved methods of food production began to influence societies. The growth of world population since then has become very clearly geometric or exponential. One wonders what the next check to increase will be. Already shortage of food is checking population growth in many parts of the world such as Ethiopia and the Sahel. AIDS is also reducing population size but kills off producers of food. A check on the birth rate would enable the increase in food production to 'catch up' and improve the standard of living of those surviving in the world.

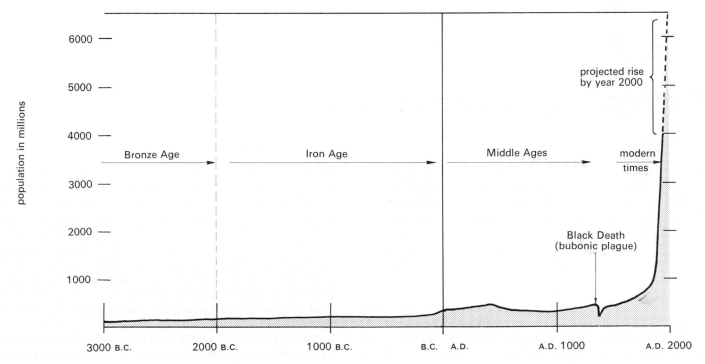

Fig. 22.4   Growth of world population over 5 000 years

Practical Work

Experiment 1   Censuses

Carry out a census among the pupils in your class. Start out with a census of family size, that is the number of children in each family. Plot the results in the form of a histogram or bar graph (*see* Fig. 22.5).

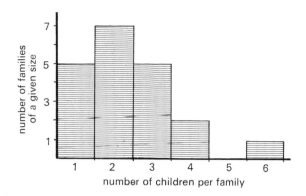

Fig. 22.5   Distribution of family size in a class of twenty-four London schoolboys

Experiment 2

The mean family size can be calculated multiplying the numbers of families in each group by the number of children in that group, adding the scores together and dividing by the total number of families. Thus:

No. of families		No. of children	
5	×	1	= 5
7	×	2	= 14
5	×	3	= 15
2	×	4	= 8
0	×	5	= 0
1	×	6	= 6

Total   48 ÷ 20 families = 2.4

The mean family size is 2.4 children for this group.

Find the mean family size from your own census.

Experiment 3

Collect information about the age structure of the population in your road or village. Don't be too ambitious and attempt to collect information about too large a group. Find out how many people there are in five-year age groups, 0–5, 6–10, 11–15, and so on, and construct either a histogram or a population pyramid. Compare your results with the surveys carried out by other students in your class. Are there more children, or more people aged 60 or over in some streets than in others?

Questions

1   Why is the infant mortality rate usually considered separately from the general death rate of a population? Outline the factors that influence infant mortality in your country.

2   The population in the United Kingdom was 38 327 000 in 1902 and had increased to 55 798 000 by 1972. Use the table on page 186 and suggest why the total number of deaths in both years is almost the same. In which age group has the reduction in death rate between 1902 and 1972 been greatest?

23
Keeping Health Records

Notifiable diseases

Health authorities in most countries require doctors and other medical workers to send in reports of certain illnesses. Consequently these illnesses, which represent serious dangers to the community, have come to be known as *notifiable diseases*. Examples of these include

cholera	scarlet fever	typhoid fever
relapsing fever	plague	yellow fever.
diphtheria	typhus	

Of these diseases, outbreaks of cholera, plague, relapsing fever, typhus and yellow fever must be reported to the World Health Organization. It is recognized that they can very easily be transmitted from one country to another and only by international co-operation can their spread be checked.

Epidemiology

Disease records are of limited use if they tell no more than the number of cases occurring in a given year. Information is needed about where the case occurred and at what time of year. The age, sex and occupation of the patient also enable a more detailed picture to be drawn up. *Epidemiology* is the study of such disease statistics and it can provide enormous support in planning preventive measures. Originally epidemiology was concerned with the study of those communicable diseases that give rise to epidemics. An *epidemic* is the occurrence of a number of cases of a disease very much in excess of what would normally be expected.

Bacillary dysentery is said to be an *endemic* disease in Britain because it is always present at least at a low level among the population. Each summer the number of reported cases

Health Records and changes in the law

Health records can be used to influence government policy. Fig. 23.4 shows the relationship between the total number of deaths and the smoke and sulphur dioxide concentrations in the air in London in the first 15 days of December, 1952. At that time most of the houses in London were heated by coal fires which produce a great deal of smoke and sulphur dioxide. On the 3rd/ 4th of December a change in temperature resulted in a dense fog which blanketed London for five days. During that time and for several days afterwards the death rate rose by 2.5 times. Doctors and hospitals recorded a striking increase in the numbers of cases of two lung diseases and a similar increase in deaths from these diseases – acute bronchitis and pneumonia.

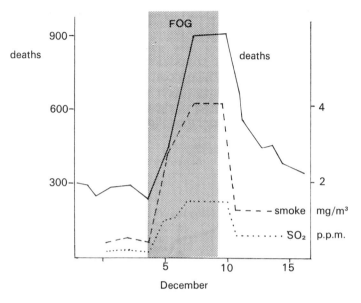

Fig. 23.4   Total deaths and levels of smoke and sulphur dioxide in London in the great fog of December 1952

During the two years that followed, a report was drawn up by the Ministry of Health, making use of medical records, which showed a correlation between the increased death rate and the presence of high concentrations of smoke and sulphur dioxide which were stationary in the atmosphere during the fog. The outcome was the introduction of new laws forbidding the burning of fuels other than smokeless fuels in London. The city became a smokeless zone. Fogs have occurred since, but they have never been accompanied by the increased death rate of 1952.

Records of behaviour of vectors

Finally, records of climate and the breeding habits of a disease vector (Chapter 31) may lead to the development of methods of controlling a disease. Romans had known for centuries that the incidence of malaria in their city was high during the hot summer months. Wealthy Romans left the city to spend that period in villas high up in the nearby hills where malaria was, if not unknown, never spread from one person to another. We now know that the particular species of anopheline mosquito that transmits malarial parasites in central Italy will only complete its breeding cycle to produce new, adult mosquitoes when the temperature of the water in which the larvae develop

is at least 15 °C. A combination of too low a water temperature for larval development and a night temperature too low for the adult mosquitoes to fly in search of blood meals keeps the vector away from the hill villages.

Sometimes man himself is the carrier of disease organisms from one part of the world to another. This is true for cholera epidemics (Chapter 29) and for influenza. In 1957 Asian 'flu' swept through Britain having originated in the Far East and in 1968, Hong Kong 'flu' followed a similar pattern. With high-speed air travel it is quite possible for a passenger to carry the virus from one part of the world to another before he has started to show signs and symptoms of the disease. Where no easy and rapid transport exists an outbreak of a disease can be contained. Thus in 1977 there was an outbreak of a virulent form of influenza in Russia. But it occurred in an isolated part of the country and there was no question of a nationwide epidemic.

Practical Work

Experiment 1

In most countries people 'catch cold' at any time of year. Keep a record of the people in your class who catch cold, noting the month (or the week) in which they develop a cold. You might be able to obtain records for people catching colds in more than one class, perhaps in the whole school, if you make it a team effort.

Having collected your records for a year construct a histogram or bar graph to show the incidence of colds over the year. You will have to be persistent to collect records for the holiday period. Fig. 23.5 shows a histogram for a group of fifty boys over one year in a London school.

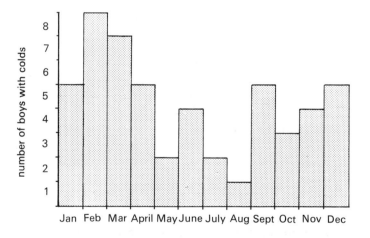

Fig. 23.5   Number of boys in a group of fifty suffering from colds each month over a period of a year

Experiment 2

You can carry out such a survey for any other illness that affects your group. You can also compare the incidence of illness in the first three weeks and the last three weeks of any term. Very often the 'score' for the first three weeks is the higher, at a time when everyone should be fresh and fit from the holidays. If this is so in your class can you suggest why this should be? Perhaps Chapter 31 will give you some ideas.

Experiment 3

Draw a twelve-month calendar, as shown in Fig. 23.6 and mark on your calendar the months in which you observe houseflies

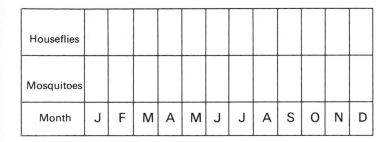

Fig. 23.6 Chart for recording the occurrence of houseflies and mosquitoes throughout one year.

and mosquitoes or gnats out of doors. Of course it will help if you catch the insects and check that you are trapping the same kind of fly or mosquito at each stage of the year. From your chart suggest when you would be most likely to encounter intestinal complaints. When might diseases spread by mosquitoes be most common?

Your chart will be more helpful if you can record whether the insects are abundant (A), frequent (F), rare (R) or absent (O).

Questions

1 Not all the people who die from lung cancer are cigarette smokers and this is sometimes used as an argument in favour of smoking. What facts justify a campaign against smoking?

2 Study Fig. 23.5. Suggest a reason why the number of boys suffering from colds is higher in September than in either August or October.

24
Health and Disease

Health

Good health is something that many of us take for granted. It is much more than an absence of disease. Good health involves all the organs of the body working properly. It also involves *feeling* well both in body and mind.

Personal health. The following are some of the steps which individuals can take to maintain themselves in a state of good health:

(a) **Diet.** Nutritional deficiency diseases have been described in Chapter 6 and it is only necessary to add that an inadequate or unbalanced diet may result in a person being more susceptible to other kinds of disease. Thus the skin eruptions and slow healing that result from a shortage of ascorbic acid, vitamin C, in the diet may lead to infection by bacteria that would not cause harm to a healthy skin.

Better feeding and improved nutrition reduce the likelihood of death from transmissible disease but, by enabling people to live longer, create situations in which the tissues of the body have time to degenerate.

(b) **Personal hygiene.** Simple precautions, such as washing your hands after going to the toilet or latrine, and before handling food can greatly reduce the chances of picking up an intestinal disease. Such habits will also prevent you from passing diseases on to other people (p. 237).

Washing your whole body and changing your clothes will help to prevent fungal infections and will remove body parasites which might be carrying disease organisms (p. 239).

(c) **Domestic hygiene.** Keeping cooking pots, plates, cups and other utensils clean will cut down the numbers of bacteria on them. Keeping flies out of the house, or at least away from food and utensils, removes one source of contamination (p. 246).

(d) **Clean food and water.** Fruit and vegetables, which are eaten without cooking, should be washed in clean water in case they are contaminated with disease organisms.

When cooking food, it is important to heat it to a high enough temperature to kill bacteria and tapeworm cysts (p. 206). Cooked food should be eaten at once or stored in a cool, fly-proof place (Chapter 38).

If you are not certain that your water is clean and free from bacteria, it is a sensible precaution to boil it if it is to be used for drinking.

(e) **Avoiding infected water.** In the tropics, unless you know that a pond, lake or river is free from disease bacteria and from the water snails which carry schistosomiasis, you should avoid paddling, bathing or washing in such places.

(f) **Not smoking.** It is now well established that smoking, especially cigarettes, causes bronchitis, emphysema, athero-sclerosis, heart attacks and lung cancer (p. 195). It is clearly sensible not to take up smoking or, if you have started, you should give it up.

(g) **Exercise and relaxation.** Regular vigorous exercise during work or recreation has a very beneficial effect on health. It seems to maintain the heart and circulatory system in good working order and also promotes a positive feeling of well-being.

Regular sleep and relaxation play an important part in maintaining health, particularly mental health.

Community health. There are many steps which can be taken by the community to maintain and improve standards of health. These include the provision of clean drinking water (Chapter 36) and efficient methods of sewage disposal (Chapter 37). The provision of clinics and health workers is also an important contribution to the health of the community (Chapter 41).

Perhaps one of the most important roles for a community health service is in education. If people in a community do not know how diseases are spread or how they may be avoided, the standard of health in the community is not likely to improve much. Matters relating to public health are discussed in Chapter 41.

Disease

Disease is not easy to define. If a person suffers from influenza then his body temperature will be above the normal 37 °C (the average daily temperature of people living in most countries) and he will probably have a headache as well as aches in his back and limbs. He will *feel* ill. A person suffering from tuberculosis, on the other hand, may have a persistent cough or may not feel very strong, but he is able to do work that does not require physical effort and he probably will not complain of feeling ill. Both influenza and tuberculosis are diseases in which parts of the body do not function normally or efficiently and both are the result of the invasion of the body by micro-organisms.

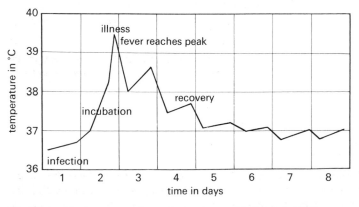

Fig. 24.1   Temperature chart of a patient suffering from influenza

Most diseases are accompanied by *signs* and *symptoms* which make you aware that you are unwell. Signs can be *seen* by the doctor or another observer. For example, body temperature in a healthy person varies between 35.8 °C in the early morning to 37.7 °C in the evening, and with exercise the temperature may rise a little higher. But if the temperature of a resting person is more than 38 °C, particularly if accompanied by a raised pulse rate, then this is a sign of illness. Diarrhoea, vomiting, swellings and rashes on the body are all visible signs of disease.

Symptoms, such as headache, nausea and lassitude, are *felt* by the patient and must be described by him.

A doctor or health worker will measure the body temperature and take into account any visible signs and a description of symptoms in order to decide whether a person has a disease or not.

Diseases which can spread from one person to another are called *infectious* or *transmissible diseases*. Diseases which cannot be spread from person to person are *non-infectious* or *non-transmissible diseases*.

Transmissible diseases

These are caused by organisms such as viruses, bacteria, round worms and flatworms which get into the body. Organisms which enter the body of another organism and feed on its living tissue are called *parasites*. The organism (e.g. man) which the parasite enters is called the *host*. Some parasites damage the host's tissue, or upset its metabolism (the normal working of its body). In other words they cause disease. Parasites which cause disease are called *pathogens*.

Most parasites are so small that they can be seen only with the aid of a microscope: they are called micro-organisms. Bacteria, viruses and single-celled animals (protozoa) are all micro-organisms. Because these parasites are so small, they are easily transmitted from one host to another, by touch, or in food, or in many other ways: thus the disease is easily spread.

Some parasites, on the other hand, can be quite large. Tapeworms, for example, can grow to several feet in length inside their hosts. The eggs or larvae of the larger parasites, however, are small, and easily transmitted.

Transmissible diseases, the organisms which cause them, the methods by which they spread and how they can be controlled are described in Chapters 25 and 30.

Non-transmissible diseases

A large number of diseases cannot be spread from one person to another. If one person has a vitamin deficiency disease such as beri-beri, then nobody can catch it from him. They can only develop the disease through being short of vitamin B_1 in their diet (p. 36). A person with an over-active thyroid gland will suffer from hyperthyroidism but this condition cannot be passed on to anyone else.

Dietary deficiencies

A gross shortage of food in children leads to the disease marasmus. This is closely associated with the disease called kwashiorkor which may be the result of a protein deficiency but other possible causes are being investigated (p. 37). A shortage of vitamins causes diseases such as pellagra and beri-beri; shortage of iron leads to anaemia. This type of deficiency disease is described on pages 34–7.

Metabolic disorders

These are disorders of the metabolism or normal working of the body. Here are two examples:

Obesity. This is the accumulation of excessive fat in the body, and is one of the common metabolic disorders. Obese people may be observed in many countries, particularly in so-called 'Western' or developed countries.

In all cases, obesity results from eating more food than is needed to supply the body's energy requirements. It can occur at any age, though it is most commonly associated with the middle years of life. The appearance of being fat is due to the excessive growth of the fat-bearing cells under the skin (subcutaneous fat, pp. 96–7). The extra weight that has to be carried by the body means that there is a strain on the heart and lungs. Both organs have to work harder to provide the energy needed to move the heavy body around. Because of this the person tires easily and has to rest, reducing energy expenditure. Heart failure is more common in obese persons than in those whose weight is normal for their height and build.

The only effective treatments are to reduce the amount of food and drink consumed so that the body slowly respires the excess fat, and to take regular exercise.

Diabetes mellitus. This is the name given to a disease in which a person is unable to regulate the concentration of sugar in the blood. One form of diabetes results from the failure of islet cells in the pancreas to produce the hormone *insulin* (p. 157). It can be serious. With too much sugar in the blood a person becomes sleepy, but with too little sugar a person may pass into a state of *coma*, very deep unconsciousness which may lead to death. Diabetes can increase the likelihood of a person developing disorders of the retina of the eye and becoming blind. It also makes people more susceptible to other diseases, such as tuberculosis.

It is not certain why diabetes develops. It is believed that some people are more *likely* to develop diabetes because of the genes that they inherit, but some other factor, such as a prolonged period of eating too much sugary food, is necessary to trigger off the disorder.

Degenerative diseases

Some diseases are called *degenerative* diseases because parts of the body degenerate and become less effective.

Thrombosis. As people grow older there is a tendency for the arteries to lose their elasticity. This is one of the factors leading to high blood pressure. If the lining of particular arteries becomes impregnated with fatty material the artery becomes narrower, making it more difficult for the blood to pass through (Fig. 24.2). If a clot forms inside the blood system then it may be trapped by the fatty material, blocking the artery. Whatever part of the body is supplied with blood by that artery is now deprived of blood. If the coronary arteries, which supply blood

to the muscles of the heart, become blocked then a *coronary thrombosis* results, leading to a heart attack (p. 79). If the blood supply to the brain is cut off in this way then a person suffers from a *cerebral thrombosis*.

Arthritis. This is a condition in which the joints become inflamed and swollen (Fig. 24.3). The joint may become permanently damaged, causing pain and difficulty in movement. The joint *degenerates* and will not recover.

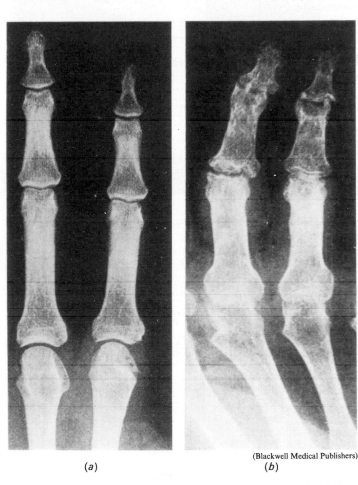

(Blackwell Medical Publishers)

(a)　　　　(b)

Fig. 24.3　(a) X-ray of healthy finger joints. (b) X-ray of finger joints swollen with arthritis. Notice that the bones of the lower joints have fused together

Cancer

Cancer is becoming a more common disease in most countries of the world. Cancer is the name given to a disease in which cells of the body reproduce at an abnormal rate and produce a swelling or **carcinoma**. Most cancers start in one of the tissues in which the normal role of the cells is to divide. Lung cancer starts in the epithelium of the bronchial tubes. These lining cells divide regularly to replace the worn-out cells that are removed by the cleaning mechanism (p. 85). This cancer is of particular concern because it seems that one of the factors causing the cells to divide more frequently is a chemical found in the tar in cigarette smoke. If a person who smokes cigarettes inhales the smoke, that is, breathes the smoke into his or her lungs, the tar settles on the lining of the tubes where it may stimulate the cells to form a cancer. This is not to say that non-smokers are

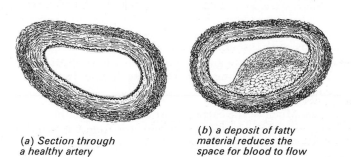

(a) Section through a healthy artery

(b) a deposit of fatty material reduces the space for blood to flow

Fig. 24.2　Thickening of the walls of an artery

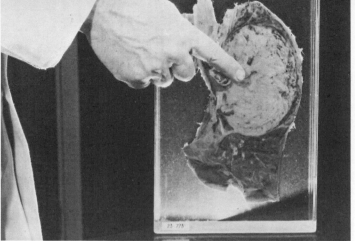

(Crown Copyright)

Fig. 24.4   Lung cancer. The cancerous growth almost fills this lung

immune to such cancers, but that they appear to be significantly less likely to develop them. The cancer-producing (carcinogenic) effect of such tars has been demonstrated on laboratory animals.

The Malpighian layer of the skin, the liver, bone marrow and the milk-producing cells of the breast are just four of the other parts of the body where cells divide naturally and where cancer may develop.

Breast cancer is the commonest cancer among women. Its occurrence increases after the menopause, the time in a woman's life when she ceases to menstruate and release eggs because of a change in the pattern of hormone production (p. 108). The alteration in the hormone balance following menopause is likely to be one of the factors inducing cancerous growths in the breasts, though other factors, such as diet, may also be involved.

A tumour can be felt as a lump in the breast tissue and, while by no means all lumps in a breast are cancerous, it is sensible to seek medical advice as soon as such a lump is discovered. Early treatment, usually by surgery, gives a good chance of recovery whereas delay in receiving treatment for any cancer increases the chance of the malignant cells spreading to other parts of the body.

The fear of cancer is very strong, particularly as people grow older, yet the record of successful treatment has grown steadily over the past few years.

Inherited disorders

Sickle cell anaemia (p. 71) is an inherited disorder in which the red cells in the blood take on an unusual shape because the haemoglobin in the cells includes the so-called S form (Fig. 24.5). This factor makes the red cells less effective in carrying oxygen. However, people with this inherited disease have greater resistance to malignant malaria.

Fig. 24.5   Sickle cells in a blood smear

(Wellcome)

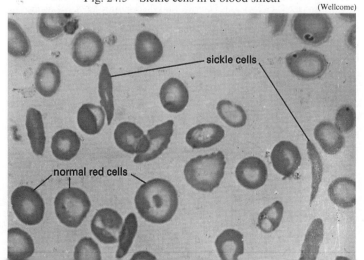

Haemophilia is a condition in which the blood takes a long time to clot. Even a small cut or wound can result in a lot of blood being lost. One kind of haemophilia results from a defect in the gene (p. 73) responsible for the production of one of the enzymes that brings about clotting. If the defective gene is carried in the sperm or the egg that fuse together at conception, then the child that is produced may inherit haemophilia.

Occupational diseases

There are several diseases, often termed occupational or industrial diseases, which result from the regular inhaling of dust or fumes associated with particular industries or industrial processes. Dusts from coal, stone, asbestos and china clay are implicated in the development of lung diseases. Irritant

(Tek Image/Science Photo Library)

Fig. 24.6   Two workers wear overalls and rubber gloves to protect their skin from chemicals. They are wearing face masks which filter the air they breathe and protect their eyes

chemicals such as ammonia and chlorine, widely used in industry, and vapours such as those of zinc and copper, released as fumes during welding, can also cause disease. In this section we consider one example.

Silicosis. Much of the earth's crust is composed either of silica (silicon dioxide) or silicates. These are found in sand and sandstone, clay and granite, all of which are widely used. When minerals are quarried or mined, when clay is made into pots and sand is used in moulds for metal casting, dust containing silica is produced. Unless workers breathe through elaborate and efficient masks, they breathe in the silica dust.

(WHO)

Fig. 24.7 Drilling rock produces a great deal of silica-containing dust which damages the lungs

Silica is more damaging to the lungs than, for example, coal dust, or flour. If these dusts are breathed into the lungs then the normal cleaning mechanism (p. 85) will usually remove the particles unless the quantity breathed in is so great as to overwhelm the cilia. Silica dust is an irritant and it is dangerous because it induces *fibrosis*. If silica dust passes the cleaning mechanism and enters the alveoli then large white blood cells, called *macrophages*, escape from the blood capillaries and ingest the dust particles. The macrophages are killed by the silica and they release poisonous substances that induce inflammation. Over a number of years the lung tissue responds by producing fibrous tissue to replace that which has been damaged by the inflammation. It is the build-up of nodules or lumps of fibrous tissue that obstructs the air passages and reduces the working efficiency of the lungs. Not only this, but sufferers from silicosis are more susceptible to such lung diseases as bronchitis and tuberculosis.

Laws have been passed in Britain and in many other countries in an attempt to regulate the conditions in which people work to reduce their exposure to silica dust. If workers become ill with lung complaints, and medical examination shows a clear connection between the illness and the working conditions then industrial compensation can be claimed.

Mental disorders

Mental disorders are discussed in Chapter 34. Mental defects may be due to failure in the normal pattern of growth and development, resulting from severe deficiency in the diet or from disturbances in the production of the hormones of the body, as in *cretinism* (p. 157). If a pregnant woman contracts rubella (German measles) then the virus causing the illness can cross the placenta and reach the foetus and may affect its brain development.

Many mental illnesses result from internal or external stress. Virus diseases sometimes leave a patient very depressed, but emotional or psychological factors may also produce depression.

The table below summarizes the different types of diseases discussed in this chapter.

Type of disease	Example
Transmissible:	
Diseases caused by micro-organisms: bacteria	cholera; tuberculosis
viruses	measles; rabies; AIDS
protozoa	malaria; trypanosomiasis
fungi	ringworm; athlete's foot
Diseases caused by larger organisms: roundworms	ascariasis; onchocerciasis; pinworm
flatworms	schistosomiasis; tapeworm
Non-transmissible:	
Nutritional deficiency diseases	rickets; scurvy; kwashiorkor
Metabolic disorders	diabetes; phenylketonuria
Degenerative diseases	coronary heart disease; arthritis
Cancer	skin cancer; breast cancer
Inherited disease	sickle cell anaemia
Occupational or industrial diseases	silicosis
Mental disorders	depression; alcoholism

Questions

1 Why are some kinds of disease described as infectious?
2 How does a degenerative disease differ from a cancer?
3 Study the sections of the book dealing with anaemia and say which forms of anaemia are (a) deficiency diseases; (b) inherited disorders; (c) metabolic disorders.
4 Suggest reasons why the *number* of people suffering from most forms of cancer is increasing throughout the world whereas the number suffering from tuberculosis is falling steadily.

25
Organisms that Cause Disease

Discovery and culture of disease-causing organisms

The possible connection between living organisms and the incidence of disease has been a matter of speculation for a long time. Fracastorius in 1546 suggested that syphilis was caused by a *contagium vivam*—a live contact.

Probably the first person to see micro-organisms was the Dutchman, van Leeuwenhoek. In about 1676, he invented a microscope for looking at very tiny objects (Fig. 25.1) and described 'little animals' which he saw in scrapings from his teeth. He noticed that these little organisms did not move if the scrapings were taken just after he had drunk hot coffee.

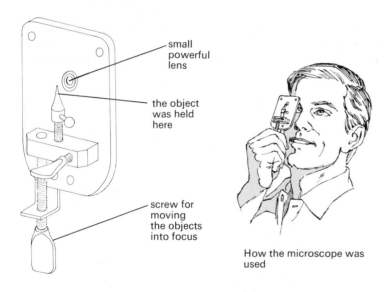

small powerful lens

the object was held here

screw for moving the objects into focus

How the microscope was used

Fig. 25.1 Van Leeuwenhoek's microscope was a very simple instrument

Both before and after van Leeuwenhoek's time, theories were put forward to explain decomposition, particularly of rotting flesh. Most people believed that dead flesh decomposed of its own accord.

Spallanzani, who lived in Italy from 1729 to 1799, is given credit for having shown that the decomposition of organic liquids depends on the introduction of living organisms from the air. Both Spallanzani and Louis Pasteur, who lived in France, nearly 100 years later, carried out experiments to show

that decay was caused by micro-organisms. If these micro-organisms were prevented from reaching the organic material, then it did not decay. At that time the theory was revolutionary and many scientists did not believe it, particularly when it was suggested that similar tiny organisms might be the cause of disease.

The connection between micro-organisms and disease *did* become accepted, slowly, particularly after Robert Koch, working in Germany, carried out investigations of the diseases called anthrax and tuberculosis. He showed that these two diseases were caused by particular organisms that could be recognized under a powerful microscope. In 1884 he summarized his findings in a series of scientific statements which are now known as *Koch's postulates*:

In order to prove that a disease is caused by a particular organism it must be possible to:

1 Observe the organism in *every* case of the disease.
2 Isolate the organism in a pure culture (*see* below).
3 Produce the disease in a suitable experimental animal by injecting it with the pure culture.
4 Recover the organism in pure culture from the diseased experimental animal.

Koch worked out methods for growing bacteria. He used jelly made from meat and bones, with nutrients in it. When a few micro-organisms were placed on the jelly they absorbed food from it and grew rapidly to produce a visible *colony* (Fig. 25.2).

One of Koch's assistants was a young scientist called Petri who invented the glass dish, known as a Petri dish, which is now widely used for growing bacteria. Today we use a jelly made from seaweed, called agar jelly, rather than the bone and meat jelly that Koch and Pasteur used. When bacteria are grown on this jelly they are said to be *cultured*. The colonies form the *culture*. Koch introduced the idea of a pure culture. A pure culture contains only one kind of micro-organism, for example, typhoid bacteria. If other organisms are found to be present, such as tuberculosis bacteria or the mould *Penicillium*, then the culture is said to be *contaminated*.

A micro-organism that causes disease is called a *pathogen*. The animal in which it causes disease is known as the *host*, because its body provides food for the pathogen.

Within the last 100 years our knowledge of the range of organisms that cause disease has increased greatly. Pasteur spoke of *microbes*—which means 'little living things'. Today we speak of 'germs', but we distinguish between several groups of micro-organisms—bacteria, viruses, fungi and protozoa, all of which are very small.

Disease-causing organisms

Bacteria

Structure. Bacteria are very small organisms, each being a living cell. Most bacteria are between 0.0005 mm and 0.002 mm long and they rarely exceed 0.01 mm in length. They are therefore visible only under a high-power microscope. A cell wall is present, made of protein and fatty substances. It encloses the cytoplasm and the nuclear or genetic material. Since the latter is not enclosed in a membrane, we cannot call it a nucleus (*see* Fig. 25.3).

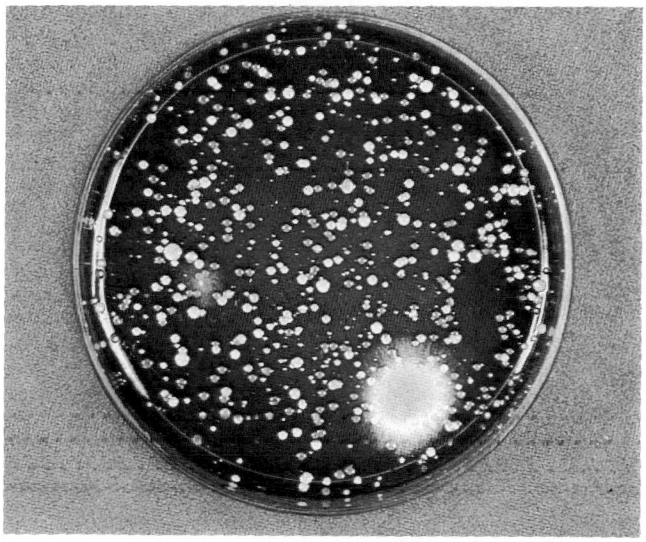

(St Mary's Hospital Medical School)

Fig. 25.2 Bacterial colonies, with one mould colony, growing on a dish of nutrient agar

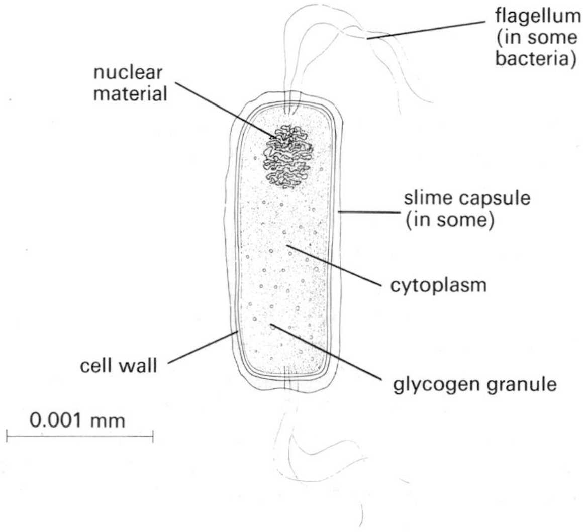

Fig. 25.3 A bacterial cell drawn from an electronmicrograph ($\times$ 2 000). There is no nuclear membrane around the chromosome. The slime layer gives colonies of bacteria their shiny appearance

Different kinds of bacteria show a range of shapes (Fig. 25.4). Certain kinds of bacteria have long thread-like structures, called flagella, with which they can move. Other kinds of bacteria without flagella, particularly spirochaetes, can also move about. All bacteria must respire. Most bacteria use up oxygen and respire aerobically, while others, including several of the pathogenic bacteria, respire anaerobically (without oxygen; p. 28).

All bacteria reproduce by fission, that is, they divide into two. If they are in a warm place, have plenty of food and can get rid of their waste products, most bacteria can reproduce every 20 minutes.

1st hour	*2nd hour*	*3rd hour*
1→2→4→8	→16→32→64	→128→256→512

In one day one bacterium can produce millions of offspring. This is one of the reasons why diseases such as bacillary dysentery (caused by bacteria) can often develop very rapidly.

The simpler kinds of bacteria may be classified as *cocci* when they are spherical, *bacilli* when they are rod-shaped or cylindrical, *spirilla* when they are spiral and do not move, and *spirochaetes* when they are spiral and are able to move of their own accord.

Activities of bacteria. Most bacteria are killed if they are heated above 50 °C or if they become dry. One group, the bacilli, are able to produce a thick protective coat which protects them against both high temperature and drying out. When surrounded by such a coat the bacterium is called a *spore*. The bacillus that causes tuberculosis (TB) can do this. When a patient spits out TB bacilli, the sputum (or spit) dries up and some of the bacteria form spores which are very small and can be blown about by the wind with dust. People may then inhale the airborne spores and become infected.

Most kinds of bacteria live in water, in the soil or else in decaying matter. In these situations the bacteria feed by secreting enzymes (p. 26) on to the organic matter to digest it.

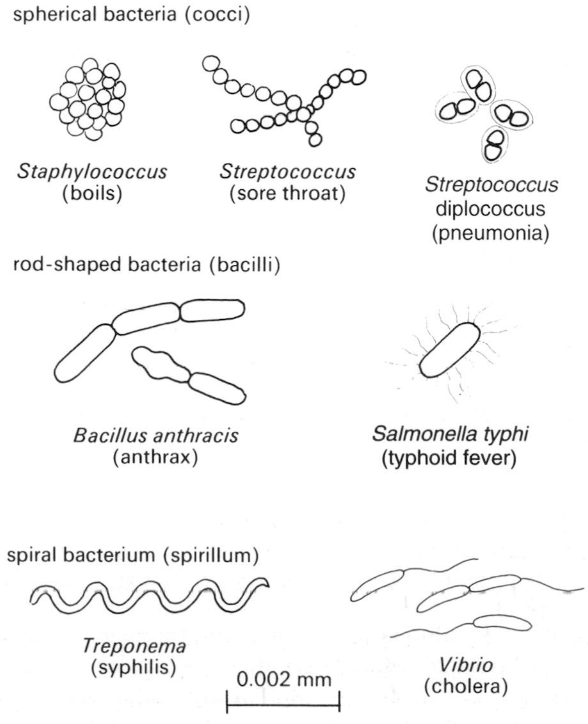

Fig. 25.4 Different types of bacteria

They then absorb the digested food through the cell wall. Inside the cell the food is used (a) to make more bacterial cytoplasm so that the bacteria can grow and reproduce, and (b) to provide them with energy through respiration. This sort of activity can be a nuisance when bacteria attack our stored food (p. 288). But in nature it is the means of breaking down dead material and recycling important chemicals. Most bacteria are harmless or even helpful.

Certain kinds of bacteria are always found in the intestines of animals, including man. Some are beneficial because they produce vitamins B_{12} and K, which we need to keep us healthy. Other bacteria simply live there without causing harm and are called *commensals*. Large numbers of bacteria are always found on our skin. Most of them are harmless unless they happen to get through the skin to the tissues underneath. While they are on the skin they are commensals. But if they get inside the body they may cause disease and then they become *pathogens*.

Pathogenic bacteria are those bacteria that cause disease. Some bacteria can affect only particular parts of the body. For example, the bacillus that causes bacillary dysentery produces a powerful poison (or *toxin*) that causes inflammation of the large intestine, resulting in loss of blood and acute diarrhoea. On the other hand, the treponema (spiral bacterium) that causes syphilis (p. 199) first causes inflammation only in or near the part of the body where it first enters, but as time passes, it can cause damage in almost any part of the body to which it can travel. Thus, although it first causes damage to the penis in men, or near the vulva in women, it can eventually damage the heart, the skeleton or the brain.

Rickettsias

These are smaller than most bacteria and vary in shape from spheres to thin rods, up to 0.002 mm in length. They are bacteria-like in that they are visible under the light microscope, and are similar in structure to the bacterium shown in Fig. 25.3, having similar genetic material. However, like viruses, they are capable of growing only within living cells and are, therefore, total parasites. Different species of *Rickettsia* cause the various forms of typhus.

Mycoplasms

These are similar in size to viruses. They are the smallest organisms that can live and reproduce outside living cells. They cause diseases of the breathing system in many domestic animals. *Mycoplasma pneumoniae* attacks lung tissue in humans and is one kind of organism that causes pneumonia. Unlike viruses, mycoplasms can be destroyed by antibiotics.

Viruses

Viruses are even smaller than bacteria, less than 0.0005 mm in diameter. Most of what we know about the structure of viruses has been learned by studying them under the electron microscope. Viruses differ from bacteria in having no cell wall and no cytoplasm (Fig. 25.5). Viruses can only reproduce inside living cells. Unlike bacteria they cannot be grown on agar jelly.

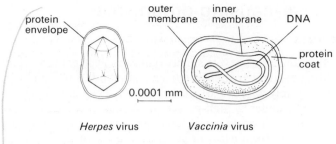

Fig. 25.5   Structure of two viruses

When a virus gets inside a suitable cell, called a *host cell*, it takes over control of the cell. It makes the host cell produce new viruses. Fig. 25.6 shows how the virus reproduces.

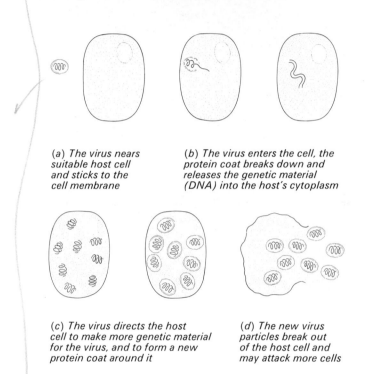

(a) The virus nears suitable host cell and sticks to the cell membrane

(b) The virus enters the cell, the protein coat breaks down and releases the genetic material (DNA) into the host's cytoplasm

(c) The virus directs the host cell to make more genetic material for the virus, and to form a new protein coat around it

(d) The new virus particles break out of the host cell and may attack more cells

Fig. 25.6   How a virus reproduces

Once a virus has entered the body it is likely to be carried around by the blood and the lymph. If the virus reaches a so-called 'target cell' then it changes the behaviour of the cell. The cells of the nasal epithelium (the lining of the inside of the nose) are the target cells for the virus of the common cold. Nerve cells of the brain are the target cells for the virus causing rabies. Measles virus affects many kinds of cell in the body but produces obvious effects in cells in the skin by causing a rash or outbreak of spots.

Fungi

Most of the fungi are *saprophytes*. These are organisms that cause decay by secreting enzymes and digesting dead organic matter. They include *Penicillium* which produces penicillin. There are not many fungi that cause diseases in man. We shall consider one group that causes infections of the skin and another group that infects the mouth, throat and vagina.

Ringworm or dhobie itch is caused by a fine, thread-like fungus called *Trichophyton* and a similar, related fungus called *Epidermophyton*. The threads or *hyphae* of the fungus (Fig.

25.7) are found in the skin where they secrete enzymes to digest the skin cells. These secretions irritate the skin cells and cause *dermatitis*, an inflammation of the skin which may either itch or be very sore. The fungus produces reproductive bodies that can be spread from one person to the skin of another person by direct contact, or on towels or clothing. Athlete's foot or tinea pedis is caused by the parasitic fungus growing between the toes, although the infection can spread over the entire foot. Athlete's foot is most common among people who wear shoes and socks which do not allow the perspiration from the feet to evaporate. The skin remains very moist, giving the right conditions for the fungus spores to germinate.

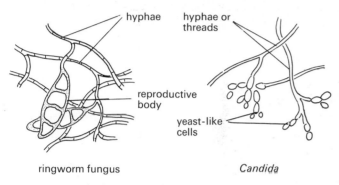

Fig. 25.7   Fungi that cause disease

A yeast-like fungus, called *Candida* (Fig. 25.7), is often present in the vagina and in the mouth where it can live as a commensal (p. 200), without doing any harm. But if a person's resistance to disease is lowered then the fungus may become more active and cause signs and symptoms of disease. *Candida* in the vagina causes itching and an unpleasant discharge of pus.

The commensal fungi are often kept in check by the commensal bacteria, living in the same place. If the bacteria are killed off by the use of antibiotics (p. 257) then the fungi, being no longer controlled, often start to grow rapidly and cause disease. They change from being commensals to being pathogens.

Protozoa

Protozoa are very small, single-celled animals. They are common in fresh water and moist soil but most of them can only be seen by using a microscope. Protozoa ingest or feed on small particles of food, such as bacteria. A few, like *Plasmodium*, the malarial parasite, absorb soluble nutrients directly through the cell membrane.

Most protozoa reproduce asexually by dividing into two, though some have more complex methods of sexual reproduction. In this chapter we are concerned only with those protozoa that are pathogenic and live as parasites in man (Fig. 25.8).

Entamoeba histolytica is visible only under a high-powered microscope. It is usually taken into the body with contaminated drinking water. Once it reaches the large intestine it begins to feed on the bacteria that are normally found there. It may remain there feeding harmlessly in this way or it may invade the epithelium of the intestine where it damages the cells and causes ulceration. This leads to bleeding. Pain and acute diarrhoea result and these are the chief symptoms of *amoebic dysentery*. If the disease goes untreated then the patient may lose so much blood that he or she becomes anaemic.

Those entamoebae that continue to feed on the bacteria in the gut may secrete a layer of material that hardens to form a case or cyst around the protozoon. When the cyst is passed out of the body with the faeces of the host, the wall of the cyst protects the entamoeba from drying up. No further activity is possible until the cyst is swallowed by another person. Then the cyst wall is dissolved away by the digestive juices and the entamoeba is free to feed and become active again.

Trichomonas vaginalis (Fig. 25.8) is even smaller than *Entamoeba* (0.015 mm long) and it may be found in the vagina where it causes inflammation and irritation, termed *vaginitis*. In men it may get into the urethra where it causes inflammation. Like *Entamoeba* it reproduces by dividing into two, but it does not form cysts. It is usually spread from one person to another during sexual intercourse. Vaginitis is, therefore, termed a *sexually transmitted disease* (see p. 266). Unlike the bacteria that cause gonorrhoea and syphilis (two other sexually transmitted diseases), *Trichomonas* can remain alive outside the human body for several days, for example in moisture in towels or in droplets on lavatory seats.

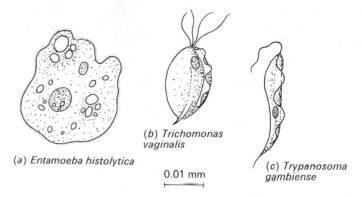

(a) *Entamoeba histolytica*

(b) *Trichomonas vaginalis*

0.01 mm

(c) *Trypanosoma gambiense*

Fig. 25.8   Some parasitic protozoa that cause disease

Trypanosoma is a protozoon with one flagellum (Fig. 25.9). It is found in the blood and is transmitted from one person to another by the bite of a blood-sucking insect, the tsetse fly. *Trypanosoma* is the cause of African sleeping sickness, known also as *trypanosomiasis*. After being sucked in with blood from a person with the disease, the parasite develops in the stomach of the tsetse and works its way to the insect's salivary glands.

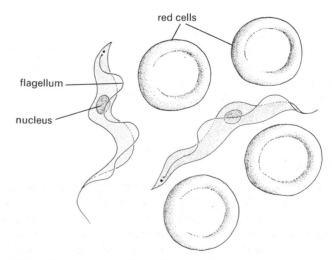

red cells

flagellum

nucleus

Fig. 25.9   *Trypanosoma* with red blood cells for comparison of size

When the tsetse next sucks blood, it injects saliva into the wound and the trypanosomes pass in with the saliva.

Occasionally parasites are transmitted on the outside of the mouthparts of biting insects after feeding on the blood of an infected person.

Chagas' disease, from which Charles Darwin is believed to have suffered, occurs in Central and South America and is caused by a trypanosome transmitted by blood-sucking bugs. This species of *Trypanosoma* leaves the blood stream, and settles and develops in muscle tissue, frequently in heart muscle.

Plasmodium, the malarial parasite, is another protozoon that lives in the blood stream and liver of infected humans. It has a much more complicated life cycle than any of the pathogenic organisms so far discussed. Like *Trypanosoma*, it is transmitted from one person to another by an insect that can pierce the skin and suck blood. *Malaria* is transmitted by the females of certain kinds of mosquitoes, and is discussed in much more detail in Chapter 29.

Roundworms (nematodes)

Most roundworms are not parasites but are free-living animals found in soil and water. But some kinds are pathogens, living in animals and plants and causing disease.

Enterobius or pin-worm is widespread among young children all over the world. Infection takes place as a result of swallowing the eggs which survive passage through the stomach and hatch in the duodenum. The larvae migrate down the intestine to the colon where they become adult and settle in the lining. The female worms are about 12 mm long while the males are only 2.5 mm in length. After mating and fertilization the female migrates from the colon to the skin around the anal opening where she deposits her eggs, up to 10 000 in number. The presence of the females around the anus induces intense itching and the child's natural response is to scratch. Eggs often lodge under the finger nails as well as on the finger tips and may then be transferred to the mouth, directly reinfecting the child. But when the child touches various objects, such as door handles, cutlery and food, the eggs are deposited and may reach the mouths of other children or adults. Since children are continually touching things as they play, the spread of the eggs, which are very resistant to desiccation, is easy. The irritation caused by the worms around the anus may be so intense as to disturb a person's sleep. Apart from this, infection is more of a nuisance than an illness.

When either acute itching or the presence of the worms in faeces indicates infection with *Enterobius* the most important measure is to keep the infected child's nails cut short and to teach it to scrub its nails and fingers after going to the toilet or after scratching the anal region. Bathing in the morning, with thorough washing of the anus and the careful washing of night clothes helps to remove eggs.

The drug piperazine is given orally. It kills the worms which are then passed out with the faeces. It is important to be watchful for signs of reinfection since the eggs can survive outside the human body for very long periods of time, surviving cold, dryness and moderate heat.

Ascaris is one of the larger roundworms, causing *ascariasis*. It lives in the small intestine where it can grow to 300 mm or more in length (Fig. 25.10). The numbers are sometimes so great that

(Wellcome)

Fig. 25.10   *Ascaris*, male and female. The male is smaller (life size)

the intestine is blocked, and this can prove fatal. It is very widespread and infection is the result of ingesting eggs which have been passed out in the faeces of an infected person. Tiny larvae are released from the eggs and these reach the small intestine and burrow through the wall of the gut into the blood vessels. The blood stream carries them to the lungs where they burrow through the walls of the alveoli into the breathing passages, to be carried up by the cilia (p. 85) to the larynx. They are then swallowed for a second time and when they reach the small intestine they settle there, feeding on digested food.

Heavy infections result in damage to the lungs, causing haemorrhage, and in the intestine may cause obstruction as well as producing toxic wastes to which the host reacts.

Ancylostoma or hookworm is a much smaller nematode, the adult female being 10 to 13 mm long. Like *Ascaris* it lives in the small intestine but it has sharp hooks at the mouth end with which it cuts through the villi to enable it to suck blood.

Fig. 25.11   Head of *Ancylostoma duodenale*, showing hooks around mouth ($\times$ 150)

(Wellcome)

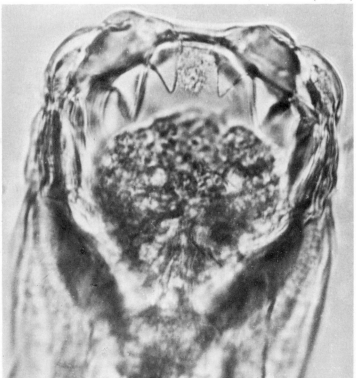

Hookworms are free-living in water as first and second stage larvae and only in the third stage are they infective. They are then able to enter the human host by boring their way through the skin of feet and ankles of people wading in infected water or walking through vegetation or on damp soil. Once inside the

Fig. 25.12 *Ancylostoma duodenale*, hookworm (× 12)

skin they enter the blood and lymphatic systems and are swept along to the lungs. Leaving the capillaries, they enter the alveoli and are carried in the mucus stream and, by coughing, to the oesophagus. If they are swallowed, they develop to become active egg-laying adults in the gut. They pierce the lining of the gut and suck blood, causing anaemia if they are numerous and the infection is unchecked. Carelessness over defaecation in fields, gardens and waterways enables the eggs to be dispersed and the life cycle to be continued. Fig. 25.11 (p. 202) shows the head of *Ancylostoma* with its jaws or hooks.

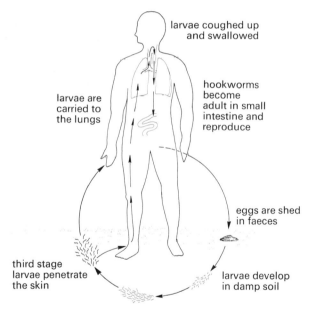

Fig. 25.13 The life cycle of hookworm

Wucheria. Some tropical roundworms are parasites in the lymphatic system. One of the most important of these is a filarial worm called *Wucheria bancrofti*. The adult worms may be 40 to 100 mm long and, instead of laying eggs, the females produce tiny living larvae. The larvae are about 0.25 mm long and, because they are so small, are called *microfilariae*. The microfilariae are carried by the lymph back into the blood stream (p. 83) but they do not develop further until they are sucked up with blood by a mosquito. Inside the mosquito each larva develops into a stage known as an infective larva. When the mosquito feeds again on human blood the larvae are released into the wound and make their way to the lymph nodes and grow to the adult stage. The worms are often so numerous that they block the nodes and prevent the flow of lymph. Since the lymph ducts can no longer remove the surplus tissue fluid

Fig. 25.14 Young man suffering from elephantiasis

(Shell)

the tissues swell up. The legs are particularly affected and in men the scrotum may also be affected. Swollen legs give rise to the term *elephantiasis* (Fig. 25.14).

Onchocerca is another important filarial worm, causing the disease known as *onchocerciasis*. This worm lives in connective tissues in the body, sometimes in the skin where it causes large lumps or nodules, usually on the scalp, elbows and knees (Fig. 25.15). The worms may be carried in the blood to the eyes,

Fig. 25.15 Feeling for nodules under the skin produced by the microfilariae of *Onchocerca*

(WHO/P. Pittet)

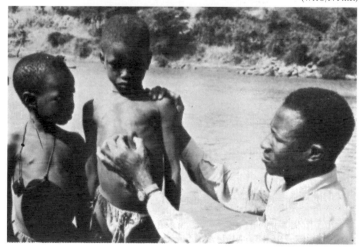

Although Fig. 25.22 shows the life cycle of the beef tapeworm (*Taenia saginata*), the life cycle of the pork tapeworm (*Taenia solium*) is very similar, with a pig taking the place of the cow. The pig or cow is called the *secondary host* because the parasite undergoes its larval development in that animal. Man is the *primary host* because the tapeworm is sexually mature and reproduces in man.

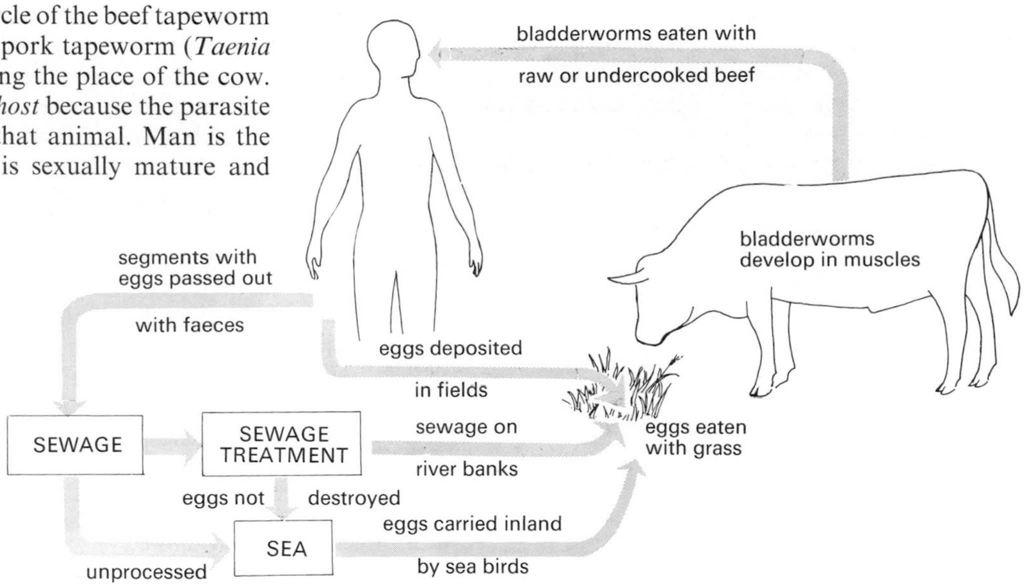

Fig. 25.22 Life cycle of *Taenia saginata*, the beef tapeworm

When the secondary host takes in tapeworm eggs it usually takes in a large number at the same time. The digestive juices of the host release the larvae from the eggs. Each larva has a ring of hooks which it uses to tear a way through the lining of the gut and enter the blood stream or a lymph vessel. In this way the larvae are carried around the body. The larvae usually settle in the muscle of the cow or pig, often near the shoulder, and there each develops into a tiny fluid-filled bladder surrounding a scolex. If the meat is heavily infested with larvae, or bladderworms as they are called, then these can be seen with the unaided eye. The bladderworms are killed by strong heat, e.g. by proper cooking, such as roasting.

Echinococcus is chiefly a parasite of dogs. While the tapeworm *larvae* will become sexually mature only in the correct host, the *eggs* will hatch and develop in the bodies of unusual hosts. When dogs become infected with *Echinococcus* the eggs are passed out in the dog faeces. Faeces get rubbed on to the fur near the dog's anus and dogs often 'nose around' their faeces. When children play with infected dogs the eggs can be transferred from the fur near the tail to their own fingers and so

to their mouths. *Echinococcus* eggs will hatch to produce larvae in the human stomach. The larvae then burrow into the gut wall and are carried in the blood to the liver, the lungs or the brain. The larvae cause a large fluid-filled swelling (hydatid cyst) to develop. If this happens in the brain then the pressure of the cyst damages the brain, and changes in behaviour result. In the liver or lungs (Fig. 25.23) the hydatid cyst may press on blood vessels and interfere with circulation.

The fish tapeworm, ***Diphyllobothrium***, differs from the pork and beef tapeworms in that its life cycle requires two intermediate hosts for its completion (Fig. 25.24, p. 207). It also differs in that it may become adult in a variety of fish-eating mammals, such as man. Thus the fact that man disposes of his faeces hygienically does not mean that there is then no chance of infection, since other mammals may contaminate the water in which the second and third hosts live. Fish tapeworm is most common in countries such as Finland where a great deal of freshwater fish are eaten, sometimes raw and often only lightly cooked.

Characteristics of parasites

All the organisms described in this chapter are parasitic in man and cause disease. To be a successful parasite, whether it be a bacterium or a tapeworm, the organism must have some means of being transferred from one host to another and in the next two chapters we shall consider the methods by which transfer takes place. In addition, it must be able to fend off the host's immune defence system. Success also depends on the ability of the parasite to obtain food from the host. You have seen that many kinds of parasites live either in the blood or in the lymph, both of which are liquids that carry soluble food to the tissues. The parasite must not take too much food, or cause such serious damage that the host dies. If the host *is* killed, there must be some means by which the parasite, or its eggs or larvae, can reach a new host.

The chance of a parasite reaching a new host is often small and most parasites have spectacular rates of reproduction. One bacterium may give rise to several million bacteria in one day and one tapeworm may release many million eggs during its life.

Fig. 25.23 X-ray, showing hydatid cyst in a lung, caused by *Echinococcus* larvae

(John Wright and Sons Ltd)

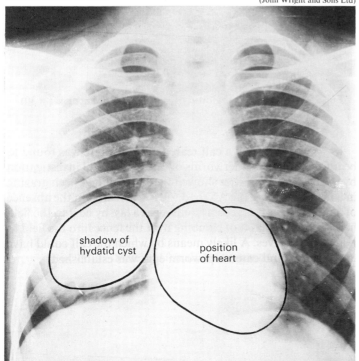

It seems that in this way the chances of *some* of the eggs hatching and giving rise to larvae that reach a new host are sufficient to keep the species from becoming extinct. While there is undoubted wastage, there is a reasonable probability of the life cycle being repeated.

The control methods are dealt with in detail in Chapter 31.

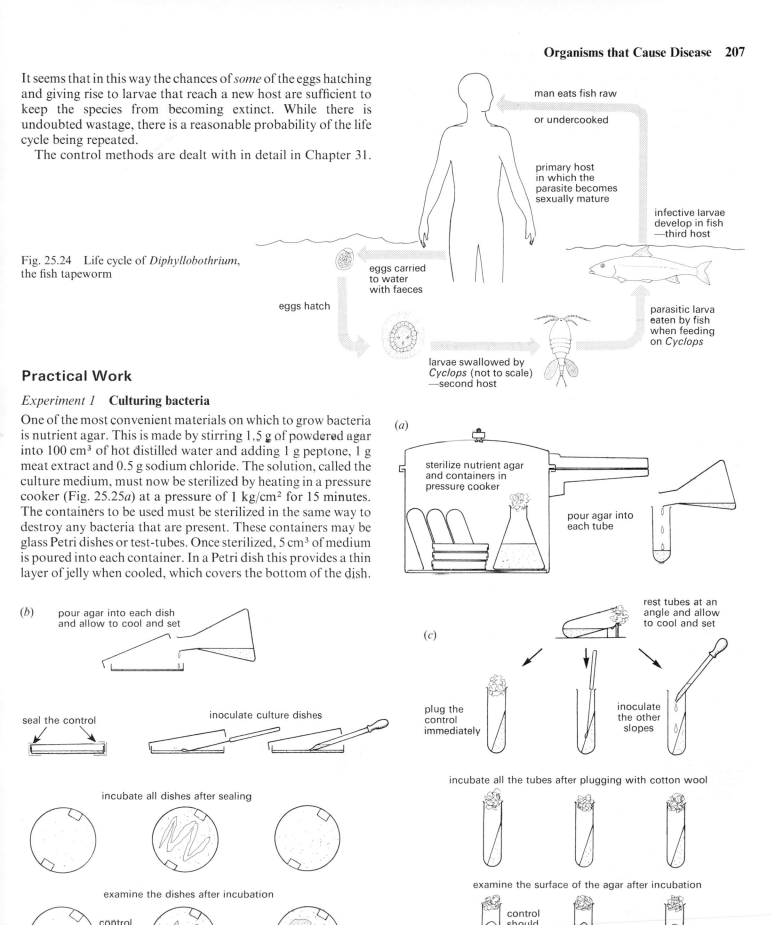

Fig. 25.24   Life cycle of *Diphyllobothrium*, the fish tapeworm

Practical Work

Experiment 1   **Culturing bacteria**

One of the most convenient materials on which to grow bacteria is nutrient agar. This is made by stirring 1.5 g of powdered agar into 100 cm³ of hot distilled water and adding 1 g peptone, 1 g meat extract and 0.5 g sodium chloride. The solution, called the culture medium, must now be sterilized by heating in a pressure cooker (Fig. 25.25a) at a pressure of 1 kg/cm² for 15 minutes. The containers to be used must be sterilized in the same way to destroy any bacteria that are present. These containers may be glass Petri dishes or test-tubes. Once sterilized, 5 cm³ of medium is poured into each container. In a Petri dish this provides a thin layer of jelly when cooled, which covers the bottom of the dish.

Fig. 25.25   Preparing and using agar plates and slopes

If a test-tube is used, it should be plugged with sterile cotton wool and laid at an angle to cool, as shown in Fig. 25.25c. When set the jelly is termed a *slope*. It can be handled at any angle. The cotton-wool plug can be sterilized by passing it swiftly through a Bunsen flame.

Bacteria can be obtained by allowing a piece of boiled potato to decompose in water. After two days' decomposition, transfer some of the water to a plate or a slope, using a wire loop (Fig. 25.26). The loop must be sterilized before and after use by heating to red heat in a flame. The method of making the culture is shown in Fig. 25.25b. As bacterial colonies grow, they show up as glistening blobs on the jelly.

Fig. 25.26 Wire loop used for subculturing bacteria

Experiment 2 **Sources of bacteria**

Collect samples of river water, pond and tap water in separate, sterile vessels and make a fourth sample by placing soil in a vessel with just sufficient water to cover it. Transfer water from each sample to each of four plates or slopes using a wire loop. Keep a fifth plate or slope, sterilized like the others, as a *control*. This provides a check that sterilization has been effective, since no bacterial colonies should appear on it. It enables you to compare the appearance of the plates on which you placed water samples with one to which no additions were made. Incubate the cultures at 37 °C for two days and then compare them. They should not be opened, since pathogenic organisms may be present. All plates and slopes should be resterilized in the pressure cooker before the containers are opened or re-used.

Experiment 3 **Nematodes or threadworms**

A nematode called *Rhabditis* is usually obtained if an earthworm, killed swiftly by dropping into boiling water, is cut open lengthwise, laid on a dish of fresh garden soil, and covered with polythene or other waterproof material for two days. The nematodes can be seen moving on the surface of the worm and may be transferred to a drop of water on a slide for examination under a hand lens or a microscope.

The kinds of micro-organisms that cause disease are characterized in the following table:

Organism	Size	Structure	Reproduction
Bacterium	0.5–10 μm	single cells with cell wall	division in two
Virus	less than 1.0 μm	nucleic acid with a protein coat	can only be made in host cell
Protozoon	5.0–50 μm	single-celled animal; no cell wall	usually by division into two, but some reproduce sexually
Fungus	thread 7.0 μm in diameter	fine threads or chains of cells	by spores or by fragmentation
Nematodes			
Enterobius	up to 12 mm		
Filaria	up to 100 mm	}multicellular worm-like animals	sexual
Microfilaria	up to 0.3 mm		
Blood fluke	up to 15 mm	multicellular animals	sexual in man; asexual in snail
Tapeworm	up to 3 m	large multicellular animals	sexual

(1 m = 1000 mm = 1,000,000 μm)

Questions

1 Draw up two tables, one showing the ways in which bacteria, viruses, protozoa and fungi resemble each other and one showing the ways in which they are different from each other, e.g. in methods of feeding and reproduction.

2 What evidence is there for the statement that it is not in the interest of a pathogenic organism to kill its host? For what organisms might this statement appear at first sight to be untrue?

3 Which of the disease-causing organisms described in this chapter may be found in (a) human urine and (b) human faeces?

4 Why is the fungus *Candida* sometimes considered as a commensal and sometimes as a pathogen?

5 Write a list of the diseases that could be greatly reduced by the hygienic disposal of sewage and the provision of pure drinking water.

26
How Disease-causing Organisms Enter the Body

Commensal bacteria which normally live harmlessly within the body may become pathogens when conditions in the body change. Here we are concerned with the ways and means by which they and other more obvious pathogens gain access to human tissues (Fig. 26.1).

Natural openings

Many micro-organisms can enter the body through the natural openings. The nasal passages, mouth, urinary passage, anus and, in females, the vagina all present pathways for the entry of pathogens. Each of these openings leads to a tube lined with a soft, moist mucous membrane. The mucus itself often contains enough food material to sustain microbial life such as the bacterium *Staphylococcus aureus*. Fortunately, if a person is in a state of good health, the cellular membranes seem able to resist penetration by most kinds of bacteria and viruses but the extent of this resistance varies from one individual to another and also within one individual from time to time. In any case,

these tubes lead to other organs that may possess less resistance. A diet that includes sufficient vitamin C helps to strengthen the membranes of the cells that form the linings of mouth, nose and throat.

The different openings are often associated with particular methods of spread (Chapter 27). Thus the nose and mouth are the principal means of entry of airborne organisms causing, for example, colds, influenza and measles. The mouth also provides access for organisms present in drinking water and food, such as the cholera vibrio, the bacterium of bacillary dysentery, and tapeworm. Pathogenic organisms can also be transferred to the lips by contaminated fingers, cups and cutlery. Entry of organisms through the urethra and vagina is associated with sexual activity, though many disorders of these passages arise from organisms that are spread by means other than sexual intercourse. Few organisms cause damage by entering through the ear passage, provided the ear drum is intact and healthy. The secretion of wax by glands in the skin lining the ear opening provides an anti-bacterial barrier.

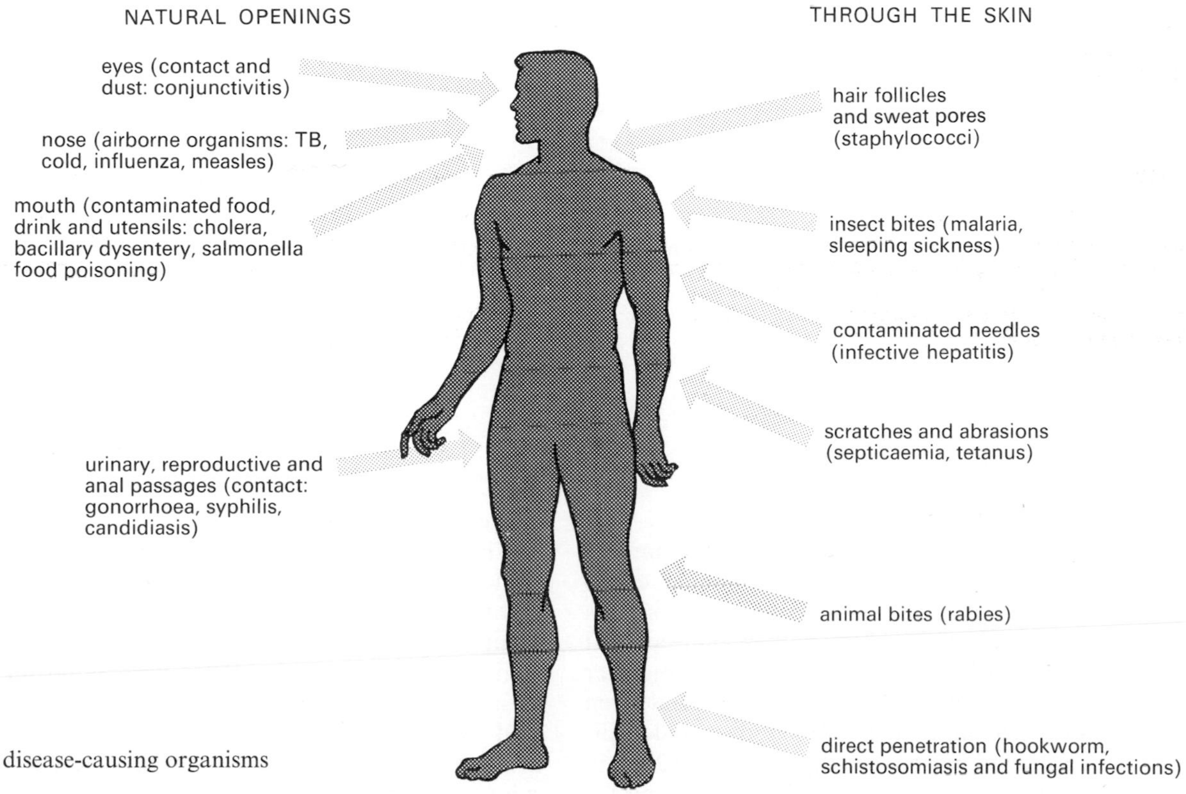

NATURAL OPENINGS

eyes (contact and dust: conjunctivitis)

nose (airborne organisms: TB, cold, influenza, measles)

mouth (contaminated food, drink and utensils: cholera, bacillary dysentery, salmonella food poisoning)

urinary, reproductive and anal passages (contact: gonorrhoea, syphilis, candidiasis)

THROUGH THE SKIN

hair follicles and sweat pores (staphylococci)

insect bites (malaria, sleeping sickness)

contaminated needles (infective hepatitis)

scratches and abrasions (septicaemia, tetanus)

animal bites (rabies)

direct penetration (hookworm, schistosomiasis and fungal infections)

Fig. 26.1   How disease-causing organisms enter the body

Wounds and breaks in the skin

Human skin provides a remarkable degree of protection against micro-organisms, provided there is no break in it. The micro-organisms that can penetrate the skin of their own accord are few, but they include the larvae of *Schistosoma* and hookworm (p. 203) which enter partly as a result of enzyme action and partly by mechanical penetration. The staphylococci that cause boils and pimples invade hair follicles and set up local infections. As the infection spreads under the skin, the tissues are broken down locally and secondary infections could result. Abrasion or grazing of the skin provides an entry for bacteria such as those that cause wounds to become septic, usually with the formation of pus. The fungi that cause ringworm and athlete's foot can also penetrate directly.

Deep wounds may provide access to muscle and other tissues such as joint capsules which are normally sterile. Disease signs and symptoms may develop rapidly in such tissues.

Creatures that bite through the skin may introduce pathogens. Blood-sucking insects and other arthropods are frequently the vectors or agents in the transmission of a range of disease-causing organisms. Thus the adult females of certain mosquito species feed on human blood (p. 213) and are the vectors of malarial parasites and yellow fever virus. These diseases cannot be contracted by having the pathogens rubbed on to the skin. It is only when the skin is punctured by the mouthparts of the insect vector that the pathogens can be

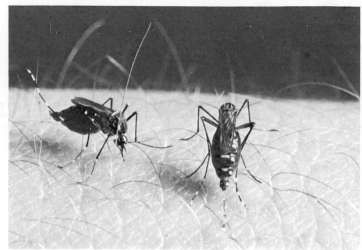

(Wellcome)

Fig. 26.2   Yellow fever mosquitoes piercing human skin (× 4)

injected into man. Horseflies and other biting flies may introduce micro-organisms on the outside of their mouthparts. Fleas that have fed on the blood of plague victims regurgitate plague bacilli on to the skin before biting. The bacteria are then carried through the skin during the act of biting. The bite of a dog infected with rabies virus usually results in the infection of the wound with virus from the saliva on its teeth.

27
How Disease-causing Organisms are Spread

Airborne infection

Many common diseases are caused by organisms that are carried in the air. This method of transmission is often called *droplet infection* because when we sneeze, cough, talk or even breathe gently, tiny droplets of liquid are carried in the air that leaves our lungs and breathing passages. When we cough or sneeze there is often a very noticeable spray of drops. But when we breathe out gently there are also droplets in the exhaled air, though these are very small indeed, usually less than 0.2 μm in diameter and so small that they can remain afloat in the air for

long periods. They can contain, and therefore carry, both bacteria and viruses.

The common cold virus is spread in this way. The virus attacks the *mucous membrane*, the moist lining of the nasal cavity which secretes a lubricating *mucus*. As the signs and symptoms of the cold develop, the membrane cells actually make more virus particles. So, as we breathe out, the movement of air over the membrane picks up some of the virus particles and they are carried out into the atmosphere. If another person is standing close to you then they are likely to breathe in quite a lot of the droplets containing the virus. Thus a new infection can be started (Fig. 27.1). Of course, by the time the air has reached a person several metres away the droplets are more widely scattered and, while some particles may be breathed in, there may not be sufficient virus to start an infection. There must be a sufficient concentration of virus to form a *minimum infective dose* before an infection will result.

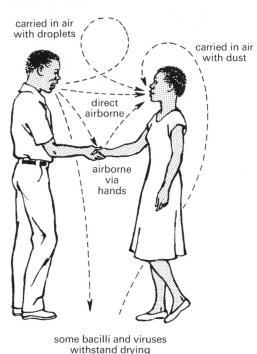

Fig. 27.1   Pathways of transmission of airborne organisms

Most of the diseases of the respiratory system (larynx, trachea, bronchi and lungs) are airborne. Those diseases, such as the common cold and measles, which are caused by viruses, can be spread in dry conditions as well as in humid conditions because the virus particles are not alive and it does not matter if they dry out when the droplets evaporate. Other airborne diseases are caused by bacteria, and bacteria are killed if they dry out unless they have formed spores. So whooping cough is more likely to be spread at damp times in the year when the droplets are less likely to evaporate. Tuberculosis bacteria, too, are killed by drying, so the disease is spread most easily when a person suffering from the illness is living in a confined space, such as one closed room, with a number of other people. This is one reason why tuberculosis patients are usually isolated from other people. Another reason is that the mycobacterium (a bacterium that forms threads, like a fungus) that causes tuberculosis is able to form spores, bodies surrounded by a tough coat so that they can resist drying. Thus, if a TB patient spits out sputum containing germs, when the spit dries up, the spores can be moved about in air currents and may be breathed in by other people.

Leprosy is spread by droplet infection. It is thought to be contracted by young children and it takes a long time for symptoms to develop. If a small child is being nursed by someone who has leprosy, the child is then held very close to that person's body and there is an easy chance of leprosy bacteria being breathed over the child's face for a long time (Fig. 27.2).

Cerebrospinal meningitis is another airborne, bacterial disease. Epidemics are most likely in parts of the world where

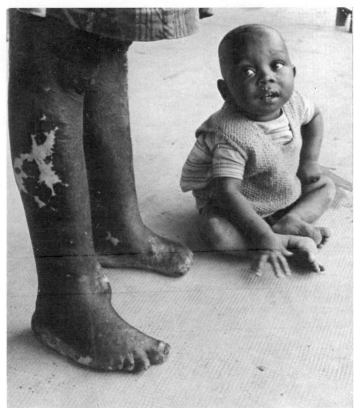

(WHO/P. Pittet)

Fig. 27.2   Feet of a woman badly affected by leprosy. Early treatment could protect her child who is probably infected

there are cold seasons, and the nights in particular are cold. Then people sleep indoors and close together to keep warm. The conditions are ideal for the spread of droplet infection. Smallpox was also spread in this way before the action of local health authorities and WHO (the World Health Organization) succeeded in eliminating it.

Transmission in water

Water is a favourable medium for the dispersal of the organisms causing gastro-intestinal infections such as amoebic dysentery, cholera and typhoid fever. When such organisms reproduce in the gut, eggs, spores or active organisms are carried out with the faeces. The sanitary habits of the infected person will determine the chances of the disease being spread.

If faeces or urine carrying disease organisms are deposited directly in water that is used for drinking, the organisms may thus infect large numbers of people. If infected faeces are left on the ground or even buried within a few metres of a stream or lake, the disease organisms may be washed by rain into the water supply and contaminate it. One person suffering from an intestinal disease such as typhoid or cholera may thus infect hundreds of others. It is important for individuals to see that their faeces are buried as far as possible from streams, rivers and lakes, and even more vital for the sewage from towns to be disposed of in such a way that water supplies cannot be contaminated (*see* pp. 275–80).

After defaecation and urination, the hands are likely to carry minute amounts of faeces and urine. If such unwashed hands touch food, utensils for handling food or even door handles and other objects, the faecal matter with any germs it contains may

(WHO)

Fig. 27.3 Contamination of drinking water is not always obvious; this well in India is protected by a raised stone surround, but the feet of the villagers carry dirt and germs on to the ropes which then contaminate the well water

be transferred to the object and later picked up by another person who eats the food or opens the door. For this reason it is essential to wash hands after each visit to the lavatory and before handling food. People handling and preparing food for others must be particularly careful in this respect.

Moving water can quickly spread micro-organisms over large distances. The sweet-water canals of Egypt have in the past been the means of spreading cholera from a point of contamination to people using the water for drinking many miles away. One of the unforeseen results of the irrigation of land by water from new dams such as those on the Nile has been the spread of the watersnail that is the host of *Schistosoma* larvae. Rivers similarly carry pathogens downstream. Even sterilized drinking water has been known to become re-contaminated through defects in pipe systems or by pollution of reservoirs used for storage of sterilized water. Severe flooding often means that sewage channels overflow, carrying raw sewage and therefore germs to places where one would not normally expect contamination to occur. In this way outbreaks of gastro-intestinal diseases may occur in cities that normally take a pride in their record of good health.

Transmission in food

Food may become contaminated with pathogenic organisms in a number of ways.

Contaminated fingers

Any food handler, a cook or a mother preparing food for her family, may suffer from an intestinal disease. If such people use the lavatory to defaecate and do not wash their hands carefully

afterwards they can spread pathogens on to food that will be eaten by other people. Typhoid and bacillary dysentery in particular are spread in this way. The eggs of *Ascaris* can be transferred from human faeces to food by being carried under the finger nails.

Flies

Houseflies are attracted to foods rich in protein such as meat and fish and they leave a trail of germs as they walk over the food. In warm conditions the germs multiply rapidly, digesting some of the protein for their own use. Houseflies are discussed in more detail on p. 214.

Washing food in contaminated water

Fresh vegetables and fruit do not usually form a breeding ground for bacteria but they may become contaminated if they are washed in water which is not pure. Human excreta are sometimes used as fertilizer for growing vegetables. If there are pathogens in the excreta then the chance of the vegetables becoming contaminated is high.

Infested meat

Tapeworm is another pathogen that is spread only in infested meat. You have read (p. 206) that the larvae in the meat can be destroyed if the meat is cooked at a high enough temperature. Thorough inspection of the meat carcasses before they are sold also helps to reduce transmission of tapeworm if the infested meat is destroyed. Salmonella is usually transmitted in contaminated meat or eggs (Fig. 27.4).

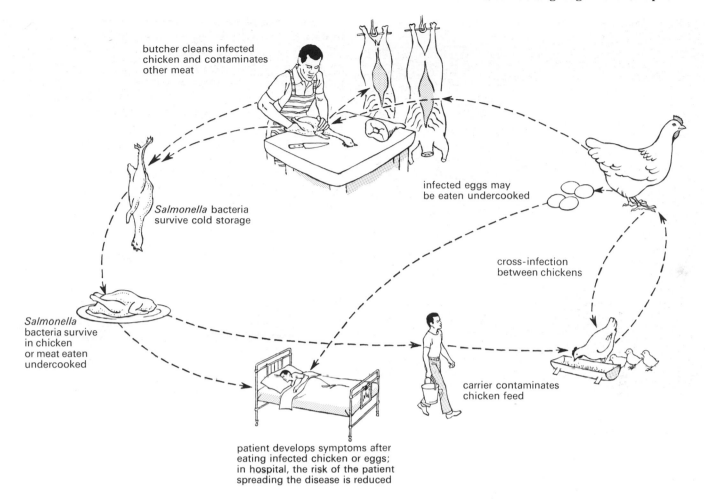

butcher cleans infected
chicken and contaminates
other meat

infected eggs may
be eaten undercooked

Salmonella bacteria
survive cold storage

cross-infection
between chickens

Salmonella
bacteria survive
in chicken
or meat eaten
undercooked

carrier contaminates
chicken feed

patient develops symptoms after
eating infected chicken or eggs;
in hospital, the risk of the patient
spreading the disease is reduced

Fig. 27.4   Transmission of *Salmonella* food poisoning

Spread by contact: contagious diseases

Some diseases are spread by person-to-person contact. For example, the spirochaete that causes yaws is thought to be spread *only* when the skin of one person comes in contact with the lesions or sores on the skin of an infected person. The bacteria that cause the venereal diseases, syphilis and gonorrhoea, are spread *only* by contact, usually during the contact of sexual intercourse, and the itch-mite causing scabies (Fig. 27.5) is also spread only by very close contact between the skins of two people.

The diseases caused by fungi, such as ring worm and tinea, may be spread by skin-to-skin contact, though they may also be spread by contact between the skin and clothing or the ground on which spores or fragments of fungus are lying. Some organisms such as the bacterium that causes tetanus can only get into the body when they come in contact with broken skin. Thus people sometimes develop tetanus after falling and scraping the skin off the knee or some other exposed part of the body, allowing the germ to penetrate the tissues.

People catch hookworm when the larval stage penetrates the skin of the foot, usually where it is soft near the ankle, or the skin of the hand if a person is working in infested water. The larva is able to bore through the skin of its own accord, but it must first come in contact with the skin, for example when a person walks barefoot over contaminated soil.

Fig. 27.5   Itch mite that causes scabies (× 55)

Diseases spread by insects and other vectors

Several kinds of insect feed by biting through the skin of mammals, including man, and then sucking blood. In Chapter 11 you saw that blood clots when blood vessels are damaged (p. 73). As the blood clots it becomes very sticky and it could fix the mouthparts of the insect so firmly in the skin that the insect could not get away. At the same time clotting would block the

(WHO)

Fig. 27.6   *Aëdes aegyptii*, the mosquito that carries yellow fever virus, piercing human skin (× 15)

feeding tube of the insect. Most blood-sucking insects secrete saliva into the wound made in the skin. The saliva contains an *anticoagulant* which prevents the blood from clotting (coagulating), so that the insect can withdraw its mouthparts after feeding (Fig. 27.6). If there were any pathogens in the saliva from the salivary glands of the insect then these would be injected into the blood stream of the host animal. Insects that transmit pathogens in this way are called *vectors*.

Mosquitoes, blackflies and tsetse flies are, perhaps, the best known vectors, but lice, fleas and ticks also feed on blood and transmit disease-causing organisms. Clearly, no blood-sucking vector is able to transmit disease-causing organisms until it has first taken them in by sucking the blood of someone suffering from the disease. Only in this way can the pathogens enter the body of the vector and be present to be injected into the body of a new host.

Houseflies

The housefly (Fig. 27.7) is one of the most widespread vectors of disease in the whole world. It is particularly important in the spread of intestinal diseases because adult houseflies are

Fig. 27.7   Head of a housefly (× 15)

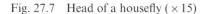

(Shell)

strongly attracted to human faeces and also to the kinds of food that we eat. When a housefly walks over faeces, the hooks and the sticky pads on its feet and the hairs on its legs and body pick up liquid from the faeces (Fig. 27.8). There are always germs in faeces. If the faeces are produced by someone suffering from an intestinal disease then some of the germs will be pathogens. If the fly, which is now carrying the pathogens on its body, walks over food which we are to eat, some of the pathogens will be scattered over the food. Fig. 27.9 shows a Petri dish with a meat jelly over which a housefly has walked on the previous day. The dish has been kept at 37 °C for 24 hours (known as *incubating* the dish) and in that time the bacteria left by the housefly have reproduced to form distinct blobs or colonies.

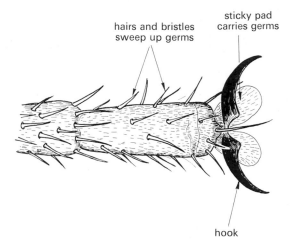

Fig. 27.8   Foot of a housefly (× 100)

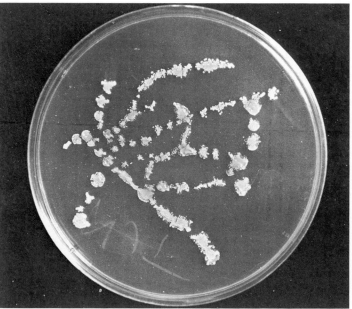

(David Broadribb)

Fig. 27.9   Trail of bacteria left on meat jelly by a housefly

In addition, the housefly feeds by extending a tube or proboscis on to the faeces and, later, on to human food. The fly can take in only liquid food, so it first squirts saliva over the food and then 'puddles' it to liquefy the food. The saliva always contains germs, and this is another way in which the housefly can spread pathogens. Finally, some pathogens survive the digestive juices of the fly after being swallowed by it and they are shed later when the fly defaecates, often on to food. You may sometimes see small black specks left behind by a fly when it has walked over something white like sugar (Fig. 27.10).

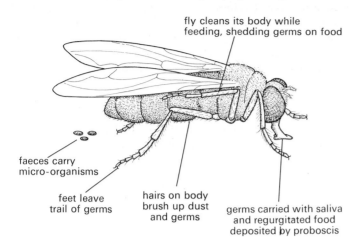

Fig. 27.10   Means by which the housefly transmits disease organisms

Houseflies are particularly dangerous because they spread the pathogens of typhoid, bacillary dysentery, cholera and bacteria such as *Salmonella* that cause food poisoning.

Mosquitoes

Mosquitoes are often divided into *anopheline mosquitoes*, of which *Anopheles* and *Aëdes* are the best known, and the *culecine mosquitoes*, represented by *Culex*. The importance of knowing the life history of these vectors is made clear in Chapter 31, dealing with the control of the vectors of disease. If you can control the insect or eliminate it, you can control the spread of the disease.

Lice

There are two species of lice that suck human blood. *Pediculus humanus* occurs in two forms, the head louse and the body louse. *Phthirus pubis* lives among the pubic hairs and is also called the crab louse because of its shape (Fig. 27.11). The long sharp claws enable all three kinds of louse to hold on to skin and hair.

Each kind of louse feeds by biting through the skin and then sucking blood. This does not result in much damage beyond causing irritation or itching, but the person who has been bitten

might then scratch the area. If he does so with dirty finger nails then bacteria may be rubbed or scratched into the tiny wound and start an infection.

However, sometimes the louse is carrying the tiny organisms called *Rickettsiae* which cause typhus. These organisms are sucked up with blood by the louse when it feeds on an infected person. The *Rickettsiae* are passed out of the louse's body in its faeces on to the skin of the host. When the host scratches his skin the *Rickettsiae* are rubbed into the bite and a new typhus infection begins. Epidemics of typhus are not very common, but where there are lice there is always the risk of infection. This may happen when travellers from a country where typhus is endemic visit another country where lice are found.

Lice spread from one person to another during close, physical contact, for example during sexual intercourse, or between children when their heads touch and when children are cuddled or nursed by grown-ups. Also, lice can live for short periods of time on clothing, but they soon leave it to search for food if the clothing is not worn. Control of lice is described on p. 249.

Fleas

It is said that every kind of mammal has its own particular flea. The flea that lives on man is *Pulex irritans* and, like the louse, its bite causes irritation that may result in the victim scratching.

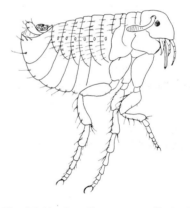

Fig. 27.12   Rat flea, *Xenopsylla* (× 15)

Occasionally the fleas from other animals such as cats and dogs may jump on to the body of man, particularly if a human is playing with and stroking such an animal. But these fleas do not usually stay long on the human body. However, the rat flea, *Xenopsylla cheopis* (Fig. 27.12), often transfers to man and if it

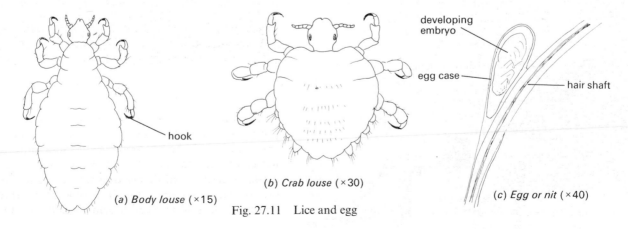

(a) *Body louse* (×15)

(b) *Crab louse* (×30)

(c) *Egg or nit* (×40)

Fig. 27.11   Lice and egg

is carrying the bacillus of bubonic plague then it can transfer these bacteria into the blood stream before it starts to suck blood. The plague bacillus multiplies in the oesophagus and stomach of the flea to such an extent that it causes a blockage. The flea is obliged to regurgitate the bacterial mass before it can swallow blood and the mouthparts become contaminated with the bacilli before the flea bites through the skin (*see* p. 232). The plague bacilli are pushed through the skin as the flea bites.

Cockroaches

Cockroaches are not always thought of as disease vectors but they are widespread and are sometimes found feeding on our food (Fig. 27.13). They carry germs on the outside of their

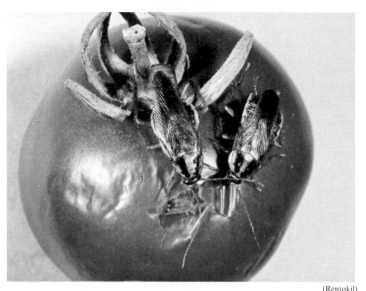

(Rentokil)

Fig. 27.13　Cockroaches feeding (× 1.5)

bodies and, if they pick up pathogens from latrines and other places where they go in search of water, they can transmit these to our food. Cockroaches are suspected of carrying the pathogens causing polio, food poisoning, typhoid, leprosy and amoebic dysentery.

Ticks

Some other small animals, which are often confused with insects, also carry pathogens. The blood-sucking ticks are related to spiders and have *four* pairs of legs (Fig. 27.14). They

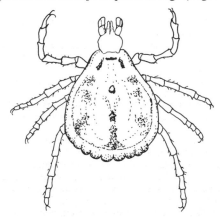

Fig. 27.14　Hard bodied tick, vector of tick-borne typhus (× 10)

are commonly found on the bodies of wild and domestic animals. The hard-bodied ticks sometimes feed on the blood of man and they are responsible for spreading the *Rickettsia* that cause tick-borne typhus.

Mammals

Some mammals are vectors of disease. Most of these are animals that bite humans. A dog suffering from rabies can transmit the virus to humans if it bites through the skin. While the disease is not very common, it is nearly always fatal. The virus is present in the saliva of the infected dog and, once having got under human skin, it travels along nerve fibres to infect the central nervous system.

Rats and mice form an important group of vectors. Since they frequent places where food is stored they often contaminate that food with pathogens carried on their fur and feet. They also carry *Salmonella*, causing food poisoning, in their urine and

(Rentokil)

Fig. 27.15　The brown rat, *Rattus norvegicus*, is also known as the sewer rat; it is a good swimmer and emerges from sewers to contaminate food with its urine and faeces, transmitting *Salmonella*, causing food poisoning, and *Leptospira*, causing Weil's disease

faeces. The spirochaete causing Weil's disease (infective jaundice), a severe and often fatal illness, is carried by rats and passed in their urine. Contact with rat urine may be made in sewers and sewage plant, rat-infested rivers and, in Australia, in sugar-cane fields. Apart from this, rats act as reservoirs for a remarkably wide range of pathogens including that of bubonic plague, the cause of the Black Death of Europe in the Middle Ages and the Great Plague of London in 1665 as well as innumerable other notorious outbreaks. We are wise to eliminate rats from human environments (*see* Chapter 37).

Reservoirs of infection

Rats are also important as *reservoirs of infection*. This means that they carry pathogens in their bodies without necessarily being able to transmit them directly to humans. Rats are also reservoirs of the *Rickettsia* that cause typhus.

The most important reservoir of human diseases is man himself. Anyone acts as a reservoir while he has disease-causing organisms in his body, whether he shows signs of illness or not.

Some people carry pathogenic organisms in their bodies without showing any signs or feeling any symptoms. Such people are called *carriers*.

A person who carries the bacillus of typhoid fever in his intestine without showing any signs and without feeling ill can release the bacilli in his faeces over a long period of time. The faeces are a source of pathogens which might be picked up by houseflies or cockroaches, but of even greater importance is the risk of the carrier transmitting the bacilli to food which other people then eat.

Various species of antelope, such as bushbuck and waterbuck, act as reservoirs for the trypanosome that cause sleeping sickness in many parts of Africa, although man himself is the chief reservoir of infection.

The virus of yellow fever kills off many kinds of monkey and chimpanzee, but the survivors form reservoirs on whose blood the *Aëdes* mosquito feeds. The mosquito is the vector that transmits the virus to man.

A person affected with HIV (human immunodeficiency virus p. 267) is a reservoir of infection and capable of transmitting the virus through sexual intercourse and by contaminating a shared hypodermic needle. A person does not have to show signs or symptoms of AIDS to be infectious.

(Rentokil)

Fig. 27.16   The black rat, *Rattus rattus*, is also called the ship rat; it is most common around seaports and is the reservoir of the bacterium causing bubonic plague. Like the brown rat, it also contaminates food

How disease organisms are spread

Method of spread		Examples of disease
contaminated water	drinking	intestinal diseases, e.g. dysentery, cholera, typhoid fever
	bathing, washing or paddling	bilharziasis, hookworm
contaminated food	eating	intestinal diseases, salmonella, tapeworm, pinworm
airborne	droplet	diseases of respiratory tract, common cold, TB
		diseases entering by respiratory tract, measles, influenza, smallpox
	dust	diseases of respiratory tract (e.g. tuberculosis) and eyes (e.g. trachoma)
contact (contagion)	skin to skin, skin to clothing to skin	smallpox, ringworm, scabies, septicaemia
	sexually transmitted	gonorrhoea, syphilis, candidiasis, AIDS
	contaminated hypodermic needle	AIDS, hepatitis
insect vector	carried externally, e.g. housefly, cockroach	dysentery, salmonella, summer diarrhoea
	carried internally, e.g. mosquito	malaria, yellow fever, elephantiasis
other animal vectors	rat (urine) dog (saliva)	Weil's disease rabies

Where disease vectors are found

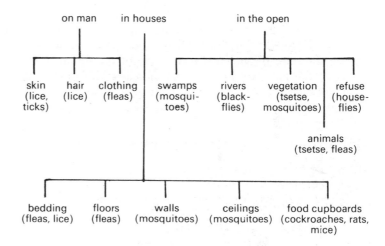

Practical Work

Experiment

Prepare two agar plates. Catch a housefly and allow it to walk over the jelly of one of the plates and allow a cockroach to walk over the other. Remove the creatures and then incubate the plates. Discuss the results, and destroy the cultures by re-sterilizing.

Questions

1   Why is it healthier to sleep in a well-ventilated room rather than with a lot of other people in a closed room?

2   Explain why leprosy is most likely to be contracted during childhood.

3   What are the essential connections between people, waterborne infections and outbreaks of illness?

4   What habits of mosquitoes and fleas make them vectors of disease organisms?

5   Why are houseflies dangerous vectors of disease organisms?

6   In what situations do you think cockroaches could become a serious threat to our health?

7   What are the differences between an animal that is a vector of disease organisms and one that is a reservoir of infection?

8   In Europe in the sixteenth and seventeenth centuries it was believed that bubonic plague was spread by touch or contact. Plague is not a contagious disease but the belief is understandable if one knows how the disease organisms really are transmitted. Suggest reasons why the belief persisted.

9   Make a list of the transmissible diseases from which you and the other people in your class have suffered. How is each of these spread?

28
The Course of an Infectious Disease

We have discussed the means by which pathogenic organisms are transmitted and gain access to the body. Infection literally takes place when the organisms first penetrate tissue, whether this be the skin, the mucous membrane of the throat or the lining of the gut. However, germs of many kinds enter the bodies of each one of us every day and usually we experience no feelings of illness and show no sign of disease.

Conditions for developing a disease

Before a disease can develop:

(a) a person must be susceptible;

(b) the causative organisms must be present in sufficient numbers;

(c) the organisms must be virulent (see below).

The susceptibility of the host, i.e. the state of his defence mechanisms, is discussed in the section on immunity on p. 253. At present it is sufficient to note that immunity to attack by pathogens takes various forms. There is genetic or innate immunity, and immunity acquired either as the result of a previous attack by a disease or through artificial immunization. Susceptibility to many diseases also varies with a person's general state of physical health.

The defence mechanisms of the body are substantial and include the action of phagocytes (p. 72) and of those white blood cells that produce antibodies (p. 74). A small number of invading pathogens, e.g. the few virus particles breathed in while talking in the open air to a friend with a cold, may be dealt with successfully. The same defences may be literally overwhelmed by the sheer quantity of virus breathed in during a train or bus journey sitting close to a fellow passenger who is breathing out virus-laden droplets into a humid atmosphere. The level of the infective dose required to produce disease varies also from one pathogen to another. Thus for *Salmonella* food poisoning it is high—the number of organisms swallowed at one time must be large—while only a small number of typhoid bacilli appear necessary to produce violent disease symptoms.

The *virulence* of an organism is a different matter. In the years 1905 to 1935 scarlet fever was regarded as a serious disease among children in western Europe. The streptococcus causing the disease is still found and recognized but it now produces only a mild reaction. Its virulence has diminished.

Virulence varies in the effect produced by pathogens in different hosts. Tuberculosis mycobacteria exist in three strains associated with specific hosts. Thus one strain is normally associated with man, a second with cattle and a third with birds. While the second and third can parasitize man, neither produces the devastating effects in man that they do in the natural hosts. Their virulence towards man seems to be diminishing.

Louis Pasteur discovered that it was possible to reduce the virulence of pathogenic bacteria by growing them in abnormal conditions such as at unusual temperatures or in unusual hosts. The use of such *attenuated* strains with reduced virulence is discussed on p. 253.

Incubation period

The period of time that elapses between the entry of pathogens into the tissues of a susceptible host and the onset of the first signs and symptoms of disease is termed the *incubation period*. The duration of this period varies greatly from one disease to another as may be seen from the table below.

Disease	Length of incubation period
Cerebro-spinal meningitis	2–7 days
Cholera	3–6 days
Bacillary dysentery	3–7 days
Gonorrhoea	5–10 days
Measles	10–14 days
Leprosy, AIDS	up to several years

A knowledge of these incubation periods is helpful in the control of the spread of infectious diseases and particularly in deciding the period of quarantine or isolation of contacts. The term quarantine originates from the belief that isolation of travellers arriving from an infected area for a period of forty days would be sufficient time for disease symptoms to develop and be recognized.

Infective period

It is even more difficult to be precise about the duration of the period in which a patient is infective, that is, capable of releasing pathogens to infect another person. As soon as the parasites begin to multiply and leave the body of a patient in airborne droplets, sputum, faeces or urine, or in insect vectors, he is *infective*. He ceases being infective only when his normal defences have destroyed all the pathogenic organisms or when these have been destroyed by antibiotics or other drugs. During the infective period the patient should be isolated from other people as far as possible, and those who must have contact in nursing him should observe strict hygiene of hands and, where possible, clothing. Occasionally a person harbours pathogens without showing symptoms of disease. If he is capable of transmitting those organisms he is said to be a *carrier*.

Signs and symptoms of disease

These are sometimes considered together as the *clinical features* of a disease. The signs can be seen sometimes by direct examination of the patient, sometimes by examining faeces, urine or a blood sample. Thus a rash, patches of yellow matter on the tonsils, diarrhoea, an enlarged spleen, or presence of the eggs of a parasite in faeces would be termed *signs of disease*.

Complaints by the patient of a feeling of nausea, of pain, headache or loss of sensation (feeling) cannot be seen by the medical observer and are termed *symptoms*. They may be invaluable to the person who is trying to determine or *diagnose* the cause and nature of a disease in order to prescribe appropriate treatment.

One sign of reaction to disease organisms that is almost universal is a rise in body temperature. Man is homoiothermic (*see* p. 98). His body temperature is usually about 37 °C and it is reduced only by long exposure to cold and increased by vigorous physical activity. Most pathogens invading the tissues and those gut parasites that excrete toxic substances are associated with a temperature rise. This may be a small rise, from 37 °C to 38 or 39 °C if the patient is suffering from the common cold. Pneumonia or acute tonsillitis may result in temperatures of 40–41 °C.

The rise in temperature seems to result from an interaction between the pathogen or its toxins and certain white cells (Fig. 11.3). This causes the release of a substance that acts on the temperature regulatory centre in the brain. Temperature rise increases the metabolic rate within the tissues, and the activity of the white cells in ingesting micro-organisms and producing antibodies is speeded up. The heartbeat similarly increases its rate.

While an increased temperature assists the body defence mechanisms, it is also the cause of some of the discomfort of disease—headache, profuse sweating and loss of appetite. The rapid reaction of a malaria sufferer to the release of merozoites (p. 261) and their toxins as the invaded red cells burst involves a cold stage followed by a dramatic rise in temperature, accompanied by headache and vomiting.

Treatment of disease

An account of the treatment of disease in general terms would be inadequate and misleading but the treatment of a number of specific diseases is described in Chapter 29.

Secondary effects of disease

Disease organisms produce a variety of effects on the body and upset the feeling of well-being of the patient. Some such organisms cause changes in the body which result in further effects such as *anaemia* (p. 71).

Anaemia is the condition that occurs when the capacity of the blood to carry oxygen is reduced, either by a reduction in the numbers of red cells or in the amount of haemoglobin they contain, or by both factors. The signs of anaemia may include pallor (paleness) of the mucous membranes, shortness of breath during exertion and a low red cell count (assuming that a normal count is 4–6 million red cells per mm³). Symptoms may include general tiredness and giddiness during exertion, experienced by the patient but not measurable by an observer.

There are many causes of anaemia, but here we are only concerned with those that result from the activities of pathogens. The strictly infectious diseases rarely result in anaemia though it occasionally results from cholera and hookworm. More frequently it is a secondary effect of attack of blood flukes, hookworms and malarial parasites. In each case heavy infestation with parasites results in destruction of large numbers of red cells; in the case of blood flukes or hookworms by the direct ingestion of blood. In chronic malaria the red cells are destroyed by the parasites living inside them.

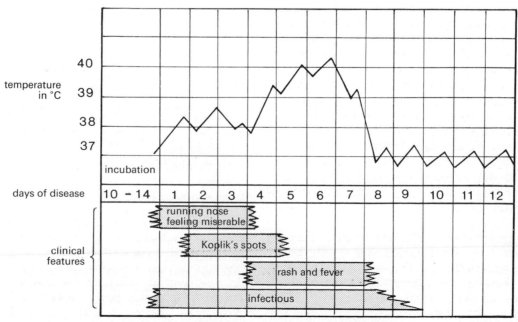

Fig. 28.1   The course of an infection with measles

29

Some Diseases of World-wide Importance

Measles

Measles is a very infectious, world-wide disease of children. It was recognized and described in London by Thomas Sydenham in 1676 and it had become common by the eighteenth century. In Europe and North America it occurs usually in children between the ages of 4 and 7 for whom it is an unpleasant but not very severe illness. By the age of 20 about 85 per cent of people in Britain will have had measles and, of the remaining 15 per cent, some will have antibodies in their blood (p. 252) suggesting that they contracted the disease without showing any signs, known as a sub-clinical infection. However, in many parts of Africa, particularly in Nigeria, measles is regarded as the most acute infection of children. Those aged 1½ to 2½ years are most at risk and there is a 5 to 10 per cent death rate. Younger children are protected by passive immunity (p. 254) gained from their mothers.

Causative agent and transmission. The disease is caused by a virus containing RNA, surrounded by a lipo-protein envelope (Fig. 29.1). The virus is airborne (*see* p. 210) and is inhaled in droplets, though it may be spread in saliva on eating utensils and on soiled clothing. From the breathing passages it enters the lymphatic system (p. 83) where it reproduces and then the new virus particles enter the blood and are carried around the body.

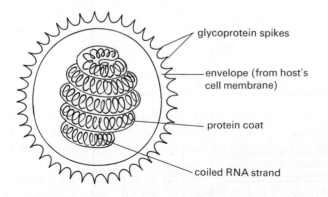

Fig. 29.1   Structure of a para-myxo virus. Measles is caused by a virus of this type

glycoprotein spikes

envelope (from host's cell membrane)

protein coat

coiled RNA strand

Course of the disease. The virus attacks the mucous membranes, the first signs being a running nose, followed by small white spots in the mouth. In 1896, Henry Koplik noticed and described these small spots which resembled grains of sugar on the mucous lining of the mouth, particularly under the tongue. Koplik's spots are now taken as a very convincing sign of the early infective stage of measles. The conjunctiva may be affected and the white of the eye looks red. Meanwhile the young patient has a rising temperature and feels miserable.

Koplik's spots disappear as the virus invades the skin. A rash develops, first on the forehead and behind the ears, spreading later to the rest of the body. The fever reaches a peak but, if there are no complications, the signs and symptoms die away after seven days (*see* Fig. 28.1). In Britain, secondary complications are not common, though the damage to the mucous membranes does open up the way to secondary infections, such as infections of the bronchial tubes and the inflammation of the middle ear by staphylococci. The disease is more serious in children suffering from malnutrition, whose resistance to disease is lower. Such children are then more prone to marasmus and kwashiorkor (p. 37).

Treatment and control. There is no specific treatment but the young patient usually prefers to remain quietly in bed during the first few days of active infection, feeling rather miserable. The child should be given plenty to drink, since the fever results in sweating and loss of body fluids. If the child has no appetite for solid food then milk is appropriate and nourishing. Antibiotics are of little help unless secondary bacterial infections develop.

By the time the disease is diagnosed there is little value in isolating the child, who will have already been spreading virus-laden droplets. The disease is truly endemic, there being *some* cases about at all times of the year. However, it becomes epidemic, with large numbers of children affected, usually between February and May in alternate years in Britain, particularly in towns.

It was not until 1960 that the virus was first cultivated successfully in chick embryos (Fig. 29.2) and a live, attenuated vaccine (p. 253) was produced. Most babies are protected by antibodies received from their mothers, giving passive immunity that lasts for about six months.

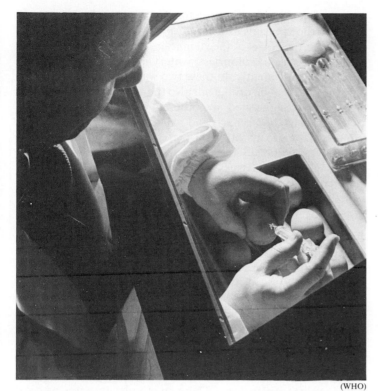

(WHO)

Fig. 29.2   Virus being injected into an egg to produce anti-measles vaccine

German measles or Rubella is another highly infectious viral disease which also produces a rash and fever, but it usually affects older children and adolescents. The symptoms are much less unpleasant than those of measles, but, if a pregnant woman contracts the disease, then there is a risk of damage to the foetus (p. 105).

Influenza

Causative agent. Influenza is an illness affecting the breathing system. It varies in severity, and during the years 1933 to 1949 three distinct strains of causative virus were isolated.

Strain A is highly infective and causes epidemics at intervals of two years.

Strain B is irregular in its appearance and outbreaks are usually localized.

Strain C is very mild in its effect and the disease is often undiagnosed.

It is spread by inhaling the virus (Fig. 29.3) in droplets in the air breathed out by infected persons.

Course of the disease. Children aged 5 to 14 years are most susceptible, and young adults aged 25 to 35 are also likely to be affected. Of course people in other age groups can and do suffer from the virus, which first attacks the ciliated epithelium of the nasal cavity, pharynx and often the lungs. Symptoms include headaches and fever (*see* Fig. 24.1) together with aches and pains and these all develop rapidly, usually within two days of infection. A person rarely remains infective for more than six days, though he may feel quite weak after an attack.

Treatment. There is no particular treatment other than allowing a person to rest, away from other people to reduce the chances of spreading the virus. However, since the virus damages the

mucous membrane of the breathing passages, secondary infections may follow from bacteria which normally exist as harmless commensals (*see* p. 200) on these membranes. One of these, *Haemophilus influenzae*, is associated with bronchitis and inflammation of the nasal sinuses, while the pneumococci causing pneumonia often become active in elderly people following influenza. Pneumonia is often fatal and was probably the most frequent cause of death in the influenza pandemic of 1918–19 when between 15 and 20 million people died.

Prophylaxis. Vaccines containing inactivated strains of the A and B virus give some protection for a short time, but new substrains of these viruses appear as a result of changes in the RNA (the genetic material) of the virus, and there is rarely any immunity to a new strain. However, medical workers and others concerned with maintaining essential services are often vaccinated when it is known that an epidemic is likely to occur.

Influenza is one of the few diseases to show a truly pandemic spread during this century. The pandemic of 1918 has already been mentioned. In 1957 a virus known as A_2 caused a pandemic of 'Asian flu'. Its spread has been traced from China to Hong Kong and from there to most parts of the world. Hong Kong is a centre of world trade with an international sea- and airport from which infected persons may carry the virus to other countries.

It is still surprising that influenza, with its maximum of six days' infectivity in man and its rapid production of symptoms, should be transmitted quite so widely and speedily. What happens to the virus between outbreaks is still uncertain. Possibly it is kept going by a chain of sporadic cases, though healthy carriers of the virus may exist. International co-operation through WHO enables outbreaks to be predicted, and some preparations to be made for dealing with a wave of infection by the preparation and administration of vaccines.

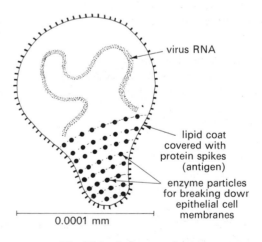

— virus RNA

— lipid coat covered with protein spikes (antigen)

enzyme particles for breaking down epithelial cell membranes

0.0001 mm

Fig. 29.3   Influenza virus

Cholera

Cholera is an acute infection of the intestines and is caused by invasion by a bacterium of the kind classed as a vibrio (p. 199). *Vibrio cholerae* and *V. el tor* are the species causing this disease in man.

Through historical time the disease has been endemic in Bangladesh and in Burma and central China. Epidemics have occurred in most parts of Asia from time to time, but it was only in the nineteenth century that, with the return of travellers, the disease spread to Africa, Europe and America. In each of these continents sporadic outbreaks may assume epidemic proportions (Fig. 29.4).

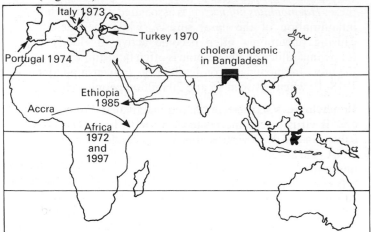

Fig. 29.4 Distribution of endemic cholera and cholera outbreaks 1970–1997

Method of spread. The spread of cholera within a population is through contamination of the water supply (Fig. 29.5). The vibrios grow well at 25 to 30 °C. Where a population draws its water from a river or canal and consumes this water without further treatment, then contamination with the faeces of a cholera victim will make the water dangerous and put that population at risk.

In the summer of 1973 an outbreak at Bari on the Adriatic coast of Italy was traced to the eating of contaminated shellfish. Untreated sewage, which included the faeces of a cholera victim, had been passed into the sea where the shellfish, in the natural process of breathing and feeding, filtered out and retained the vibrios from the seawater. Presumably the shellfish were not adequately cooked to destroy the vibrios before they were eaten.

Both natural disasters, such as floods, and man-made disasters, such as war, result in people fleeing to safety. Large numbers may be living in overcrowded conditions in camps with no effective sanitation and no safe water supply. Intestinal diseases, with diarrhoea as one of the signs, can be spread very rapidly in refugee camps. In 1996 and 97 cholera outbreaks among refugees in Rwanda and Zaire in central Africa have resulted in thousands of deaths.

Houseflies, which as adults are attracted to and feed on human faeces, may also spread the vibrios when they subsequently walk over food intended for human consumption.

Course of the disease. After an incubation period of two to six days diarrhoea begins as a result of the cholera toxin. Having cleared out the bowel of faecal matter, the diarrhoea continues with the passage of colourless, watery material. The patient is likely to vomit violently. In these two ways the patient loses not only the fluid that he drinks but also the considerable volume of digestive juices secreted by different parts of the digestive system such as the stomach and pancreas. Water is withdrawn from the tissues to maintain the secretion of juices and, since no water is reabsorbed by the colon, there is a net water loss from the body and the tissues become dehydrated. When death occurs it is not caused directly by the pathogen or its toxins but by dehydration of the tissues and loss of potassium. In mild cases recovery takes two to three weeks.

Treatment. It is necessary to kill the vibrios but it is also important to replace the water and salts lost from the tissues as a result of the action of the toxin on the gut lining. This is done ideally by the use of a solution of glucose with sodium and potassium salts, taken by mouth (Fig. 29.6). UNICEF (the United Nations Children's Fund) has distributed packets of 'oral rehydration salt' containing glucose and salts of sodium and potassium for use in time of epidemics. Alternatively rehydration can be achieved by the careful injection of biologically balanced salt solution into a vein or by the use of a saline drip transfusion. These methods require medical supervision.

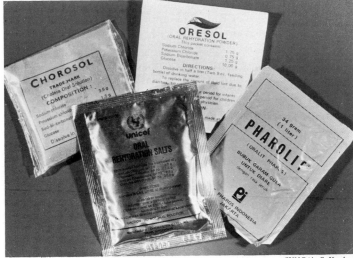

(WHO/A. S. Kochar)

Fig. 29.6 Sterile packages of oral rehydration salts can be used by trained health workers to help patients suffering from dehydration

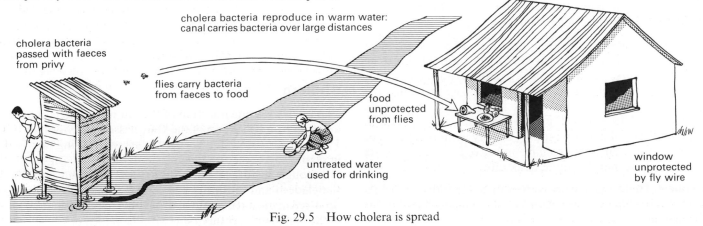

Fig. 29.5 How cholera is spread

The swallowing of a sulphathiazole drug or tetracycline to kill the vibrios is necessary not only for obvious cholera victims but also in the treatment of those people, termed carriers, in whose faeces vibrios are identified but who show no signs or symptoms of the disease.

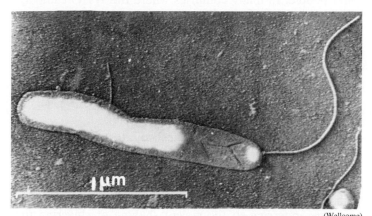

(Wellcome)
Fig. 29.7 Electronmicrograph of the vibrio of cholera (× 4 000)

Control. To control the spread of cholera, patients must be isolated in a place where neither vomit nor faeces will contaminate water supplies, food or even clothing, or be reached by flies. Both vomit and faeces carry large numbers of vibrios.

People who have been in close contact with patients may be given the drug tetracycline, and other people in an infected area may be inoculated against the disease (p. 256). Inoculation gives protection for a period of at least three months and possibly up to a year, in which time an epidemic is likely to have died down.

Unfortunately, while people who have been inoculated do not develop signs and symptoms of cholera, active vibrios can still be swallowed and, since they do not come in contact with the blood stream, they are not killed. Such people are therefore carriers who pass out pathogens in their faeces. Good sanitary habits remain very important.

Thorough control of houseflies near hospitals, lavatories, kitchens and eating places must be maintained and it is usual to chlorinate drinking water supplies more heavily than normal.

Outbreaks of cholera should be notified to local health authorities, whose duty it is to inform the World Health Organization. Restrictions on travel from infected areas and strict quarantine of those travellers who do enter a country from an infected area help enormously in checking the spread of a cholera epidemic.

Prophylaxis. Effective prevention is achieved by ensuring that people drink only filtered and chlorinated water, free from vibrios, and by controlling houseflies.

The vaccine made from killed vibrios, used to immunize people gives only short-term protection. Protection for the resident population of a country normally free from the disease, when an epidemic breaks out elsewhere, is best achieved by very vigorous quarantine measures. Thus when holiday travellers return to Britain from regions where cholera outbreaks have occurred, the health authorities endeavour to trace these people, and doctors are alerted to be on the watch for patients showing signs of cholera. A water-borne epidemic would be unlikely in Britain now, though we should remember that it was still endemic in this country only 100 years ago.

Tuberculosis or TB

Tuberculosis is caused by a very small bacillus, known as *Mycobacterium*. The micro-organism, which varies from 1 μm to 5 μm in length and 0.2 μm to 0.6 μm in breadth, was first isolated by Robert Koch in 1882. There are several species of *Mycobacterium* and the one most commonly causing tuberculosis of the lung is *M. tuberculosis*. In countries where a great deal of unpasteurized milk is drunk *M. bovis* may be the commoner organism, and infection having started in the alimentary canal may spread to other parts of the body. Since children drink more milk than adults as a rule, the risk to them from infected, unpasteurized milk is greater.

Spread and course of the disease. *Mycobacterium tuberculosis* is spread chiefly by droplet infection (p. 210). The bacteria are inhaled and the first lesions or signs of damage to tissues develop in the lungs. The pathogen develops slowly and quite often is destroyed by the antibodies produced by the infected person. Small scars may be left in the lung and can be detected by careful X-ray examination.

However, if resistance to the pathogen is low then it may do extensive damage to lung tissue over many years. This damage includes the formation of masses of epithelial cells which, unlike the alveoli, offer no surface for gaseous exchange. Breathing efficiency is reduced and irritation set up by the *Mycobacterium* results in violent coughing. Pathogens are blown out of the lung at each cough. If blood vessels are burst by the violence of the coughing then blood droplets may also be scattered with each cough. This stage of the disease is exhausting and the patient loses weight. In Victorian times the disease was known as consumption since it seemed to consume or burn away the patient's body.

The pathogen may invade lymphatic tissue and be spread to other parts of the body. This is more likely in the case of *Mycobacterium bovis*, present in milk from tubercular cows, which penetrates the body through the alimentary canal. In the 1930s children with tuberculosis of bone tissue and joints were commonly found in Britain, but pasteurization of milk together with improved standards of hygiene in dairies and amongst cow handlers have reduced this form of tuberculosis. The introduction of new and effective drugs has helped to cure those cases that have developed.

Treatment. Treatment for pulmonary tuberculosis involves complete rest, usually in hospital, together with a course of a drug such as streptomycin. This destroys the bacterium, but careful nursing is necessary to restore the patient to full physical vigour. Low resistance to the pathogen is often associated with inadequate diet, so good food with sufficient protein is essential for recovery. Patients are kept in very well-ventilated wards, partly to reduce the risk of infecting patients suffering from other complaints and partly because fresh air and sense of space seem to improve the patients' morale, an important factor in recovery.

Prophylaxis. Since sufferers from tuberculosis may be infectious for years before signs and symptoms are noticed, it is important to apply hygienic principles in order to prevent the spread of the pathogens, particularly in those areas where the disease is known to exist. As with all diseases spread by droplet infection, living in overcrowded, humid conditions increases enormously the risk of infection. Conversely, good ventilation and space, in schools and factories as well as in the home, play a considerable part in reducing the spread of this disease.

In the 1950s and early 1960s the Ministry of Health and local Health Authorities sent out teams of radiologists with mobile X-ray equipment (Fig. 29.8) to visit schools, offices, factories and villages. Chest X-ray photographs were taken of large numbers of people and the procedure became known as Mass X-ray. The film could be fed through a projector to give enlarged pictures of the lungs on which any lesions or damage caused by tuberculosis showed up as patches. The people with such patches were then recalled for thorough examination and treatment. In this way TB sufferers were detected often before they were aware of having the disease. Not only was treatment likely to be more effective but measures could be taken to prevent these persons from spreading the pathogens.

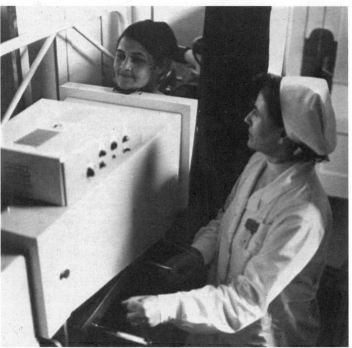

(WHO/P. Almasy)

Fig. 29.8   With between 15 and 20 million people in the world suffering from TB, many countries still use mass X-ray to identify patients for treatment at an early stage

In 1920 two French medical scientists, Calmette and Guérin, used a strain of *Mycobacterium bovis* to make a vaccine which is still referred to as BCG (Bacille Calmette–Guérin). Before vaccinating, a skin test is carried out. Antigens from a preparation of dead mycobacteria are injected under the skin. If the person has previously contracted and recovered from TB (and many do without knowing it) then the antibodies resulting from this infection react with the antigen to give a red swelling in two to four days. If there has been no previous infection then there is no antibody–antigen reaction and such a person can usefully be given the BCG vaccination to encourage the development of antibodies. Although this is active immunity it does not last very long and re-vaccination is recommended after three to five years, particularly for young people.

During the 1990s new strains of *Mycobacterium tuberculosis* have developed which are more virulent and resistant to the antibiotics at present in use. Nevertheless the use of *combinations* of antibiotics is proving effective.

Studies by the World Health Organisation have shown that failure in their campaign to eliminate TB is often due to the failure of patients to complete their course of treatment once they begin to feel better. It is at this stage that the more resistant bacteria can reproduce and be spread, infecting other people. In many countries where the WHO campaign is operating, patients are being supervised by health workers to ensure that the complete course of drugs is taken.

The use of BCG vaccine continues to be an important part of the WHO programme of prevention.

Malaria

Malaria is endemic in many tropical countries and until recent years it was widespread throughout many European countries. It occurred in Britain until the last years of the nineteenth century under the name of *fen ague*, because of its association with the *fens*, marshy breeding grounds of the mosquito vector, and *ague*, referring to the violent shaking or shivering that accompanies one stage of the disease.

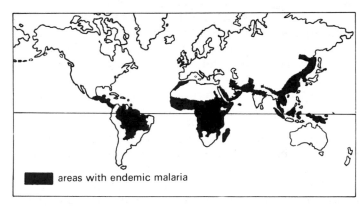

areas with endemic malaria

Fig. 29.9   World-wide distribution of malaria in 1997

Causative agent and vector. The parasite causing the disease is a microscopic single-celled animal (protozoan) belonging to the genus *Plasmodium*. It lives in the red blood cells and liver cells of man. There are several species of *Plasmodium* but only four affect humans and these are listed in the table below.

Four types of malaria

Type of malaria	Causative organism	Duration of cycle in blood stream	Geographical distribution
benign tertian	*Plasmodium vivax*	48 hours	Europe, N. America, Argentina, Australia
quartan	*Plasmodium malariae*	72 hours	commonest form throughout tropics
subtertian	*Plasmodium falciparum*	24 or 48 hours (irregular)	Central America, Brazil, Sri Lanka
ovale tertian	*Plasmodium ovale*	48 hours	mostly West Africa

Infection is normally the result of being bitten by a female anopheline mosquito that has already sucked blood from a malaria victim and is therefore carrying the parasites (*see* Fig. 29.10).

A complex development, involving sexual fusion followed by rapid increase in numbers of the parasites, takes place in the mosquito and a stage known as a *sporozoite* develops in its salivary glands. The mosquito thrusts its feeding tube or *stylets* through the human skin, and injects saliva containing an *anticoagulant*, a chemical that prevents blood from clotting. Thus it can suck blood freely and also withdraw its stylets after feeding. When an infected mosquito injects saliva into its human host it introduces hundreds of sporozoites into his blood stream.

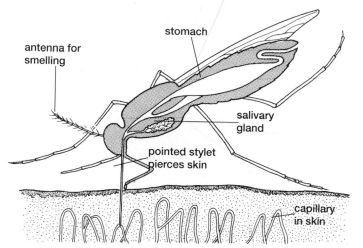

Fig. 29.10 The head and mouthparts with which the female Anopheles mosquito injects saliva and sucks blood (× 10)

The sporozoites are carried around the body in the blood stream but they can develop further only in cells in the liver, where they multiply over a period of about one week. Some remain multiplying in the liver while others are released in their thousands into the blood stream, where they invade red blood cells. Inside the red cells the parasites feed, grow and again reproduce asexually by dividing, usually to form sixteen offspring called *merozoites*. This reproduction is completed about 48 hours (72 hours for *P. malariae*) after the red cell was first invaded, and the cell now collapses. The sixteen merozoites together with their waste products are released into the blood plasma, but by some curious mechanism not clearly understood, this release of parasites seems to occur almost simultaneously from all the affected red cells. The numbers of parasites released into the blood plasma may number many

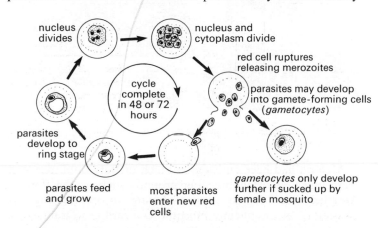

Fig. 29.11 The development of malarial parasites in red blood cells

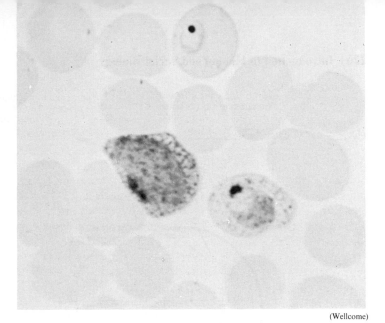

(Wellcome)

Fig. 29.12 Three stages in the development of malarial parasites in red blood cells (× 2 000)

millions, and Sir Ronald Ross, a pioneer in research into malaria, reckoned that at least 150 million must be present before a patient would show signs of fever (Fig. 29.3).

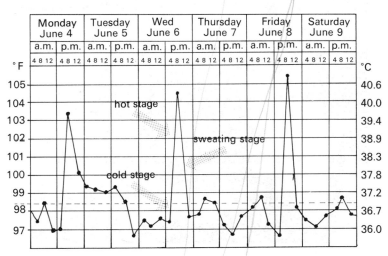

Fig. 29.13 Temperature chart of a patient suffering from benign tertian malaria

Course of the disease (Fig. 29.13). The first attack usually takes place ten to fourteen days after infection and is often preceded by tiredness, aching and sometimes by vomiting.

Cold stage. This is the ague, characterized by shivering and chattering of the teeth, and it lasts for one to two hours. It follows the release of the parasites from the red cells. The parasites quickly enter fresh red cells and the disease passes to the next stage.

Hot stage. The patient feels hot and his temperature may rise to 40 °C (104 °F) or higher. Rapid breathing and pulse rate are accompanied by headache and general discomfort for three to four hours.

Sweating stage. Sweating is profuse over a period of two to four hours, but the patient's temperature falls to below normal and eventually he experiences a feeling of relief. He is likely to feel exhausted, but is capable of rising from his bed and moving about until the onset of the next fever, two or three days later, according to the type of malaria.

225

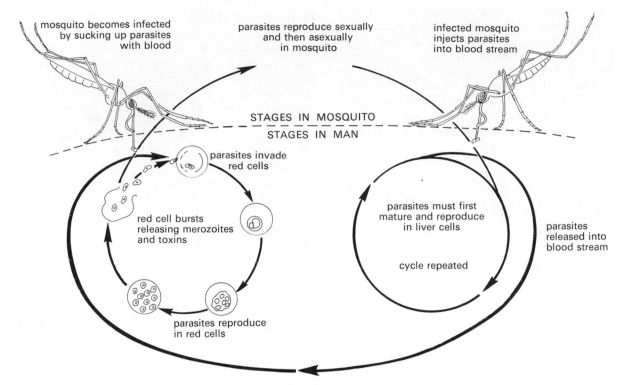

mosquito becomes infected
by sucking up parasites
with blood

parasites reproduce sexually
and then asexually
in mosquito

infected mosquito
injects parasites
into blood stream

STAGES IN MOSQUITO

STAGES IN MAN

parasites invade
red cells

red cell bursts
releasing merozoites
and toxins

parasites must first
mature and reproduce
in liver cells

cycle repeated

parasites
released into
blood stream

parasites reproduce
in red cells

Fig. 29.14   Life cycle of the malarial parasite

Control of malaria

In the 1960s the World Health Organization spent a great deal of time, effort and money on trying to eradicate malaria. In more recent years it has been realized that, while the disease can be eliminated from confined areas, world-wide eradication is not yet possible. Resources are now devoted to the control of the spread of the disease, research into new anti-malarial drugs and vaccines, and techniques of mosquito destruction.

Principles of control. Malaria is a community disease, large numbers of people usually being affected. Application of a combination of the following principles is essential if the disease is to be controlled.

(i) Prevention of the mosquitoes from breeding.
(ii) Use of insecticides to prevent mosquitoes from becoming effective carriers.
(iii) Prevention of mosquitoes from gaining access both to patients and uninfected individuals.
(iv) Use of drugs to kill malarial parasites in the blood and liver of infected persons.
(v) Use of drugs in a healthy person so that parasites are killed as soon as they are injected into the blood stream.

These preventative measures are called *prophylaxis*.

Control of mosquitoes. Mosquitoes lay their eggs in still water in lakes, ponds, swamps, roof gutters, drains and even in water-filled pots and tin cans. The eggs hatch into larvae which live and feed in the water for several days before turning into adults. Measures to prevent the adults from breeding include drainage of swamps and the spraying of lakes and ponds with oil and insecticide. Oil forms a thin film on the surface of water. An insecticide in the oil kills both larvae and egg-laying adults that settle on the water. Roof gutters and drains should be treated with insecticides, as should any other places containing still water. Old jars, pots and tins should be destroyed or buried so

that they will not contain even the small amount of water needed for eggs to be laid.

While it is still important to spray such places as drains in which mosquitoes breed it is now realized that in swamps and open water so many breeding 'pockets' exist that total control is impossible. Environmentalists in particular are reluctant to accept widespread use of insecticides (p. 305).

(Wellcome)

Fig. 29.15   Spraying the breeding grounds of anopheline mosquitoes with insecticide

Malaria had been endemic in much of Italy until measures were taken to drain breeding grounds such as the Pontine Marshes near Rome, thus reducing or eliminating the possibility of anopheline mosquitoes breeding. Drainage of areas of stagnant water has been effective in reducing malaria in Malaysia. The introduction of fish that feed on mosquito larvae, so reducing the numbers of vectors, is an interesting feature of biological control. *Gambusia affinis* is used increasingly in the tropics in situations where it is not possible to eliminate water and, while the fish do not eat all the mosquito larvae, the number of adult mosquitoes that emerge is much smaller than would otherwise be.

(WHO/P. Almasy)

Fig. 29.16   A small fish, called *Gambusia*, can be reared in ponds and reservoirs to feed on mosquito larvae

One of the most promising methods of biological control to be introduced is the use of a bacterium, *Bacillus thuringiensis*, into the water in which the larvae live. This organism produces a protein which is highly poisonous to the larvae of the mosquito.

The measures so far described reduce the numbers of mosquito larvae but they do not ensure that no adults emerge to act as vectors.

Fig. 29.17   Concrete-lined irrigation canal in Turkey boosts agriculture but does not provide breeding grounds for mosquitoes

(WHO)

Destruction of adult mosquitoes. The introduction and use of insecticides first began on a large scale in 1935 and the material used was pyrethrum. The newer synthetic insecticides such as DDT, gamma HHC and dieldrin had an incomparably greater success and the massive reduction in the world-wide incidence of malaria in the 1960s owed more to the systematic use of insecticides than to all other methods. The aim was not to kill all mosquitoes at once but to prevent the adults from living for more than twelve or fourteen days, the time taken for the parasite to complete its sexual reproduction in the mosquito and migrate to the salivary glands. The adult female mosquitoes which transmit malaria are most active at night and rest during the day. Some rest on the walls of houses and under the roof.

(WHO/R. da Silva)

Fig. 29.18   Spraying where mosquitoes rest

When their resting places are known, these surfaces can be sprayed with residual insecticides (Fig. 29.18). 'Residual' means insecticides that stick to a surface and remain active for many days. The mosquito had only to land on and walk over the sprayed surface to be affected. DDT was once used. It was cheap and lasted for a long time. It still kills many mosquitoes, but organophosphates, such as malathion and senitrothion, and carbamates (which are more expensive) are now increasingly used for house spraying. By trapping mosquitoes and counting them, it is possible to test how effective a spray is (Fig. 29.19). If a spray is effective then fewer mosquitoes are trapped on successive days. However, reports from many of the malarial regions of the world suggest that mosquitoes have developed resistance to some of the older and cheaper insecticides.

Fig. 29.19   These simple traps, made of wire and muslin, are used to test the effectiveness of different spraying techniques against mosquitoes

(WHO/R. da Silva)

Penetration through human skin can occur when people bathe or wade through or simply get their feet wet with infected water. The larvae may also penetrate the body from the gut if they are swallowed in untreated drinking water. Partly by the secretion of enzymes and partly by a wriggling movement the cercariae make their way through, for example, the thin skin at the ankle, and shed their tails. Inside the body, the larvae enter lymph vessels and are then carried slowly to one of the large veins near the heart and so into the circulatory system (*see* p. 83). They appear in the liver after a few days and develop there into adult worms which then re-enter the blood stream. *S. mansoni* and *S. japonicum*, after a total of twenty-three days in the body, establish themselves in branches of the hepatic portal vein near the intestine; while *S. haematobium*, after a similar period, is found in the blood vessels of the bladder. Egg production soon begins and the cycle is resumed.

Course of the disease. Apart from some itching caused when large numbers of cercariae pierce the skin, little is felt until about a month after infection. Then, for five or six weeks the patient may feel ill and have a fever. Heavy infection with *S. mansoni* or *S. japonicum* is accompanied by diarrhoea with blood and mucus appearing in the faeces together with eggs, though these can only be found by examining faeces under a microscope. Over a period of time eggs may be carried in the blood stream to other parts of the body and although they do not develop into larvae, they may cause damage by obstructing small blood vessels, particularly in the brain and spinal cord.

S. haematobium by damaging blood vessels around the bladder causes blood to appear in the urine. Urination becomes uncomfortable and, over a period of years, the ureters are often scarred and obstructed with build-up of pressure in the urine back to the kidneys which may, in turn, be damaged.

Treatment. Diagnosis is usually made by the recognition of eggs in either faeces or urine. The usual treatment is to inject a compound containing antimony into a vein. Such a compound must, of course, be toxic to the blood flukes but there is a risk that the patient might react also. One new antimony compound can be injected into muscle, which is less uncomfortable for the patient but not as effective as those injected into a vein. Drugs known as Nilodin and Miracil D can be taken by mouth. One of the problems is that treatment and cure of individuals does not prevent re-infection.

Control and prophylaxis. If control of schistosomiasis is to be effective then it is necessary to know where infection is coming from. This means mapping an area carefully to show all canals, streams, ponds and other wet areas that may harbour the snails. Then the places where people bathe, wash clothes, obtain drinking water, plant rice, or come into contact with water used for irrigation must be noted on the map. Where the snails are found, they are examined for the presence of schistosome larvae, taking precautions against infection while handling them, for example by wearing rubber gloves.

When this preparatory work has been done, the following measures are put into operation.

(a) Treatment of drinking and washing water. Filtration of water followed by chlorination kills cercariae, but if treated water is to be stored then it is essential that the tanks and cisterns be snail-free, to prevent re-contamination. Boiling destroys these parasites as effectively as it destroys other micro-organisms.

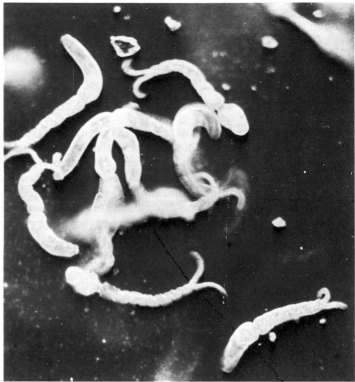

(Dr M. Blair)

Fig. 29.25   Cercaria larvae of *Schistosoma* (× 80)

(b) Preventing the eggs from reaching fresh water. It is not sufficient to provide latrines or an effective sewage disposal system if people suffering from schistosomiasis still urinate or defaecate in or near fresh water. People living in infected areas must be educated always to use latrines and this is not an easy or obviously practical goal to achieve. Children in particular should be told about the hazards of snail-infested water and taught the need for hygienic behaviour, since they often play near and splash in and out of water, exposing themselves to the risk of infection and of contaminating the water (Fig. 29.26).

(WHO)

Fig. 29.26   Children paddling in this sweet-water canal in Egypt are exposed to the risk of infection with *Schistosoma*

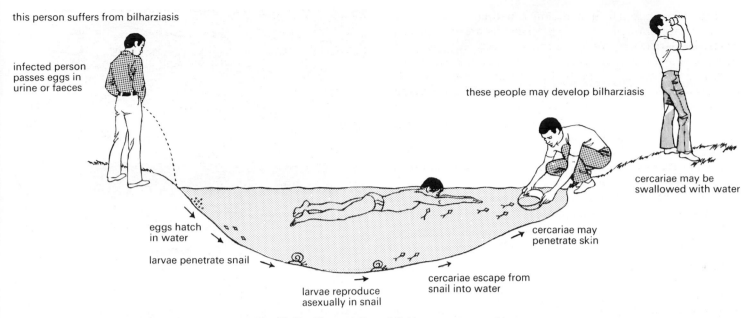

this person suffers from bilharziasis

infected person passes eggs in urine or faeces

these people may develop bilharziasis

cercariae may be swallowed with water

cercariae may penetrate skin

eggs hatch in water

larvae penetrate snail

cercariae escape from snail into water

larvae reproduce asexually in snail

Fig. 29.27   Transmission of *Schistosoma mansoni*

(c) Cercariae can be prevented from reaching the skin either by keeping out of water or by wearing protective boots. Building bridges over streams and ditches is also helpful.

(d) Control of the snails, the secondary hosts, can be achieved by regulating the flow of irrigation canals. A rapid flow with few obstructions discourages the growth of the water-weeds on which the snails feed. If the water can be cut off to allow ditches to dry out periodically then large numbers of the snails die. This has proved effective in Ghana where the periodic drying out of rice paddy fields has reduced the numbers of snails carrying larvae. Another method is to destroy the snails by the use of copper salts in the water. This requires careful control because copper molluscicides (snail killers) are expensive and can also be dangerous to humans, domestic animals and crops if used at too high a concentration.

In Sharkiah Province in Egypt a project was successfully launched following an investigation of the spread of schistoso-miasis that accompanied a new irrigation scheme in the area. Irrigation of the land led to a rise in productivity and prosperity, and the money was used to provide piped, sterilized water and a simple but effective latrine system. Then the part of the canal where village children played was lined with stone and a control dam built at either end. In the hot season the water in this section is treated with a molluscicide to destroy snails and with bleaching powder to control bacteria and other micro-organisms. At the same time, to provide a check on the effectiveness of these measures and the accompanying health education programme, the children's faeces and urine were checked for schistosome eggs. The egg count is falling and the incidence of schistosome in this community is being reduced.

In 1964 WHO estimated that 200 million people suffered from schistosomiasis and that the number was rising. This is particularly so where new irrigation schemes designed to make arid lands in the Philippines, Iraq, Egypt and West Africa more fertile also provided a habitat for snails whose previous absence meant freedom from blood flukes. Awareness of the biology of the schistosome life cycle and determination to break this cycle are leading now to a slow reduction in numbers of patients.

Rats, fleas and plague

Bubonic plague is a disease caused by bacteria. It appeared in epidemic proportions throughout most of Europe in the thirteenth to nineteenth centuries and was often referred to as the Black Death, because of the appearance of dark areas of diseased tissue on the bodies of victims, particularly after death. An epidemic in the fourteenth century killed about a quarter of the population of Europe.

The spread of plague throughout the tropics has been more recent and has followed the transport of rats, the principal reservoir of the causative agent, from one country to another with grain shipments, particularly as movements of grain foods

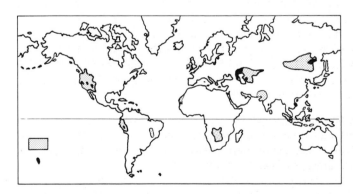

Fig. 29.28   Areas where plague was endemic in 1997

between countries have greatly increased. A major outbreak in India in 1896 lasted for fifteen years and killed over eight million people. The epidemic followed the extension of rail and sea transport facilities. It soon spread into East Africa but it was only in 1903 that it was first recorded on the west coast of South America where it is now endemic (Fig. 29.28).

Scientists believe that the disease was originally confined to wild animals which, being the hosts in which the pathogen normally lives, still form the reservoir of infection; from these animals it may be introduced or re-introduced to humans. An

Fig. 29.29 The black rat, *Rattus rattus*, is the reservoir for *Yersinia pestis*, the plague bacillus

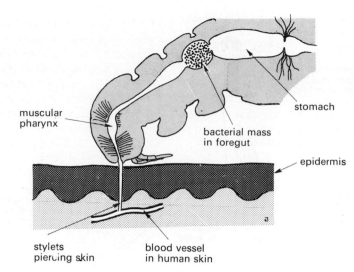

Fig. 29.31 Section through flea showing plug of bacteria in foregut

infection of this kind, originating in another vertebrate, is called a *zoonosis*. To understand its relationship to man one has first to know the ecology (life pattern in relation to its environment) of the reservoir animal and also of the vector, in this case the rat flea.

For plague to become endemic there must be a suitable host animal with a high resistance to disease. Rats such as the black rat, *Rattus rattus*, and the brown or sewer rat, *Rattus norvegicus*, certainly harbour the plague bacillus in such countries as India, Pakistan and Burma, though in other parts of the world to which plague has been introduced the rats have had a low resistance to the bacteria and have died. Marmots, mice, gerbils and other rodents are among the mammals that act as reservoirs in different parts of the world.

Causative agent and method of spread. The micro-organism causing plague is a bacillus with rounded ends called *Yersinia pestis*, first identified in 1894. Experiments carried out in Hong Kong in 1898 showed that these bacilli were present in fleas that had fed on plague rats, and that these fleas could infect healthy rats. The details of transmission to humans were not discovered until 1914 when it was noted in India that only certain kinds of rat flea were capable of transmitting the bacilli. Fleas that digest the bacilli as well as the blood are not able to act as vectors. However, when the rat flea, *Xenopsylla cheopis* (Fig. 29.32), has fed on infected blood the bacillus multiplies in the flea's foregut and forms a mass which blocks the gut and prevents further feeding until the flea, by now hungry, regurgitates the bacterial mass on to the host's skin (Fig. 29.31). As it pierces the skin to feed, some of the bacteria are introduced into the wound and, provided the numbers of bacteria are sufficient, a new case of plague results. When feeding, a flea can take about 0.5 mm³ of blood into its stomach. This volume of blood from an infected rat may contain large numbers of bacilli,

but 0.5 mm³ of human blood, unless the host is nearly dying of plague, will not contain sufficient numbers of bacilli to provide a minimum dose (*see* p. 211) so that no infection results. It was only in 1953 that the human flea, *Pulex irritans*, was finally shown to be capable of transmitting plague bacilli from one human host to another and then only when infection was very heavy and when the fleas were numerous. When many fleas feed together the number of bacilli entering the blood exceeds the minimum needed to produce an infection (Fig. 29.30).

Course of the disease. Plague bacilli produce a powerful antigen (*see* p. 74). In addition they inhibit the phagocytic activity

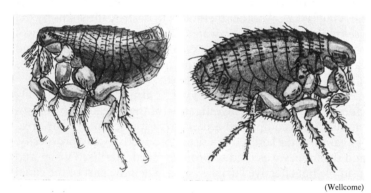

(Wellcome)

Fig. 29.32 Rat flea, *Xenopsylla cheopsis* (×15), male (left) and female (right)

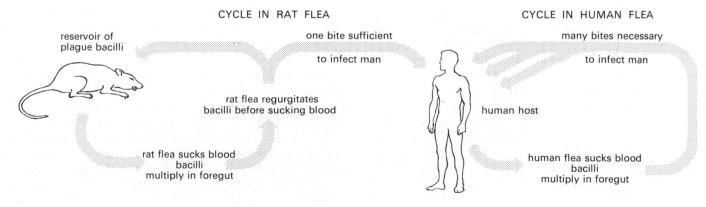

Fig. 29.30 Transmission of bubonic plague

CYCLE IN RAT FLEA

reservoir of plague bacilli

one bite sufficient to infect man

rat flea regurgitates bacilli before sucking blood

rat flea sucks blood bacilli multiply in foregut

CYCLE IN HUMAN FLEA

many bites necessary to infect man

human host

human flea sucks blood bacilli multiply in foregut

of leucocytes. Symptoms develop after an incubation period of two to four days, occasionally longer, and include fever of 39–40 °C (103–104 °F) or higher, and the patient feels very ill and weak. The white cells of the blood stream combat the invasion of fast-breeding bacilli and the lymph nodes become choked with dead cells and bacteria. Lymph nodes in the groin and armpit in particular become swollen, sometimes to the size of a hen's egg, and the name 'bubonic plague' is a recognition of these buboes or swollen lymph nodes. Mortality from plague is high, particularly at the start of an epidemic, but the antibiotic drug streptomycin destroys the bacilli.

Prophylaxis. The most effective way of preventing plague epidemics is to destroy the reservoirs of the disease, usually rats, and the fleas that act as vectors. To destroy rats successfully one must know their habits and where they feed and breed.

Stored food, particularly grain, is attractive to rats. Granaries and rice stores should be made rat-proof (*see* p. 250) and so should domestic food stores. Rats are good climbers, and rat-proofing a building at all levels is not easy and is often expensive. Rats also frequent rubbish tips, drains and sewers, each of which provides food for the animals. Methods of making these less accessible to rats are described in Chapter 39.

Where rats have already established themselves and are breeding they can be destroyed by poisons, laid in a bait, usually crushed grain. Warfarin is still widely used. It reduces the clotting properties of the blood, and rats that eat it die from internal haemorrhages. However strains of rats have evolved that are resistant to warfarin and the search for a new rodent destroyer is being intensified.

Attention to personal hygiene, particularly the wearing of clean underclothing (p. 239), eliminates or at least reduces the numbers of fleas on the body and therefore reduces the chances of a minimum infective dose of bacilli penetrating the skin. DDT powder on floors and DDT spray on walls of rooms where fleas may be present reduce the numbers of vectors.

Public health authorities must act quickly when plague outbreaks do occur but the fact that epidemics are now rare is

a reflection of the attention given to control of rats. The introduction of plague to a healthy country is prevented by such measures as the use of anti-rat shields on the mooring ropes of ships and by drawing up gangways when these are not being used (Fig. 29.33).

Dysentery

Dysentery is a disease of the large intestine, with diarrhoea as one of the main signs. But we must remember that diarrhoea is a sign of several other infections, such as cholera and typhoid. We must also distinguish between *bacillary dysentery*, which is caused by the bacterium *Shigella*, and *amoebic dysentery*, caused by *Entamoeba histolytica*, which is a protozoon (Fig. 25.8, p. 201). Amoebic dysentery is often called *amoebiasis*. Both kinds of dysentery are spread when the causative organisms are transferred from faeces to the mouth. The course of each disease and the treatment are different but the methods of prevention are similar.

Bacillary dysentery. Outbreaks of bacillary dysentery are not uncommon in Britain in the summer months. The *Shigella* bacterium directly affects the large intestine. Two to seven days after the causative organism has entered the digestive system diarrhoea develops. In mild cases the stools or faeces are passed frequently but the quantity is small. The stools may carry flecks of blood and the patient may have a slight fever, but he usually recovers in a few days. The stools, of course, contain the bacteria. In severe cases the stools consist of mucus, stained bright red with blood from the damaged lining of the colon, and the patient has a high fever. Dehydration of the tissues occurs.

Method of spread. Bacillary dysentery is a disease associated with poor sanitation and hygiene (Fig. 29.5). If a person's fingers come in contact with infected faeces, for example when cleaning the anal region after defaecating, then the bacilli can be transmitted directly to food or to objects such as lavatory handles and toilet roll holders that other people could touch.

Since diarrhoea is one of the signs of dysentery and a person may be 'taken short', or feel an urgent need to defaecate, faeces are sometimes deposited behind hedges or in other places where flies can reach them easily. Houseflies pick up the bacteria on their feet while walking over infected faeces and later on may deposit them on our food.

Treatment. A few people carry the bacilli in their intestines without showing any signs of disease and, therefore, without knowing that they are carriers. If they are cooks or food handlers then they can spread the causative organisms if they do not wash their hands thoroughly after using the lavatory.

The patient should rest and eat no food for a day or two. It is important to drink plenty of pure water, preferably with sugar and a little salt added, to prevent dehydration and to replace sodium (see oral rehydration, p. 222). Sulphonamide drugs are used to treat severe cases.

Control. Personal hygiene is important, particularly washing the hands in clean water after using the lavatory. Good sanitation and a supply of safe drinking water (Chapters 36 and 37) are particularly important. Food, whether uncooked, or particularly if it has been cooked and allowed to cool, must be protected from houseflies and from the hands of carriers. Careful, sanitary disposal of the patient's faeces is very important.

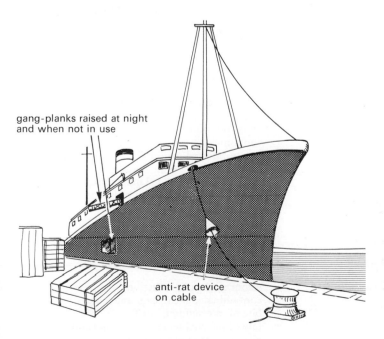

gang-planks raised at night and when not in use

anti-rat device on cable

Fig. 29.33 Precautions against the spread of rats by ship

Amoebic dysentery (amoebiasis)

The microscopic animal or protozoon *Entamoeba histolytica* can live harmlessly in the caecum, feeding on the bacteria that are normally found in the gut. The protozoa can form cysts which are passed out with the faeces. But if a person becomes ill, particularly with some other disease of the large intestine, then the *Entamoeba* can become more active. The organisms invade the lining of the colon and cause ulcers to develop. When the ulcers break, blood escapes and stains the faeces. But the blood stains are dark and there is a strong smell and the stools look very different from the bright red stools of acute bacillary dysentery. Amoebiasis is rare in Britain and patients are usually found to have contracted the disease while on a visit abroad.

Method of spread. Amoebic dysentery, like bacillary dysentery, is spread by transmission on the fingers following contact with infected faeces. Again, young children often receive the organisms in food handled by their mothers. The cysts of *Entamoeba* may lie on the skin of the hands or under the finger nails after a person has used the lavatory. It is possible for the organisms to be spread in drinking water, but this is unusual. However, houseflies and cockroaches, both of which visit lavatories and feed on faeces, can carry the cysts on their feet and transfer them to food.

The disease is endemic in many tropical countries, with a low frequency of infection. Epidemics, with rapid spread to large numbers of people, are unusual.

Treatment. Various drugs can be used which destroy the amoebae. Many of these drugs destroy the cysts as well as the active amoebae. Metronidazole is a modern drug that is very effective and which produces few side-effects.

Control. The same methods of control that apply to bacterial (bacillary) dysentery are also effective for amoebic dysentery. The sanitary disposal of faeces where flies and cockroaches cannot reach them is essential. Care in washing the hands after defaecation and after cleaning up a patient is important.

High temperature kills the cysts of *Entamoeba*, so food that is freshly cooked is safe. But food that is eaten raw, such as fruits and vegetables, should be washed carefully in clean, safe water.

Typhoid

Typhoid fever is caused by the flagellate bacillus *Salmonella typhi* while the similar but milder paratyphoid fever is caused by *S. paratyphi*. Typhoid is found world-wide and is one of the commonest fevers in tropical countries.

Method of spread. There are no natural animal reservoirs of *Salmonella typhi* other than man, though some other kinds of *Salmonella*, causing food poisoning, may be carried by ducks and poultry (*see* opposite). Patients suffering from the disease pass the bacilli in their faeces, vomit and, sometimes, in their urine. Care over the sanitary disposal of the faeces and vomit of known sufferers prevents the spread of the pathogens, but there are many cases on record of so called 'healthy carriers', people who have the bacilli reproducing in their bodies and who excrete them but who show no signs or symptoms of the disease. If such a carrier is employed as a cook or a food handler then, unless that person is scrupulously clean in his habits, there is a real risk

Some widespread diseases and their causes

Disease	Causative organism	Method of spread
AIDS	virus	sexual intercourse, sharing needles
Ascariasis	large nematode or roundworm (*Ascaris*)	eggs swallowed with food or water
Cerebro-spinal meningitis	bacterium (*Neisseria*)	airborne, droplet infection
Cholera	bacterium (*Vibrio cholerae*)	contaminated drinking water or food
Dysentery (bacillary)	bacterium (*Shigella*)	contaminated drinking water or food
(amoebic)	protozoon (*Entamoeba*)	contaminated drinking water
Elephantiasis (filariasis)	nematode, filarial worm (*Wucheria*)	night-biting mosquitoes
Hookworm	nematodes (*Ancylostoma, Necator*)	direct penetration of skin by larvae
Jigger (Chigger)	flea (*Tunga*)	direct penetration of skin
Leprosy (Hansen's disease)	bacterium (*Mycobacterium leprae*)	airborne, droplet infection
Malaria	protozoon (species of *Plasmodium*)	bite of infected female anopheline mosquito
Measles	virus	airborne droplet infection
Pin-worm	nematode (*Enterobius*)	anus-to-mouth
Poliomyelitis	virus	contaminated water
Rabies	virus	bite of infected mammal
Ringworm	fungus (*Trichophyton*)	skin contact
Scabies	mite (*Sarcoptes*)	direct penetration of skin by female mite
Schistosomiasis (Bilharzia)	blood fluke (*Schistosoma*)	direct penetration of skin by larvae, or swallowed in water
Sleeping sickness (trypanosomiasis)	protozoon (*Trypanosoma*)	bite of infected female tsetse fly
Smallpox	virus	airborne, droplet infection
Tapeworm	platyhelminth (species of *Taenia*)	eating raw or undercooked infected meat
Tetanus	bacterium (*Clostridium tetanae*)	enters through breaks in skin
Tinea	fungus (*Trichophyton*)	skin contact
Tuberculosis	bacterium (*Mycobacterium tuberculosis*)	airborne, droplet infection
Typhoid	bacterium (*Salmonella*)	contaminated meat, eggs and water
Typhus	virus	bite of infected louse or flea
Yellow fever	virus	bite of infected female *Aëdes* mosquito
Whooping cough	bacterium (*Bordetella*)	airborne, droplet infection

of bacilli being transferred onto the fingers from the anus while cleaning himself after defaecating.

In much of the tropical world the principal method of spread is in contaminated drinking water and, occasionally, by flies (*see* Fig. 29.5). If untreated sewage reaches a river or other source from which water is consumed without being chlorinated, then the pathogens are likely to be ingested. Holidaymakers from Britain occasionally take in the pathogens while overseas by eating shellfish, particularly oysters. These molluscs filter seawater and extract minute organisms on which they feed. If the sea is contaminated by raw sewage then there is a risk of the pathogens being filtered by the shellfish.

Infections that originate in Britain usually originate from eating contaminated food. An outbreak of typhoid in Aberdeen in 1964 was traced to a can of imported corned beef. It is possible that the meat had not been properly sterilized before being canned, though more likely that the can was imperfectly sealed before being cooled in unsterilized water.

Course of the disease. After the pathogens have been ingested they are carried to the intestine and, from there, to the lymphatic system associated with the gut where they reproduce. They are then returned to the blood stream and carried around the body, setting up infections particularly in the liver, spleen and bone marrow. The toxins released provoke the high fever which is accompanied by severe abdominal discomfort. If not correctly diagnosed and treated the illness may result in death, though treatment with antibiotics usually results in recovery.

Control. Countries in which people use a WC or a lavatory for the sanitary disposal of faeces, where sewage treatment is efficient and in which drinking water is effectively treated are usually free from typhoid. The disease, which was common in Britain in the nineteenth century, has been reduced in this country by these measures. Travellers to parts of the world where the disease still exists and sanitary arrangements are inadequate can be given a TAB vaccination. This should be given at least three weeks before travelling to allow the level of antibodies to build up adequately (Fig. 32.4).

When extensive outbreaks of typhoid occur it is necessary to seek out 'healthy carriers', particularly among cooks and food handlers.

Food poisoning

Bacterial food poisoning affects the stomach and intestines and is usually accompanied by pain and diarrhoea and, sometimes, by vomiting. In Britain the commonest cause is infection with *Salmonella*, but several species are involved and *S. typhimurum* and *S. enteriditis* are the most widespread.

Illness only follows when the bacteria are ingested in large numbers, although age or another infection may reduce a person's resistance. Rapid reproduction of the bacteria in warm, cooked, protein-rich foods builds up the numbers of salmonellae to the danger level and provides one of the common sources of infection. Under-cooked eggs or raw eggs,

used for example in making mayonnaise, provide another source (Fig. 27.4, p. 213).

Staphylococcus aureus, a common bacterium in the nose and in infected cuts, for example on the hand, may get into food where it produces a powerful toxin, particularly when food is kept at warm temperatures before being served. When the food is eaten the toxin, being already present, quickly produces symptoms of food poisoning, often in a matter of hours.

Food poisoning is not uncommon among the population but when it occurs among patients and staff in hospitals, or among people using a restaurant or canteen, then the effects attract wide publicity, and rightly so. Control and prevention of food poisoning is discussed further on p. 293.

Questions

1 Study Fig. 28.1 and explain why a child with measles will spread the virus before developing a rash.
2 Why are measles epidemics most common in the winter months during the school term?
3 In what ways are influenza and measles (a) similar and (b) different?
4 Why do outbreaks of influenza frequently reach epidemic and sometimes pandemic proportions?
5 How is cholera most likely to be introduced into an area?
6 Why must particular care be taken over the disposal of faeces during a cholera epidemic?
7 What are the two important methods used to treat and cure cholera?
8 Why was cholera an epidemic disease in London in the nineteenth century? (see Chapter 23)
9 Explain why people who have been inoculated against cholera can still transmit the disease.
10 Why was tuberculosis a much more common disease in Britain in the 1920s than it is in the 1980s?
11 What are the chief means by which the incidence of tuberculosis is kept low in Britain?
12 Why do you think the world-wide death rate from malaria decreased between 1955 and 1970?
13 Why is the incidence of malaria rising at the present time?
14 Digging holes near towns and villages in malaria regions and then leaving them to fill with rain water can be dangerous. Why?
15 The malarial parasite does not cause fever while it is in the liver. Why is it necessary to kill the parasites in the liver if a person is to be cured of the disease?
16 Once a person has contracted malaria, why is it a good idea to prevent mosquitoes biting him?
17 What is the point of taking anti-malarial drugs when you do not have malaria, e.g. when visiting a tropical country?
18 In what situations are insecticides used against mosquitoes?
19 What are the characteristics of malarial fever that distinguish it from other fevers?
20 List the places near your school or home where water collects and mosquitoes breed.
21 What are the signs and symptoms of schistosomiasis?
22 Why is schistosomiasis found only in parts of the country where freshwater snails live?
23 Why is it necessary to map the places where snails are found and the places where people come in contact with water if schistosomiasis is to be controlled?
24 What steps can individuals take to avoid catching schistosomiasis?
25 What reasons can you give to account for the numerous epidemics of bubonic plague in Britain and Europe during the fifteenth and sixteenth centuries?
26 Plague is now endemic in only a few areas of the world (Fig. 29.30). What precautions are taken to prevent it becoming more widespread?
27 Why is diarrhoea described as a sign, rather than a disease?
28 Why is bacillary dysentery sometimes thought of as a contact or contagious disease?
29 Why is typhoid now increasing in Britain?
30 Suggest ways by which typhoid could be further controlled.

30

The Control and Prevention of Disease by Personal Hygiene

The best protection against disease is good health. This is not such an obvious statement as it may sound, because good health involves a person's way of life: a person who is well-fed, clean and not overtired usually resists infection better than one who is undernourished, dirty and tired. Hygiene is the science and practice of maintaining good health.

The importance of an adequate, balanced diet has already been discussed in Chapter 6. The body's first lines of defence against invasion by bacteria and viruses are the skin and the mucous membranes lining the breathing system and alimentary canal. If these are to be effective barriers then the diet must contain adequate protein, fat and the appropriate minerals and vitamins for healthy cell growth.

Hygiene of hands and skin

Much of the food that we eat is handled at several stages between its production and the time when we eat it. Some aspects of food hygiene are dealt with in Chapter 38, but here we are concerned with the part played by our own hands in contaminating food.

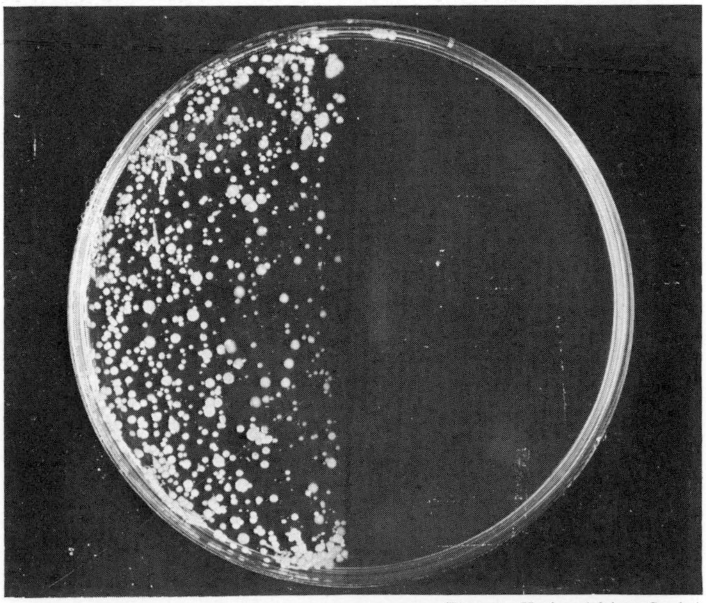

We are continually handling things such as furniture, implements and door handles that have been touched by other people, and many of us handle domestic animals and stroke our pets. These all carry micro-organisms that, in turn, stick to the skin of our hands, as is demonstrated in the practical work on p. 207. Our fingers touch our food, our mouths and other parts of our bodies and the micro-organisms are again transferred. We build up a resistance to many of the germs spread in this way (*see* p. 253) and for much of our lives we are unaware of any risk of infection being spread, especially if our hands do not *look* dirty. Nevertheless there are in Britain each year outbreaks of a mild form of dysentery, particularly among school-children, that probably spreads because they are not taught to use the facilities for washing their hands after using the toilet (Fig. 30.1).

The use of soap or a similar detergent when we wash is important. Detergents have the effect of breaking the bonds between fats and our skin and, since many kinds of micro-organisms are held by oils and fats, these too are removed by detergent action.

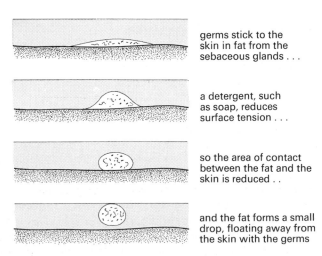

germs stick to the skin in fat from the sebaceous glands . . .

a detergent, such as soap, reduces surface tension . . .

so the area of contact between the fat and the skin is reduced . .

and the fat forms a small drop, floating away from the skin with the germs

Fig. 30.2 How a detergent removes fat and germs from the skin

Antiseptics and disinfectants are not usually used on the hands unless we know we have been in contact with pathogenic organisms. Thus, during an epidemic of cholera or any other disease for which diarrhoea is a sign, it is wise to wash the hands thoroughly with an antiseptic after touching a WC seat (toilet seat) or indeed after touching the walls or any other part of a lavatory.

(Domestos Hygiene Advisory Service)

Fig. 30.1 A sterile swab was rubbed over a lavatory seat and then over the left-hand side of the plate. A second swab was rubbed over the seat after it has been cleaned with disinfectant and applied to the right-hand side of the plate. The photograph shows the growth of bacterial colonies after incubation for forty-eight hours

The fingers can transmit other kinds of micro-organisms such as the bacteria causing septic spots, and if these are touched on breaks in the skin like cuts and grazes—and we do touch them when they itch or irritate—then infection can be caused. Clean hands make this less likely.

Many kinds of organisms, even those the size of *Ascaris* eggs, can be carried under our finger nails. Keeping the nails cut short and scrubbing them, particularly after using the lavatory, helps to keep down the number of germs under the nails.

The micro-organisms on most of the skin of our bodies are commensals (p. 200) and they are harmless or even mildly helpful in resisting colonization by pathogenic organisms (Fig. 30.3). So washing all over with soap and water too frequently —say several times in one day—is not necessarily a wise practice, unless our bodies become very dirty, as a result, for example, of the kind of work we do.

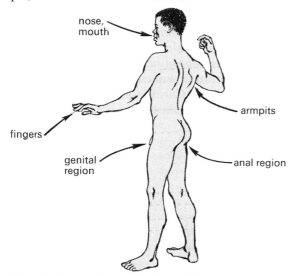

Fig. 30.3   Regions of the body where micro-organisms are found in greatest numbers

Hygiene of the eyes and ears

The conjunctiva (*see* p. 133) and epithelium, the thin layer of cells around the cornea of the eye, have an important protective function similar to that of the skin covering the soft tissues of the body. If the cornea is damaged, for example by the scratching action of sharp dust blown into the eye, then a break is made in the protection, and pathogenic organisms may invade the surface tissues of the eye. Such pathogens include the rickettsia-like organisms causing *trachoma*, a form of blindness common in some tropical countries, in which the cornea becomes opaque and unable to transmit regular patterns of light rays. The pathogen is present in the discharge from an infected eye and is transferred on towels and by flies which alight on the eyelids to feed on the liquid film over the eye. The same methods of spread are responsible for transmission of a virus causing epidemic conjunctivitis (inflammation of the conjunctiva) and a bacterial conjunctivitis which is highly infectious (Fig. 30.4).

A continuous secretion of watery liquid from the tear glands bathes each eye. This lubricates the eyeball so that the eyelids can move freely over it and it washes away fine dust particles and bacteria. Sir Alexander Fleming first described an enzyme that he discovered in tears which is capable of *lysing* or breaking down the outer coat of bacteria, and which he called lysozyme.

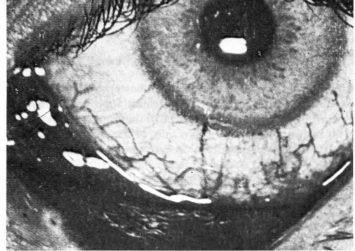

(Blackwell Scientific Publications)

Fig. 30.4   Conjunctivitis, inflammation of the membrane around the cornea

Clearly this enzyme does not destroy all pathogenic organisms but it is important in controlling many of them, just as a steady flow of fluid washes away many organisms.

The importance of vitamin A in the maintenance of good vision has already been described on p. 36 and the common defects in vision are dealt with on p. 138.

The middle and inner ear are encased in bone which provides protection against mechanical damage. The tube of the outer ear is less well protected. Wax is secreted from cells in the skin lining the tube, and its function, apart from being mildly antiseptic, is uncertain. If too much accumulates in the tube then hearing is less distinct, but care should be taken when trying to remove it since the careless use of any implement such as a matchstick could easily damage the ear drum, with the possibility of infection and deafness. Wax near the entrance to the ear passage should be removed in the course of washing; wax deep in the passage should be removed only by a qualified medical worker.

As far as possible the ear tube should be dried after swimming or bathing to prevent the germination of fungal spores, causing a condition more common in tropical than in temperate countries.

A short tube, the Eustachian tube, connects the middle ear to the pharynx (back of the mouth) and its function is to permit movement of air to equalize the pressure on each side of the ear drum (Fig. 30.5). The drum then vibrates with maximum

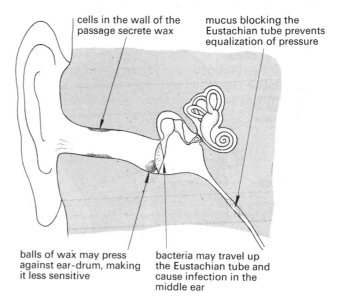

cells in the wall of the passage secrete wax

mucus blocking the Eustachian tube prevents equalization of pressure

balls of wax may press against ear-drum, making it less sensitive

bacteria may travel up the Eustachian tube and cause infection in the middle ear

Fig. 30.5   Causes of temporary deafness

31

Control of the Vectors of Disease

Many disease organisms are spread by insects and other animals, called *vectors* of disease. If the pathogenic organisms are spread *only* by a vector then destruction of the vector will stop the spread of the disease. This is an important principle in the attempts to eradicate several diseases.

Sometimes a vector transfers disease-causing organisms in a mechanical way. The housefly carries micro-organisms on its legs and feet, on its body and mouthparts (p. 214). The cockroach can transmit the pathogens of food poisoning in similar ways. But many vectors have a more complex role. The pathogenic organisms are sucked up, usually with blood, from one host and then reproduce and develop in the body of the vector before being transmitted to a new host. This is illustrated by the anopheline mosquito in whose body the malarial parasites reproduce and move to the salivary glands before they can be injected into the body of a new host (p. 225).

Control of vectors

Insecticides

Not all the vectors of disease are insects, but since the majority are either insects or other *arthropods* (animals which, like insects, have segmented bodies and jointed legs) the chemicals that are used to kill them are called *insecticides*, which means 'insect killers'.

The insecticides in common use against insect vectors are either *residual*, that is they are sprayed on to a surface, such as a wall, and they remain there in an active state, killing insects that land on them over a long period of time; or they are *non-residual*, usually sprayed into the air where they lose their effect quite quickly.

Non-residual insecticides. One of the most widely used non-residual insecticides is *pyrethrum*. This is made from the petals of the *Pyrethrum* flower, which are dried and made into a powder, then either dusted on to surfaces or else made into a liquid extract by dissolving the active substance, *pyrethrin*, in kerosene (paraffin). The extract can then be used as a spray. This is the insecticide that is most widely used in domestic sprays and in aerosols (Fig. 31.1).

Pyrethrum sprayed in a closed space kills most flying insects very quickly. However, when the sprayed space, for example a room, is reopened other insects can fly in and will not be affected. Since pyrethrum is non-residual, these insects can only be destroyed by re-spraying. Pyrethrum is a natural material and can be decomposed or broken down chemically. It is said to be *biodegradable*, which means that organisms such as fungi and bacteria can break it down. This is why it is non-residual.

Residual insecticides. These are stable chemicals that are sprayed or dusted over surfaces on which insects land or crawl. The active insecticide dissolves in the waxy outer layer of the cuticle of the insect's feet and slowly penetrates the body. It reaches the nervous system of the insect and causes paralysis and death. These chemicals are called residual because they remain active for weeks or even months.

Most of the residual insecticides are synthetic (man-made) compounds called chlorinated hydrocarbons. They are more familiar under such names as DDT and Gamma HHC (hexachlorocyclohexane, which used to be known as Gamma BHC) which is often sold as Gammexane or Lindane. Gamma HHC is very poisonous to insects. It vaporizes slowly and is not so long lasting as DDT.

All these chemicals are poisonous to man and to other animals if they enter the body in sufficient quantity, so great care has to be taken over where and when the insecticides are sprayed. Particular care has to be taken by the people who are doing the spraying to make sure that they do not breathe it in or get it on their skin.

DDT is not only stable in the atmosphere but it resists the normal process of detoxication in the liver (p. 69) if it does happen to get into the body. It is soluble in fat and so it accumulates in the fat in the body of any creature that takes it in. Animals feeding on crops or vegetation that have been sprayed with DDT may accumulate sufficient of the chemical in their bodies to cause damage or even death.

A further group of insecticides, the *organophosphorus* compounds, is widely used. Some of the early organophosphorus compounds were very poisonous to man and their use has been discontinued. Even those that are now in use are very dangerous, but they are very powerful insecticides. People concerned with their use take rigorous precautions to avoid

Fig. 31.1   Using an aerosol spray indoors

(Shell)

inhaling spray or dust or getting them on to their skin. Research to produce safer, longer-acting and more powerful insecticides continues.

Control of mosquitoes

Mosquitoes are found in most parts of the world. As adults they all feed on plant juices though the females of most species will also suck the blood of mammals and birds before producing eggs. It is these blood-sucking adults that are potential vectors for disease organisms. Whether they actually do transmit pathogenic micro-organisms depends on the capacity of the micro-organisms to survive inside the body of the mosquito as well as in the blood of the host animal from which they were sucked. A particular pathogen will usually be transmitted by one of a small group of related mosquito species, but not by all species.

Although there are some hundreds of species of mosquito the main features of their life cycles are surprisingly uniform and provide a basis on which the general principles of control can be established (Fig. 31.2).

The females lay their eggs on water either singly or in groups termed 'egg rafts' (Fig. 31.3). Hatching occurs after two or three days in the tropics, though it may take longer in cooler climates. The larvae feed by filtering minute algae from the water and, with this protein-rich diet, they grow rapidly and wriggle actively. They breathe atmospheric air through a tube or siphon near the tail end. After a period of ten days to several weeks the larva changes into an active pupa which still breathes air through a breathing tube, but at the head end. Metamorphosis, a dramatic change in body shape, takes place inside the pupa case which is eventually split to let the young adult emerge. While its wings expand and harden the newly emerged adult stands on the floating pupal case and then flies off in search of vegetation from which it proceeds to suck its first meal. For this purpose the mouthparts are arranged to form two channels with sharp, piercing ends. Saliva can be pumped through the smaller channel while liquefied food is sucked up through the larger. The table shows four important diseases and the more important kinds of mosquitoes that transmit them.

Disease	Causative organism	Mosquito vectors	Geographical distribution of disease
Malaria	protozoon	*Anopheles*	potentially world-wide
Dengue	virus	*Aëdes*	S.E. Europe, Middle East, North & West Africa, S.E. Asia, North Australia, Central & S. America
Filariasis	filarial worm	*Aëdes Culex Mansonia Anopheles*	Africa, Caribbean, India & Sri Lanka, S.E. Asia, North Australia, Polynesia
Yellow fever	virus	*Aëdes Haemogogus Taenorhyncus*	Africa, S. America, Caribbean

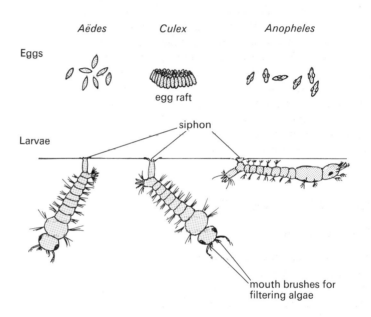

Fig. 31.3   The eggs and larvae of *Aëdes, Culex* and *Anopheles*

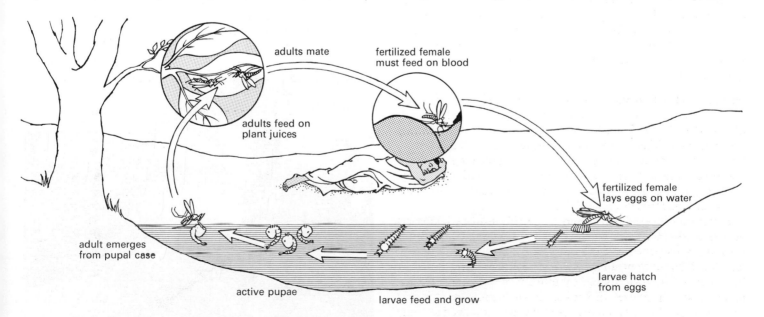

Fig. 31.2   Life cycle of a culecine mosquito

Although the life cycles of all these mosquitoes are similar, precise breeding sites vary greatly as do their feeding habits. For example *Anopheles gambiae*, the most widely distributed malaria carrier in Africa, will lay eggs in almost any sunlit water such as roadside puddles, wells, pools in river beds and even slightly salted water. *Anopheles claviger*, an urban mosquito found in Syria and Israel, lays eggs in wells and cisterns, while *A. plumbeus*, found in Britain and Northern Europe, lays eggs in tree holes. *A. punctimaculata*, which spreads malaria in Central America, will only lay in pools or sluggish water in shade.

Information about where eggs are laid, when the adult females feed on blood, and where they rest when not feeding, is of importance in planning control measures. So is knowledge of the length of time that the adults spend between taking their first blood meal and the next meal, and the frequency with which they feed.

With a carefully planned campaign based largely on the use of insecticides, as described on pp. 226–7, malaria and yellow fever have been brought under control in some areas of the world where they were once endemic. Success in controlling filariasis has not been so successful, partly because species of *Culex* transmitting this disease are less susceptible to insecticides. However, in Sri Lanka (Ceylon), where filariasis is transmitted by *Mansonia*, control has been achieved by the use of a herbicide to destroy the water plant *Pistia stratiodes* to which the larvae attach themselves and on which they are dependent.

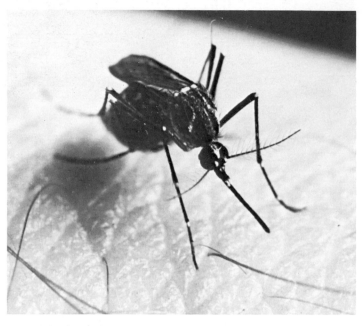

(*a*) *Adult preparing to feed on human arm*

(*c*) *Larvae* (× 20)

(*b*) *Egg raft* (× 10)

Fig. 31.4 Stages in the life cycle of *Culex*

(*d*) *Pupae* (× 10)

(Shell)

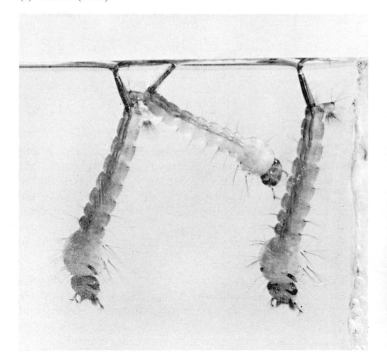

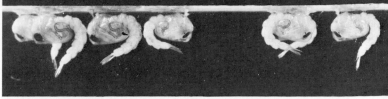

Fig. 31.5 Stages in the life cycle of the *Anopheles* mosquito

(a) Eggs and larvae (× 20)

(Shell)

(b) Pupae (× 5)

(c) Adult female (× 5)

Once a vector has been destroyed in an area it is important to prevent its re-introduction. For example, since the eradication of *Anopheles* from the Mediterranean island of Cyprus, health authorities check and spray with insecticide all aircraft arriving from malarious countries.

In Guyana, S. America, *Anopheles darlingi* was systematically destroyed along the coastal belt by spraying houses with residual DDT to kill the adults. It is no longer necessary to maintain this spraying programme, provided that a narrow barrier zone is continuously treated between the cultivated and populated coastal zone and the high bush of the interior where the mosquito still breeds. People visiting the country from known malarious areas have to provide a drop of blood to be examined for parasites (Fig. 31.6).

Fig. 31.6 Taking a blood sample to test for the presence of malarial parasites

(Shell)

Where extermination is not possible, use is made of the knowledge that most mosquitoes bite at night. By reducing the area of exposed skin by wearing slacks and long sleeves, the risk for those people who must be out after dark is reduced. Indoors, spraying with an insecticide destroys mosquitoes that have entered the building during the day. Keeping windows covered with fly wire, while allowing ventilation, excludes the vectors.

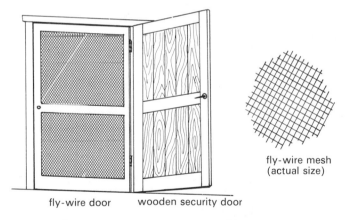

fly-wire mesh
(actual size)

fly-wire door   wooden security door

Fig. 31.7 Fly-wire is placed over windows and doorways to keep out mosquitoes

Sleeping under a mosquito net is important in such circumstances, not only to keep mosquitoes from healthy persons but particularly from malarial or other patients. If such patients are bitten by the appropriate mosquito then the life cycle of the parasite is continued and other people are at risk. The principles of control of the *Culex* mosquito that spreads the filaria worm causing elephantiasis are similar to those that apply to the control of the *Anopheles* mosquito.

Eradication or control? Eradication means that all the mosquitoes in an area are destroyed. This is often not possible for a number of reasons, such as the difficulty of finding and treating all the breeding grounds, or the ease with which mosquitoes can re-invade the area from other places.

243

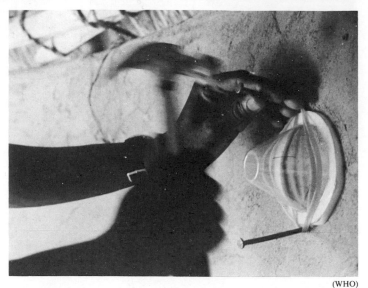

Fig. 31.8 A plastic cone, nailed to a hut wall, to test insecticides: mosquitoes that enter the cone are killed and can be counted

(WHO)

(UN/WHO/R. Witlin)

Fig. 31.9 A helicopter spreads ABATE on breeding grounds

Control, on the other hand, involves the reduction of the number of mosquitoes in an area without wiping them out completely. If the number of mosquitoes is reduced then the incidence or number of cases of malaria will also be reduced.

Every campaign to eradicate or to control mosquitoes in an area must be carefully planned. The first stage or *preparatory stage* involves making a geographical survey of the area. A map is drawn, showing where human dwellings lie and where there is water in which the mosquitoes breed. It also shows the direction of the prevailing wind which might carry in mosquitoes from another area. When these facts are known the necessary supplies of insecticide and the means of spreading it can be organized. The extent to which walls and buildings will have to be sprayed with residual sprays is worked out and the methods of dealing with the breeding grounds are decided. Sometimes the drainage of swampy areas or the filling in of wet regions to raise the ground level may be the best long-term approach, since both these methods destroy permanently the places where the mosquitoes breed.

The campaign moves to the *attack stage* in which the decisions to spray or drain are put into effect. The teams of men who form the spraying squads go into action, wearing protective clothing.

The *consolidation stage* follows, in which the effectiveness of the attack stage is studied. If eradication was the objective, then areas where mosquitoes have survived are likely to receive further spraying, usually at intervals of from three to six months. Where eradication has been successful a *maintenance stage* involves checks at intervals to find out whether there has been any return of vectors.

If the aim was control, to reduce but not eradicate malaria, then checks must also be made at intervals to find out how effective the action is (Fig. 29.19).

Larvicides. DDT is widely used to kill mosquito larvae. The DDT is dissolved in fuel oil and sprayed on to standing water. This is satisfactory for still water that is not to be used for drinking, but insecticides such as ABATE are being used increasingly. ABATE is an organophosphorus compound that is not very poisonous to man. It is also used in the control of blackfly larvae in tropical rivers. Paris Green, a compound containing copper and arsenic is used as a powder, scattered on water. Mosquito larvae take it in as they filter the water. It is

cheaper than many other insecticides and does not make the water unusable. The insecticide dieldrin can be mixed with sand and cement to form bricks, which are used to control larvae in water tanks and cisterns. The insecticide dissolves slowly out of the bricks over a long period of time.

Most recently a bacterium, *Bacillus thuringiensis*, has been used to destroy the larvae. This is an example of biological control and it promises to be effective. The bacteria are filtered from the water as the larvae feed. Once inside the digestive tract the bacteria reproduce and kill the larvae which then decompose, releasing the bacilli into the water.

Controlling adult mosquitoes. Since adult mosquitoes spend part of their lives feeding on the juices of plants, it is sensible in a heavily infested area to reduce the number of bushes near huts and houses. But a more effective method of control is possible

Fig. 31.10 Spraying a doorway with residual insecticide to control mosquitoes in Taiwan

(Shell)

through knowing the habits of the adult mosquitoes. *Anopheles* in particular is a night-flying insect which often spends the day perched on walls of houses. Spraying these surfaces with a very even spray of a residual insecticide, such as DDT or HHC, is effective since the insects remain in contact with the chemical long enough for it to penetrate their bodies. It may take several days to kill the mosquitoes, but less time than it takes for the malarial parasites to develop into sporozoites (p. 225) and move into the salivary glands of the mosquitoes.

In recent years it has been noticed that certain strains of mosquitoes have developed resistance to DDT. This means that other, sometimes more costly insecticides have to be used.

Control of houseflies

Houseflies of the genus *Musca* are world-wide in their distribution. A connection between flies and the spread of disease has been assumed since very ancient times. Moses' instruction to the children of Israel to bury their excrement suggests an awareness of this, and the Romans supposed that dysentery was spread by flies falling into or walking over food. Adult flies are attracted to and walk over and feed on foetid, decomposing material and, particularly, human faeces. More than a hundred different pathogens have been identified from the bodies of houseflies at one time or another, including those causing cholera, bacterial and amoebic dysentery and typhoid fever.

Adult flies are capable of feeding only on liquid food, that is, on food already in liquid form or else food which can be dissolved in the plentiful saliva that the fly secretes. In order to suck liquid the fly uses its proboscis which has a flap or labellum at the end. Bacteria can be shown to be present in the liquid on this flap and also in the saliva that the fly produces while feeding. When it is feeding on food destined for human consumption then contamination occurs. Sometimes partially digested food is regurgitated with the saliva, carrying still more micro-organisms (Fig. 31.12).

Flies drop their faeces as and when necessary and this frequently occurs as they walk over food. Many kinds of pathogens are unaffected by passage through the alimentary canal of the fly and are still capable of causing disease if ingested by humans.

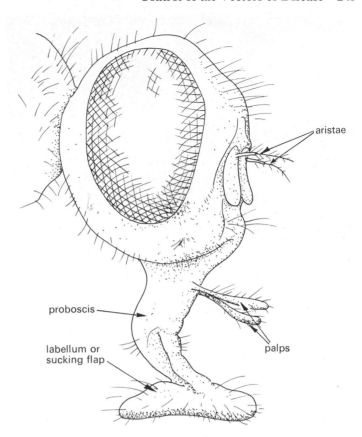

Fig. 31.12   Head and proboscis of housefly

Careful examination of an adult fly shows the body and legs to be covered with hairs. These are able to hold small particles of dust and bacteria in the same way that the bristles of a brush do, and they may be scattered from these hairs on to food as they walk over it.

In order to control this vector of disease it is necessary to know its life cycle (Fig. 31.11). During mating the female fly receives large numbers of sperms into her body and these will fertilize the eggs. Before laying she seeks out decomposing organic matter such as faeces, a manure heap or a dustbin. Flies are attracted to this material by its smell. Bacterial decomposition converts it into a liquefied state on which the larva will be able to feed easily after hatching. Hatching occurs ten to seventy-two hours after laying, depending on temperature.

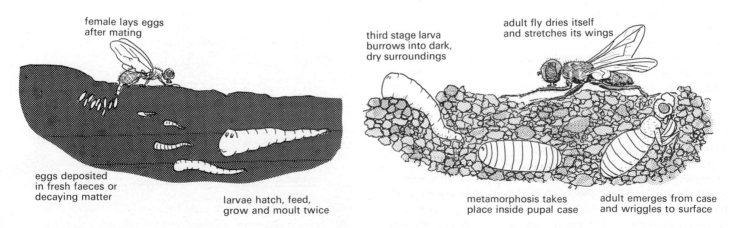

Fig. 31.11   Life cycle of the housefly

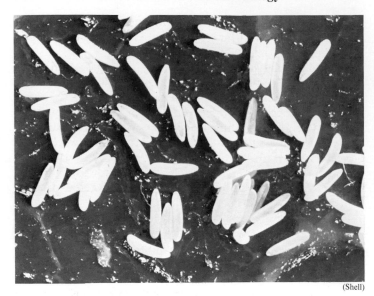

(Shell)

(a) Housefly eggs (× 10)

(Shell)

(b) Housefly larvae (× 5)

Fig. 31.13

Bacterial decomposition results in release of heat energy and this accelerates development of the embryo within the egg.

Effect of temperature on the development of the housefly

Stage	Length of each stage	
	Warm climate 30 °C	Cool climate 10 °C
egg	10 hours	3 days
larva	3 days	8 weeks
pupa	3 days	4 weeks
adult	4 weeks	12 weeks

The larva has a white body, almost transparent when newly hatched, tapering from a blunt posterior to a small head (Fig. 31.13*b*). Although it has no eyes the larva is sensitive to light and burrows into the decomposing mass away from light, thereby avoiding the risk of death through drying out. It grows through two moults to a length of about 1 cm and then changes its habits, seeking a dry situation in which it can pupate. The larval cuticle forms the pupal case inside which metamorphosis, or change to the adult shape, takes place. After three or four days the adult fly emerges, its wings expand and harden, and the adult then flies off to feed and mate.

Control can be gained by (i) preventing flies from breeding and producing more flies, (ii) killing flies, and (iii) preventing them from having access to food for human consumption.

1   Preventing flies from breeding. We have seen that flies are attracted to foetid, decomposing matter in which to lay their eggs. If faeces are buried or disposed of through a sewage system or in deep latrines then flies will not lay eggs in them.

Kitchen waste, which usually is wet and contains organic matter that can quickly decompose, should be kept covered, for example in a bin with a tight-fitting lid, to prevent flies reaching it. Bins should be emptied regularly and frequently (Fig. 31.14).

When rubbish or garbage is collected for disposal it should be taken to a point well away from human habitation. Experiments in which flies marked with drops of paint have been released and later recaptured show that they can travel distances of over 2 km, so that while accepting that it is difficult

to make a rubbish tip free from flies, the chances of their reaching an urban area should be kept to a minimum.

Where there is no local authority to collect and remove refuse, such waste can be either buried or else burned in a fire or an incinerator (a small furnace for reducing combustible material to ashes).

2   Killing flies can be effectively done by destroying them with a fly swat but insecticides, sensibly used, will destroy more flies in less time. At the present time, flies throughout the world are developing resistance to DDT, which had the advantage of being a residual insecticide (*see* p. 240). Similarly resistance to HHC and dieldrin is growing. For spraying rooms and other enclosed spaces a combination of pyrethrums and piperonyl butoxide is most effective, and there is no sign yet of resistance to these chemicals developing in houseflies. However, since it is not a residual spray applied to surfaces this treatment has to be repeated frequently. Trials have been made of other substances that destroy flies, but in spray form there is the risk that the chemical may reach and affect other kinds of animals, including humans. Some of these more dangerous compounds have been impregnated into cards and strips of absorbent material to be hung from ceilings. The best remain toxic to flies for up to six months and are useful in buildings such as dairies and lavatories where flies are attracted. But care must be taken while handling the strips when putting them in place!

flies can reach refuse

Fig. 31.14   A dustbin (garbage bin) for kitchen waste must have a tight-fitting lid to keep out houseflies and rats

3 **Protection of food.** A shopkeeper wishing to sell goods will usually display them to his customers. Meat, wet fish and most perishable foods attract flies which may have come from feeding on faeces or putrefying matter. It is important to keep food on display covered, for example with butter muslin or gauze or, better, behind glass or transparent plastic, to prevent flies from settling on it (*see* Fig. 38.5).

Similarly when such food reaches the home it is necessary to keep flies away. Wrapping in polythene or other flyproof material is not necessarily the best solution since impenetrable material of that kind also excludes air and thus provides conditions for any anaerobic bacteria already present to multiply to dangerous numbers. Where a refrigerator is not available perishable food should be stored in a ventilated cupboard or a meat safe, ventilation being provided through fine mesh gauze which excludes flies. (*See* Chapter 38.)

Methods which prevent houseflies from breeding are the most important control measures because since they reduce the chances of flies breeding, there will be fewer adults to act as vectors. No country has eliminated houseflies but many have greatly reduced even the hot-season fly population, when breeding is most rapid.

Control of tsetse flies

Tsetse flies, which transmit the trypanosome causing sleeping sickness (*see* p. 201), are confined to tropical Africa. There are many species, though they all belong to one genus, *Glossina*. All have mouthparts for piercing and sucking, though the males feed on plant juices while the females suck blood. Only a few species transmit trypanosomes to humans (Fig. 31.15).

(Shell)

Fig. 31.15   Tsetse fly (× 3)

The fertilized eggs hatch within the female. Each larva develops rapidly, being nourished by a secretion from special glands in the abdomen of the mother. They are quite large when they are deposited by the mother in a place where they can burrow into dry sandy soil or under dry leaves to pupate. This stage takes three to four weeks. As adults, the flies live from three to six months during which time the female will produce between six and twelve larvae. Compared with mosquitoes and houseflies, they are slow breeders.

Certain species of tsetse breed only among vegetation in shady, humid conditions along river banks, while others breed in thickets of vegetation in open savannah country. Of the twenty or so species of *Glossina* only five or six species have been shown to be vectors of human sleeping sickness, so the control of trypanosomiasis requires careful investigation work to be done if it is to be effective and not wasteful. In 1972 it was found in Nigeria that *Glossina morsitans* moves into the foliage at the tops of trees and stays there all night so it can be controlled by spraying from the air in the early morning.

For controlling those species breeding on river banks the technique most used is total clearance of trees, though elimination of only the type of tree on which the males feed would be as effective and would be less damaging to the environment. In savannah land it is possible to eliminate small patches of thorn scrub but where the thorn scrub is extensive another approach can be made. By removing scrub along a band two kilometres wide the tsetse can be contained in an infested area while development of grazing land takes place on the 'safe' side of the band. The band of cleared ground can be extended further into the thorn scrub as resources become available.

Barriers and bands of cleared ground give protection around villages and water-holes but, with increased use of road transport, care has to be taken to avoid bringing tsetses into a community in or on cars or, indeed, simply following foot travellers, as flies do.

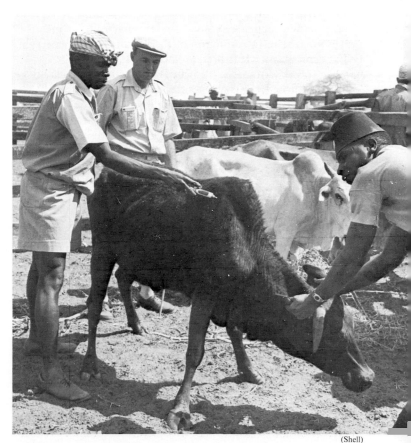

(Shell)

Fig. 31.16   Domestic cattle infected with trypanosomes, being treated by injection

Burning of grassland is discouraged because, although it may drive out the flies temporarily, it is usually followed by the growth of a variety of plants which may include some that encourage the adult flies to re-enter the area.

No tsetse feeds exclusively on human blood and the destruction of game animals has been advocated as a means of eliminating alternative hosts in which trypanosomes may be present. In Uganda and Zimbabwe reduction of game populations as a tsetse control measure has been carried out on a large scale. Conservationists are reluctant to accept this approach and a solution in which game is confined to game reserves may be possible.

A modern method of control involves breeding tsetse flies in captivity and exposing the adult males to heavy doses of radiation. This renders the males sterile but they can still mate with females. The irradiated males are released among wild tsetses at a time in the year when the proportion of males in the natural population is low. Mating takes place and sterile seminal fluid is passed. The females rarely mate more than once and so, their eggs being infertile, produce no young. The method is as yet in use on an experimental rather than a widespread scale.

Control with insecticides introduces problems. Tsetse flies are very susceptible to the effects of chlorinated hydrocarbons such as DDT, but spraying from the air, essential if large areas are to be treated, must be done during daylight, and at this time most adult flies are resting on the *undersides* of leaves. In tall, dense forest, penetration of spray is inadequate anyway but small-scale experiments in riverside villages show that spraying dieldrin emulsion from the ground on to vegetation around the village reduces the fly population.

Control of traffic and the spraying of the insides of cars and aircraft with insecticide should be re-emphasized as a means of preventing the re-introduction of the vector, possibly carrying trypanosomes, into a cleared area (Fig. 31.17).

(Shell)

Fig. 31.17   Spraying a car to destroy tsetse flies which could be carried to a fly-cleared area

(FAO/WHO/T. Land)

Fig. 31.18   Challier traps are cheap and can be made locally. They are effective in attracting tsetse flies, so reducing the transmission of sleeping sickness

Recently it has been discovered that tsetse flies are attracted by particular shapes and certain colours. Fig. 31.18 shows a Challier trap which is used increasingly in West Africa. Once inside the trap the flies crawl up the inner cone and are then killed by a powerful insecticide. The traps can be made locally and are quite cheap, particularly when compared with the cost of aerial spraying. Also the insecticide is not dispersed over an area. It is a selective method.

Body parasites

Bugs, fleas, chiggers (or jiggers), lice and ticks are all blood-sucking arthropods that live in close contact with the human body for at least part of their lives. Because they bite through human skin to suck blood they represent a means by which pathogenic organisms can gain access to the blood stream. In some cases pathogenic organisms complete a part of their life cycle in the body of the vector, while in other cases the blood sucker transmits pathogens mechanically on the outside of its mouthparts.

Bedbugs (Fig. 31.19) cause irritation through their bites but they are not known to be vectors of disease. Scratching the inflamed region around their bites often breaks the skin and

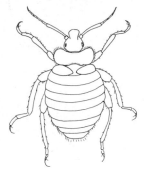

Fig. 31.19   Bedbug, *Cimex* ($\times 6$)

allows pathogenic organisms to enter. The 'assassin bugs' of S. America (Fig. 31.20) transmit *Trypanosoma cruzi*, the pathogen causing Chagas disease, but it is not the mouthparts that introduce the pathogens; the trypanosomes live in the insect's gut and are shed onto the wound in a drop of fluid from the hindgut, expelled after the bug has finished feeding on blood. Bedbugs hide in bedding and in cracks in floors and walls during the day. Thorough spraying of floors and walls with DDT or dieldrin–HHC in oil kills them. Bedding may be sprayed, but in this case pyrethrum is a safer insecticide to use on material that is likely to be in close contact with the skin.

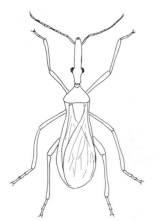 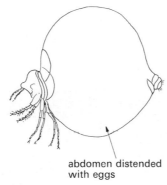

abdomen distended
with eggs

Fig. 31.20   Assassin bug, *Rhodnius* (×2)

Fig. 31.21   Female chigger, *Tunga* (×15)

The assassin or triatomid bugs are controlled by house spraying with gamma HHC. They are resistant to DDT.

There are many kinds of fleas, wingless insects whose laterally compressed bodies permit them to crawl between hairs and through fur. *Xenopsylla* the rat flea is a tropical flea that can survive in heated buildings in temperate countries and plays a part in the transmission of bubonic plague (*see* p. 232), as does the common human flea, *Pulex irritans*. Eradication of fleas is difficult because although fleas in clothing can be killed with DDT powder the insects do not spend all their time with the host. Fleas may be found in bedding and furniture and on floors as well as on alternative hosts, such as dogs. Because they can live for some time away from the host's body it is easy for a flea from an infested person to reach the body of a person who is very clean in his habits.

Flea larvae spend ten days to two weeks feeding and growing, living in cracks in floors and similar places. Keeping the house well swept and the floors washed regularly helps to remove the larvae. Since domestic animals carry fleas it helps if they are kept out of living quarters. Cats and dogs can be washed with carbolic soap and dusted with DDT powder, or sprayed with a pyrethrin aerosol if a person particularly wants these animals in the house.

Chiggers (Fig. 31.21) are a small group of fleas originating in S. America but now found also in West, South and East Africa and parts of India. The chigger is of importance not as a vector but because the female, after mating, burrows into human skin using her powerful, toothed mandibles. Her body swells as the eggs inside her develop and her presence irritates the skin which become inflamed. When the eggs have been laid the skin breaks and ulcerates and secondary infections are common. Before mating, the female chigger is similar to any other kind of flea and spends time in the dust and debris on floors. She is not a good jumper and is most likely to reach a naked foot. Wearing shoes reduces the chances of acquiring chiggers.

Chiggers in their mobile state are controlled by the same methods used for fleas, but once under the skin they should be removed with a sharp, sterilized needle, the small wound that results being dressed with an antiseptic and covered to prevent the entry of dirt and germs.

Unlike the animals so far described in this section, the louse (*Pediculus*) is obliged to spend most of its life in contact with its human host, and each species possesses curved claws for ensuring a firm hold on the skin (*see* Fig. 27.11*a*). The two varieties, the head louse and the body louse, can interbreed and the females in each case attach eggs to the shafts of hairs on the head or the body. The eggs are commonly termed 'nits'. The adults are vectors of the rickettsiae that cause epidemic typhus and relapsing fever, which are spread when the lice move from the body of one host to another. Transfer takes place when people sleep huddled together, during sexual intercourse, and sometimes as a result of changing clothes. A third louse, the crab louse, which lives mostly among pubic hair, is also transferred during sexual intercourse (*see* Fig. 27.11*b*).

Lice can be killed by Gamma HHC (BHC) and by Dicophane (a form of DDT), which is effective against head lice and pubic lice. When there is a threat of a typhus epidemic it is particularly important to control body lice which are vectors of the rickettsia (p. 200) that causes typhus. Gamma HHC powder can be dusted over the person's body and clothing, though it should be kept away from the face. Body lice live in the underclothing next to a person's body when they are not feeding and that is where they lay their eggs. It may take up to ten days for eggs to

(London School of Hygiene and Tropical Medicine)

  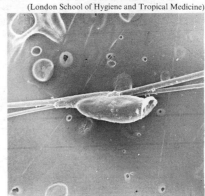

(*a*) *Adult louse* (×20)

(*b*) *Egg of louse attached to hair* (×100)

(*c*) *Empty egg case or nit* (×40)

Fig. 31.22   Electronmicrographs of the head louse and its egg

32
Immunity and Immunization

It has been noticed for a long time that certain people in a population never develop particular diseases. We say they are *immune* to these diseases. In the eighteenth century Edward Jenner, a country doctor in England, noticed that people who looked after cattle and milked the cows rarely caught smallpox. At that time smallpox was quite common and it was often fatal. Jenner noticed that the cow handlers often developed *cowpox*, which is a much milder and less serious disease than smallpox. It seemed possible that a person who had caught cowpox might be immune to the more serious smallpox. Dr Jenner tested his hypothesis by infecting a young boy with cowpox. The boy developed a slight fever and a rash, but soon recovered. Jenner then took scrapings from the pustules or scabs on a smallpox victim and rubbed them into a scratch on the skin of the boy's body. The boy stayed healthy and showed no signs or symptoms of smallpox, yet we now know that these scrapings must have contained the virus that causes smallpox (Fig. 32.1). The process of either rubbing or injecting organisms into the body became known as *inoculation*.

(Wellcome)

Fig. 32.1   Dr Jenner inoculating James Phipps. Note the dairymaid, who has given the cowpox serum, binding her wrist

Jenner thought that cowpox was simply a mild form of smallpox. This is not true, but the two diseases are caused by viruses that are very similar, particularly the outer coat (p. 200). When the white cells in the blood 'learn' to produce the antibody to counteract cowpox virus, the same antibody also counteracts smallpox virus. Since the cowpox virus came from a cow, for which the Latin name is *vacca*, the procedure became known as *vaccination* and the extract of cowpox virus is called a *vaccine*. However, we should remember that it is very unusual to find that immunity to one disease-causing organism gives us immunity to another, different disease-causing organism.

Within ten years Jenner's technique of vaccination was being tried in Europe though there was still plenty of mistrust to be overcome at home! However, by 1840 a Vaccination Act had been passed in England making vaccination compulsory. New and improved methods of producing vaccine were developed.

Innate immunity

In studying immunity to pathogenic organisms we must consider again the ways in which the body can resist invasion by pathogens. Some of these ways are *innate*, i.e. present when we are born. These are listed below.

Innate immunity

Physical barriers to entry	*See page/s*
Skin	96
Mucous membranes	–
Hairs (e.g. in nose)	238
Secretions (e.g. sweat, tears)	98, 133

Active protection inside the body	
Secretions containing active agents (e.g. gastric juice)	63
Cilia in the trachea and bronchioles	85
Interferon in cells	255
Phagocytes in blood	72
Antibodies made by lymphocytes	74, 252

The effectiveness of these barriers varies from one person to another. All microbes carry on their surfaces, or else secrete, large protein molecules which are called *antigens*. These foreign protein molecules come in contact with lymphocytes (p. 72). The lymphocytes form the second line of defence in the blood of the host. The lymphocytes produce *antibodies* called immunoglobulins and release them into the blood plasma. A particular antibody will react with a particular antigen, forming a chemical link. We say that each antibody is *specific* to each antigen.

The chemical links formed between antibody and antigen may destroy the pathogenic organism, or it may neutralize the poisonous substances (toxins) that the organism has produced. So the host resists the invading organism. The antibody has made the host immune. This ability to produce antibodies against certain disease-causing organisms is present from birth and results in the pathogens being inactivated before they can cause disease symptoms.

Precipitation is the name given to the reaction between an antibody and a soluble toxin as a result of which the toxin is precipitated and made inactive.

Agglutination takes place when the antibody makes the pathogenic organisms stick together and so become ineffective.

Opsonization is the name given to the process of coating invading bacteria with antibody in a way that makes them more attractive to phagocytes which then ingest them.

Acquired immunity

Naturally acquired immunity

Immunity may be acquired when disease organisms invade the body and the signs and symptoms of disease develop. The lymphocytes 'learn' to make the appropriate antibody. When the patient recovers, the lymphocytes retain this ability to make the particular antibody. If the patient is infected with the same kind of germs on another occasion, the antibodies are produced very quickly and the pathogens are destroyed rapidly. The patient does not usually show any signs of disease. This way of acquiring immunity applies in particular to diseases caused by viruses. Unfortunately it does not seem to work for illnesses caused by protozoa. Immunity to bacterial diseases does not usually last very long.

Artificially acquired immunity

Dr Jenner discovered the possibility of immunizing or protecting a person from a possible disease attack by introducing the pathogen into the body artificially, so causing a mild form of the disease and inducing the lymphocytes to produce antibodies. The techniques of immunization have developed a great deal since Jenner's day and nowadays three kinds of antigen preparations or vaccines are used:

Toxoids. These are extracts of the toxins secreted by the bacteria such as those causing tetanus. The poisons are made harmless by addition of the chemical *formalin* and when an extract from the bacteria is injected into the body, it stimulates the lymphocytes to produce antibodies. This kind of immunity is *active* because the white cells make the antibody and retain the ability to do so. Active immunity lasts for months or even years.

Killed vaccines. These are made from cultures of bacteria or from viruses that have been grown in suitable host cells. Again, chemicals such as formalin are used to kill the organisms and the extracts are usually given by injection. Short-term protection against cholera is given in this way. Salk anti-polio vaccine is prepared by killing polio virus with formalin.

Attenuated living vaccines. Neither of the methods just described involves a vaccine containing living pathogens. There is therefore no risk of the person developing the disease as a result of vaccination or inoculation. In some cases only the germ itself will stimulate the lymphocytes to produce antibodies. So cultures of the organisms have to be made that are less *virulent* (p. 218) than those that are caught at the start of an infection. This can be done by growing the pathogens in an unusual host, for example, by growing them in a horse. The strain or kind of the organism that grows in the horse survives there without causing disease symptoms. When these *attenuated* or weakened organisms are taken from the horse and injected into man (Fig. 32.2), they can no longer cause disease. But they *do* cause the lymphocytes to manufacture the antibody, and so immunity is acquired. This is the method used to produce the BCG (Bacille Calmette Guérin) vaccine. The two French bacteriologists, Calmette and Guérin, found that while the killed mycobacterium of tuberculosis would not induce the formation of, antibodies, the attenuated, live bacillus would do so.

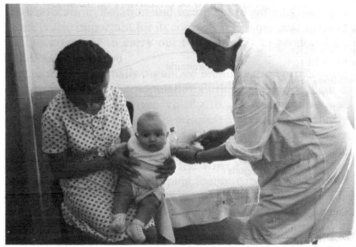

(WHO/H. Herioud)

Fig. 32.2   This six-month-old child is being given an anti-measles vaccination

Immunity to yellow fever is produced by injecting an attenuated virus, grown in a tissue culture in which the virus will multiply. The Sabin polio vaccine, given by mouth, is also made from an attenuated virus (Fig. 32.3).

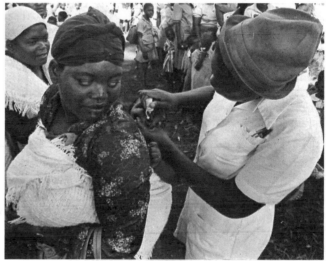

(Aspect Picture Library)

Fig. 32.3   A health worker is gently squeezing this baby's cheeks to open his mouth. A few drops of Sabin vaccine are given orally on three occasions during his first year and will provide life-long protection against polio

Some vaccines in common use

Purpose	Type of vaccine	Method of introduction
Vaccines for large-scale use:		
BCG (antituberculosis)	attenuated bacillus	injected
antitetanus	toxoid	injected
antipoliomyelitis: Sabin	attenuated virus	oral
Salk	killed virus	injected
antidiphtheria	toxoid	injected
antimeasles	attenuated virus	injected
anti whooping cough	killed bacteria	injected
Vaccines for use with persons who are exposed to particular risk:		
typhoid	killed bacteria	injected
yellow fever	attenuated virus	injected
meningitis	killed bacteria	injected
measles	attenuated virus	injected
rabies	attenuated virus	injected
cholera	killed bacteria	injected
Serum for use after a person has been exposed to infection:		
antirabies	both contain ready-made	injected
antitetanus	antibodies	injected

Questions

1 In what ways is the body provided with innate immunity?

2 Explain what is meant by a vaccine. How does a vaccine bring about immunity?

3 Distinguish between active and passive immunity.

4 Why are new-born babies often immune to certain diseases although they become susceptible to the same diseases as they grow older?

5 Why is surveillance particularly important when an immunization programme is showing signs of being effective?

33

Drugs, Antibiotics and Antiseptics

When diseases are treated and cured by the use of chemicals, such treatment is termed *chemotherapy*. The chemicals used must be able to destroy pathogenic micro-organisms without destroying or damaging human tissues. Up to the year 1930 most of the materials used in treating infections were obtained from plants, some of which contain very powerful drugs. The modern antibiotics are extracted from soil bacteria and fungi.

Synthetic drugs

By 1900 many of the organisms that cause disease had been seen under the microscope and classified. At the same time advances had been made in the understanding of the chemistry of carbon compounds (organic chemistry). A German scientist, Paul Ehrlich, who had knowledge of both chemistry and medicine was, at this time, investigating the use of coloured dyes in the treatment and control of the trypanosomes that cause sleeping sickness. He shifted his research to the study of certain compounds of arsenic and, in 1910, published his discovery of a chemical compound that he called *salvarsan*. This would kill trypanosomes not only in mice kept in his laboratory but also in man. He also discovered that salvarsan would kill the treponeme bacterium that causes syphilis. It had to be used with care, since large doses proved poisonous to the patient. Thus he recognized the need to calculate carefully the size of dose that would kill the pathogen without harming the patient. In this way the measurement of the *safe dose* was put on a scientific basis.

In the 1930s, intensive research led to the discovery of a group of drugs called the *sulphonamides*. These proved effective against many kinds of bacteria, including the diplococci that cause pneumonia, a disease that, until then, had often been fatal. Sulphonamides are still in wide use, particularly for the treatment of meningitis and infections of the urinary tract.

Antibiotics

In 1929, Alexander Fleming was working in a London hospital, culturing bacteria on agar jelly in Petri dishes (p. 207). Some of his plates became contaminated by colonies of a blue-green mould, called *Penicillium*. Fleming noticed that there was an area of clear jelly around each mould colony, with no bacteria growing. *Penicillium*, it seemed, stopped the growth of bacteria (Fig. 33.1).

(Beecham Research Laboratories)

Fig. 33.2   The fungus, *Penicillium* (× 400)

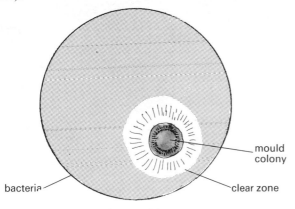

Fig. 33.1   Clear zone around a mould colony growing on a plate of bacteria

He found that a substance could be extracted from the mould fungus which would still check the growth of bacteria. The extract was called *penicillin*. Tests carried out on animals in the laboratory showed that penicillin could destroy pathogens without damaging the tissues of the animal itself. Eventually penicillin was used to treat some desperately ill patients and was shown to cause them no harm.

Chemicals that are extracted from one living organism and which will destroy pathogens in another living organism are called *antibiotics*. Streptomycin was one of the other antibiotics to be discovered soon after penicillin. Now there are many more, including tetracycline, aureomycin and cycloserine. New substances from fungi and other sources are being tested every year. Many of the antibiotics are made more stable by chemical treatment to alter their molecular structure. For example, patients are not treated with raw penicillin extract. The antibiotic is supplied as benzyl penicillin, which will destroy most of the cocci bacteria that cause certain diseases. Benzyl penicillin will not kill *Salmonella* (cause of food poisoning) or *Shigella* (cause of bacillary dysentery) but ampicillin, also made from penicillin, kills both these dangerous pathogens. However, ampicillin is less than half as powerful as benzyl penicillin in killing other pathogens.

New antibiotics are sometimes the result of careful planning and prediction by scientists, though some of them follow intelligent observations, made by chance, as in the case of

Fleming noticing the effect of the mould on the growth of his colonies of bacteria. Most antibiotics are effective against a variety of bacteria and are termed *broad-spectrum antibiotics*. Benzyl penicillin is a good example. The drug isoniazid is used exclusively to destroy the mycobacterium of tuberculosis and is called a *narrow-spectrum* drug.

Care in the use of antibiotics

When industrial methods of producing antibiotics on a large scale were developed, many people assumed that the drugs could be used at the slightest sign of illness. In fact, even 'safe' antibiotics will cause harm to the patient if taken in too large an amount. There are also more subtle effects. Broad-spectrum drugs taken orally (by mouth) may be absorbed through the gut wall into the blood stream to attack the bacteria causing tonsilitis in the throat. The same antibiotic will affect many of the bacteria living normally in the gut. Most of these are commensals, organisms that live harmlessly inside us. Some of them perform a useful task of making riboflavin and vitamins B_{12} (p. 35) and K. If our diet contains an adequate amount of these vitamins then the destruction of the bacteria will not matter. For a person on an inadequate diet, killing these gut bacteria by the antibiotics used to treat an illness may result in a vitamin deficiency disease.

Fig. 33.3   Equipment for producing penicillin on an industrial scale

(Beecham Research Laboratories)

We should note at the same time that many of the bacteria in the gut use up vitamins from our diet and perhaps the destruction of gut bacteria by antibiotics is of less importance than was once thought.

It is important to note that penicillin taken orally kills off *susceptible* strains of bacteria, while leaving *resistant* strains to increase in number. Resistance is possible when the microbe can digest the antibiotic. These resistant strains are capable of transmitting to other bacteria the genetic material (DNA) which gives them resistance. These other bacteria then become resistant to penicillin. The use of diluted antibiotics, or doses that are too small, is most likely to have this effect. In the same way, it is important when a person has been given a certain amount of antibiotic by the doctor, that *all* of the prescribed dose is taken. The person should not stop taking the medicine simply because the *symptoms* have disappeared: enough antibiotic must be taken to kill *all* the pathogens.

Some people are more sensitive to the side-effects of drugs than others, and this is one of the reasons why it is necessary for antibiotics to be made available only through qualified doctors. Very often an alternative drug can be prescribed to which the patient will not be sensitive.

Herbal medicines

Man has used plants or herbs to treat diseases for at least 2 500 years and many of our present-day drugs, although they are made by chemical processes, are based on substances obtained from plant materials. The extract from boiled cinchona bark has been used for the treatment of malaria for hundreds of years. The drug, quinine, was isolated from the bark in 1820. An extract was made from willow bark in a similar way to relieve pain and treat rheumatism. The willow tree is called *Salix* and the drug extracted from it is called salicylic acid, better known as aspirin. Nowadays it is more satisfactory to synthesize salicylic acid in a chemical factory. Nevertheless, 30 per cent of the drugs used today are still obtained either from plants or from plant products.

Fig. 33.4   Drawing of a medicinal plant from a book of herbal remedies printed in 1491. It would be difficult to identify the plant from this drawing

Most drugs, whether synthetic or made from plants, are poisonous if too much is taken. With modern medicines it is possible to work with pure substances and it is possible to measure how much of the drug is in each tablet or dose. It is not so easy to do this with herbal medicines, and there is the additional problem that the user must be sure that the correct plant is being used. For example, the leaves of the plant *Commelina diffusa* are used to reduce swelling and to cure sores on the skin. But this plant has at least eight different local names in different parts of West Africa.

(WHO)

Fig. 33.5   This is an accurate drawing of a fern used in the treatment of worm infestations. It comes from a Chinese medical book

Medical people have realized for a long time that the accurate identification of medicinal plants is important. In 1735 the Swedish naturalist, Linnaeus, worked out a system of naming plants, known as the binomial system. Each plant has a name in two parts. A record of the description of the plant, together with the two parts of its name is kept at Kew, in London. The names in the catalogue or *index* at Kew are accepted and used all over the world. The WHO list contains the botanical names of over 20 000 medicinal plants, with an indication of the countries in which each plant is used and a note of the synonyms or alternative names by which plants are known locally. Most 'herbals' or books about medicinal plants contain illustrations of the plants to help in identification. Fig. 33.4 shows a picture from an old herbal, not very helpful for identification, and Fig. 33.5 shows an accurately made drawing from a modern book.

Drug testing

In some patients the act of *taking* the drug makes them feel better. For example, suppose that 40 patients complain of headache; 20 are given aspirin and 20 are given chalk tablets which look like aspirin tablets but contain no active chemical. It is quite likely that several people in the second group will claim that the tablet has cured their headache. The effect must have been psychological. The inactive, fake medicine is called a *placebo* and the apparent cure is known as the *placebo effect*.

It is thus impossible to know whether a remedy is effective because of the action of particular chemicals, or whether it is simply the act of taking anything thought to be a drug, or even the 'magic' that goes with it, which makes the patient feel better. The only way to test the remedy is by carefully controlled trials with a large number of animals and then people.

One of the other purposes of such trials is to discover whether the active drug has any side-effects. That is to say, while the drug may cure one particular disease it may produce undesirable effects in other parts of the body. Aspirin is useful in reducing the symptoms of headache. If one or two tablets are taken then there is no apparent ill effect. When five to ten tablets are taken to reduce the pain in joints affected by arthritis, bleeding may occur from the lining of the stomach. This is an undesirable side-effect, leading to anaemia. Certain anti-malarial drugs are very effective in killing malarial parasites but produce feelings of nausea in some people.

Disinfectants and sterilization

Micro-organisms on utensils, clothes and other surfaces can be killed by chemical means. The chemicals that are used for this purpose are called disinfectants. They are powerful substances and must be used with care, since they may kill the tissues of our bodies if they come in contact with them. Calcium chlorate(I) (chloride of lime) and sodium chlorate(I) (sold as 'Domestos' and under other proprietary names) produce an acid in water. This acid is unstable (easily decomposed) and it breaks down to release oxygen in a very active state. This newly released oxygen kills bacteria, viruses, fungi and protozoa, but it takes a little time to work. Materials to be sterilized must be left in contact with the disinfectant for 20 minutes or more. Hypochlorite solutions are used for sterilizing babies' feeding bottles, soiled clothes, lavatory seats and lavatory pans (WC pans) and to sterilize equipment used in the large-scale preparation of food.

There are many other disinfectants in use, many of them based on the chemical called phenol, which was used by Lord Lister to reduce infection in the early days of antiseptic surgery. It helped to reduce infection, but there was a risk that the phenol itself, in the wrong concentration, would cause destruction of the tissues as well as the bacteria. Hexachlorophene is a chemical, derived from phenol, that is widely used to control bacteria on the skin, for example on a surgeon's hands and on the patient's skin. But it is much milder than a disinfectant, safer to use, and is one of a range of *antiseptics*.

Antiseptics

If we say that a wound is septic we mean that bacteria are growing there and causing infection. An *antiseptic* is a chemical that prevents the growth and reproduction of bacteria without necessarily killing them. Such chemicals are sometimes called *bacteriostats* because they keep the bacterial numbers static or steady. Antiseptics can be used for washing wounds and as mouthwashes, without damaging the tissues. But if they are used constantly they will destroy the harmless bacteria (the commensals) on the skin (p. 200) which have a useful effect in controlling pathogenic micro-organisms. Hexachlorophene is a very good example of an antiseptic which is extremely useful in

modern medicine but which has to be used with knowledge of its effects. 'Dettol' is another widely used antiseptic which is safe for washing wounds and for use during childbirth, washing around the vulva before the baby is born.

Asepsis and aseptic surgery

Asepsis is the condition in which no live bacteria are present. In the operating theatre in a hospital it is desirable that there should be no bacteria to infect the wound that a doctor makes when he operates on a patient. You have read that the surgeon's instruments are sterilized by heat treatment. Disinfectants are used to sterilize all the equipment in the operating theatre. The surgeon and all his assistants wear sterile clothing and take care to control the bacteria on their own bodies. It is very difficult to destroy *all* the bacteria and other micro-organisms, but the aim is *aseptic* (bacteria-free) surgery.

Practical Work

Experiment 1 Hands and bacteria

Prepare four agar plates (p. 207). Rub your fingers on the floor and then press them firmly on the agar jelly of the first plate and mark it 'Dirty fingers'. Now wash your hands with soap and water. Dry them by waving them in the air and *not* with a towel (why not?) and make a set of finger marks on the second plate, labelling it 'Washed fingers'. Scrub your fingers thoroughly for not less than half a minute and make a third set of marks. Label this 'Scrubbed fingers', and the fourth plate 'Control'. Seal the lids with sticky tape and do not remove the lids when you examine the colonies of bacteria.

Incubate the four plates at 37 °C for not less than two days and then compare and discuss the growth of bacterial colonies on them. Destroy the cultures by sterilizing when you have observed the bacterial colonies.

Experiment 2 The action of antibiotics

Prepare a suspension of soil bacteria by just covering a quantity of fresh soil in a sterile beaker with distilled water. Pour off the first 5 cm³ of water into a sterile test-tube and use a sterile pipette to transfer 1 cm³ of this soil water to each of four agar plates or slopes. Tilt the jelly so that the soil water wets the entire surface and then pour off the surplus water. Incubate the cultures for one day.

Meanwhile collect moulds from four different sources such as the skins of ripe fruit and the rind of cheese. 'Blue' cheese is coloured by the growth of *Penicillium* and is a particularly interesting source of mould. If the cultures show bacterial growth, use a sterilized wire loop to transfer a small quantity of one mould to one plate, another to the second and so on. Incubate at room temperature for two days and observe whether there is any interaction between the mould and the bacteria. Remember that you are likely to have the same kinds of bacteria in each culture. Examine the results without opening the cultures, which should be re-sterilized after the experiment.

Experiment 3 Extraction of an antibiotic

Penicillium mould can be grown well in a liquid medium, made by dissolving 1 g of meat extract and 4 g of dextrose in 200 cm³ of distilled water and sterilizing it in a pressure cooker for 15 minutes. When cool the medium is inoculated with *Penicillium*

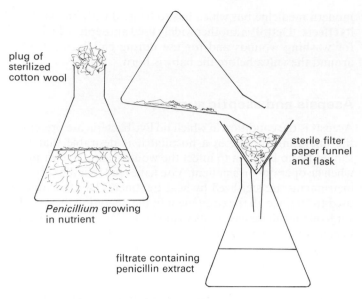

plug of sterilized cotton wool

sterile filter paper funnel and flask

Penicillium growing in nutrient

filtrate containing penicillin extract

Fig. 33.6   Penicillin production in the school laboratory

spores and incubated at room temperature until a strong growth of mould is seen through the liquid, usually after one week. The medium is now filtered into a sterile vessel using sterile filter paper and funnel. Any antibiotic formed will be in the filtrate (Fig. 33.6).

Experiment 4   **Tests with an antibiotic extract**

Prepare a fresh plate of soil bacteria as in Experiment 2. Cut out and remove a strip of agar to form a trough (*see* Fig. 33.7). Use a sterile pipette to transfer about 0.5 cm³ of the antibiotic extract from Experiment 3 to the trough and incubate at room temperature for one day. Observe whether there is any interaction between the bacteria and the extract.

Experiment 5   **Response of different bacteria to an antibiotic**

You will need two or three different species of bacteria for this experiment, The following cultures can be obtained from biological supply firms:

Chromobacterium lividum   *Escherichia coli*
Bacillus subtilis   *Staphylococcus albus*

Prepare a Petri dish with nutrient agar as described on p. 207 and cut out a strip as in Experiment 4. Use a sterile loop to streak a sample from each bacterial culture up to the trough, sterilising the loop between each transfer (Fig. 33.7). Place

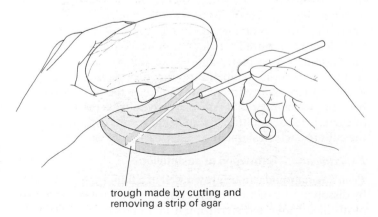

trough made by cutting and removing a strip of agar

Fig. 33.7   Inoculating a prepared plate with bacteria

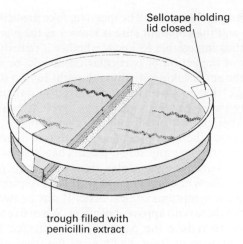

Sellotape holding lid closed

trough filled with penicillin extract

Fig. 33.8   Plate with bacterial streaks interacting with penicillin extract

antibiotic extract in the trough, tape the lid securely on the Petri dish and incubate as in Experiment 4. After one or two days observe (without removing the lid) the extent to which bacterial growth has been prevented by the antibiotic.

Sterilize the plates again before washing them.

N.B. Although the micro-organisms used in the experiments are regarded as non-pathogenic, all bacteria should be treated as if they were potentially harmful.

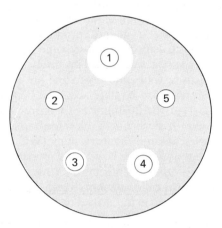

Fig. 33.9   A bacterial culture was spread over the nutrient agar in a Petri dish. Five discs, each impregnated with a different antibiotic were placed on the agar, each numbered disc containing the same quantity of antibiotic

Questions

1   What important principle did Ehrlich discover, arising from the early use of salvarsan in treating syphilis?
2   Fig. 33.9 shows the effect on the growth of bacterial colonies of five different antibiotics. Each of the numbered discs contains the same quantity of antibiotic. Which one of the antibiotics is most effective in destroying the bacteria? How can you tell?
3   How could you use this technique to see whether the antibiotic marked '3' was a broad-spectrum antibiotic?
4   Several of the drugs used in the treatment of disease at the present time contain the same or very similar chemical substances to those found in the plants used in herbal remedies. What do you consider to be the most important advances in the use of modern drug treatment over the use of old herbal medicines?
5   What advantages might the use of chemical sterilizing agents have over the use of boiling water?
6   Why is it unwise to use antibiotics to treat trivial complaints?
7   What is the difference between antiseptic and aseptic?

34
Mental Health

A healthy mind is just as important to a person as a healthy body. But it is just as difficult to define mental or emotional health as it is to explain the positive meaning of physical health.

We recognize that our bodies suffer from minor disorders from time to time, such as a high temperature, or pain or diarrhoea. Recovery is usually quite rapid. In much the same way our nervous systems suffer from the effects of stresses. The results are changes in behaviour. Thus the death of someone whom we love, or a deep disappointment or a sense of frustration may each result in depression, the commonest form of mental or emotional ill-health. If we are in a state of physical and emotional good health then we adjust to the stress and soon recover. On the other hand, a person who is already emotionally unwell does *not* adjust to stress so quickly or easily. Bereavement or frustration tend to produce much more dramatic changes in behaviour and these signs indicate that the person is very unwell.

Mental and emotional balance

A person who is mentally and emotionally 'whole' or well is able to:

1 Form and sustain relationships with other people.
2 Tolerate behaviour patterns in other people that are different from their own.
3 Look after his or her own welfare, including personal hygiene and feeding.
4 Make rational decisions and plan for the future.
5 Adjust to stress resulting from changes in the environment, whether physical, such as changes in climate, or social, such as changes in attitudes and behaviour in other people.

Most of the time we assume that we ourselves are mentally well-balanced and healthy. We may be so used to signs of minor emotional upsets in ourselves that we scarcely notice them, though we may be much more aware of such signs in other people. One of the difficulties lies in deciding whether a change in behaviour is trivial and temporary, leading to quick recovery, or whether it is more serious and long lasting, requiring medical help.

Mental health and physical health are closely connected and are not two distinct, different things. A person who is suffering from a physical illness is less able to deal with the problems of everyday life and may become anxious and irritable. Anxiety in turn often leads to depression which is one of the commoner forms of mental illness. Irritability and bad temper are often connected with stress, but they can also disguise depression. On the other hand, a person may be short tempered and incapable of making sensible decisions because he is very tired. Tiredness is a stress and sleep relieves the stress and allows the mind to recover. But the person who is always short-tempered may be very stressed indeed.

Mental and emotional illness

We become aware that a person is emotionally or mentally sick when we notice changes in his or her behaviour. We say that their behaviour has become abnormal. For example, to be able to cry or weep is a means of relieving emotional stress, even in an adult. Nonetheless, most adults do *not* weep very often. So if an adult is seen to weep frequently, for no obvious reason, this behaviour is *abnormal*. It is an indication that the person is

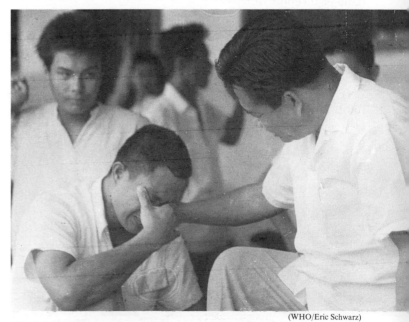

(WHO/Eric Schwarz)

Fig. 34.1   Facing the stresses of life. Some people reach breaking point

261

reacting to stress that may be emotionally or, perhaps, physically painful. Weeping is helpful if it relieves the stress, but weeping that continues is frequently a warning signal that the stress is continuing and may produce more serious effects.

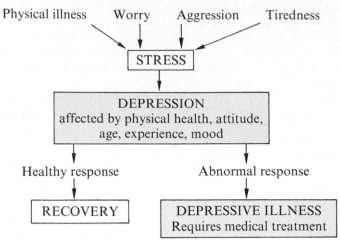

Most mental illness is *not* inherited, though many forms of mental handicap and mental retardation are hereditary. A child may be born suffering from phenylketonuria, a disorder in which the body is unable to metabolize the amino acid called phenylalanine (p. 169). The accumulation of this amino acid in the blood stream of a young child affects the development of the brain, resulting in serious retardation or slowing down of mental development.

Down's syndrome, sometimes termed mongolism because of the angle at which the eyes are set, results from the presence of an extra chromosome (p. 168) in the body cells. This arises in the formation of the egg cell before fertilization. Mental development is seriously affected.

Temperament and personality vary a great deal from one person to another, and some people are more likely than others to react to stress by becoming ill. Mental illness, unlike mental handicap or retardation, is likely to be triggered off by factors in the environment, particularly by the attitudes and actions of

(WHO/H. Henrioud)

Fig. 34.2   Children suffering from Down's syndrome are encouraged to express themselves and develop skills

other people. Such illness can be treated and often cured, so that the person behaves normally again and talks to and works with other people. Because mental illness involves *feelings* as well as behaviour it is often called 'emotional' illness. A young child does not automatically control its feelings. When it is happy it will laugh or sing. If it is frustrated it becomes cross and loses its temper. We may speak of 'childish tantrums', meaning uncontrolled outbursts of temper, as shown by a child. As we grow older we continue to feel pleasure and anger, sadness and joy. But we are expected to show some control over the way in which we express these feelings, though sometimes we may express anger in a violent and uncontrolled way, as a young child would.

Patterns of mental illness

Psychiatrists, the medical doctors who are trained to help people who are mentally ill, frequently find it takes time to diagnose a patient's illness. The causes of mental illness are less easy to identify than, say, the cause of diarrhoea or anaemia. The doctor has to listen to the patient describing his symptoms, which are often an account of his feelings, and he has to study the patient's behaviour. There are no causative organisms to identify, though the psychiatrist may be able to identify events that have affected the feelings and behaviour of the patient.

Psychosis

Psychosis (plural = psychoses) is the term used to describe mental illness in which a person loses touch with reality. It is a serious kind of illness that results when part of the brain is not functioning properly (dysfunction), sometimes as a result of physical or chemical damage.

Schizophrenia is the name given to one of the most serious psychoses. There are many signs and symptoms of schizophrenia. The patient may see or hear things that are not really there. We say that the patient suffers from hallucinations. His way of thinking about the world is affected. Schizophrenics often lose interest in their appearance, and this is one of the signs of their withdrawal from the real world.

Affective psychosis is a very severe form of depression in which the patient feels both helpless and hopeless. It is not the same as the feeling of being low, or fed up, that most of us experience from time to time, but from which we recover quite quickly. Most of us recover from slight depressions while sleeping, and wake up feeling well.

In schizophrenia and affective psychosis there are thought to be changes in the way in which brain cells work. One widely held theory suggests that, when these disorders occur, there is a change in the balance of chemicals produced at the synapses between nerve cells (p. 147). There are difficulties in investigating the working of a live brain without a serious risk of damaging it, and this is one of the reasons why medical scientists are still uncertain about the absolute cause of psychosis.

Neurosis

Neurosis is a less severe form of illness than psychosis and it is psychological in origin. That is to say, it starts in the mind. The patient may be affected by events, such as the actions of other people towards him or her. A *neurotic* person becomes very tense, or excitable, or shows unreasonable fear. He or she may not be able to make decisions but, unlike the schizophrenic, does not lose touch with the real world.

Anxiety is an emotional reaction to the fear of someone or something. There is nothing unusual about it. We all feel anxious from time to time. But anxiety can become a neurosis or an *anxiety state* when the fear gets out of proportion and cannot be controlled. Anxiety may temporarily become so intense that a person vomits, or suffers diarrhoea or loses control of the bladder. Children, and even adults, sometimes wet themselves when they are very frightened. Intense fear of a real situation is not the same as an anxiety neurosis, which is fear of a situation which offers no real threat.

Phobias are unreasonable fears that affect the way people live. For example, many people are afraid of dogs, but this is not irrational and would not prevent most people from going out. A person with a phobia about dogs would take extreme measures to avoid meeting even a small dog, perhaps to the extent of not going out of doors. Some people are afraid of open spaces and will not venture out. They are said to suffer from *agoraphobia*.

Anorexia nervosa is most often seen in teenage girls and young women who develop a reaction against eating, particularly food rich in carbohydrate. It is often associated with a desire to lose weight and, since the patient starves herself, the loss of weight is often dramatic. Menstruation stops (amenorrhoea) and the patient often has no interest in sex.

There is no clear-cut cause of the disorder but it may result from the interaction of several factors. These may include excessive shyness, fear of sexual activity, reaction to a domineering parent and, sometimes, a wish to avoid growing up. The causes are psychological but the effects are most markedly physical.

Treatment includes persuading the patient to eat; this calls for patience on the part of doctor and nurses in charge. An attempt is made to identify factors that may have triggered the condition so that these can either be eliminated or the patient enabled to come to terms with them.

Alcoholism and drug addiction

Alcohol is a powerful drug that affects the working of the brain. Muslims forbid its use and many other societies recognize its dangers. If it is consumed in small quantities it can make people socially more relaxed. But it removes inhibition and a person is then more likely to commit antisocial acts such as stealing or assaulting people. Drinking a lot of alcohol makes a person less aware of the world around him and some people get into the habit of consuming alcohol for this reason. They feel that it helps them to get through the day when they are frustrated or depressed. In fact alcohol makes them *less* able to solve their problems but they do not *feel* the effect of the problems so acutely. When a person has become dependent on alcohol to get through the day we say he is an *alcoholic*. He is an emotionally sick person. For this reason alcoholics are often treated in psychiatric hospitals, that is, in hospitals for the mentally ill.

Some people use other drugs to help them to escape from the unpleasant realities of life. If life seems boring or depressing then it is understandable that a person should want to escape.

But escape does not remove the *cause* of the boredom or depression. The kinds of drugs used are termed *psychotropic* drugs, because they alter a person's state of mind and, in particular, their perception of the world and people around them. Thus the drug LSD may affect the brain cells in such a way that a person sees a common flower, such as a daisy, as very beautiful and radiating light. The same drug may affect that person's judgement in hazardous situations so that he or she ignores traffic dangers, or steps out of a window high up in a building with no fear.

Dependence on drugs can be *psychological* or *physiological*. Psychological dependence means that the person has become used to taking the drug and feels that he or she really *needs* it and must have it. People may become psychologically dependent upon alcohol, or on tobacco or coffee, as well as other drugs. When a drug is used over a long period of time, the body may develop tolerance. This is a physiological reaction and means that the body can put up with increasing quantities of the drug without being too seriously affected. It also means that the person needs to take increasing amounts of the drug in order to produce the desired effect.

Physiological dependence is often termed *addiction*. It means that the use of a drug has brought about changes in the way one or more organs in the body work. The drug may have been used originally in order to feel 'high' but there comes a time when the person feels very ill if he does *not* take the drug. The drug no longer makes the person feel high, but is needed to prevent him or her feeling bad. Without the drug the person may suffer stomach cramp, as when the drug heroin is withdrawn. The stomach cramp is called a *withdrawal symptom*. An alcoholic may suffer from *delirium tremens* (frightening hallucinations) when he is *not* able to take alcohol.

Prevention of emotional illness

If we all lived in secure surroundings with pleasant, considerate people then there might be very little emotional illness. In fact we have to learn to live with other people as they are—sometimes kind and thoughtful, but often selfish, jealous and sometimes malicious and cruel.

Your family provides your first experience of other people. As a small child you were entirely dependent on your mother for food and protection (Fig. 34.3). As you grew and learned to

Fig. 34.3   A baby is entirely dependent on its mother in whatever part of the world it is born

(Barnaby's Picture Library)

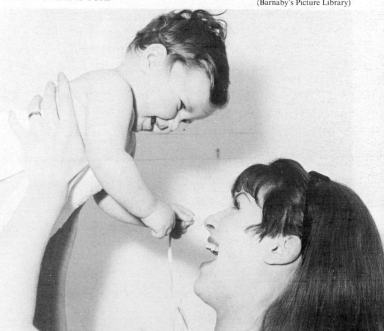

walk, your contacts with other people increased and you became less dependent on your mother. But most of us continue to depend on our families to some extent for support, encouragement and help, from parents, brothers and sisters, uncles and aunts. This support helps us to build up mental and emotional stability.

There is often rivalry between brothers and sisters, starting frequently when a new baby is born. A young child sees that the new baby is taking a lot more of its mother's time and attention than the child now receives. Jealousies may develop. At this stage a mother can help the older child to overcome jealousy.

In large families the older children may help to look after the younger children and do jobs around the home. They learn to take responsibility. Children in a family usually 'stick together' as a defence against older people. But children imitate grown-ups and in this way they learn the skill of living with other people and adapting to them. They learn how to develop their emotions and how to use them, as well as learning how to exercise some degree of control over the way in which they express them. The example set by parents at this learning stage is very important and has a great influence on the development of the child's personality. A stable home environment, with a background of care and affection, is a good basis for preventing emotional disorders.

If parents provide affection for their children as well as giving protection then there is a good chance that those children will grow up to be stable, well-balanced adults. Children who are deprived of affection are more likely to be antisocial in their attitudes and behaviour as they grow older. It is as if, in some respects, they do not grow up.

Fear leads some people to be aggressive. They attack other people either physically or with words to show that they are strong when really they are trying to convince themselves that they are not afraid. Often they do not recognize the cause of their fear, so they cannot overcome it. Providing affection and giving people a sense of being valued are two of the most important ways of preventing emotional illness.

Treatment of mental illness

If a person is feeling slightly depressed then he or she may be able to do some activity, usually physical, that distracts them from thinking about the depression and becoming preoccupied with it. Often the feeling of depression then disappears. But when a person is suffering from a psychosis or a severe neurosis, treatment by another person, usually a medically trained person, is necessary.

Psychotherapy

Psychotherapy means 'treatment of the mind'. There are many approaches to psychotherapy, some of which are scientifically based while others are not easily assessed by scientific tests. The attitude of mind of the patient towards himself or herself and towards the person who is trying to help is of great importance. In European society there is some unwillingness to believe that anything can cure illness other than medicinal drugs. So there is a reluctance to accept treatment by psychotherapy. Indeed many sick people, when told that drugs are not needed for their cure, refuse to believe that they are really ill.

Psychotherapy is a slow treatment, involving patience and care, and it is the principal treatment for neuroses. It involves helping the patient to understand and come to terms with his or her problems. When a person is suffering from a psychosis, then treatment is more likely to involve the use of drugs.

Drug therapy

Drug therapy depends on the fact that many chemicals affect the way in which the brain works.

Tranquillizers are used to slow down the brain activity in very disturbed patients so that psychotherapy can be started. *Chlorpromazine*, for example, is a tranquillizer used in the treatment of psychoses. In recent years the carefully monitored

Fig. 34.4   A group therapy session for homeless, alcoholic people in London encourages them to talk and recognise their problems

(DAVID HOFFMAN)

use of this drug has enabled many patients to return to their homes, sometimes after many years in mental hospitals. They are not necessarily cured but the psychoses are controlled.

Antidepressants raise the feelings and change the mood of very depressed patients and make life more acceptable. All drug therapy must be carried out only under the supervision of a trained doctor since, if too much of the drug is given, side-effects may result.

Attitudes to health and illness

Many people are embarrassed by mental illness, and in many societies the mentally ill are shut away in institutions where most of us forget about them. Success in treating both mentally ill and physically handicapped people is influenced very strongly by the attitudes of the rest of society. Mental and physical health are inseparable from each other and from the social, economic and political situations in which people live.

We do not regard a person with dysentery as an oddity, though he is obviously affected by the disease. But a person whose body is partly paralysed and wasted away as a result of poliomyelitis and who cannot move about easily may be regarded as 'different'. We may be more ready to tolerate an alcoholic than a schizophrenic who cries out and suffers from hallucinations, or who is so withdrawn that he will not make contact with other people. Our attitudes towards the people concerned make a big difference to their feelings of being accepted or not. If they do *not* feel accepted then there is much less likelihood of recovery. They may not even want to recover.

When we *know* people, we are usually more tolerant of differences between them and us. In a village there is often a greater chance of a mentally ill person being accepted by the community than in a large town where many people will not know him and may be embarrassed by his behaviour.

This question of recognition, of being known and therefore feeling that we belong to a society is important for every one of us. In order to belong, we must communicate and much of this is done by talking. The ability to talk, to express feelings and ideas is one of the features that distinguishes humans from other animals. But we also communicate by gestures, with our eyes, our faces and our hands, indeed often by using our whole bodies. By communicating friendly and sympathetic feelings to those who are ill, whether physically or mentally or emotionally ill, we help them to feel accepted. This will affect their chance of recovery and, if recovery is not possible, it will help them to cope with their disability.

The patient's own attitudes both to himself and his illness and to the person treating him is also of great importance. If the patient does not believe that he can be cured, or does not have faith in the ability and skill of the person treating him, then cure is less certain and sometimes impossible. Consequently doctors and nurses must cultivate the ability to reassure their patients and build up their confidence.

Very young children are usually quite tolerant of disability in other people, including the disabilities of old age, when people become slower in their movements, sometimes confused in their minds and often conservative in their attitudes. As we ourselves grow older there is a tendency to become less tolerant and this is not helped by the fact that sick people are often shut away from society. We get used to the view that society is made up of people who at least *appear* to be normal. Becoming aware of the fact that society is made up of very varied people, some well and some ill, is an important part of our development if our society as a whole is to be strong and healthy.

(WHO/T. Farkas)

Fig. 34.5 Physical and emotional contact continue to be important as we grow old

Questions

1 In what circumstances would you regard an outburst of anger as (a) normal and (b) abnormal?

2 Which of the following would you regard as an example of (a) normal and (b) abnormal behaviour? State your reasons.
 (i) A person who shows no anxiety about examinations.
 (ii) A person who feels tense and anxious before examinations.
 (iii) A person who feels sick and is absent from school every time there is an examination.

3 What would you regard as (a) normal and (b) abnormal behaviour towards snakes, or spiders?

4 Why is alcoholism often seen as a sign of mental or emotional illness?

5 Discuss the differences between psychological and physiological dependence on drugs.

6 What factors in our upbringing are likely to reduce the chances of mental illness?

35
Socially Significant Diseases

Sexually transmitted diseases

The term sexually transmitted diseases, or STDs, covers the many infections whose causative organisms are spread by contact during sexual intercourse. Three of these diseases, gonorrhoea, non-specific urethritis and syphilis, are termed *venereal diseases* (from Venus, goddess of love and love-making). These are spread only through sexual contact which may be between male and female (heterosexual) or between members of the same sex (homosexual). The common fantasy or 'old wives' tale' that venereal diseases can be caught from towels, lavatory seats and cups is simply not true. Other STDs such as candidiasis and cystitis may be spread by contact between fingers and genital organs, or by soiled towels or clothing, although by far the commonest means of catching any STD is through sexual intercourse.

The chance of becoming infected with a sexually transmitted disease is, therefore, related to sexual behaviour. A couple who make love together and with no one else are most unlikely to contract the three venereal diseases. It would be possible, though not very likely, for them to contract candidiasis from a contaminated towel. Cystitis, an inflammation of the bladder, could also be caught by non-sexual transfer.

The risk of infection with STDs increases when a person makes love with a number of sexual partners, and increases more if these partners also sleep around. A person who has sex with a number of partners, with no regular relationship, is said to be promiscuous.

Venereal diseases

Gonorrhoea has been known and described for many hundreds of years. It is caused by a small bacterium, *Neisseria gonorrhoeae*, which is easily killed by drying, so that it rarely survives outside the human body. Hence transmission from person to person is almost exclusively through sexual intercourse. In males there is an incubation period of 3 to 5 days, followed by discomfort in the urethra. The urethra becomes inflamed and a greeny-yellow discharge appears at the end of the penis. The acidity of the urine produces a burning sensation in the infected urethra during urination. If the infection spreads to the prostate gland (p. 101) then the patient finds it more difficult to pass water. The bladder may be infected, causing cystitis, and infection of the sperm duct or the epididymis may result in sterility.

Less than half the females infected with *Neisseria* show any signs or symptoms, so that they may not know they have the disease and are infecting their sexual partner. Signs of the disease which may appear include inflammation of the urethra and a burning sensation when passing urine. The bacteria in the vagina reproduce readily and can be transferred to the penis during intercourse. In both sexes there is the risk of the infection spreading more deeply into the reproductive system and causing sterility.

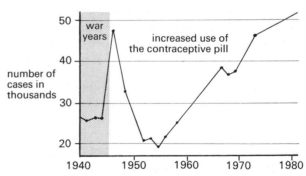

Fig. 35.1 Changes in the number of people seeking treatment for gonorrhoea in Britain

Treatment and control. In 1981 about 55 000 cases of gonorrhoea were reported in Britain. It is important that any persons showing signs and feeling symptoms of the disease, or who know that they have exposed themselves to risk of infection, should seek medical treatment quickly. Delay allows the organisms to spread further into the reproductive system, causing more damage. It also means that the person remains infectious, with the possibility of transmitting the disease to other people. Most strains of *Neisseria gonorrhoeae* are destroyed by one or more of the various forms of penicillin but there is concern that resistant strains are finding their way into the population at risk. At present even these resistant strains are susceptible to other antibiotics, but these alternative drugs tend to be very costly and, if they are used indiscriminately, resistance to these drugs, too, could develop. In all cases it is desirable to check up after treatment to ensure that it has been successful.

Unfortunately no immunity to the disease seems to develop, and while a patient who has been treated is no longer infectious there is nothing to stop him or her from catching the disease again from an infected partner.

Non-specific urethritis, often abbreviated to NSU, means inflammation of the urethra without any readily identifiable cause. There is still speculation as to whether a virus or a mycoplasm (a bacterium with no cell wall) might be the causative agent or, indeed, whether a number of different organisms might produce similar signs and symptoms. Generally these are similar to the signs and symptoms of gonorrhoea, appearing in males after an incubation period of 3 to 5 days. An urethral discharge, painful urination and a frequent need to empty the bladder may be followed by damage to the prostate gland and the epididymis. Females rarely show signs, though they certainly transmit the infection.
Treatment is less certain than for gonorrhoea, though tetracycline, given over a number of days, is often effective.

Syphilis is caused by a spiral, motile bacterium called *Treponema pallidum* which, in males, produces a painless ulcer or *chancre* on the penis after an incubation period of about 9 days. Though the chancre looks red and unpleasant it causes no discomfort and after a few weeks it clears up. In females a similar chancre develops, sometimes on the labia (p. 101) but often out of sight in the vagina or on the cervix. In both sexes it is possible for the chancre to develop on the lips or on another part of the skin.

A secondary rash develops over the body after a delay of from 6 weeks to 6 months, but this, too, disappears in time. In the long term, if untreated, the *Treponema* may spread to any part of the body, producing tertiary signs and symptoms in the heart, liver, bones or brain. Today people infected with the disease rarely go untreated for long enough to develop tertiary signs and symptoms.
Treatment. Like *Neisseria*, *Treponema* is killed by penicillin. Early treatment is most effective since this prevents the development of the irreversible tertiary symptoms while at the same time making the patient non-infectious.

AIDS stands for *acquired immune deficiency syndrome*, a disease that hit the headlines in the 1980s. It is caused by a virus that attacks the body's immune system, making a person much more susceptible to common infections and to certain cancers. These include a skin cancer, known as Kaposi's sarcoma.

At one time it seemed that the disease was most common among homosexual males. In anal intercourse the epithelium lining the rectum is frequently broken, making a pathway by which the virus can both enter and leave the blood stream. The virus is also found in semen and in vaginal fluid and can be transmitted during sexual intercourse between males and females. It can also pass through the placenta and infect an unborn foetus.

Since the virus is carried in the blood, there is a risk of its being spread from an infected donor during a blood transfusion, though techniques of screening donated blood now make this unlikely. However, when drug addicts share needles there is a real risk of this and other viruses being transferred from an infected person to a healthy user.

After infection a person is said to be HIV positive, and a sensitive test can confirm this. However, it may take months or years before that person shows symptoms of AIDS. During this time they are infectious, though they may be unaware of the fact. In Third World countries and particularly among prostitutes, who may not attempt to find out whether they are HIV positive, and therefore carriers, the risk of spread is great and WHO statistics show an accelerating increase in the number of cases of AIDS.

Other sexually transmitted diseases are listed in the table in Fig. 35.2. They are not to be dismissed as trivial, though they rarely have the far-reaching consequences of gonorrhoea, syphilis, NSU and AIDS.

Name	Causative agent	Signs and symptoms	Frequency
Gonorrhoea ('clap')	Bacterium *Neisseria gonorrhoeae*	Discharge from urethra, pain during urination, symptoms more common in males than females	Widespread
Non specific urethritis (NSU)	Uncertain	Discharge from urethra, pain during urination, symptoms more common in males than females	Widespread
Syphilis ('pox')	Bacterium *Treponema pallidum*	Sore or chancre on genital organs. Can spread through whole body	Not very common
AIDS (Acquired immune deficiency syndrome)	Virus	Loss of immunity to diseases, including cancers	Increasing in male and female population
Candidiasis (vaginal thrush)	Yeast-like fungus *Candida*	Itching, soreness, discharge from vagina in females; signs rare in males	females; males carry the fungus
Cystitis	Bacterium *Escherichia coli*	Inflammation of bladder, frequent and painful urination	Common in females; not often diagnosed in males
Trichomoniasis	Protozoan *Trichomonas*	Itching and discharge from vagina	Common in females; rarely diagnosed in males
Genital herpes	Virus	Weeping sore in genital region, like cold-sore on mouth	Becoming more widespread
Genital warts	Virus	Warts on penis or labia, rough and similar to warts on the hand	Fairly common

Fig. 35.2 Some of the more frequent sexually transmitted diseases

Social significance of sexually transmitted diseases

The social significance of all these diseases lies in their effect on the quality of life of individuals and is related to the pattern of spread in society. With widespread use of contraceptives, particularly the contraceptive pill, the fear of unwanted pregnancy which in the past deterred many people from sexual activity has been reduced. Diseases which half a century ago mostly affected men and female prostitutes are now affecting men and women in roughly equal numbers. STDs affect people in all sections of society, though in Britain the most noticeable increase in numbers of patients is among teenagers.

Sexual intercourse is pleasurable and the sex urge is strong in most people. Yet many countries around the world impose strict rules, condemning and punishing people who have sexual intercourse without being married to each other. In most western countries the once powerful social constraints on sex outside marriage, particularly from parents, are now weak or non-existent. Television programmes, magazines and newspapers confront us with examples of people making love without a background of a constructive relationship. It seems sometimes as if random or promiscuous love-making were approved.

While the act of sexual intercourse is physical, love-making is a skill that involves the building up of a relationship between two people, each feeling and showing concern for the pleasure and well-being of the other.

Control of sexually transmitted disease

Even in countries where strong sanctions are imposed on people who have sexual intercourse outside marriage, people still break the law. Attempts to control sexual activity by legislation have been made in the western world on many occasions, with no more than partial success. Increasing awareness of the signs and symptoms of the various diseases, coupled with a much greater encouragement for people to seek early treatment seems the most promising line of control at present. Clinics for the treatment of STDs are found in many large hospitals. Even more important, the sensitivity and embarrassment of patients is respected and in many clinics the patients remain anonymous, being identified only by a code number. Patients are encouraged to persuade their sexual contacts to seek treatment. This is important since, as Fig. 35.2 shows, both males and females may carry the organisms without realizing that they are infected, since they show no signs and feel no symptoms. There is then an understandable reluctance to seek treatment. Many clinics employ carefully trained health workers to trace contacts and persuade them to visit a clinic. This requires tact and skill, but in some parts of London, for example, it is believed to be one of the more important factors in reducing the level of infection with STDs. With total co-operation between patients and contacts, health authorities and those whose job it is to provide information, whether in the media or in schools or elsewhere, there is no doubt that these diseases could be controlled, perhaps even eliminated.

Meanwhile, considerable protection against the spread of most STDs, including AIDS, is provided by the regular and correct use of a condom (p. 109) whch prevents body fluids being exchanged.

Diseases associated with tobacco smoking

Smoking tobacco is a social custom that has a considerable effect on a person's health. The leaves of the tobacco plant contain nicotine, which is both a stimulant drug and a powerful poison. A small quantity of nicotine in the blood stream speeds up the activity of the nervous system, but a very little more of the drug produces a poisoning effect. About 0.07 g of nicotine injected into the blood stream would kill a grown man. But people do not inject nicotine into their blood streams and the amount of nicotine in the smoke from even 20 cigarettes is far less than 0.07 g. Nevertheless it remains a poison.

Tobacco smoke is composed of gases, vapours and very tiny solid particles that make the smoke visible. The gases include carbon monoxide and hydrogen cyanide, both of which are poisonous, while the vapours include nicotine and tar. The amount and proportion of these materials present in smoke depends on the way in which the tobacco is burned. A cigarette burns at a high temperature and the smoke from cigarettes contains the highest proportion of nicotine and tar. Tobacco smoked in a pipe or as a cigar burns at a lower temperature and the smoke contains less tar.

The *way* in which a person smokes determines how much of each chemical gets into the lungs and, from there, into the blood stream. Most cigarette smokers inhale the smoke, that is, they draw the smoke into their lungs. The alveoli in the lungs provide a large surface area for the exchange of oxygen and carbon dioxide (p. 86). This surface will absorb nicotine and carbon monoxide as well. The solid particles in the smoke settle on the surface of the alveoli and on the lining of the bronchial tubes where they act as an irritant, making the surfaces more easily attacked by pathogenic organisms. The tar in the smoke settles on the lining of the tubes where it often provokes the cells to react by dividing in an uncontrolled way, producing a cancer.

The lungs have a natural cleaning mechanism, provided by the cilia that line the bronchial tubes (p. 85). Tobacco smoke is quite acidic and one of the effects of this acid, particularly of the hydrogen cyanide, is to inhibit the action of the cilia, causing them to stop beating. With the cleaning mechanism switched off, there is no means of removing the harmful tar and smoke particles from the lung surfaces. The effect of inhaling the smoke from *one* cigarette is to stop the action of the cilia for 20 minutes. If a person is a heavy smoker who smokes with only a short interval between one cigarette and the next, the cleaning mechanism may be out of action for a long time.

Long-term effects of smoking

Lung cancer. In the early years of this century, cigarette smoking was an uncommon habit and lung cancer was a rare disease. Now cigarette smoking is widespread throughout most countries of the developed world, including Britain, where over half the deaths from cancer among men are due to lung cancer.

Some people who smoke heavily do not develop lung cancer, but the risk of dying from the disease is very much greater among cigarette smokers than among non-smokers, as Fig. 35.3 shows.

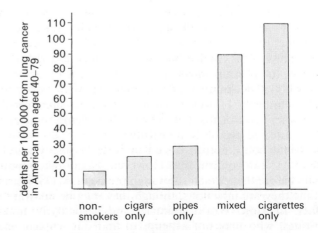

Fig. 35.3   Smoking and lung cancer

You will see that people who smoke cigars or pipes are less at risk than people who smoke cigarettes. This is probably because cigar and pipe smokers do not inhale the smoke as a rule, though they are likely to breathe in some of the smoke from the air around them. Nevertheless, they still have a higher death-rate from lung cancer than non-smokers.

The table below shows how the risk of developing lung cancer is related to the number of cigarettes a person smokes each day.

Cigarettes and lung cancer

Number of cigarettes smoked each day	Increased risk of developing lung cancer
1–14	× 8
15–24	× 13
over 25	× 25

It is now clear that passive smoking, inhaling the smoke from other people's cigarettes, puts non-smokers at risk. No Smoking areas at work, in public transport and in recreational places give protection.

Heart disease

Coronary heart disease is one of the major causes of death in the developed countries of the world. Lack of exercise, excessive eating and wrong diet, together with high blood pressure, are all contributory factors leading to heart attacks, but a significantly high proportion of people who die from coronary heart disease are heavy smokers.

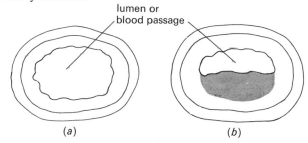

lumen or blood passage

(a)   (b)

Fig. 35.4   Section through (a) a healthy artery, with an open lumen allowing easy passage of blood, and (b) an artery in which fatty material and fibre have been deposited in the lining, reducing the blood flow and causing high blood pressure

Bronchitis

Bronchitis is an inflammation of the bronchial tubes, particularly the finer tubes leading into the alveoli. This results in the production of considerable quantities of thick mucus or phlegm which the patient coughs up frequently, particularly in the morning after getting up.

There are several causes of bronchitis, one of the commoner causes being infection by air-borne bacteria. Cigarette smoke does not in itself cause the disease, but it aggravates the condition and makes it often become chronic (long lasting). Chemicals in the cigarette smoke inactivate the cilia that form the natural cleaning mechanism of the lungs (p. 85) and the phlegm can then be removed only by coughing. Failure to remove the phlegm reduces the ability of the lungs to carry out gaseous exchange and death from bronchitis is by no means uncommon. The disease is sometimes referred to, euphemistically, as 'smokers' cough'.

Emphysema

When the build-up of phlegm makes it difficult to breathe out air from the alveoli and fine bronchioles, these become stretched to an abnormal degree. Coughing weakens the walls of the alveoli, causing them to break down, and holes appear where the lung should have a fine, spongy texture. Also, abnormal cell metabolism causes disruption of the cell walls. These factors greatly reduce the amount of alveolar surface through which gaseous exchange can take place and the efficiency of the breathing system is impaired (Fig. 35.5). Eventually so much lung tissue is destroyed that the patient dies.

Emphysema and bronchitis are very commonly associated and there is a far higher incidence of both conditions among smokers than non-smokers.

(Philip Harris Biological)

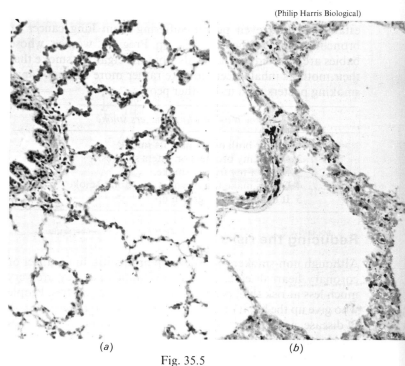

(a)   (b)

Fig. 35.5

(a) Healthy lung tissue showing part of a bronchiole and about 20 alveoli (× 200)
(b) Lung tissue from a person with emphysema. This is the same magnification as (a) but the alveoli have broken down leaving only five air sacs in the same space. This provides a much smaller surface for gaseous exchange

Smoking and pregnancy

One of the measurable effects of smoking during pregnancy is on the birth-weight of the baby. Taking the minimum desirable weight for a baby as 2 500 g, one large-scale survey in Britain showed that 5.4 per cent of non-smoking mothers gave birth to babies of less than this weight, but 9.2 per cent of smoking mothers produced babies of under 2 500 g. Women who do not smoke, or who give up smoking during pregnancy, tend to produce heavier babies.

Of at least as much importance as the reduction of birth-weight is the effect of the chemicals in the smoke on other aspects of the baby's development. Remember that cigarette smoke contains carbon monoxide which is absorbed into the mother's blood. This poisonous gas forms carboxy-haemoglobin more readily in the baby's blood than the mother's and this reduces the capacity of that blood to carry oxygen. Oxygen is needed not only for the growth of the baby

(Metropolitan Police)

Fig. 35.10   The level of alcohol in the blood can be measured using a breathalyzer

drinking driver. The breathalyzer provides a quick way of measuring blood alcohol since a proportion of this is excreted in the breath through the lungs.

Use and misuse of alcohol

The majority of people who consume alcoholic drinks are termed social drinkers because they drink in company at a pub, a party or at a meal. Their behaviour is relaxed and does not overstep accepted social conventions.

Excessive drinkers have developed the habit of consuming alcohol to the point at which they are psychologically dependent and they are irritated if deprived of drink. The excessive drinker will often carry a hip flask or have a bottle handy in case there is not as much to drink at a party as he would like. He often drinks alone and may feel ashamed to be seen to be dependent. Such drinking habits often start because a high level of alcohol produces oblivion—a form of escape from unpleasant reality.

Alcoholics are unable to control their drinking. They may consume no more than the excessive drinker but the effect of alcohol and of the deprivation of alcohol on mind and body is most marked. Amnesia or loss of memory is common and many alcoholics lose their self respect, ceasing to care about their appearance and other people's opinion. The withdrawal of alcohol leads to the 'shakes', when patients—for they are sick persons—need alcohol to steady their nerves and reduce their fears and anxieties.

Non-drinker → Social drinker → Excessive drinker

↓

ALCOHOLIC

There is nothing inevitable about the transition from non-drinker to alcoholic. Most social drinkers are moderate in their drinking habits and do not feel the need or urge to drink to excess. If they have problems in their lives then they solve them instead of trying to escape from them by drinking.

Prevention of misuse of alcohol

Clearly if people never started to drink alcohol then there would be neither excessive drinkers nor alcoholics. But since in the western world social drinking is acceptable then a different strategy must be found for preventing excessive drinking.

In Britain a person may not legally buy or drink alcohol in a pub under the age of 18. It is clear that the law is not strictly

adhered to, and most under-age drinkers never become alcoholics, though for a number of reasons they may drink enough to make their behaviour generally unacceptable. This may be part of the teenage rebellion against the rules laid down by authority. Arguments are put forward for reducing the age at which it is legal to drink in pubs. Would this increase the number of people with drink problems or would it make under-age drinking less daring and less glamorous?

Happy, friendly people, satisfied with life, are rarely excessive drinkers or alcoholics. How can we ensure that we are part of this section of society and do not inflict damage on our bodies and minds by excessive drinking? Clearly we are faced with some important decisions.

Legislation against the sale of alcohol works only when a large majority of people support the law. Prohibition in the United States in the 1920s had little effect on alcoholism and generated an era of racketeering in the illegal sale of alcohol. Enforcement in Britain of the laws affecting the sale of alcohol is neither easy nor very widespread.

However, the laws forbidding people to drive with more than the legal limit of alcohol in the blood are gaining widespread support, probably because the punishments are severe and also because detection using a modern breathalyzer is quick and legally acceptable.

Drug dependence and drug misuse

Many of the drugs associated with the terms drug addiction and drug dependence have recognized medical uses. Morphine, for example, is one of the most effective pain killers known and it gives relief to many people suffering the pain of terminal cancer. Other drugs and chemicals, including cannabis or 'pot', and the solvents used by glue sniffers have no medical use.

Some commonly misused drugs

Type of drug	Example	Effect
Narcotic	Morphine, heroin	Pain killer, euphoria
Stimulant	Amphetamines, e.g: benzedrine	Prevents sleep
	Caffeine (in coffee)	
	Nicotine	See p. 268
Sedative	Barbiturates e.g.: phenobarbitone, Valium, Librium	Reduces tension and anxiety, and promotes sleep
	Ethanol (alcohol)	See p. 271
Hallucinogens	LSD (lysergic acid diethylamide)	Changes perception
	Cannabis ('pot')	Changes perception

We are not concerned here with the medical uses but with the fact that many drugs, including alcohol, affect the working of the brain and alter people's perception of the world. Some drugs produce a sense of euphoria or well-being, feeling 'good' or 'high'. Others dull the senses and make a person less aware of the world and the unpleasantnesses of life. The mood-changing drugs do not create new feelings; they modify or alter existing feelings. Valium, for example, may make the user feel less worried, but it does not create a feeling of positive happiness. Valium, like other depressants, including alcohol, will, if taken to excess, so depress the part of the brain controlling breathing as to cause coma and death.

272

Heroin

Morphine, extracted from the juices of the opium poppy, can be modified chemically to produce heroin. Unlike morphine, heroin is not used medically but is taken by people seeking escape from the unacceptable realities of life or wanting a 'high'. It is usually injected in solution, first of all under the skin (hypodermic). Since tolerance to the drug develops over a few days, that is, the drug produces less effect, most users then inject into a vein (main-lining) to speed up the effect. Heroin users often have no access to sterile needles and may share a syringe and needle. This creates the possibility of transferring infections from one person to another (*see* AIDS, p. 267). The 'high' is a strong incentive to go on using the drug.

Withdrawal. Heroin is one of the drugs associated with powerful withdrawal symptoms so that, if the user does not get a 'fix' or injection, he may suffer violent vomiting, diarrhoea, shivering and feel very ill indeed. He needs to take the drug simply to prevent the unpleasant reactions and the psychological pain. He no longer gets a 'high'. We say he is addicted or 'hooked'.

Obtaining the drug. Most heroin is obtained illegally. It is smuggled from country to country and sold at considerable profit, most of the money going to the organizers of the illegal trade. Finding the money to pay for the drug is just one of the problems facing the user who, as he becomes more dependent on heroin, is less able to sustain a job and obtain money legally.

Treatment. Some users decide to get off the drug and most of these need medical help. Some units for the treatment of addiction use methadone as a substitute for heroin in the early stages of treatment since it relieves the withdrawal symptoms without itself being addictive. The person coming off heroin also needs psychotherapy to help build up self-confidence and enable him to face the world without using drugs.

Glue sniffing

Synthetic glues contain volatile solvents which evaporate when the glue is exposed to the air. If these fumes are inhaled they enter the blood stream and are carried in solution to the brain. There they produce changes in mood or feeling, some of them producing a 'high'. They also depress brain activity, particularly in the breathing control centre, causing shortage of oxygen which leads to death.

Solvents of this kind are widely used in the chemical industry and in the manufacture of glues and resins. The Health and Safety at Work regulations, governing the conditions in which people work in such industries, are very strict with regard to safeguards against the poisonous effects of solvent fumes. The deliberate glue sniffer is not protected by such safeguards and many young sniffers are probably unaware that they are risking death.

Control of drug misuse

Many people use drugs to get relief from anxiety as well as to get 'high' and others find that drugs alter their awareness or perception of the natural world. Still others take drugs to gain mystical experience and many take them to keep up with their peers, that is, to keep in with the group.

How can people be persuaded to stop using drugs or, better still, never to start using them in the first place? Laws, unless very ruthlessly and vigorously enforced, are not by themselves very effective. Besides, happy and contented, active people do not, as a rule, become misusers of drugs. Alcohol (*see* p. 271) is socially acceptable in moderation and tobacco smoking, though no longer so widely accepted in society, is not illegal. Yet even these accepted drugs can lead to ill-health and death. Why are attitudes different towards heroin, cannabis and glue sniffing? While cannabis may not kill people, it increases blood pressure, and misuse of both heroin and glue solvents is frequently fatal. But the difference in attitude towards these three and to alcohol and tobacco is cultural and not rational. In Moslem countries the attitude to alcohol is to forbid its use, though some Moslem states accept the use of cannabis.

(DAVID HOFFMAN)

Fig. 35.11   Why do people get hooked on drugs?

Dependence on any drug means denying one's ability to cope with life. This is a hard statement, but true. There are many reasons why daily life may seem dull, tedious or unbearable but the reasons or causes are not changed by drugs. The alternatives come from what we and the people with whom we associate actually do about life and living conditions. Even when work (or the lack of it) seems depressing we have the capacity to develop friendships, hobbies, take up sport and do things which direct our thoughts outwards from ourselves. By maintaining our respect for ourselves as persons we reduce the risk of becoming dependent on drugs.

Questions

1 Why are many diseases described as sexually transmitted diseases but only a few are termed venereal diseases? Is the distinction a sensible one?
2 What are some of the reasons for the increase in the number of cases of sexually transmitted diseases in recent years?
3 How can the spread of sexually transmitted diseases be prevented?
4 What is meant by *dependence on tobacco*?
5 Suggest ways in which advertising may work against efforts to develop good health in society.
6 In a public place, does the cigarette smoker have rights as well as the non-smoker? Think out your reasons.
7 How does one's family influence smoking habits?
8 In what ways do you think the anti-smoking posters shown in Fig. 35.7 are trying to persuade people to stop smoking or not to start?

9 Why do you suppose the number of non-smoking compartments on trains and in aircraft is increasing?
10 Why is alcohol described as a *depressant drug*? Do you think it should be described in this way?
11 Find advertisements for alcoholic drinks in magazines. What is the aim of each advertisement? How does it achieve its aim? Whom is it trying to reach?
12 Do you support the use of the breathalyzer? Write down your reasons for or against its use.
13 From your own experience, do you think there is much drinking among teenagers? What age group drink most? Why do you think this is so?
14 Discuss the differences between psychological and physiological dependence on drugs.
15 Many intelligent people smoke, drink to excess and misuse drugs, knowing the dangers to their health. Why do you think this is so?
16 How would you try to dissuade a friend from becoming hooked on drugs?

36
Safe Drinking Water

Sources of water

The water cycle

We all need to drink water to keep healthy. A grown person in a hot country may drink 5.5 litres of water a day, and this water must be free of pathogenic organisms. All the water that we drink comes from the clouds, usually as rain. Fresh rainwater is usually very pure and safe to drink, because the clouds from

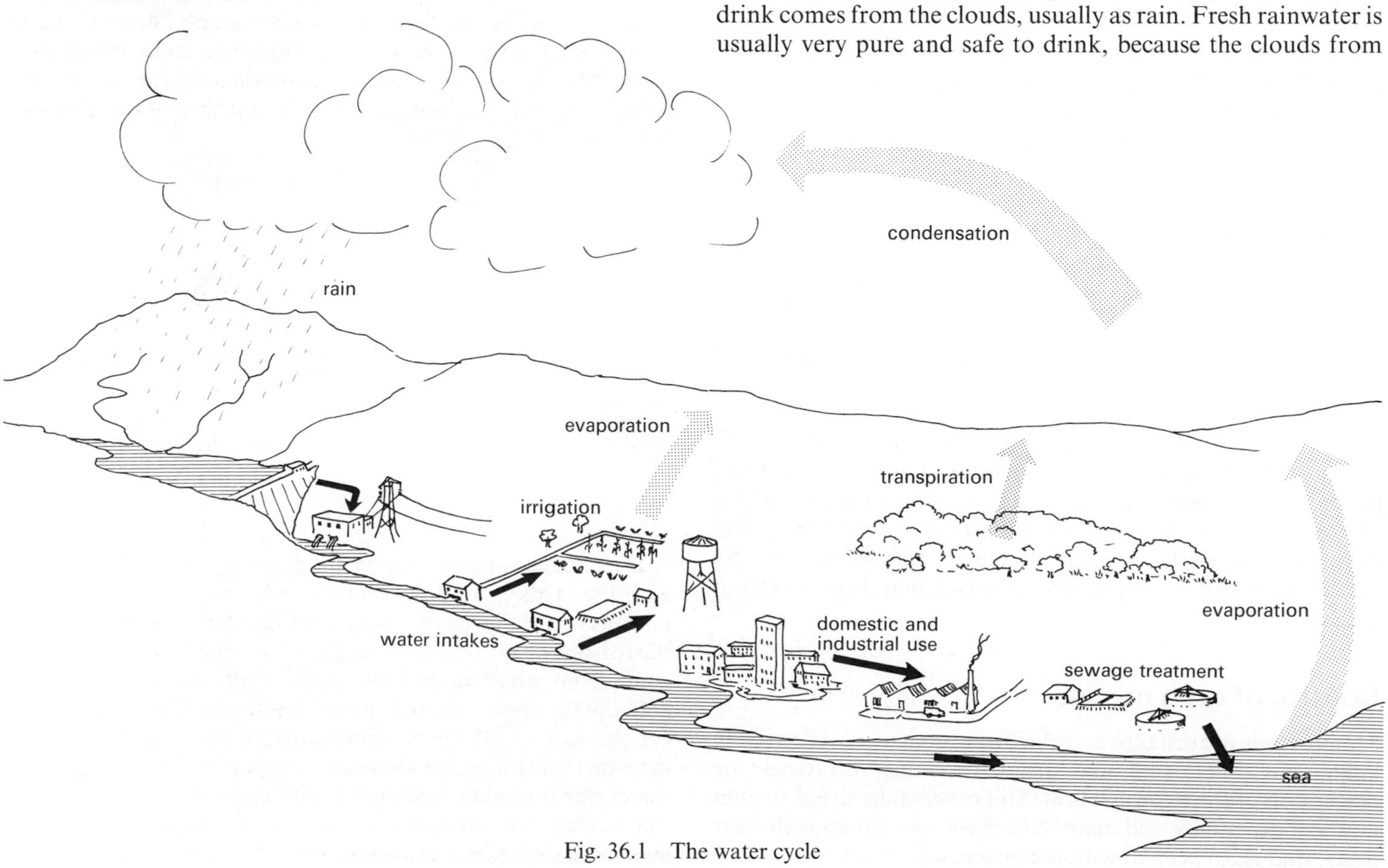

Fig. 36.1   The water cycle

which it comes are formed by the condensation of water vapour. When rainwater falls from the clouds to the earth it may collect in streams and rivers and run into lakes or into the sea. Water from these sources, and from plants, evaporates again to form more clouds as it condenses. This sequence is called the *water cycle* (Fig. 36.1).

As rainwater runs over the surface of the ground many substances dissolve in it. It is unusual for these dissolved substances to be present in a high enough concentration to be harmful to people drinking the water, but they enable all sorts of micro-organisms to live in it. These organisms include bacteria, algae, and protozoa. Some of these organisms are pathogens and many of them get into the water as a result of people urinating or defaecating into or near the water.

In Chapter 28 you can read that a minimum number of pathogenic organisms must be taken into the body in order to produce signs and symptoms of disease. If the pathogens in the excreta from a small village pass into a large, fast-flowing river then they are so diluted that they will be unlikely to cause disease to people drinking the water lower downstream. But if the sewage from a large town runs, untreated, into a river then the risk to people drinking that water is much greater. The means of treating sewage to make it safe enough to put into a river are discussed in Chapter 37.

Choosing a source of water

Whatever the source of water, it must supply a sufficient quantity to meet the needs of the community. It is helpful if the source provides water that is naturally pure. Rainwater collected from a roof is naturally pure, but there may not be enough of it to meet daily needs except in regions where there is a high rainfall on most days. Water from a stream or river that has not passed any other human community and is therefore not polluted may be safe to drink, but if people are careless about where they urinate and defaecate then the organisms of waterborne diseases may easily get into river water (Fig. 36.2).

Water collected from roofs (Fig. 36.3) and rivers is called *surface water*.

(UNICEF/WHO/Thorning)

Fig. 36.3   Rainwater collected from a roof is usually very pure

Water may also be obtained from underground. As water soaks through the soil it reaches the underlying rock. If this rock is impervious or waterproof, then the water collects above the rock surface (Fig. 36.4), and either forms an underground store called an *aquifer* or, if the rock surface slopes, flows along to appear as a spring. The water is filtered to some extent as it passes through the soil, but such filtration usually removes only mineral particles while many bacteria pass through. Such water is not pure.

Fig. 36.2   River water is often made unsafe to drink as a result of people urinating, defaecating or simply washing in it

In some parts of the world the underlying rock is limestone and this is porous and therefore pervious to water. However, the pores are very small and water that has passed through limestone is usually free from bacteria and safe to drink, though it contains dissolved calcium salts. Ways of obtaining water from underground sources are illustrated in Figs. 36.4 and 36.5.

A spring is a very convenient source of water because no pump or other mechanism is needed to bring the water to the surface. More often a well has to be sunk to reach the water which has then to be raised to the surface. Shallow wells are usually dug out by hand if the water is not more than 6 metres below the surface. They do not form the best source since the water is not likely to have been effectively filtered and there is a serious risk of contamination from the surface.

If the soil is deep then it is possible to drive a tube down to reach the water, but where the water lies under rock it is usual to sink a *borehole*. This can be quite expensive since, although a shallow borehole can be sunk using hand-operated equipment, a deep borehole through rock can only be driven by using a drilling machine. Because of the cost, a survey is usually made before drilling starts to find out how likely it is that water will be found. Water from deep wells and boreholes is usually pure.

Sometimes after a borehole has been sunk the water comes to the surface of its own accord because it is under pressure. Such a well is called an *artesian well* and it is always found where a layer of porous rock, such as limestone, is sandwiched between two layers of impervious rock (Fig. 36.4). Water seeps into the porous rock where it is exposed on high ground and passes through the sloping rock-layer to form an underground reservoir or aquifer. Artesian well water is thoroughly filtered and very pure.

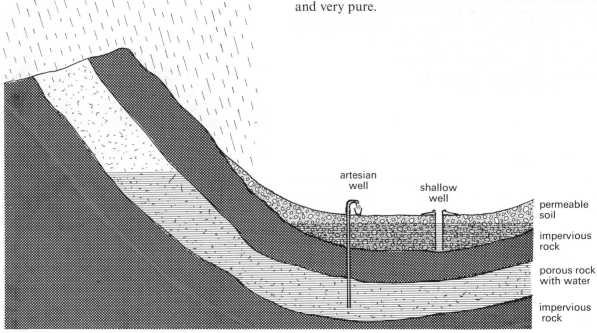

Fig. 36.4   A shallow well and an artesian well compared. Water from the artesian well has been filtered through porous rock and comes to the surface under pressure

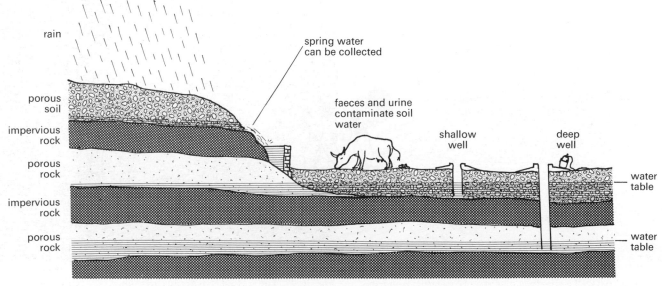

Fig. 36.5   A spring, a shallow well and a deep well compared

Fig. 36.6   This well is dangerously close to a latrine

(WHO/A. S. Kochar)

Protecting a well from contamination

Particular care must be taken over the siting of shallow wells. If a shallow well is close to where people live then there is a risk of contamination. If there are pit latrines close by, then bacteria may be carried by liquids seeping from the pit through the soil (Fig. 36.6). If the well is on sloping ground then it is important that the latrine is dug lower down the slope.

If the water is taken from a deep well or borehole (Fig. 36.7) then there is little risk of contamination, provided that germs do not percolate into the upper parts of the shaft. It is sensible to draw the water up through a pipe and to take steps to ensure

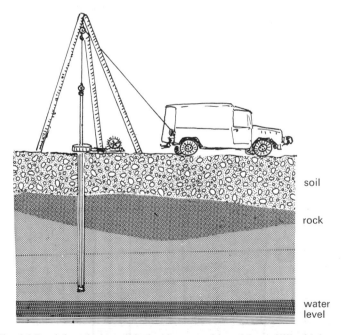

Fig. 36.7   A borehole well is dug by a machine-driven drill which can bore through rock to reach water

that no contamination can pass down around the outside of the pipe. The same precaution applies to all other kinds of well. Lining the upper part of a shallow well with concrete or other impervious material helps to stop contaminated water getting in. A concrete or stone surround at the top of the well prevents people's feet making mud which could trickle into the well. A

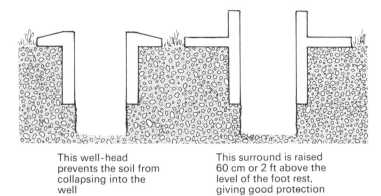

This well-head prevents the soil from collapsing into the well

This surround is raised 60 cm or 2 ft above the level of the foot rest, giving good protection to the well

Fig. 36.8   Two methods of protecting well-heads

raised well-head (Figs. 36.8 and 36.9) gives still greater protection, but the wall should be narrow enough to prevent people from standing on it. A head-wall also reduces the chances of leaves, dirt and other debris falling into the well, both polluting it and reducing the volume of available water. An open well should be covered when not in use.

(WHO/A. Holbrooke)

Fig. 36.9   The raised edge prevents mud, carrying germs, from slipping into this well

Chlorination. The water that comes from the filter looks clear and sparkling but it may still contain dangerous pathogens. Chlorine gas is bubbled into the water, producing an unstable acid, as described on p. 278. The amount of chlorine needed to

Emergency supplies. Sometimes we do not have access to pure water. Water from an unreliable source can be made safe to drink by boiling, which kills most kinds of pathogens. When there are outbreaks of cholera, typhoid or other water-borne diseases it is sensible to take the precaution of boiling drinking water. At such times, water authorities usually add extra chlorine to piped supplies, as a precaution.

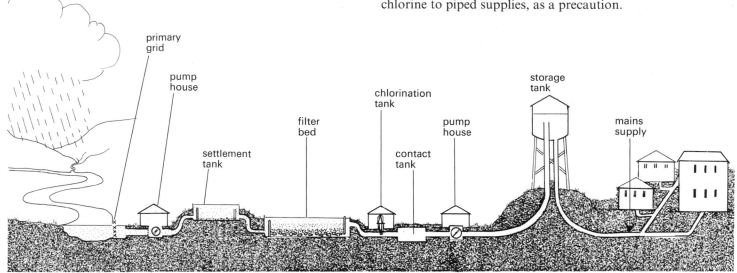

Fig. 36.16   Stages in the large-scale treatment of drinking water

destroy all the bacteria can be worked out very precisely, so waste is avoided. However, a small excess of residual chlorine is allowed. Chlorine is a very poisonous gas and great care has to be exercised in handling it.

The water now passes to a storage tank where the chlorine is given time to act. It has cost a lot of money to make the water safe to drink and it would be foolish to allow it to become re-contaminated. So the storage tank is covered to keep out sources of infection such as wind-blown dust, bird droppings and insects and the water is then pumped through pipes to houses and hospitals and other users (Fig. 36.16). The pipes, if they are properly looked after, protect the water from further contamination.

Questions

1   Why is water which has been collected high up in hill country more likely to be safe to drink than water from a river passing through a town?
2   Explain why water from a deep well is more likely to be safe to drink than water from a shallow well.
3   Make a list of the kinds of pathogens that are transmitted in untreated water.
4   Why are water-treatment works more likely to be associated with large towns rather than small villages?
5   In what ways are slow and rapid sand filters similar? In what ways are they different?
6   List the ways in which pathogens get into drinking water.
7   Why do you think the cholera epidemics in London in the 1840s and 1850s were associated with drinking-water? Why do you think they occurred in the summer months?

37

The Disposal of Sewage and Refuse

'Sewage' is the name given to human urine and faeces when collected together. Very often it is diluted with the water used to flush WCs and with water carrying organic waste from kitchens and washrooms. It may also contain industrial waste or effluent.

'Refuse' and 'garbage' are names given to solid waste that accumulates as a result of domestic and industrial activity. It includes paper, cartons and cans, and may also include unused food material that can harbour micro-organisms and feed creatures such as flies and rats which act as disease vectors. The methods of disposing of sewage and refuse differ.

Sewage disposal and treatment

As explained on p. 211, urine and faeces carry bacteria and these organisms may be pathogens. By urinating and defaecating carelessly about the place, people can make it easy for such vectors as houseflies, cockroaches and rats to become infected with pathogens which can then be spread to other, healthy people.

Carelessness over these functions also results in the spread of the cysts of amoebic dysentery, the eggs of tapeworm and blood fluke, and the larvae of hookworm.

In country districts where the population is widely dispersed, the risk of the spread of intestinal disease by vectors is less than in a densely populated town. When people live close together in large numbers, flies and other vectors have only a short journey from a source of disease-causing organisms to the food which may be eaten by a large number of people. Thus one person suffering from dysentery, thoughtlessly defaecating near a food market, could result in the spread of dysentery amoebae by flies to the food, so infecting a considerable population.

The number of people travelling from one town to another, as traders or casual visitors, increases the risk of the introduction of a disease to a healthy community. This can be seen clearly in the spread of cholera (p. 222). However, the arrival of an infected visitor will result in the spread of cholera only if he is careless over the disposal of his faeces or if the sewage system is faulty.

Domestic disposal

Pit latrines. In isolated houses and small villages both urine and faeces may be disposed of safely in a pit latrine (Fig. 37.1), provided that the soil is permeable to water and that there are no wells within 30 metres that are likely to be contaminated by seepage. The latrine should never be sited uphill from a well. A pit latrine should be dug as deep as the soil will allow, provided

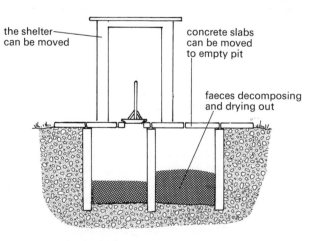

(a) A simple and inexpensive pit latrine

(b) A more elaborate pit latrine which can be used where soil is shallow. One pit is in use. The faeces in the other pit are allowed to decompose and dry out. They can then be removed and the pit re-used

Fig. 37.1 Pit latrines

that there is no danger of the walls of the pit collapsing inwards. A floor of concrete, brick or stone that can be cleaned easily and which will not collect puddles should be laid around the top of the pit. If the upper part of the pit is lined with concrete or brick then it will be more difficult for rats to burrow their way in. A lid or cover must be placed over the opening of the pit when it is not being used, to keep out flies and cockroaches which are attracted by the faeces. These creatures gather micro-organisms as they feed in the latrine and then transfer them to other places.

Chemical closets. These are useful in caravans and on temporary camping sites. A bucket with a capacity of between 25 and 40 litres receives the faeces and urine. It must be of a shape that is easy to clean and made of a non-corrodable material. The bucket is placed in a plastic container with a toilet seat and a non-spill lid.

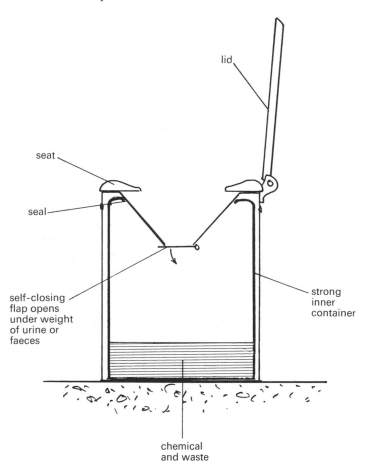

Fig. 37.2   One type of chemical closet

The chemical put in the bucket to kill germs is usually based either on phenol or on formaldehyde, both of which are powerful disinfectants. They have the added advantage of producing fumes which tend to discourage flies. The bucket has to be emptied from time to time, in a pit dug in the ground or in a lavatory linked to a main sewage disposal system. It should not be emptied into a system connected to a septic tank since the disinfectant will upset the normal decomposition processes in the tank.

Water carriage systems. When a good water supply is available then a flush privy or WC can be installed and connected to a water carriage system. This means that a flow of water is used to carry the excrement to the sewage disposal point. The lavatory pan is made of earthenware or china with a shiny, glazed surface that makes it easy to clean. Urine and faeces are flushed away into the drainpipe and sewer by the rapid release of 10–15 litres of water from a cistern. Some of this water remains in the pan, forming a seal that prevents unpleasant-smelling gases from the sewer from passing back into the house. A simple pattern of squatting plate is shown in Fig. 37.3, and an alternative design of lavatory pan is shown in Fig. 37.4.

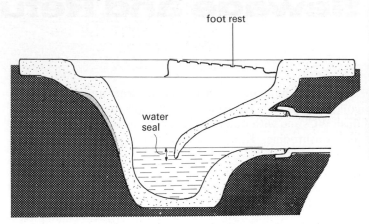

Fig. 37.3   Section through a squatting plate, showing how the water seal is formed

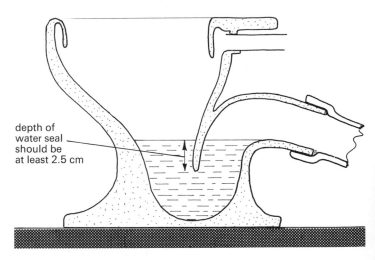

Fig. 37.4   Section through a lavatory pan or WC

The operation of both pieces of equipment depends on the rapid flow or flushing action of the water from the cistern. This flushing generates a fine spray which is not easily seen, and most people are unaware of it. Like any fine spray or aerosol, these droplets float in the air and settle on the seat, on any fitting in the lavatory and on the person who just used it. Normally the danger from this is not great, but when the WC is used by a person suffering from diarrhoea, germs from the liquid faeces are spread very easily by the spray. It is because one cannot avoid touching things such as handles which are contaminated by the spray-borne bacteria that so much importance is attached to washing the hands carefully after using the lavatory.

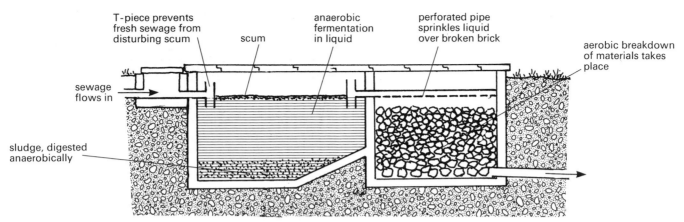

Fig. 37.5   Section through a septic tank

Septic tanks. The sewage from an isolated house or group of houses may be carried by water through pipes to a septic tank (Fig. 37.5). The name 'septic tank' means that the tank contains active bacteria. These bacteria are necessary for the working of the tank and it is *very* important that they are *not* destroyed by, for example, putting disinfectant into the lavatory pan, or into the sinks or drains that run into the tank.

The septic tank consists of two large chambers lined with waterproof concrete. Solid matter settles out in the first chamber and a thick scum forms on the surface of the liquid. Anaerobic bacteria are active in the water and they slowly decompose the solid waste. Liquid flows out of the first tank into a pipe perforated with holes, through which the liquid is sprinkled on to broken stone or brick. The surface of the bits of stone or brick becomes covered with a jelly containing large numbers of one-celled animals called *ciliates* (*see* Fig. 37.10) together with aerobic bacteria. The ciliates feed on bacteria carried by the water from the first chamber, while the bacteria in the jelly break down organic chemicals, such as urea, in the water. The water that runs out of the second tank should be quite safe and it is allowed to soak away underground.

From time to time the sludge or sediment in the first chamber has to be cleared out. It can be pumped out into a mobile tank if the local health authority provides such a service. Otherwise it has to be emptied by hand and the sludge disposed of in pits or trenches. The sludge is sometimes used as a fertilizer for crops, but this is not very sensible since it often contains pathogens.

Septic tanks are now widely used in rural Britain and have replaced bucket latrines and the more unsatisfactory custom of building a 'little house' with the lavatory seat over a stream. Regulations demand that sewage is disposed of in a sanitary way, even in the most remote areas.

Cesspits. A cesspit is a much simpler structure, built of brick or concrete. The walls should be impermeable to water to prevent seepage. It is a temporary storage place for faeces and urine and must be emptied from time to time when it becomes full. It must be covered to keep out flies and other vectors, such as rats, which are attracted by the smell.

Community sewage disposal and treatment

In towns and cities it is usual to find a system of sewage disposal run by the health authority or another branch of the local administration. Waste is flushed from houses and other buildings into *sewers* (Fig. 37.6). The pipes carrying the waste from the lavatories or latrines are usually made of glazed earthenware or strong plastic and they connect up with much larger pipes or, in a large city, tunnels made of brick or concrete.

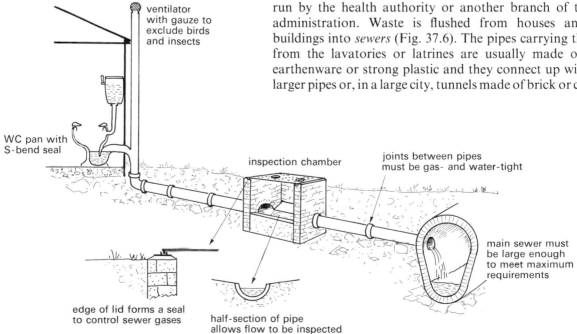

Fig. 37.6   Connection between WC and sewer

Fig. 38.2 An abattoir should have a concrete or tiled floor that can be washed easily, good drainage, fly-proof ventilation, containers for removing animal waste and skins, and a place where the slaughtermen can keep their clothes and wash

disposal of the intestines and other waste from the slaughtered animals is important in this respect. The skins or hides from the animals, which are valuable for the making of leather, also attract flies and they must be stored in a way that provides the least attraction to flies. Both hide and intestines carry germs that get on to the hands of the men cutting up the carcasses.

There must be provision for the men slaughtering and cutting up the animals to clean both themselves and their equipment. The knives and other instruments they use get covered with blood which is a very good food in which bacteria breed. Unless they are kept clean, the men's hands and the knives they use can be means of spreading bacteria to meat.

Storage of meat. Whatever precautions are taken, some bacteria are always present in meat and will bring about decomposition. The enzymes that the bacteria produce to break down the meat proteins are more active at high temperatures than at low temperatures. In a small village, meat which has been killed is usually sold and eaten in a short space of time. In a large town or city, animals are usually killed some distance away and it may be days before meat reaches the shops and markets and is sold. Storing meat in a cool place or, better still, in a cold refrigerated place becomes more important. Also, the longer the time between slaughter and the meat being eaten, the greater the chance of flies settling on the meat and spreading pathogens. Meat which is kept in a cold-store is also protected from flies landing on it and contaminating it. (See also pp. 290–92, 'Freezing and storing food'.)

Fish. Fish is an important food for communities living near the sea or a river. Freshly caught fish does not usually represent a health risk, provided it does not come from water that is contaminated or polluted (Chapter 40). But if fish is to be sold in a market some distance from where it was caught, then it must be remembered that, like fresh meat, it is perishable. Bacteria already in the fish start to decompose it though decomposition is slowed down if the fish is frozen or kept cold. In some parts of the world, fish may carry tapeworm larvae (p. 206) and it is particularly important that such fish are properly cooked before they are eaten. High temperature kills the tapeworm larvae and also kills any bacteria that are present.

Fruit and vegetables

Most fruit and vegetables contain much less protein than meat and fish and they do not provide such a rich food for bacteria to feed on. There is less risk of disease being caused by organisms in the food, but there are two chief dangers: decomposition of the food, and contamination from other sources.

Decomposition. Fruit and vegetables are often gathered in large quantity, partly because a crop ripens or is ready for gathering at the same time and partly because it is economical to take a large quantity to market at one time. Fungi and insect pests, such as weevils (a kind of beetle), can spread through a large basketful or a truck-load of fruit or a sack of beans.

Many fruits produce a gas called ethene (ethylene) as they ripen. In a closed space the ethene from one fruit can quickly affect all the other fruit around it and, once ripe, the fruit attracts small flies which often carry fungus spores. The fungi are usually the kinds we call 'moulds' and they can feed on the fruit by secreting enzymes. This makes the food 'go bad' and it becomes uneatable in a very short time.

Well-ventilated conditions disperse the ethylene, so preventing too rapid ripening, and also reduce the humidity around the fruit, so discouraging the growth of the moulds.

Contamination. While fruits are usually gathered from trees and bushes, vegetables often grow on or near the ground. In districts where manure is used as a fertilizer for growing plants, there is always a risk that the outsides of the vegetables may become contaminated by pathogens and pesticides (Fig. 38.3). It is wise, therefore, always to wash the outsides of vegetables very thoroughly before eating them.

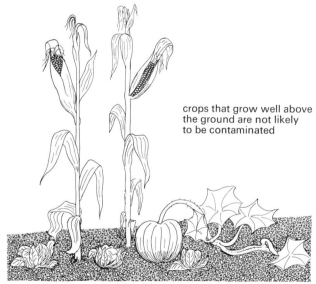

crops that grow well above the ground are not likely to be contaminated

Fig. 38.3 Vegetables that are grown close to the ground may be contaminated by manure and soil organisms

Many crops such as maize, rice and beans are harvested when they are ripe and dry. Provided they are kept dry there is not much risk of them being attacked by micro-organisms, though they may go mouldy in damp conditions.

The reason we grow such crops is that they provide us with very useful food. But they also provide food for rats and mice and many kinds of beetles, particularly weevils which can bore through the hard outer coverings. Making grain stores rat-proof is described on p. 300 but it is not very easy. Protecting dry crops against weevils is even more difficult and it is usually wise to try not to keep grains and beans too long between harvesting and eating.

Since the risk of fruit and vegetables carrying pathogens is less than the risk with meat and fish, the precautions taken when selling them are less strict. But a high standard of cleanliness should always be maintained.

When it is necessary to store such crops in large quantities, then apart from maintaining dry conditions it is often necessary to use chemicals to deter insects and rats and mice. Care has to be taken to see that such chemicals will not harm the people who eventually eat the food. The use of such chemicals is usually under the supervision of a trained person who knows when to use them and how much to use.

Food shops and markets

If food is produced in ways that ensure that it is safe for people to eat, then it is important to make sure that it is sold in hygienic conditions. If meat or fish is sold on an open stall (Fig. 38.4) where flies can get at it, then it can be easily contaminated with many kinds of bacteria, including pathogens. Meat can be covered with a thin cloth, called muslin, in which there are fine holes that allow air through but which prevent flies from getting at the meat. The oxygen inhibits the dangerous anaerobic bacteria.

In a town, where it is possible to spend quite a lot of money on building shops, meat is usually kept in glass-fronted cabinets (Fig. 38.5) that are refrigerated, that is, kept cool by a

(J. Sainsbury plc)

Fig. 38.5 Joints of meat to be sold are prepared under hygienic conditions and then placed in a refrigerated cabinet

refrigerator unit. The same approach applies to the sale of fresh fish.

One of the responsibilities of a local health authority is to inspect markets and food shops to check that proper standards of hygiene are being maintained. This includes inspecting the surroundings to see whether there are places where flies are breeding, such as open rubbish dumps. They should also ensure that there are lavatories, so that people can dispose of urine and faeces in a sanitary way and wash their hands thoroughly before returning to the food counters.

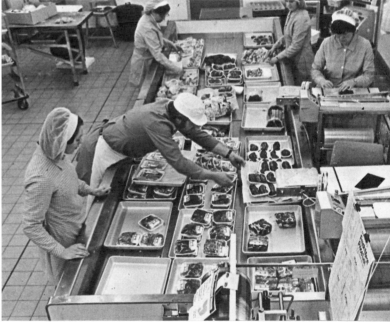
(J. Sainsbury plc)

Fig. 38.6 Prepacking food for sale in sterile containers reduces the chance of contamination through handling by customers

Methods of preservation

Methods of food preservation are of two general kinds. The first is *bactericidal*, in which all micro-organisms are killed. This includes canning and any other process involving sterilization. The second group of methods is termed *bacteriostatic* since the bacterial activities are stopped, but without actually killing the organisms. Freezing, dehydration and pickling come in this group.

(FAO/F. Mattioli)

Fig. 38.4 This fish market is clean and not too crowded. There are no dogs about

289

1 Canning

This was first introduced in America during the Civil War and was widely used in Europe by the time the First World War broke out in 1914. The principle was developed over a hundred years earlier by Appert in Paris. He discovered that if food is adequately cooked and then sealed while hot, in jars or cans, then it will remain safe to eat for a long time. In fact, roast veal canned in London in 1823 for a naval expedition was opened in 1958 and the meat found to be in good, wholesome condition.

Canning is applied on a commercial scale to a wide range of foods, including meat, vegetables, fruit and fish. Heat treatment at 90 °C (or blanching) is essential to destroy natural enzymes as well as to kill micro-organisms and is a form of sterilization. The food is then placed while it is still hot in cans made of tin plate (sheet steel coated with tin) that has been coated with lacquer to prevent the metal causing discolouration of the food. After sealing and cooling, the contents contract and a vacuum develops, causing the ends of the can to curve in slightly (Fig. 38.7). This vacuum is maintained as long as the food is free from micro-organisms and the seal is intact. On piercing a can to open it, one should hear a slight rush of air into the can. If the ends of a can bulge then there is a positive pressure inside and this usually means that the anaerobic micro-organisms (p. 283) have been respiring, reproducing and feeding. The food is suspect and should be destroyed. Cans in this condition are said to be 'blown'.

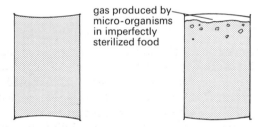

gas produced by micro-organisms in imperfectly sterilized food

Fig. 38.7   Sections through perfectly sterilized and 'blown' cans

Although the record for successful storage of veal in a can is 135 years, most meat products tend to deteriorate after 7 years and some acid fruits such as pineapple are likely to corrode the tin plate from the inside in less than 2 years, thereby destroying the seal and making contamination possible. It is wise therefore not to rely on keeping canned foods for indefinite periods of time.

2 Freezing

An increasing amount of food around the world is preserved by freezing. At sub-zero temperatures, bacteria and most moulds are unable to digest and decompose food or to reproduce. Natural enzymes in food are inactive also. For over a hundred years, fishermen have preserved fish by packing it in ice or in brine (salt water) at sub-zero temperatures, and refrigerated ships have carried meat safely through the tropics to be eaten thousands of miles away from where it was produced.

In recent years the technique of quick freezing has been applied to a wide range of foods from meat, fruit and vegetables to ready-prepared complete meals. The material to be preserved must be cooled very rapidly to a temperature of −18 °C (0 °F). The food must be spread thinly on shelves or a conveyor belt in close contact with brine pipes at a temperature a little below −18 °C. Strong brine does not freeze at 0 °C like pure water.

Thus brine at −20 °C can be circulated between the refrigerator unit and the chamber containing the food, so extracting heat from the food. An alternative method of freezing involves blowing very cold air from a refrigerator unit through the food on a conveyor belt in a tunnel. Once frozen, such food must then be stored at temperatures below 0 °C, for example, in a deep freeze. If the food is allowed to thaw out then bacteria and moulds that have been inactive at the low temperature become active once more.

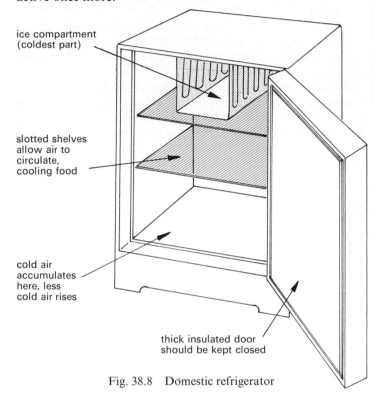

ice compartment (coldest part)

slotted shelves allow air to circulate, cooling food

cold air accumulates here, less cold air rises

thick insulated door should be kept closed

Fig. 38.8   Domestic refrigerator

Domestic refrigerators do not run at temperatures as low as those we have discussed. The icebox or freezer compartment will be at a temperature lower than 0 °C, but much of the space in the refrigerator will have a temperature between 0 °C and 5 °C. At such temperatures bacterial activity is slowed down but not completely stopped, and certain moulds seem to grow quite actively at such temperatures. This does not mean that refrigerators are useless, but it does mean that we should use our knowledge and sense in deciding what sorts of food can be kept safely in a 'fridge' and for how long they can be kept.

3 Dehydration

One of the reasons that perishable foods are so readily used for nutrition by micro-organisms is that they contain plenty of water. If this water were removed then the food would be preserved against the digestive action of enzymes, whether bacterial or those natural to the food.

The simplest method of dehydration is air drying. Fish, for example, can be split, the internal organs removed, and then hung in the sun and wind to dry. Wind speeds up the rate of evaporation. The sun's rays provide the heat necessary and may kill bacteria through the effect of ultraviolet radiation. Vegetable foods such as peas, beans and split coconut can be dried in this way as can many fruits such as dates and plums. Dried foods are lighter and less bulky than fresh foods and, provided they are protected from humid conditions, may be

(a) *Peas have been arranged on shallow trays in the freeze-drying cabinet*

(b) *The cabinet is sealed before pressure is reduced*

(Unilever Educational Publications)

Fig. 38.9   Accelerated freeze drying

kept for long periods. When placed in water they absorb liquid, swell up again, and may then be eaten.

Vacuum drying is a more modern method of preservation by dehydration. Use is made of the fact that water evaporates quickly at temperatures well below boiling point if the air pressure around it is reduced. Perishable foods such as eggs (taken from their shells), milk and several proprietary foods are treated in this way to reduce them to dry solids which can be powdered and packaged. If the water had been removed by boiling then the foods would have been cooked and changed in flavour. It is a curious fact that many kinds of bacteria can survive vacuum drying, even though the degree of vacuum would cause the death of all other kinds of organisms. Bacteria will not multiply in the food while it is kept dry, although they will resume activity when water is added.

Freeze drying (Fig. 38.9) is an even more up-to-date process in which food is first quick frozen, as described on p. 290, and then subjected to reduced pressure. The ice crystals evaporate directly, a process called sublimation, instead of first melting. As with any other dried food, material dried by this method can be stored at the normal temperature of the environment.

4 Curing and salting

These are very old methods of preserving meat and vegetables. Common salt, sodium chloride, dissociates in water (*see* ionization, p. 21) and the sodium and chloride ions attract water around themselves. If sufficient salt is added then all the available water in a piece of meat may be drawn out in this way. The resulting solution is osmotically active (p. 22), and if bacteria are present then water may be drawn out through the bacterial membrane by the salt solution, thus killing the organisms. Many kinds of fish are preserved in this way and so are beef and pig meat (e.g. bacon and ham).

5 Pasteurization

This is the name given to a process devised by Louis Pasteur to prevent the souring of wine by bacterial action. The French scientist observed that when wine was exposed to the atmosphere, microbes (micro-organisms) appeared in it, which he believed were the cause of chemical changes that spoiled the wine. He discovered that boiling the wine killed these microbes, but boiling also destroyed the fragrance (or bouquet) and removed the alcohol. Further experiments showed that the microbes could be killed by heating the wine to a temperature of 50–60 °C for thirty minutes. At this temperature the quality of the wine was not very much altered. This treatment is a partial sterilization and does not kill all possible kinds of micro-organisms.

The term *pasteurization* was first used in connection with the partial sterilization of beer, but nowadays we think of it in connection with the treatment of milk produced for human use.

Fig. 38.10   Pasteurization plant. Milk is passed through stainless-steel pipes to heater units, one of which is seen in the lower right-hand corner. From these the milk passes to rapid cooling units

(National Dairy Council)

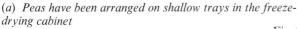

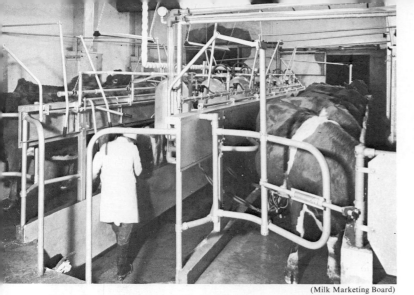

(Milk Marketing Board)

Fig. 38.11 Cows being milked in hygienic conditions. The milk is drawn by vacuum along sterilized, stainless-steel pipes to a chilled collecting tank

Raw milk from a cow is an excellent medium for bacterial growth. Most of the kinds of bacteria commonly found in milk do not cause disease but several of them alter the flavour of the milk and others cause it to curdle. However clean the cows and the dairy are kept it is not possible to exclude all bacteria from the milk. Some come from within the mammary glands or udder of the cow. The milk is either heated to 63 °C for thirty minutes and then cooled, or else passed between metal plates heated to 80 °C for about half a minute and then cooled rapidly. This kills most of the contaminating bacteria. We now know that this treatment destroys *Mycobacterium tuberculosis*, the pathogen causing TB which may be transmitted in milk.

Pasteurized milk is safe to drink and should 'keep' for several days if kept cool; but with pasteurized milk, as with any other preserved food, once fresh micro-organisms have been allowed into it, deterioration may take place and further sterilization may become necessary.

Storing food in the home

Many cases of food poisoning, bacillary dysentery and gastro-enteritis result from contamination of food after it has been brought into the home. Cleanliness is important at all stages in handling and storing food (see below) but even if food has become contaminated, the risks of infection can be kept to a minimum with a little care.

Bacteria and fungi grow much more rapidly in warm conditions than in cold, so it is wise to keep food in the coolest place in the home. Most homes have refrigerators in which perishable food can be kept safely for two or three days. Most domestic refrigerators have a shelf temperature of 4 or 5 °C. While bacteria grow and reproduce only slowly at such temperatures, it is important to remember that these micro-organisms are *not killed*.

In a home without a refrigerator, or when camping, meat and other perishable foods should be kept in a well-ventilated meat safe (*see* Fig. 38.16). Most of the kinds of bacteria that cause serious food poisoning are anaerobic. They grow best if there is no oxygen. If there is a flow of air around the meat, with the 20 per cent oxygen usually found in fresh air, the anaerobic bacteria are inhibited and do not grow.

A closed food store will be unlikely to allow any movement of air and will soon become saturated with water vapour,

evaporated from the food. In such saturated air, moulds in particular grow rapidly and start to decompose food and make it taste unpleasant. A store with ventilation that allows air to circulate may not stop the growth of moulds completely, but it makes it less likely (Fig. 38.12).

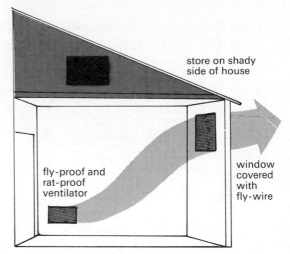

Fig. 38.12 A food store should have good ventilation with circulating or moving air

The size of a food store depends on the space available as well as on the quantity of food to be stored. It may range from the meat safe (Fig. 38.16) to a cupboard or a room with air intakes and outlets. Whatever the type of store, it is desirable to keep out flies, cockroaches, ants, rats and mice.

(Rentokil)

Fig. 38.13 Common cockroach. This female is carrying an egg case which she will deposit in a crevice (× 2)

Fig. 38.14 Cockroaches are tropical insects. In temperate countries they seek warm protected environments such as this passage for hot-water pipes in a hospital basement

(Rentokil)

Houseflies and other flies contaminate food with bacteria carried on their bodies and in their saliva and faeces (p. 245). Cockroaches (Fig. 38.13) contaminate food in similar ways and damage packages and leave their droppings in dry foods, such as flour and rice, giving them an unpleasant taste. They also leave a trail of bacteria.

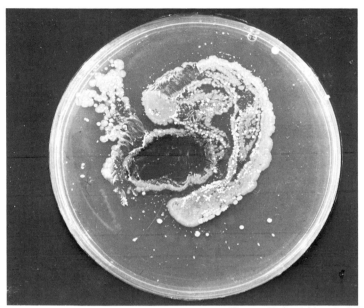

(Domestos Hygiene Advisory Service)

Fig. 38.15   Growth of bacteria on nutrient agar twenty-four hours after a cockroach had walked over the plate

Ants are usually so numerous that, once having found their way to stored food, they can carry away and eat quite a lot, particularly of sweet foods such as sugar. Rats (p. 231) carry germs on their bodies as well as in their urine and faeces and spread these to food, contaminating it.

Keeping out pests. Flies can be kept out by covering ventilation openings with fine wire gauze (p. 242) and keeping the doors of the store closed. Cockroaches and ants are crawling insects that get through small cracks, and it is very difficult to exclude them by purely physical methods. Sealing up cracks helps, and spraying a residual, contact insecticide around the gap under the door and around the window also helps. Small storage boxes and meat safes can be stood with the legs in dishes of kerosene (paraffin) with insecticide, or water containing a strong disinfectant to act as a barrier to crawling insects (Fig. 38.16). This is a useful precaution in tropical countries and when camping.

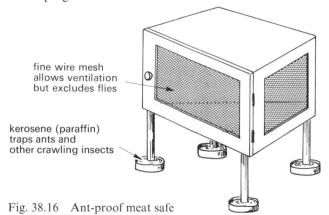

fine wire mesh
allows ventilation
but excludes flies

kerosene (paraffin)
traps ants and
other crawling insects

Fig. 38.16   Ant-proof meat safe

Rats are able to gnaw their way through wood, so keeping them out of food stores is not easy. But dry food such as flour, rice and beans can be stored in jars, tins or earthenware containers with close fitting lids, so that rats and mice do not get at the food inside.

Preparing and cooking food

While preparing meat or fish for cooking it is important to look for signs of bacterial activity. Heavy growths of bacteria show up as shiny patches or smears, often whitish or creamy in colour on the surface of the meat. Usually there is a distinctive smell, though unfortunately, some of the most dangerous bacteria, such as *Salmonella typhi*, causing typhoid fever, and *Clostridium botulinum*, causing botulism, do not produce very much smell. Both of these forms of food poisoning result usually from people with contaminated fingers handling the food. *Salmonella* is spread on the fingers of a person who either is carrying the organism in their intestine and who has been to the latrine and touched the bacteria from their own faeces or who has touched faecal material from another person. This is why at all times, and particularly when there is an outbreak of typhoid fever or any other intestinal disease, great care should be taken after using the lavatory to wash one's hands before touching or preparing food. The carelessness of just one person in this matter may result in many people contracting the disease.

Clostridium welchii is found in soil and in the intestines of animals and, also, in the intestines of infected persons. A person who has been working in the fields or handling manure can therefore spread these bacteria to food.

The *Staphylococci* that cause pimples, boils and septic wounds can also cause food poisoning and people who have septic cuts on their hands should not handle food for their own or other people's use. Putting a bandage or a plaster over a septic cut is no real protection against contamination of food by the bacteria from the wound (Fig. 38.17).

Fruit and vegetables should always be washed in pure water before being eaten either raw or cooked. This will not remove all micro-organisms but it will reduce the numbers of micro-organisms present and, therefore, reduce the risk of infection.

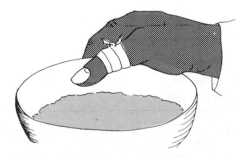

Fig. 38.17   A person with an infected finger should not handle or prepare food. Once a bandage is wet it does not stop germs reaching the food

Cooking food

So far we have discussed some of the more important methods of preserving and storing wholesome food. The ultimate aim of storing food is to eat it, and while primitive man no doubt ate all his food raw there are good reasons for cooking much of the

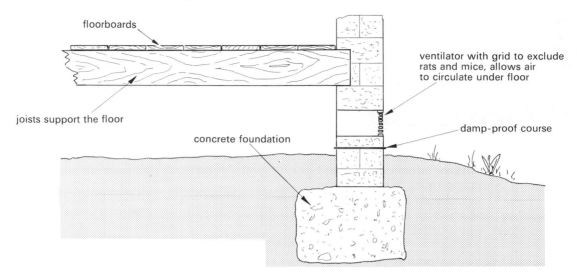

Fig. 39.3   Foundations and damp course

The avoidance of damp is important since moisture favours the growth of fungi which destroy wood and many other materials that we put inside a house.

Floors. Floors may be made of wood resting on joists which are secured in the walls above the level of the damp course (Fig. 39.3). An alternative method is to level the earth between the wall foundations and put down broken stone or brick (termed *hard core*) covered with sufficient sand to fill the spaces. A continuous sheet of waterproof polythene is placed over the sand and a concrete floor is poured over this, levelled and smoothed. This makes a very strong, dry and durable floor (Fig. 39.4). Wood can be laid over the concrete to give heat insulation.

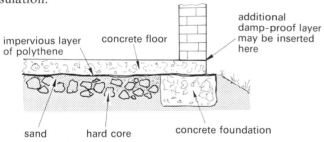

Fig. 39.4   Alternative construction for damp-proof ground floor

Roof. A strong, durable and watertight roof is essential. The pitch or slope of the roof is determined often by tradition but really should take account of climate. A steep pitch is desirable in countries experiencing heavy snowfall since a thick layer of snow may add a great weight to the roof supports. On the other hand, snow is quite a good heat insulator and in regions of moderate snowfall a strong roof with a snow 'blanket' on top may lose less heat to the atmosphere than a steep roof with no snow. Rainfall also has to be taken into account. In countries with low or sporadic rainfall a flat roof may serve to collect rainwater which can be led off to storage tanks. A gutter around the edge of any roof is desirable to prevent rainwater running on to the walls and making them damp. In Europe, roofs have traditionally been made waterproof by laying overlapping slates or tiles on the roof timbers. Asbestos sheeting and corrugated galvanized steel may be used. Sheet aluminium is particularly useful in tropical countries because it reflects the heat of the sun. Although less durable, a thatch of palms or other leaves, skilfully laid, gives a waterproof roof with good insulating properties. The roof frame and the fixing of the covering should be sufficiently strong to resist wind action.

The space under the roof is important. In the temperate zone, heat is lost through the ceiling into the loft in cold weather, and a layer of insulating material such as glass-fibre wadding or granulated polystyrene will reduce heat wastage (Fig. 39.5). In the tropics the space under the roof is often kept open to give extra height and a greater volume of air. This helps to produce a circulation of air, with a cooling effect. Where the under-roof space is closed it can be a dead space in which pests may breed, such as wasps or mice. There should be a trap-door giving access for inspection of this space.

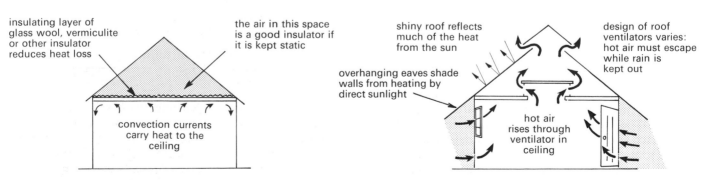

Fig. 39.5   Roof design in temperate and tropical regions

Fig. 39.6   Termite damage to the timber frame of a house

(Shell)

Timber construction and termites

If a house is made of timber, then it is important to choose a dry site to prevent the timber foundations from becoming permanently damp, since this would encourage wood-destroying fungi. Alternatively, the timber can be treated by soaking it with a wood preservative chemical that kills fungi. Timber has some definite advantages. It is often the cheapest and most abundant local building material and it is a good insulator. Unfortunately rats and other pests can gnaw their way through wood.

In many tropical countries termites cause extensive damage to buildings constructed of timber (Fig. 39.6). These insects use their powerful jaws to bite into wood, chewing it to a pulp. This pulp is digested, not by the termites themselves but by protozoa living in their gut. One kind of termite, *Macrotermes*, differs in that the wood pulp is laid down as food for a fungus, cultivated in the underground nests of the termite.

Because of the large numbers of termites in a colony and their enormous capacity to chew wood, these creatures can do immense damage to wooden buildings, eating timbers such as pillars, door and window frames and floor joists, resulting in the collapse of the structure.

Complete protection against damage is difficult to achieve. Concrete piles or pillars help by raising a timber building clear of the ground, but many species of termite can build covered, tunnel-like runs over the concrete to the timber above. Impregnation of timber with proprietary insect-killing chemicals such as Rentokil fluid gives long-lasting protection. Alternatively the ground around the house may be treated with HHC or a similar persistent insecticide to prevent the insects from reaching the house.

Ventilation and draughts

Fresh air is desirable for the comfort and well-being of the people living in a house. In hot weather in any country a free movement of air also aids the evaporation of perspiration or sweat. The cooling effect thus produced adds to one's feeling of comfort.

In hot countries louvre shutters permit movement of air in and out of a room while preventing the direct rays of the sun from overheating the room. They also provide privacy (Fig. 39.7). In cooler climates ventilation must be achieved without simply admitting rushes of cold air and losing warm air. Adjustable ventilation is preferred and this may be obtained by having different-sized sections in a window that can be opened

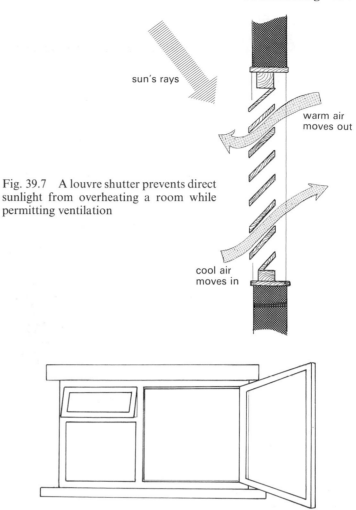

Fig. 39.7   A louvre shutter prevents direct sunlight from overheating a room while permitting ventilation

Fig. 39.8   A window should have sections that open separately to allow ventilation to be controlled

to varying extents (Fig. 39.8). Ventilator or air bricks in an outside wall allow air in but should carry an adjustable shutter or grille inside the room. In rooms with an open fire or a gas fire connected to a chimney the flue ensures a natural movement of air, as shown in Fig. 39.9.

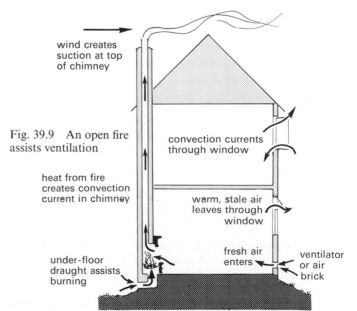

Fig. 39.9   An open fire assists ventilation

Where central heating radiators are used, however, more positive action has to be taken to ensure ventilation. Extractor fans may be used to draw out excessively humid (steamy) air from a kitchen while a meal is being cooked, and they can also be used to ventilate other rooms such as bathrooms and lavatories where there is no natural air movement.

Good ventilation should ensure a slow, gentle movement of air through a room without any direct, concentrated air stream. Anybody sitting in a draught or stream of air is likely to be chilled either because the air is cold or, more usually, because moving air accelerates the evaporation of perspiration from the skin. Excessive cooling could result in hypothermia—lowering of body temperature below the normal range (p. 99). People do not *catch* cold through sitting in a draught, but chilling reduces the body's resistance to disease organisms.

Whenever a person breathes out, micro-organisms are exhaled in droplets into the atmosphere (p. 210). In a closed room the concentration of micro-organisms builds up rapidly to a level at which the risk of infecting other people in the room becomes much higher. Sensible ventilation reduces the concentration of micro-organisms to a safe level.

Heating

In cold climates a great deal of money can be spent on heating a house, so wastage of heat should be avoided. The influence of the choice of building materials on insulation has already been mentioned (p. 295) and so has the effect of excessive ventilation on heat conservation.

What constitutes an ideal living temperature is a matter of opinion, related to feelings of comfort. This is related to the rate at which the body loses heat (*see* Chapter 14). People are usually less active physically inside a house than out-of-doors and therefore generate less heat. Loss of heat from the body to the surroundings is reduced by wearing clothes that act as insulators. A living-room temperature between 15 °C and 20 °C (60 °F and 69 °F) is considered comfortable in temperate regions and light clothing provides adequate insulation.

The sources of heat include open fires, burning coal or a smokeless fuel; central heating, in which heat from the burning of fuel in a boiler is carried by water (which has a high specific heat capacity) through pipes to radiators; electric heaters, in which electric current either heats a coil to such a high temperature that it glows and throws out radiant heat, or else heats a different type of coil enclosed in special bricks—a storage heater, so called because, once heated, the bricks emit heat over a long period; gas fires, in which burning coal gas or natural gas (methane) heats clay elements to a high temperature so that they glow and radiate heat (bottled gas is now widely used); and finally oil heaters, in which the hot products of the burning of kerosene (paraffin) circulate in the room.

Many kinds of gas and kerosene heaters are not connected to chimneys or any other means of removing fumes and the products of combustion. For every litre of kerosene burned, about one litre of water is produced which will condense on cold surfaces, creating problems with dampness.

All of these methods involve some heating by convection, as well as by radiation. The proportion of heat transferred by the two methods varies with the design of the heating equipment used. In a fan heater, for example, most of the heat from an electric coil is dispersed by air currents, increased by the action of a fan. Most of the heat from a central heating radiator is

Fig. 39.10 Most of the heat from a central heating 'radiator' is dispersed by convection

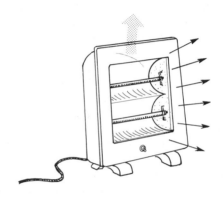

Fig. 39.11 Most of the heat from an electric fire is radiated while some is spread by convection

dispersed by convection, while a high proportion of heat from an electric fire is radiated (Figs. 39.10 and 39.11).

There is a tendency for the humidity of the atmosphere to be reduced by electric and central heating because the air temperature is increased without moisture being added.

While reduced humidity may be an advantage when drying clothes or when the air temperature is very high, it is not always an advantage where health is concerned—lung disorders such as bronchitis are accentuated by very dry air—and some means of increasing atmospheric humidity may be desirable. This can be achieved by placing an open container of water on a radiator or near an electric fire.

In tropical countries and in high summer in temperate regions it may be desirable to cool the air in rooms rather than to heat it. Increased ventilation often produces this effect but sometimes air-conditioning equipment is installed. This usually involves passing air over the cooling coils of a refrigeration unit and then blowing it into the room. While cool air can be a pleasant contrast to the heat out-of-doors, it can produce rapid cooling of the body with risks of chilling already mentioned on p. 99. Acceptable cooling can be obtained in all but the most humid climates by good ventilation, the circulation of air cooling the body by speeding up the evaporation of sweat.

Lighting

Natural lighting is desirable whenever possible because it is more intense than all but the most sophisticated and expensive forms of artificial lighting. The human eye is able to adapt to seeing under a wide range of light intensities (*see* p. 135), but many of the activities that depend on acute vision—reading, writing and fine manipulative skills—are often carried on indoors and under artificial lighting. In designing housing, windows should occupy not less then one-tenth of the wall surface. In practice window areas may be much greater than this, but in the tropics large unshaded windows can certainly lead to overheating of a room. Even in temperate areas such as Britain a large window area can be a source of heat intake during sunny weather. Of course, a window may be a considerable source of heat loss in cold weather.

Electric lighting is the accepted source of artificial lighting in many parts of the world and, in homes, is provided either by filament lamps or by fluorescent tubes. Filament lamps depend on the emission of light energy from a coil or filament, heated by the electric current. Such lamps are cheap but in terms of light output for the energy consumed are not very efficient. Fluorescent tubes are more expensive but have a longer working life and are certainly more efficient. They also provide a more even or diffused light. The siting of lamps depends very much on the use to be made of a particular room. For reading, it is helpful to have lamps so sited that light falls directly on the pages of the book. For general lighting, a lamp suspended from the ceiling is often adequate though a softer illumination is obtained when light is reflected from the walls and ceiling. Such light is termed indirect light.

Whatever the source of light, natural or artificial, it should produce no sense of strain or discomfort in the person using it. It should not produce intense shadows in the room. Although shadows look interesting in photographs, they produce an undue strain on the eyes when the gaze is shifted from a brightly lit object to one in shade. Fluorescent lamps sometimes flicker as they age and they should be replaced, since this flicker is harmful to the retinal cells and irritating to the nervous system.

Lavatories and washing facilities

When lavatories can be connected to a water carriage system (Chapter 37) it is usual to install WCs or lavatory pans inside the house. Ideally the WC should be located in a small room used for no other purpose, though quite often a WC is located in a bathroom. Since, however careful the users are, a lavatory is always a source of germs, it is wise to locate it well away from the kitchen and food store. Building regulations in Britain require that a lavatory shall not lead directly off a kitchen or living room. There must be a passageway and two doors between the two. The WC should be sited where there is a natural fall from the lavatory pan through the pipes to the main drainage system. The same applies to the siting of the bathroom or washroom. By modern standards a house should have a bath or a shower so that the whole body can be washed easily.

Kitchen facilities

The kitchen should be a light, airy, well-ventilated room, and the surfaces on which food is prepared, whether a table or the tops of cupboards, should be made of a material that can be cleaned easily and thoroughly. Modern plastic laminates covering wood have a hard, smooth surface that can be kept free from germs provided there are no cracks (Fig. 39.12). Wooden surfaces eventually become porous and may carry large populations of germs. If there is no alternative, then wooden work surfaces should be washed daily with a powerful (but non-tainting) disinfectant. Bleach solution, suitably diluted, is one of the best.

The type of cooking facilities and whether a refrigerator is provided or not will depend on circumstances, but a cool, airy food store is necessary (*see* p. 292).

Larders and store-rooms are best placed on the north side of a house in northern temperate countries so that they are cooler. In the tropics such rooms should also be away from direct sunlight but accessible to ventilating air currents (*see* Fig. 38.12).

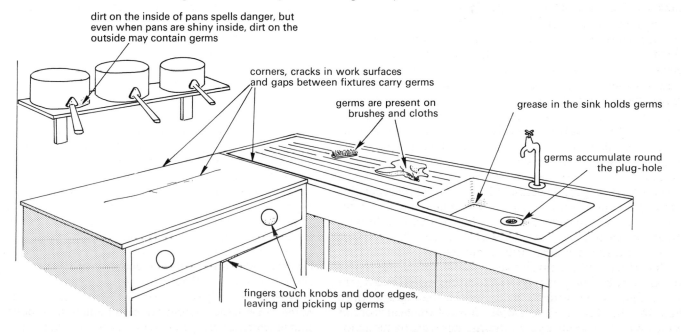

Fig. 39.12   Some of the sources of germs in a kitchen

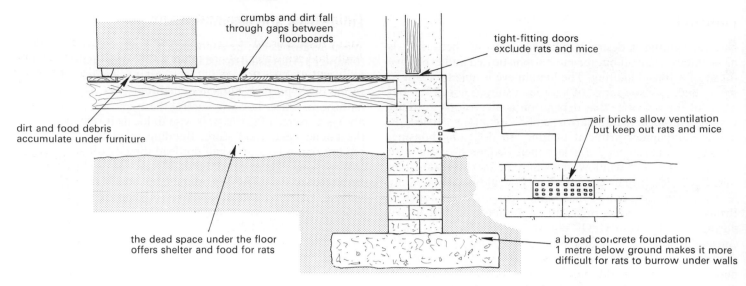

Fig. 39.13 Where rats are numerous, special precautions must be taken to make houses rat-proof

Keeping out pests

Care in the design and equipping of a kitchen and food store may be of no avail if vermin such as rats, mice, cockroaches, ants or houseflies invade the place. If any of these creatures is found in the neighbourhood then excluding them from houses can be surprisingly difficult. Rats and mice can enter through open doors or windows even if a house is otherwise rat-proofed.

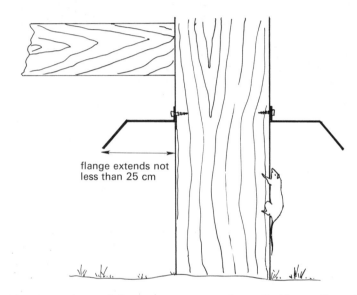

Fig. 39.15 A metal flange prevents rats from climbing walls and wooden piers

(Rentokil)

Fig. 39.14 Rats are good climbers and use roof timbers and beams as routes from one part of a building to another; routes are stained with oil from the rats' fur

Rat-proofing involves making ground floors of concrete, continuous with the walls, or else sinking the foundations sufficiently deep into the ground to prevent rats burrowing underneath (Fig. 39.13).

Tight-fitting, well-maintained doors are essential and, in regions where rats are numerous, additional precautions may be necessary. A metal flange will form a barrier to prevent rats and mice from climbing walls and gaining access to open windows (Fig. 39.15). Such a flange is particularly important where houses are made of wood. When houses are built on piles or stilts a flange should be fitted around each pile. In many parts

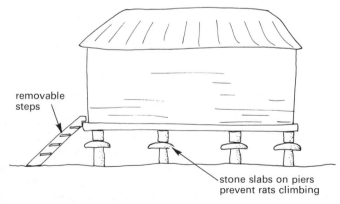

Fig. 39.16 Grain stores require special protection against rats and mice

of the world, grain stores have been built on piers with slabs of stone to prevent rats climbing up. The steps used for entering the granary are removed when not is use (Fig. 39.16).

Rainwater gutters and drains

Many insects breed in surprisingly small amounts of water. A badly laid rainwater gutter around a roof may hold water for a sufficient number of days after rainfall for hatching and larval development of an insect to be completed.

(WHO/D. Deriaz)

Fig. 39.17   The grain stores in this village are raised above the ground on stone pillars to give protection from rats and damp

Cockroaches and ants are primarily crawling insects, but whereas ants form their nests usually in the earth under or near a house and crawl in to find food, cockroaches live and breed inside houses. In Britain they are often mistakenly called black beetles. They spend the hours of daylight under floorboards, behind cupboards and in the gaps where hot-water pipes pass through walls. They often gain access to houses in boxes and baskets and apart from killing them with DDT or pyrethrum powder, the principal preventative measure is to reduce the number of cracks in floors and other small gaps in which they can hide. The dead space under a wooden floor provides an ideal breeding ground, with crumbs and other food accumulating through the cracks between floorboards (*see* Fig. 39.13), and this space is usually difficult to disinfest.

Ants are rarely a serious problem in temperate regions but in the tropics they can be a considerable nuisance. The design of a house rarely prevents their entry and when the path of entry is noticeable it should be sprayed with a residual DDT spray.

Flies and mosquitoes can be excluded by covering window openings with fine wire mesh, and doors that are left open for ventilation in hot weather can also be protected by a second, light frame door covered with wire mesh. Kitchens and food stores in particular should be protected in this way since free air movement is desirable, especially in hot weather. It is also helpful to make bedrooms or sleeping rooms free from insects in the same way, partly because they can be irritating and particularly because most biting insects, many of which are night fliers, are also vectors of disease organisms.

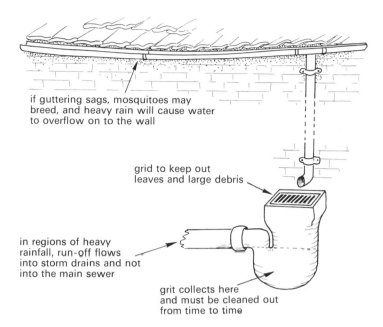

if guttering sags, mosquitoes may breed, and heavy rain will cause water to overflow on to the wall

grid to keep out leaves and large debris

in regions of heavy rainfall, run-off flows into storm drains and not into the main sewer

grit collects here and must be cleaned out from time to time

Fig. 39.18   Design of rainwater guttering and drain

Where possible, drains around the house should be of the type shown in Fig. 39.18. The grid allows water to pass through but excludes leaves and other large pieces of material that might block the drainpipes. The grid also deters flies and mosquitoes from entering the drain, though it is not always 100 per cent effective in this respect. The space below the grid traps grit and also provides a seal to prevent gases from the drainpipe seeping back. The grit trap needs to be cleaned out at intervals, depending on the amount of water passing through the drain and the amount of grit and sediment deposited.

Practical Work

Experiment 1   Heat transfer

Take two jam jars or two tin cans of equal size. Paint one black or cover it with black paper and paint the other white or cover it with white material. Place an equal amount of cold water at the same temperature in each. Read the temperature on a thermometer and record it.

Now place the two containers either in strong sunlight or in the rays from an electric fire. Measure the temperature of the water after ten minutes and after twenty minutes. Which container warms up more rapidly? What does this result suggest about the choice of colour for the outside of a house in a hot, sunny climate?

Now place equal quantities of hot water in each of the two containers and put them in a cool place. Which *loses* heat more rapidly? What does the result suggest about the choice of colour for the outside walls of a house in a cold climate?

Exhaust fumes

All motor cars and lorries produce poisonous exhaust fumes (Figs. 40.3 and 40.4). One of the gases present is carbon monoxide which forms a stable compound, carboxyhaemoglobin, with the haemoglobin in the red blood cells. This prevents the red cells from doing their job of carrying oxygen from the lungs to the tissues. The exhaust from the few cars passing down a minor road in the open country is quickly diluted in the atmosphere and does no damage. However, in a large city with high traffic density (a large number of cars in a small area) the build-up of exhaust fumes can be serious, particularly for those

Fig. 40.3   The exhaust from motorcycles, cars and trucks contains carbon monoxide and carcinogens

people who spend most of the day in the streets, such as street traders and police. Most people do not spend long periods each day in an exhaust-polluted atmosphere, but even for these people the effects over a number of years can be serious. Statistics show that there is a higher incidence of lung cancer in people who live in large, traffic-lined cities than among people living in the country. This may be linked with the fact that car exhaust contains carcinogens, chemicals that induce cancer.

Fig. 40.4   In Hungary spot checks are made on exhaust emission and heavy fines are imposed if excessive fumes are produced

(WHO/K. Hemzö)

Cigarette smoke

People are becoming more aware of the harmful effects of cigarette smoke (p. 268). While the most damaging effects are to the person who smokes cigarettes, it is recognized that, if several people are smoking in a confined space such as a closed room or a bus, then the non-smokers must inhale the smoke as well. For this reason there are regulations in many countries which forbid smoking of any kind in certain sections of buses, railway carriages and in restaurants (Fig. 40.5). Attempts, therefore, are being made to prevent pollution of the atmosphere by cigarette smoke.

Fig. 40.5   'No smoking' areas protect non-smokers from the harmful pollutants in other people's cigarette smoke

Water pollution

Sewage

When raw sewage is discharged in quantity into a river it causes pollution in several ways. The presence of pathogenic organisms in the sewage is a danger to people using the river water for washing or bathing as well as to those who drink it. The organic matter in the sewage forms food for many kinds of organisms in the water, in particular bacteria. If these decomposing bacteria reproduce in enormous numbers then they use up the dissolved oxygen in the river water so effectively that there is not sufficient oxygen left for fish to breathe or for the water-fleas and insect larvae that form part of the food chain feeding the fish.

Nitrates and phosphates in sewage effluent may stimulate the growth of water plants to such an extent that they become a nuisance. When the plants die, the organisms that decompose them deplete the oxygen supply.

Modern detergents are very effective for cleaning things, from clothes and kitchen utensils to industrial products. Most detergents produce foam when mixed with water and sometimes this foam is very difficult to destroy. This can be a nuisance in a sewage works since the treatment processes do not destroy the detergent chemical. When the treated sewage is discharged into a river more foam is produced which forms a blanket layer over river water that prevents oxygen reaching the water and dissolving in it. Fish and other animals in the water may die from lack of oxygen. The phosphates in modern detergents may also make water plants grow rapidly, causing blockage of rivers and lakes.

Industrial waste

Factories of many kinds produce soluble waste materials which they have to dispose of. It is easy to discharge such soluble waste into a river but the effects may be disastrous. Organic waste has a similar effect to raw sewage, described above. The wastes from chemical processes involving heavy metals such as copper and lead can be very poisonous and in countries with such industries, regulations may have to be enforced to prevent rivers being polluted by poisonous chemicals.

The leakage of highly poisonous chemicals into the Rhine as firemen fought a blaze at a factory at Basel, Switzerland, in 1986 resulted in the death of all fish and other aquatic life in the river over 200 miles downstream. Towns in Germany which filtered and chlorinated drinking water from the Rhine could not remove the toxic chemical. Alternative emergency supplies of water had to be used. By the time the water reached Holland the poison was very much diluted, but still caused alarm and presented a risk to water users.

Fertilizers

Fertilizers that are spread on the land to produce better crops must be soluble if they are to be absorbed through the roots of the plants. In areas of heavy rainfall fertilizers, particularly nitrates and phosphates, may be washed off the surface of the soil or washed through the soil, to appear in spring water and rivers (Fig. 40.6). Such fertilizers increase the population of the algae and other plants growing in the water and may result in much more material having to be removed by the filtration plant treating drinking water. Large growths of microscopic algae sometimes produce chemicals which taint the water, giving it an unpleasant taste.

Nitrate is a very effective fertilizer for producing increased yields of green crops; however, recent research has shown that even quite small quantities of nitrate taken in with drinking water over a number of years can damage the haemoglobin of young children. The excess nitrate and phosphate in lakes give rise to *eutrophication*, described on p. 52.

Pesticides and insectides

Many chemicals are used to kill insects and prevent them doing damage to crop plants. DDT is probably the best known of these chemicals. When DDT was first used it was very successful in destroying aphids and caterpillars that feed on crops. What was not realized at the time was that the animals, mostly birds, that feed on insects accumulate DDT in their bodies from the insects they have eaten. Birds are more easily poisoned by DDT

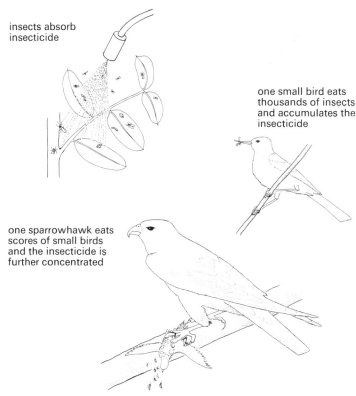

Fig. 40.8   How a pesticide becomes more concentrated as it is carried along a food chain

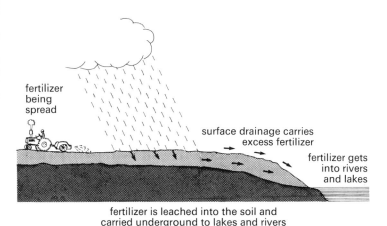

Fig. 40.6   How fertilizers get into river water

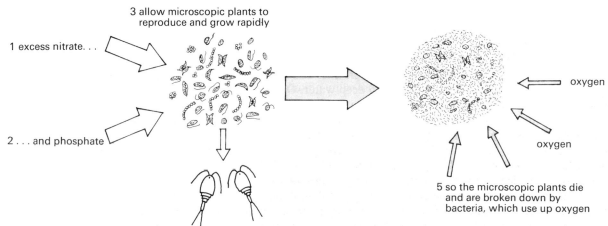

Fig. 40.7   Effects of eutrophication

Ozone is present throughout the atmosphere but reaches a peak at about 25 km above the Earth's surface, where it forms what is called the **'ozone layer'**. This layer filters out much of the ultraviolet radiation in sunlight.

The chlorine from the CFCs reacts with ozone and reduces its concentration in the ozone layer. As a result, more ultraviolet (UV) radiation reaches the Earth's surface. Higher levels of UV radiation can lead to an increase in skin cancer. A survey in Australia showed a threefold increase in certain forms of skin cancer between 1992 and 1995. Increased UV radiation can also affect crops, damage marine plankton and even distort weather patterns.

The reactions between chlorine and ozone are complex. There are natural processes which restore ozone, but these do not keep pace with the rate of destruction.

CFCs are not the only ozone-depleting compounds. Others include methyl chloride, (used for destroying pests in soil) and halons, (for use in fire extinguishers).

The greatest destruction of ozone takes place over the North and South Poles. This is a result of the very low temperatures and other climatic conditions. The thinning of the ozone layer over the Arctic and Antarctic regions has led to the formation of 'ozone holes'. Their formation is seasonal but, nevertheless, the ozone holes get deeper and more extensive year by year. Between 1969 and 1993 there was a 14 per cent reduction in the ozone layer over North America and Europe in the winter. Figure 40.11 shows the decrease in ozone levels in the Antarctic over a 40-year period.

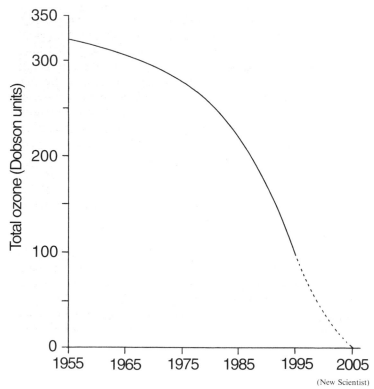

Fig. 40.11   Spring levels of ozone in the Antarctic. At current rates the ozone layer might disappear by 2005

Protecting the ozone layer

The appearance of 'ozone holes' in the Arctic and Antarctic and the thinning of the ozone layer elsewhere has spurred countries to get together and agree to reduce production and use of CFCs and other ozone-damaging chemicals.

1987 saw the first Montreal protocol which sets targets for the reduction and phasing out of these chemicals. In 1990, nearly 100 countries, including Britain, agreed to the next stage of the treaty, which committed them to reduce production of CFCs by 85 per cent in 1994 and phase them out completely by 1996. In 1995, 150 governments were party to the ozone treaty.

The use of CFCs is now banned in the industrialised countries but world production still stood at 360 000 tonnes in 1995 because developing countries are still permitted to produce and import CFCs until, at least, 2010. They are used mainly for refrigeration and air conditioning equipment.

Third World countries, quite reasonably, point out that 90 per cent of CFCs in the atmosphere come from the rich countries and these should agree to meet the cost of replacing them with less harmful chemicals. So far, the industrialised countries are not providing sufficient funds to meet even the 2010 deadline.

Less harmful replacements include hydrochlorofluorocarbons (HCFCs) which still contain chlorine but cause only one third of the damage of CFCs, and hydrofluorocarbons (HFCs) which contain no chlorine and do not react with ozone. The ozone treaty includes a programme for phasing out HCFCs but unless funds are provided for the production of alternatives, the targets are unlikely to be met.

Radioactivity

High doses of radiation will kill people, yet there is always *some* radioactivity in our surroundings, all the time. This is known as *background radiation*. Man and other animals seem to live successfully in spite of it, though we cannot be sure that it has no effect on us.

Higher levels of radioactivity cause some cells to divide more frequently than is usual. This leads to a form of cancer. At least one kind of leukaemia (p. 171) is caused by radiation. Radiation may also cause changes in the genetic message carried in the nucleus of cells, particularly of sperms and egg cells. This means that the changes are transmitted to the children who are produced from these gametes. The children may be deformed as a result.

In many parts of the world *nuclear power* is being used increasingly as a means of generating electricity. This means that energy from radioactive materials, such as uranium, is used to produce heat to drive generators. Eventually the radioactive fuel will produce no more useful energy. However, it is still radioactive and there is the problem of removing it from the power station and disposing of it safely. Some countries do this by putting the used or spent fuel in strong steel containers and sinking these in deep parts of the sea. In other countries the radioactive waste is buried deep underground. Provided that care is taken in handling radioactive material then the risk to human health is probably very small indeed. Problems arise when things go wrong. Damage in a nuclear power station in the United States in 1980 resulted in the escape of radioactive material into the atmosphere. People living in a wide area around the power station had to leave their homes until the damage had been brought under control and the level of radiation was declared safe.

The explosion at the nuclear reactor at Chernobyl in Russia in 1986 led to a high level of radioactivity around the plant and several people died as a direct consequence. But air movements high in the atmosphere resulted in radioactive material being carried to Scandinavia and Britain, 2 000 miles away.

In 1945 the first atomic bombs were exploded over two cities in Japan. Thousands of people were killed and many thousands more were badly injured by radiation. Since that time there have been many 'tests' of bombs producing much more radioactivity than the original atomic bombs. The radioactive material from these bombs is shot up into the air to great heights, and then slowly settles back on to the earth as radioactive fallout (Fig. 40.12). This represents a risk to people living in a wide area, though the risk is smaller and much less immediate than to those people who live in a 20 km radius of the explosion (Fig. 40.13). However, the radioactive fallout contaminates drinking water and crop plants, and may be taken into the body.

Summarizing, pollution is a risk to health that has only been recognized fairly recently and the ways of measuring or even estimating the risks to human health are still being worked out. This does not mean that we should ignore pollution or stop trying to assess the ways in which it affects us.

Questions

1 Why are car exhaust fumes more likely to be damaging to health in a large city than in the open countryside?
2 In what ways can people avoid the effects of cigarette smoke produced by other people?
3 How does raw sewage affect river water, even when the sewage contains no pathogens?
4 Describe the ways in which you see the river nearest to your home becoming polluted.
5 How might excessive use of agricultural fertilizers lead to pollution of a river?
6 List the gases which contribute to a) the greenhouse effect and b) the depletion of the ozone layer. Does any gas appear in both lists?
7 Which of the greenhouse gases cannot be controlled by human action?
8 In what way do we benefit from the greenhouse effect and in what ways could it be harmful?
9 Why should industrialised countries bear most of the costs of avoiding ozone depletion and global warming?
10 What contribution does the motor car make to atmospheric pollution?
11 In what two ways might global warming contribute to a rise in sea level?
12 Suggest reasons why it is difficult to get international agreement on reductions of polluting gases.
13 How does radiation affect the body?
14 Why was a rapid increase in the radioactivity recorded in the meat from sheep in North Wales and the Lake District in England linked to the explosion at the Russian nuclear power station at Chernobyl?

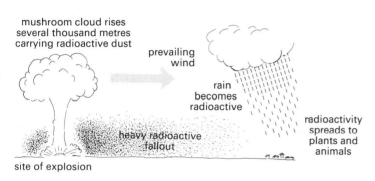

Fig. 40.12   Radioactive fallout can spread over a great distance

(UN/Sygma)

Fig. 40.13   Nuclear explosions produce a characteristic 'mushroom' cloud

41
Community Health

Primary health care

With the increase in the scientific understanding of the causes of illness and disease it is necessary to train people to bring new knowledge of health care to everyone. The training of health workers is the responsibility of government, both at a national level as well as local. Training costs money and this must be provided chiefly from taxes, though some money is provided by voluntary agencies and by the World Health Organization (p. 312), who also provide some of the teachers to train the health workers.

The World Health Organization, or WHO as we shall call it from now on, states in its constitution that 'Governments have a responsibility for the health of their peoples which can be fulfilled only by the provision of adequate health and social measures.' The same constitution also states that 'Informed opinion and the active co-operation of the public are of the utmost importance in the improvement of the health of the people.' In Third World countries this last aim is achieved largely through primary health care. The aim of primary health care is that it should be available to everybody and efficient and cheap to run, so that the community can afford it. Primary health care workers set out to improve the nutrition of the people they serve. This may be a matter of teaching people how to produce more food, though more often it is a matter of showing how a better balance in the diet can be obtained. Well-nourished people are better able to fight off disease (Fig. 41.1).

Primary health care workers (PHC workers) have the task of showing the community how to obtain and maintain an adequate supply of safe drinking water and how to dispose of sewage safely. This may include teaching children how to use latrines properly and why they should develop the habit of using the latrine, rather than urinating and defaecating around the place. Co-operation from the community is essential if safe water and good sanitation are to bring the improvements to health that they should do.

Immunization of the population against those infectious diseases that are common in the district is another function of the PHC workers. They also have a knowledge of local endemic diseases, such as measles or malaria, and can give a lead in establishing control measures. All of these aspects of their work are linked to health education which, if it is to be effective, must

Fig. 41.1   Training primary health care workers in aspects of nutrition. They will advise mothers on how to feed their children

(WHO/R. da Silva)

gain the active and enthusiastic support of the population. PHC workers can identify the serious cases of illness that require more specialized treatment, perhaps in a hospital. They may also provide care during pregnancy (Fig. 41.2).

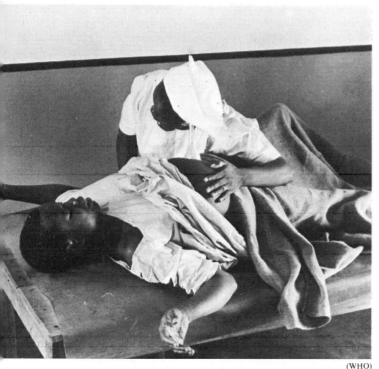

(WHO)

Fig. 41.2   A nurse listens to the heartbeat of an unborn baby

Clinics

Both at home and overseas, clinics are usually concerned with one particular aspect of health, for example with dental health (p. 128), or baby care.

Antenatal clinics   Only pregnant women will receive advice and treatment at an antenatal clinic. The health of the mother-to-be and her developing baby will be checked at each visit. For example, a simple test will show whether there is sugar in the mother's urine. If there is, she may be diabetic and this could affect the development of the unborn child. Blood pressure, too, is checked regularly. High blood pressure can be harmful to both mother and baby. A blood test is taken early in pregnancy and if disease organisms, e.g. trypanosomes or malarial parasites, are found to be present then remedies can be applied. The blood sample enables the mother's blood group to be identified and also whether she is rhesus positive (Rh+) or rhesus negative (*see* Chapter 11).

Advice is given about diet, for although the mother does not, literally, have to eat for two people, her choice of diet can have a considerable effect on the health and development of the foetus. She is also given advice on feeding and caring for the baby after its birth.

A midwife normally supervises the birth, whether this takes place at home or in hospital, and she is able to judge whether the assistance of a doctor is necessary. A health visitor may take over duties a few weeks after the baby has been born, visiting the mother in her home and reminding the mother at the

appropriate time about immunizing the baby against diphtheria, whooping-cough and polio.

Family planning clinics. These provide information about contraception (*see* p. 109), ways of preventing fertilization, but they also give positive guidance on planning a family, particularly the time spacing between one baby and the next, as well as explaining the stages in life when it is safest for the woman to have babies. Fig. 41.3 shows that the health risk to a mother through having a baby is lowest at 25. The risks to the mother and her child are greater below the age of 20 and above 30 years of age.

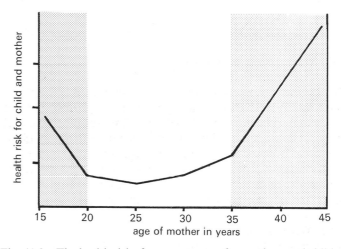

Fig. 41.3   The health risks from pregnancy for mothers and children are lowest when the mother is between 20 and 30

Special clinics. These are sometimes set up to deal with particular situations. In a region where sleeping sickness (trypanosomiasis) is endemic it is sensible to provide a clinic where people can come to check whether they need treatment and to check whether the treatment is being effective (Fig. 41.4).

Fig. 41.4   A sample of blood is given to be tested to see whether treatment for trypanosomiasis is effective

(WHO/P. Pittet)

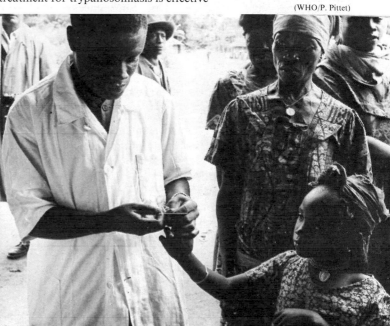

Child health

Healthy children who have been educated to look after their health are likely to become healthy adults. Most countries, therefore, are spending time, effort and resources on improving child health (Fig. 41.5).

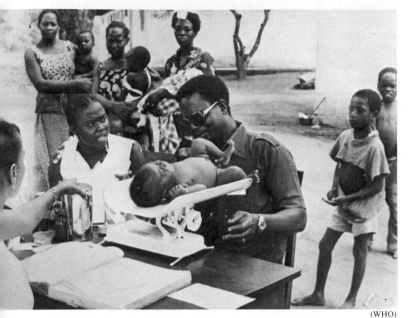

(WHO)

Fig. 41.5 Primary health care. Mothers and children waiting at this baby clinic can hear advice being given to the mother sitting next to the health worker

At school, children are seen by the school medical officer or by a health worker. How frequently they are seen varies, but it is sensible for a child to be medically examined when he or she first begins schooling. He or she may be suffering from a disease which could be transmitted to other children. A child may suffer from a disability such as deafness or partial blindness that should be recognized by the teachers if the child is to benefit from its education. Sight and hearing should be checked regularly while a child is at school and so should the child's posture and physical development. Such checks enable the trained health worker or medical officer to notice defects in teeth, skin disorders and signs of malnutrition (which is distinct from being underfed: *see* p. 37).

Local Health Services in Britain

The health of a community cannot be left to chance. In Britain the Government, through the Department of Health and Social Security, has organized a number of Regional Health Authorities as part of the National Health Service. Under them a number of Area Health Authorities control the hospitals, public health services and the provision of doctors to serve the community. At a more local level District Management Teams are responsible for family doctors, dentists, pharmaceutical services to provide drugs and medicines, and ophthalmic services to see to the care of eyes. These Management Committees have to co-ordinate the services under their control and one of the key people is the District Community Physician (commonly referred to as the DCP). He has largely taken over the duties once carried out by the Medical Officer of Health or MOH, and he organizes the services in his district. His depart-

ment analyses health statistics (pp. 186 and 190) to see whether patterns of health are changing and to ensure that the medical services are being put to best use. He takes charge of the co-ordination of resources in the event of an epidemic and this may include the vaccination of the local population, the tracing of contacts and the arrangement of emergency hospital services. He is expected to notify the National Health Organization of the nature and extent of the outbreak. Health Care Planning Teams are an important feature of the Health Service. Under the DCP they are responsible for providing care for expectant mothers, schoolchildren, the elderly and the people who are chronically ill.

The provision of safe drinking water (Chapter 36) and the efficient removal of sewage and domestic refuse (Chapter 37) are also matters of public health and in Britain they are the concern of the Environmental Health Officers who also arrange for the inspection of food processing plant, slaughterhouses and catering establishments as well as for the control of rats, mice and other pests.

Public health inspectors

Public health inspectors (sometimes known as environmental health inspectors) are responsible for supervising and inspecting many aspects of public hygiene (Figs. 41.6 and 41.7). They check that shops and markets are kept clean and free from rats and flies and they also inspect places where meals are cooked and served to the public and, if an outbreak of food poisoning occurs, they will take away samples of food for testing to try to trace the source of the infection. They are responsible for checking the standard of hygiene in slaughterhouses and abattoirs (p. 287). Where standards of inspection are slack or where there are no inspectors at all, the frequency of outbreaks of food-borne disease increases and the health of the community is at risk.

(WHO)

Fig. 41.6 An environmental health inspector in Guatemala checks the food on a street vendor's stall

(J. Sainsbury plc)

Fig. 41.7 A meat inspector wearing sterile overalls examines sheep carcasses before their distribution to shops; each carcass is labelled with its origin

In order to help the health inspectors, most countries have laws enacted either at national level (statutes) or at local community level (bye-laws) which require food dealers and caterers to observe certain minimum standards of hygiene.

The community health service has as another of its duties the extermination of rats. The breeding and feeding areas have to be discovered and then methods devised for the destruction of rats that do not put the human population at risk. Thus gassing of rats may be appropriate in a warehouse that can be sealed off and isolated, but it is a dangerous method in houses or flats.

Hospitals

The provision of hospitals is usually a national rather than local responsibility, but the pattern varies throughout the world. In some countries hospitals are run by private concerns such as religious orders. Certain hospitals cater for the training of doctors as well as nurses and are known as teaching hospitals. They form part of a medical school and may be very large with several hundred beds.

Whether a hospital is large or small, it is usual to treat the patients in wards—large, airy rooms with several beds. In order to organize the treatment efficiently the patients are separated into *surgical* and *medical* cases, according to whether a surgical operation is necessary or treatment with drugs and medicines alone is required. Patients suffering from infectious diseases are usually isolated in rooms with a single bed or, in large communities, in special *isolation hospitals*. Apart from these divisions it is usual to separate male patients from female, and children are usually treated in children's wards or in children's hospitals since the patterns of medical care they require are often quite different from those appropriate to adults.

In addition a hospital will usually have a *casualty unit* for the immediate treatment of injured persons and, close to this, an X-ray unit so that the state of a patient's bones can be studied. Using modern techniques it is also possible to study many of the soft tissues and organs of the body by X-ray. One or more operating theatres are located near to the surgical wards and close to the casualty department. Fig. 41.8 shows the layout of a small modern hospital in which the distances that patients have to be moved and nursing staff have to travel between one department and another have been kept to a minimum through thoughtful planning.

In many countries the provision of medicines prescribed by a doctor to out-patients is made from the hospital. The

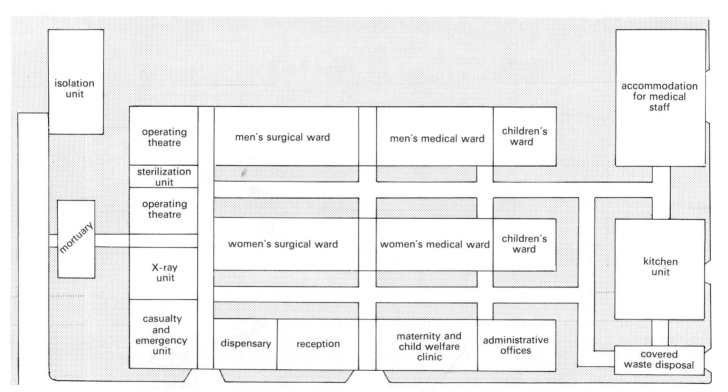

Fig. 41.8 Plan of a small, single-storey hospital

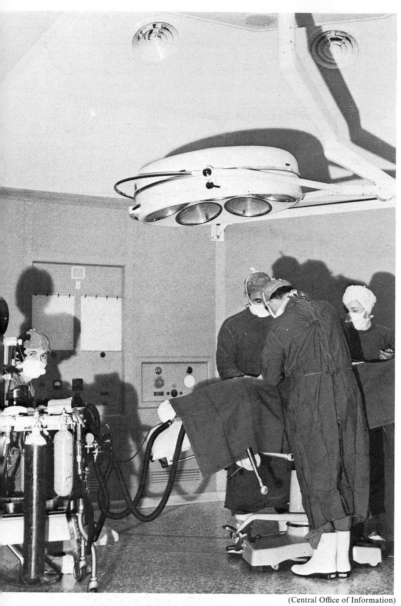

(Central Office of Information)

Fig. 41.9 A small operating theatre, designed for easy construction in remote areas

department concerned is called the *dispensary* and it also provides and checks the medicines used for treating the in-patients, those patients being treated in the wards.

The organization of health services is an expensive business and it is an unfortunate fact that the resources available both in terms of material facilities—hospitals and health centres—and trained medical personnel vary a great deal between one country and another. The money to pay for medical services is usually provided through taxation and, in some countries, by levying a contribution from all employed people and also from their employers, as happens in Britain. The introduction of the National Health Service into Britain in 1948 resulted in a general improvement in the health of the population. It marked a further step forward in a pattern of progress that began more than a hundred years ago with the discoveries and energetic action of men like Jenner (p. 252), Pasteur and Snow (p. 190). As a result, most of us can look forward to longer and healthier lives than our great-grandparents.

World health

Mutual help between countries in tackling health problems is by no means new. In the nineteenth century the discoveries of Pasteur in France, Koch in Germany and Lister in Britain were being made known throughout Europe and through much of the rest of the world. Quarantine was already operating between many countries. This involved the isolation or quarantine of travellers for a period of time, originally forty days (quarante is French for 40), in which any disease organism would incubate so that the person concerned would show recognizable signs of the disease. There were, however, only sporadic attempts to standardize methods of promoting health.

In 1948 the first World Health Assembly met in Geneva, Switzerland and established the headquarters of what is now known as the World Health Organization or WHO. The activities of WHO are many and are organized under divisions, for example the Divisions of

> Education and Training
> Public Health Services
> Health Protection and Promotion
> Communicable Diseases
> Malaria Control
> Environmental Health
> Health Statistics

The Division of Public Health Services co-ordinates developments in such fields as maternal and child health, nursing, organization of medical care, and public health administration. It makes funds available for the setting up of health services in countries where resources are limited, and provides expert knowledge on the methods of organizing such services. The influence of WHO in the field of maternal and child health has led to a very noticeable reduction in the infant mortality rate (*see* p. 186) in all the countries in which the Organization has been active. This is one of the major achievements of the campaign to train primary health care workers (p. 310).

The Division of Communicable Diseases runs the Epidemiological Intelligence Service. This grand name implies that information is collected about the spread of disease around the world. In particular, reports are collected in Geneva about outbreaks of cholera, plague, relapsing fever, typhus, and yellow fever—the internationally *notifiable diseases* (p. 189). Member states of WHO are bound by the International Sanitary Regulations to send in this information. The division publishes quarantine regulations which, if put into practice, help to check the spread of these diseases to other parts of the world. The same division also publishes vaccination requirements for travellers who have to carry a certificate of vaccination when journeying from countries where a given disease is endemic (Fig. 41.10). For example, visitors to the Caribbean islands from South America and people travelling from other parts of the world to the Caribbean via South America must carry a valid certificate of vaccination against yellow fever, signed by a qualified doctor. This is checked on entry to the country.

The present campaign to control TB throughout the world and the successful campaign to eliminate smallpox (p. 255) were sponsored by this division of WHO.

Malaria is a disease that kills millions of people every year as well as making many more millions seriously ill. The disease is

The Division of Environmental Health is concerned with such matters as community water supply, sanitation and environmental pollution. Much has been achieved in the matter of providing safe drinking water and co-operation is growing. When a new dam is being built for the purpose of providing

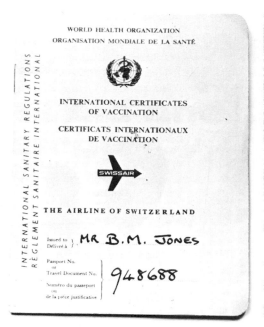

Fig. 41.10   International Certificate of Vaccination

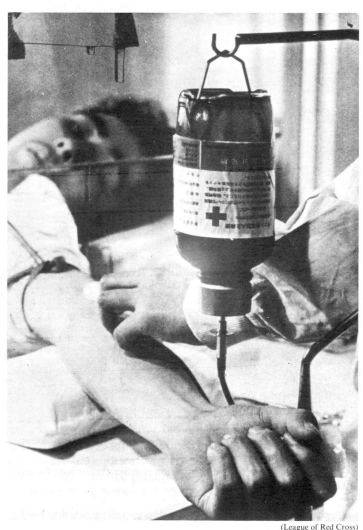

(League of Red Cross)

Fig. 41.12   The International Red Cross ensures supplies of blood are available for transfusion in many parts of the world

considered to be of such worldwide significance that a special Division of Malaria Eradication (now Malaria Control) was set up in 1955. By providing an international scientific team to study the parasite and its mosquito vector and, in particular, the ways in which the vector can be controlled by using insecticides, many former malarial regions have now been made free from the disease. By 1970 over 60 per cent of the population of areas formerly with malaria were no longer at risk. However, the situation has become worse as mosquitoes have developed resistance to DDT and several other insecticides. Attempts are still being made at an international level to control the vector of malaria. More important, new antimalarial drugs are being tested to control the parasite inside the human body. However, as long as the disease exists anywhere in the world and people travel from one country to another in which a suitable vector exists, there is the chance of the disease being re-established. One of the tasks of this Division therefore is to check all reports that indicate the possible spread of the disease and to maintain mosquito control measures, even when endemic malaria has been eliminated from a country (Fig. 41.11).

Fig. 41.11   A stewardess sprays the passenger compartment of an international aircraft in a WHO-sponsored campaign to reduce the transmission of insect-borne disease

(WHO/A. Kochar)

Minor injuries

Cuts. Cuts which do not require the emergency treatment described under 'Arrest of bleeding' should be cleaned by carefully washing the skin round the damaged area using a clean cloth. Wherever possible, avoid using antiseptics; they slow down the healing process. Cover the wound with a dry bandage or plaster and change this if it gets wet. Leave the wound exposed to the air as soon as possible.

Sprains. If the ankle or wrist joints are forced in the wrong direction by a fall, the ligaments holding the bones together (p. 116) get torn, resulting in a sprain. This is accompanied by pain and swelling. If the patient cannot move the fingers or toes, you may suspect a fracture or dislocation. Otherwise, the treatment for a sprain is to wrap the joint firmly with a bandage soaked in cold water and wrung out. Soak the bandage again as it dries out to keep it cold.

Particles in the eye

In the eye really means 'on the cornea or sclera'. The eyes will water and the person will blink. These two actions may wash the particle into one corner of the eye or on to the lower lid. If the object is on the lower lid, in the corner of the eye or free to move on the sclera (white of the eye) it can usually be brushed gently away with the corner of a clean handkerchief or piece of cotton wool soaked in water. If this does not work, the casualty should put his or her face in a bowl of clean water and open and close the eyelids. The object will usually float away.

If the particle is on the cornea or if it appears to be embedded in the surface of the eye and does not move, or cannot be seen clearly, no attempt should be made to dislodge it. The patient must be dissuaded from rubbing the eye. The eyelid should be kept closed and covered with a soft pad held lightly in place with a bandage. Help should then be sought from a clinic or from a health worker.

Wasp and bee stings

Stings can be painful and may cause alarm. They are seldom serious unless they affect the lips or the eyes, when the sufferer should receive professional medical treatment. However, for stings on other parts of the body, some relief can be gained from the application of an antihistamine cream or ointment. A person who receives a large number of stings, e.g. as a result of disturbing a wasps' nest, would be wise to receive medical attention.

Further information

This chapter gives only an outline of emergency treatment. For more information consult the First Aid Manual of the Red Cross and St John Ambulance or apply to the local branch of either of these organizations.

Questions

1 Suppose you arrive at the scene of an accident. There are several people looking on but they are doing nothing. What would you do?

2 You find a man unconscious in his room. There is a half-empty bottle of pills on the floor. Say what you would do.

3 Some children are playing on a boat in a river when the boat capsizes and sinks. When you arrive on the scene, one child is on the bank screaming and holding his arm, one has pulled a limp and inactive companion from the water and a fourth is struggling in the water near the bank. What are your priorities?

4 Explain why a first-aider cannot prevent shock but can stop it getting worse.

5 If treating a bleeding wound, in what circumstances would your first concern be for the cleanliness of the wound and dressing?

6 If you were confronted with two unconscious people, how would you decide which to treat first?

7 Why do you think it is unwise to move an injured person unless you have expert help?

8 Why is it necessary to tilt an unconscious person's head back (a) while giving resuscitation, (b) when leaving them in the recovery position.

9 Why is it unwise to stand around on a windy day after sweating during vigorous exercise?

Reagents for practical work

Ascorbic acid (0.1 per cent) (from tablets). One 50 mg tablet per 50 cm³ distilled water will give a 0.1 per cent solution. Crush the tablets in a mortar while adding successive portions of water and decanting into a flask through glass wool. The solution deteriorates rapidly and should be prepared shortly before the lesson.

Benedict's solution. To make 1 litre, dissolve 170 g sodium citrate crystals and 100 g sodium carbonate crystals in 800 cm³ warm distilled water. Dissolve separately 17 g copper(II) sulphate crystals in 200 cm³ cold distilled water. Add the copper(II) sulphate solution to the first solution with constant stirring.

Ethanoic (acetic) acid (M/10). Place 6 cm³ glacial ethanoic (acetic) acid in a graduated flask and make up the volume to 1 litre with distilled water.

Hydrochloric acid (M/10). Dilute 10 cm³ concentrated hydrochloric acid with 990 cm³ distilled water.

Hydrochloric acid (2 M). Dilute 100 cm³ concentrated hydrochloric acid with 400 cm³ tap water.

Hydrogencarbonate (bicarbonate) indicator. Dissolve 0.2 g thymol blue and 0.1 g cresol red powders in 20 cm³ ethanol. Dissolve 0.84 g 'Analar' sodium hydrogencarbonate (sodium bicarbonate) in 900 cm³ distilled water. Add the alcoholic solution to the hydrogencarbonate solution and make the volume up to 1 litre with distilled water. Shortly before use, dilute the appropriate amount of this solution 10 times, i.e. add 9 times its own volume of distilled water.

To bring the solution into equilibrium with atmospheric air, bubble air from outside the laboratory through the diluted indicator using a filter pump or aquarium pump. After about ten minutes, the dye should be red.

Iodine solution. To make 100 cm³ stock solution, grind 1 g iodine and 1 g potassium iodide in a mortar while adding distilled water. Pour the solution into a measuring cylinder and dilute to 100 cm³. Do not store in polythene bottles because the solution will become decolourized. For experiments with enzymes, dilute 5 cm³ of the stock solution with 100 cm³ water.

Lime water. Shake tap water with excess of calcium hydroxide and allow to settle overnight. Decant the clear liquid.

Methylene blue. Dissolve 0.3 g methylene blue in 30 cm³ ethanol and add 100 cm³ distilled water.

Phenolphthalein solution. Dissolve 1 g phenolphthalein in 200 cm³ ethanol.

Sodium carbonate solution (M/20). Dissolve 5.3 g anhydrous sodium carbonate in 1 litre of distilled water.

Starch solution. Shake the appropriate quantities of starch powder and cold water together until the powder is dispersed completely. Heat the mixture while stirring constantly, until the liquid becomes translucent.

Water cultures (Experiment 4, p. 48). Either purchase the Sach's water culture tablets or prepare the solutions as follows, using chemicals of the highest purity available and freshly distilled or deionized water. Store the stock solutions in stoppered Pyrex flasks and mix and dilute them when required as described in Table 2. The three solutions selected have been found to give the most consistent and reliable results with wheat.

Table 1: stock solutions†

Solution	Compound	Formula	Mass/ g	Volume of water/ cm³
A	calcium nitrate	$Ca(NO_3)_2 \cdot 4H_2O$	9.5	100*
B	calcium chloride	$CaCl_2 \cdot 2H_2O$	4.5	100
C	magnesium sulphate	$MgSO_4 \cdot 7H_2O$	15.5	300
D	sodium nitrate	$NaNO_3$	6.9	100
E	potassium dihydrogen-phosphate(v)	KH_2PO_4	8.5	300
F	iron(II) chloride	$FeCl_2 \cdot 4H_2O$	1	200

*Add one drop of concentrated nitric acid to prevent the carbonate forming.

To make up 500 cm³ of each culture solution (enough for about 16 experiments) add the solutions in the volumes indicated in Table 2 to 485 cm³ distilled water. If the mixed culture solutions are to be kept for more than a week, add a further 0.5 cm³ iron(II) chloride solution to each 500 cm³ just before use. Set out the culture solutions in labelled containers and also a container of distilled water.

Allow 30 cm³ of each solution per experiment.

Table 2: culture solutions†

	Full culture	Lacking calcium	Lacking nitrogen
Solution A calcium nitrate	5 cm³	no calcium	no nitrate
Solution B calcium chloride			5 cm³
Solution C magnesium sulphate	5 cm³	5 cm³	5 cm³
Solution D sodium nitrate		5 cm³	
Solution E potassium phosphate(v)	5 cm³	5 cm³	5 cm³
Solution F iron(II) chloride	0.5 cm³	0.5 cm³	0.5 cm³

†These tables have been selected from *Wellington Hydroponics Solutions* worked out by D. J. Angwin of Wellington College.

76 (a) Many thousands of people in Britain suffer from chronic bronchitis. How does the disease develop, and how does it affect the sufferer? (b) What measures could be taken to reduce the number of new cases of bronchitis? (c) How does asthma differ from bronchitis? **[CE]**

77 (a) How does an adult housefly differ from an adult mosquito in the way in which it feeds and in the way in which it spreads disease organisms? (b) Copy this simple diagram and use it to illustrate the life cycle of the housefly by labelling each box with the name of one of the stages.

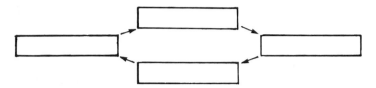

Write the word *metamorphosis* alongside the box for the stage at which metamorphosis takes place. (c) Use your knowledge of the life cycle and habits of the housefly to explain how this pest can be controlled. **[CO]**

78 (a) Briefly describe what is meant by the term parasite. (b) Write an account of the transmission of typhoid from man to man. (c) What steps can be taken to break the chain of transmission of typhoid? **[L]**

79 (a) What is a fungus? Your answer should include reference to the main biological features of fungi. (b) Describe how fungi may affect man. Your answer should make specific reference to *two named* beneficial examples and *one named* harmful example. **[L]**

80 (a) Give an account of the reproduction of viruses. (b) How may the spread of the common cold be prevented? **[L]**

81 (a) Explain clearly what is meant by the terms: (i) parasite, (ii) secondary host. (b) Describe those features of the structure, physiology and life cycle of a tapeworm *or* a fluke which enable it to be an efficient parasite of man. **[L]**

82 (a) Distinguish between bacteria and viruses. (b) How may the common cold be communicated from person to person and what steps should be taken to prevent its spread? **[L]**

83 (a) Describe *five* ways by which micro-organisms enter the body. For each way which you describe, name *one* micro-organism that enters normally by that pathway. (b) Why must patients suffering from tuberculosis (TB) be discouraged from spitting onto the ground if the spread of the disease is to be controlled? (c) Explain what is meant by airborne or droplet infection. **[CO]**

84 (a) Given a pure culture of a bacterium, describe an experiment you could conduct to find out the effectiveness of washing-up liquid as a disinfectant against it. (b) In what ways is the body normally protected against invasion by micro-organisms and what measures can the individual take to increase this protection? **[O]**

85 (a) Outline the stages in the life cycle of an anopheline mosquito, explaining where each stage lives and how it obtains food. (b) How is knowledge of the life cycle of the anopheline mosquito used in attempts to control the spread of malaria? (c) How do mosquitoes spread malaria? **[CO]**

86 (a) What is meant by the course of an infectious disease? (b) Describe fully the ways in which a person may become immune to a disease. (c) What is the biological or medical reason for quarantine regulations? (d) How do antiseptics differ from antibiotics? **[CO]**

87 Consider the following statements carefully and explain them as far as possible, dealing with each of the diseases mentioned separately. (a) Diphtheria and cholera are only very rarely encountered in Britain today. (b) The incidence of tuberculosis has dropped dramatically during the last fifty years. (c) Large numbers of people still suffer from the common cold. (d) The incidence of heart disease and venereal disease has risen considerably during the last twenty years. **[W]**

88 (a) Describe the features of the structure, life cycle and habits of the body louse which enable it to successfully parasitize man. (b) The louse is a vector of disease in man. Explain carefully what this means, and describe the effects of the disease concerned. **[W]**

Immunity and Immunization

89 (a) State briefly what you understand by the term *immunity to disease*. (b) What is meant by *artificial passive immunity*? (c) How may a person develop artificial *active* immunity, and what changes in the body provide this immunity? (d) What is meant by 'a healthy carrier of typhoid'? Why should such a person not be employed as a cook? **[CO]**

90 (a) Explain what is meant by active immunity to disease. (b) How can active immunity to tuberculosis be acquired? (c) In what ways does a vaccine differ from an antiserum? (d) Give an account of the ways by which the body is protected against disease-causing organisms, other than by immunity. **[CO]**

91 (a) Why is a person who has recovered from diphtheria unlikely to show such severe symptoms of the disease ever again? (b) Someone can suffer from the common cold as many as three times in one month. Explain carefully why this can occur. (c) Name *two* types of disease other than those caused by infective organisms and give *one* example of each. **[L]**

Socially Significant Diseases and Mental Health

92 (a) What are the signs and symptoms of gonorrhoea in (i) men and (ii) women? (b) Why is it desirable to control the spread of gonorrhoea? (c) What are the most effective means of controlling the spread of the disease? (d) Name two sexually transmitted diseases, other than gonorrhoea and syphilis. **[CO]**

93 (a) Explain why a person drives a car less safely after drinking alcohol. (b) People who feel depressed, often drink alcohol to 'cheer themselves up'. Explain why alcohol is unlikely to cure depression. **[CO]**

94 (a) How may a *named* venereal disease be spread in a community? (b) What measures may be taken to prevent this spread? **[L]**

95 Richard, aged 48, is an overstressed top-level sales manager who smokes heavily, entertains clients several times each week and gets little physical exercise. John, his son, is a normal 12-year-old who plays regularly for a school football team. Compare the two individuals with reference to: (a) heart and circulatory system, (b) lungs, (c) musculo-skeletal system, (d) mental health. **[L]**

Drinking Water and Sewage Disposal

96 (a) Use a diagram to help you to explain how water from a polluted river can be made safe for the people of a large town to drink. (b) If you were camping in the country, with many other campers around, and the only source of water to drink were a small stream, how would you make sure that you did not contract any infection from that water? **[CO]**

97 (a) List the main sources of water supply. (b) Describe in detail the way in which the water supply to a small town may be treated before it is made available through the taps. (c) In what ways may impure water affect the health of the community? **[O]**

98 Give (a) three uses of water to the body, and (b) three ways in which drinking water may be contaminated/polluted. Explain how tap water is made safe to drink for homes in our cities. **[L]**

99 What are the dangers of faulty sewage disposal? Describe a water-carriage system of sewage disposal and include simple diagrams of house drainage and public processing. What use can be made of the final products? **[O]**

100 Various methods are used to ensure a clean and safe water supply to housing in an urban area. Give an account of the principles and methods involved in: (a) slow filtration, (b) fast filtration, (c) chlorination. **[L]**

101 (a) Why is water from a deep well more likely to be pure and safe to drink than water from a shallow well? (b) Use a drawing to show how water from a river can be made safe for drinking by the people of a large town. (c) Why is it important for water to be filtered before it is chlorinated and not afterwards? (d) Explain why a person in a hot climate must drink about 2 litres of water each day in order to stay healthy. **[CA]**

102 Distinguish between cesspits, septic tanks and main sewers. Outline the way in which domestic sewage is treated in a large city. [O]

103 Why is it necessary for local authorities to make regulations about sewage disposal in towns while in country districts such regulations often do not exist? Explain briefly a system for the disposal and treatment of town sewage, once it has been collected. [CO]

104 Describe the various methods in use for the disposal of kitchen waste. What are the dangers from improper disposal? What measures should a housewife take to maintain a hygienic kitchen? [O]

105 Discuss the problems associated with living in large cities, using the following as a guide: (a) pollution; (b) disposal of human waste; (c) disposal of refuse; (d) effect of overcrowding. [W]

106 Describe either the aerated sludge method of purifying sewage or the percolating filter method. Explain what happens to the sludge and to the purified liquid. [W]

107 (a) Make a large, clearly labelled drawing of a section through a septic tank showing how water is disposed of after it has been treated. (b) Why is a septic tank given that name? (c) Make a list of the advantages of a septic tank over a pit latrine as a method of sewage disposal. [CO]

108 A sludge digestion unit is a major part of a modern sewage works. (a) What is sludge? What happens in a sludge digestion unit? (*Your answer should make reference to by-products of the sludge digestion process.*) (b) What is activated sludge? Briefly describe its function in effluent treatment. [L]

Safe Food

109 What are the chief kinds of organisms that cause the decomposition of food? Choose examples to show that some of the organisms that decompose food are pathogens while others are saprophytes deliberately used by man. Name one kind of food, rich in protein. Describe the methods that are used to preserve this food and explain the biological principles involved. [CO]

110 Describe the various ways of preserving and storing foods and indicate the danger of faulty food storage in the home. For five named methods of preservation, state one food suitable for each. [O]

111 Devise a code of hygiene that would be practical for the worker in a school kitchen. Explain briefly why you have included each point. [L]

112 Imagine that you have been asked to inspect a restaurant. State and explain ten considerations that would guide you in deciding whether the restaurant provided for the health and safety of the customers and staff. [J]

113 (a) Giving one good example for each, list the main ways in which food is preserved. (b) Using suitable examples give an account of the dangers of eating food which has been contaminated. (c) In what ways can food be prevented from contamination by houseflies? [O]

114 (a) Explain why each of the following procedures helps to reduce the spread of disease through food or to reduce food spoilage by micro-organisms: (i) washing hands thoroughly before touching food, (ii) adding salt, (iii) not allowing raw and cooked meats to be in contact, (iv) using a hard smooth surface for kitchen worktops, (v) quickly raising then lowering the temperature of food, as in the pasteurization of milk, (vi) allowing washed and rinsed dishes to dry by evaporation instead of wiping dry with a cloth. [A]

115 What are the main ways in which food can be treated to prevent it from going bad? In what different ways may food become contaminated and how may such contaminated food be dangerous to health if consumed? [O]

116 (a) Explain the biological reason for each of the following statements. (i) Food scraps should be wrapped before being placed in a dustbin. (ii) A frozen chicken should be thoroughly thawed before cooking. (iii) All prepared food should be kept covered. (b) Food spoilage can be caused by micro-organisms. Explain how high temperature treatment can extensively prolong the storage period of food while refrigeration is far less effective. (c) Explain the difference between pasteurization of milk and sterilization of milk. [A]

117 (a) Why is a person more likely to develop food poisoning after eating unsterilized meat or fish than after eating raw fruit or vegetables? (b) Explain how cooking makes food safer to eat. [CO]

118 Give an account of the method and scientific basis of each of the following methods of food preservation: (a) canning, (b) salting, (c) pasteurization. [L]

119 Explain why the following help to maintain standards of personal and community health: (a) Regular hair washing. (b) Pre-packing foods such as cheese. (c) Use of warm-air machines to dry hands in public conveniences. (d) Including bran or any other form of roughage in the diet. (e) Breathing through the nose, not through the mouth. [W]

Good Housing

120 In designing and building a house, care should be taken to make it difficult for rats to gain entry. (a) Use drawings to help to explain how this can be done. (b) How can food stores in the house be protected from rats that do enter the house? (c) Why are rats considered a danger in the spread of disease? [CO]

121 What is the importance of sunlight to housing? Discuss the principles involved in lighting, ventilation and heating of either an office building or a school or a hospital. [O]

122 What factors which would influence your health would you consider when buying a house? Explain your choice. [L]

123 State five major deficiencies common in poor housing and describe their probable effects on the residents. [L]

124 Describe, with its interior fittings, a house which would provide a healthy environment for a small family in your area. Use the following headings: (a) structure; (b) temperature control; (c) light; (d) water; (e) sewage disposal; (f) ventilation. [L]

125 Outline the main ways in which new houses may be: (a) built to keep out damp and insulated against heat loss; (b) sited in urban developments for maximum safety, amenity and health of the occupants. [O]

126 Discuss the hazards associated with living in housing which is below standard. Your answer should make reference to: (a) damp, (b) cold, (c) poor ventilation, (d) overcrowding, (e) poor lighting. [L]

127 (a) What factors should be considered when the site for a new house is being chosen? (b) Why should a pit latrine not be dug close to a well? (c) Draw a large, clear diagram of a section through a pit latrine and write notes explaining the precautions taken in building the latrine. [CO]

Pollution

128 Hazards to health may result from radiation and air pollution. (a) Name the common forms of radiation and their natural and artificial sources. (b) Describe the benefits and hazards associated with radiation. (c) Briefly describe the health hazards associated with (i) carbon monoxide, (ii) lead and (iii) inorganic dust. [L]

129 (a) What are the origins and characteristics of the major air pollutants? (b) Briefly describe some controls of air pollution that have proved effective. [L]

Community Health

130 Explain how the following have helped to improve the health and well being of the community: (a) Refrigeration of foodstuffs, e.g. meat. (b) Non-smoking areas on public transport. (c) Outdoor sports in school programmes. (d) Post-natal clinics.

131 Explain the importance of the following: (a) The provision and use of clean overalls and headwear in food departments in shops. (b) The use of ventilator fans in kitchens. (c) Sterilizing milk foods and bottles for feeding babies. (d) The presence of aerobic organisms in sewage filter beds. [W]

132 (*a*) Explain the biological reasons for each of the following practices: (i) storage of fresh food in a cool place, (ii) washing hands after defaecation, (iii) constructing kitchen work surfaces of hard, smooth materials. (*b*) (i) Explain the methods used at refuse tips to limit the housefly population. (ii) Why should houseflies be excluded from places where food is prepared or stored?   [A]

133 Describe the adverse effects on the body caused by: (*a*) bad posture, (*b*) a diet rich in animal fats, (*c*) frequent consumption of sweets, (*d*) smoking tobacco.   [L]

134 How does the community benefit from the following? (*a*) Child Welfare Clinics. (*b*) Nursery Schools. (*c*) Special (Venereal Disease) Clinics. (*d*) Playing fields.   [L]

Glossary

The explanations of terms given in this glossary are meant to be reminders rather than formal definitions, and are restricted to the context in which these terms are used in the book. The numbers in parentheses are page references.

abdomen the part of the body below the diaphragm and above the legs.

absorption (63) uptake of a substance, usually in solution, by a tissue or organ.

accommodation (136) an automatic adjustment made by the eye enabling it to focus effectively on either close or distant objects. The curvature of the lens is altered in doing this.

active transport (29) a hypothetical method by which substances are taken up or expelled by living cells other than by simple diffusion. It is an energy-consuming process.

aerobic respiration (28) the production of energy in cells by removing hydrogen atoms from food substances and eventually combining them with oxygen to make water.

agglutination (75) clumping together of cells such as bacteria or red blood cells.

allele (175) one of a pair of genes controlling the same characteristics but not necessarily producing the same effect. Genes *B* and *b* are alleles both affecting coat colour in mice but *B* results in black fur and *b* in brown fur. The alleles occupy corresponding positions on homologous chromosomes.

amino acid (18) a chemical compound containing a —COOH (acid) and an —NH$_2$ (amino) group both attached to the same carbon atom. Proteins consist of many amino acids joined together in long chains.

anaerobic respiration (28) a process in which energy is made available by breaking down food inside a cell, but oxygen is not used in the reactions.

antibiotics (257) chemical substances produced by some bacteria and fungi, which destroy certain other bacteria, but do not harm man. The antibiotic chemicals are extracted from the organisms and used to treat disease in man and animals.

antibodies (74) chemicals made in the blood and tissues which counteract harmful organisms or substances.

antigen (252) a foreign chemical, tissue or micro-organism which, if it gains access to the body, causes it to produce antibodies. The antibodies act against the antigen.

biological control (227) the use of one living organism to reduce the numbers of another, harmful organism.

calcification (111) the deposition of calcium salts in a living tissue, usually cartilage, altering its physical properties and making it harder.

capillary action (attraction) (295) the force that causes water to be drawn between surfaces which are close together, e.g. making it spread into blotting paper or between soil particles.

carcinogen (196) a substance capable of causing cancer if eaten, inhaled or applied to the skin over a period of time.

catalyst (26) a chemical which alters the rate of a chemical reaction (usually speeding it up) without being used up in the reaction.

centriole (159) a small body in an animal cell which participates in cell division. It divides into two and the two parts separate early in cell division, contributing to the spindle. The term *centrosome* is sometimes used to describe the paired centrioles before they separate.

centromere (160) the part of the chromosome which becomes attached to the spindle and by which the chromatids appear to be pulled apart at cell division.

centrum (117) the solid cylindrical part of a vertebra.

cerebellum (155) the large outgrowth from the roof of the hindbrain which plays an important part in controlling co-ordinated movement.

cerebral hemisphere (153) a large outgrowth from the roof of the forebrain which is associated with intelligent behaviour, memory and consciousness.

chiasma (170) a region of close contact between homologous chromosomes in the early stages of meiosis. When first formed the chiasma indicates where the exchange of portions of chromatids is taking place.

chlorofluorocarbons (308) chemicals used as refrigerants and propellants in aerosol cans, associated with ozone depletion.

chromatid (160) the product of replication of a chromosome. During cell division the chromatids are separated into the daughter cells.

chromosomes (162) structures which appear in the nucleus at cell division. They carry the genetic information (genes) which controls the activity of the cells and the development of the entire organism.

collagen (125) a substance produced by the body in the form of tough fibres. These fibres contribute to many tissues throughout the body, providing strength and flexibility.

combustion (57) the chemical term for burning, in which a substance combines with oxygen from the air and releases energy as heat and light.

commensal (200) one organism living in close association with another, sometimes actually inside it. The commensal organism is harmless and may even benefit its host.

concentration (20) the quantity of a substance that is present in a given volume, e.g. how much oxygen is dissolved in a fixed volume of water.

conservation (45) the preservation of a stable environment. It does not imply an unchanging environment, but one in which organisms and resources are not totally destroyed in the course of change.

contagion (213) transmission of disease organisms by contact with an infected person, his clothing or bedding, etc.

contamination (212) the introduction of harmful chemicals or organisms into food or water.

contraception (109) the prevention of conception. Sperms are prevented from fertilizing an ovum, or the fertilized ovum is prevented from implanting in the lining of the uterus.

convection (296) when a gas or liquid is heated it expands, becomes less dense and consequently rises. This movement is called 'convection'.

co-ordination (145) the way in which the organs and systems of the body work together to maintain life and efficient activity.

cortex (93) the outer layer of certain organs in the body.

crossing over (179) the exchange of portions of chromatids which takes place during the cell division giving rise to reproductive cells. It results in new combinations of characteristics in the offspring.

culture (207) the provision of ideal conditions for the growth of selected organisms.

deamination (69) the removal of the amino (—NH$_2$) group from an amino acid, leaving a compound which can be used to provide energy or changed to a carbohydrate.

deficiency disease (34) an illness which results from lack of a substance in the diet. It cannot be 'caught' or transmitted, and can usually be cured by adding the missing substance to the diet.

dehydration (66) excessive loss of water from tissues.

demography (185) a study of changes in population.

denaturation (18) an irreversible chemical change undergone by proteins when heated above 50 °C or treated with certain chemicals.

dialysis (24) a controlled process of diffusion by which dissolved substances can be extracted from a mixture through a membrane.

diffusion (22) the natural movement of a gas or dissolved substance from a region of high concentration to a region of low concentration.

diploid (163) the number of chromosomes in non-reproductive body cells. The name reflects the fact that each chromosome is represented twice, one of each pair coming from the male and one from the female parent.

dissociation (21) the separation of the components of a compound into ions, usually when it is dissolved in water.

DNA (166) the abbreviation for deoxyribonucleic acid, a chemical present in the nucleus of all cells and which controls the cell's activities.

dominant (169) when two contrasting genes (alleles) are present in an organism, the one which produces observable effects is called 'dominant'.

droplet infection (210) the inhaling of disease organisms in droplets of water from the exhaled air of an infected person.

drug (272) a chemical taken into the body to alter the metabolism medically, in such a way as to counteract an illness or relieve its symptoms.

effector (149) an organ which does something in response to a nervous or hormonal stimulus.

embryo (104) the developmental stage of an organism from a single cell to the stage when all its organs and systems are functioning.

endemic (189) the continuous presence of a disease in a population at a low level.

endocrine system (156) a number of glands in the body which secrete chemicals (hormones) directly into the blood stream. These chemicals regulate the level of activity of the body's systems.

enzyme (16) a protein made in the protoplasm of cells that speeds up the rate of chemical reactions in them.

epidemic (189) a rise in the number of cases of a disease in a population far above the level normally expected.

epithelium (10) the layer of cells lining the surface of internal organs.

erosion (53) the removal of topsoil by the action of wind or rain.

eutrophication (52) an excessive growth of plant life in inland waters as a result of a high level of nutrients. The eventual decay of the plants removes nearly all the oxygen from the water.

excretion (2) removal of harmful or excess substances from cells and from the body.

F₁ generation (175) the offspring resulting from the mating between two individuals. The first filial generation.

F₂ generation (175) the offspring resulting from a mating between the individuals of the F₁ generation. The second filial generation.

feedback (141) information about the functioning of the body's system which is sent to the central nervous system or endocrine system, helping to maintain precise control over the bodily functions.

fertility rate (186) the numbers of offspring produced by an individual or a population.

fertilization (100) the joining together of reproductive cells, e.g. sperm and ovum, to produce a zygote which can develop into a new individual.

fetus (104) the developmental stage of a mammal from the time its organs are formed to the time it is born.

gamete (100) a reproductive cell. Gametes from opposite sexes must meet and join together in order to produce a new individual.

gastric (63) any structure or function related to the stomach.

gene (164) a sequence of DNA molecules on a chromosome, inside the nucleus of a cell. It controls the chemical processes in the cell and the development of the organism produced from a zygote. The genes are hereditary units passed on from parents to offspring, and determine the characteristics of the offspring.

genetics (159) the study of the way in which characteristics are inherited.

genotype (159) the genetic constitution of an organism, i.e. all the genes present in a cell of that individual, whether or not they have an observable effect.

gland (10) a group of cells, some or all of which produce a chemical which is released and used by other parts of the body.

global warming (306) an increase in the Earth's temperature.

greenhouse effect (306) the effect caused by the Earth's atmosphere letting in heat and light from the Sun, but reducing the amount of heat which escapes.

haploid (170) the number of chromosomes in the gametes. This is half the number present in the other cells of the body.

hepatic (65) any structure or function related to the liver.

herbivore (58) an animal that feeds on vegetation.

heterozygous (175) carrying a pair of contrasting genes (alleles) for any one characteristic, e.g. *Bb*, where *B* determines black fur and *b* determines brown fur.

homeostasis (92) the maintenance of stable conditions inside the body. The temperature and composition of the body fluids are not allowed to fluctuate outside certain limits.

homoiothermic (98) the maintenance by an organism of a constant internal temperature independent of the temperature of its surroundings.

homologous chromosomes (163) the corresponding chromosomes of a pair, alike in shape and size. One is derived from the male and the other from female parent.

homozygous (175) carrying a pair of identical genes (alleles) for a given characteristic. Breeding true for this character.

host (194) the animal in or on which a parasite is living or feeding.

hydrolysis (16) the breaking down of a compound by reacting with water.

immunity (74) the ability to resist disease organisms as a result of possessing chemicals (antibodies) made by the body, which counteract the organisms.

implantation (103) after an ovum has been fertilized it sinks into the lining of the uterus. This is implantation, and the zygote will undergo all further development in this position.

inoculation (74) introduction of antigens, usually by injection, into the body to stimulate it to produce antibodies. This gives the body immunity to a particular disease.

insecticide (240) a chemical that kills insects.

insulation (296) a method of reducing the transfer of heat or electricity.

involuntary action (118) an action that is not consciously controlled, such as blinking, sneezing or swallowing.

ion (21) an atom or small group of atoms carrying an electrical charge and having different properties from the uncharged atoms.

ionization (21) the splitting up of a compound into ions, usually when it dissolves in water to make a solution.

irrigation (42) the artificial supply of controlled amounts of water to agricultural land.

larva (241) an immature stage in the development of certain animals from the egg. It is independent of its parents but different from them in appearance and activities and, usually, in feeding habits.

latent heat (99) the amount of energy (number of joules) needed to turn 1g of a liquid into a vapour.

leguminous plant (51) a plant whose fruits are pods, e.g. beans and peas.

Leguminous plants have nodules on their roots containing nitrogen-fixing bacteria.

linkage (179) the presence of genes close together on the same chromosome so that they tend to be passed on together to the offspring.

lipids (26) fats and fat-like substances; often a fat combined with other chemicals, e.g. phospho-lipids.

lymph (83) a body fluid derived from tissue fluid and returned to the circulatory system in lymphatic vessels.

lymphatic (83) a thin-walled vessel in the body which returns lymph from the tissues to the circulatory system.

macrophage (74) type of white cell present in most kinds of connective tissue. It can ingest foreign particles.

malnutrition (36) general term for illnesses resulting from inadequate feeding. Usually due to lack of one or more essential components of the diet.

meiosis (170) type of cell division which results in the production of gametes. The chromosome number after meiosis is halved (haploid).

menstruation (108) the breakdown of the lining of the uterus at intervals of about twenty-eight days. This occurs about fourteen days after an ovum has been released (but not fertilized), and is recognized by the loss of a small amount of blood through the vagina.

metabolism (30) all the chemical changes in the body that contribute to the living processes.

metamorphosis (241) the profound changes that take place when the larval form of an animal changes into an adult, e.g. caterpillar to butterfly.

micro-organisms (198) very small animals and plants such as protozoa, bacteria and moulds.

mitosis (160) the events taking place at cell division which result in an equal distribution of chromosomes to the daughter cells and a maintenance of the diploid number of chromosomes.

motor (146) in biology, the term implies some positive action in response to a stimulus, e.g. a muscle contraction or secretion by a gland. *Motor* nerve fibres carry the impulses from the central nervous system to the organ that produces such action.

mucous membrane (60) the layer composed of epithelium, mucous glands and connective tissue, which lines the alimentary canal and breathing passages.

mucus (9) the viscous lubricating fluid produced by glands in the food canal, breathing passages and vagina.

mutation (168) a spontaneous change in a gene or chromosome which alters the metabolism of an organism.

nephron (93) one of the microscopic units in the kidney which filters the blood and reabsorbs some of the products.

neural (153) a term relating to structures associated with the nervous system.

neurone (146) a nerve cell. The cell has two or more processes which can conduct nervous impulses.

nitrogen-fixing (51) the ability, possessed by certain bacteria, to use atmospheric nitrogen to make organic compounds of nitrogen such as amino acids.

nitrogenous (73) containing compounds of nitrogen.

nutrient (20) a substance having food value, or able to be used by plants to make food.

nutrition (32) the intake, digestion, absorption and effective use of food.

obesity (37) being overweight. Caused by accumulating fat to the point where health is likely to be affected.

olfactory (132) a term describing structures and activities to do with the sense of smell.

oogenesis (171) the production of mature ova (eggs).

optic (133) biological term describing structures and functions related to the eye.

optimum (27) literally, the best. Usually refers to the best conditions for some biological activity.

osmo-regulation (94) adjustments made to the concentration of the body fluids, usually via the blood, to maintain their concentrations within narrow limits.

osmosis (22) the diffusion of water through a membrane from a weaker to a stronger solution.

osmotic pressure (23) the pressure built up as a result of the diffusion of water into a system by osmosis.

ossification (111) gradual replacement of cartilage by bone.

ovulation (102) the release of a mature ovum (egg) from the ovary.

ovum (102) the female reproductive cell; the egg.

ozone layer (308) the peak in ozone concentration in the atmosphere about 25 km above the Earth's surface.

pandemic (190) the spread of an epidemic disease from one country to a number of other countries.

parasite (206) an animal or plant living and feeding in or on another living organism without necessarily killing it.

pathogen (198) a micro-organism that lives in or on another plant or animal, causing disease.

permeable (23) allowing a gas or liquid to diffuse through.

pesticides (58) chemicals that destroy plant or animal pests which threaten crop plants or the health of communities.

pH (21) a measure of how acid something is. When accompanied by a figure it indicates the concentration of hydrogen ions in a solution.

phagocyte (71) a type of white blood cell which can ingest foreign particles such as bacteria.

phenotype (175) the observable characteristics of an organism. Usually contrasted with *genotype*. A genotype may contain genes for both black fur, *B*, and brown fur, *b*, but the phenotype is black fur.

placenta (104) a structure formed between an embryo and the uterine lining which enables the embryo to obtain food and oxygen from its mother's blood.

plankton (59) microscopic organisms living in the surface waters of ponds, lakes and oceans.

poikilothermic (98) having a body temperature the same as or a little above that of the surroundings. The body temperature rises and falls with that of the environment.

pollution (303) accumulation of harmful substances in the environment.

polymerization (41) the formation of very large molecules by combining together numerous small molecules of the same type.

predator (56) an animal that kills and eats other animals (prey).

prophylaxis (214) taking measures to reduce the chances of catching a particular disease.

proprioceptor (131) an internal sensory organ, usually in a muscle. It is sensitive to stretching and sends impulses to the brain which enable one to judge the position of limbs and the tensions in muscles.

protein (17) complex chemicals used in building protoplasm, and hence cells, tissues and organs. Consequently it is essential that the diet contains some protein.

protoplasm (8) the living contents of a cell. The cytoplasm and the nucleus.

protozoa (201) single-celled animals.

psychosis (262) mental illness in which the patient loses touch with reality.

recessive (167) the gene which, in the presence of its contrasting partner (allele) is not expressed in the observable characteristics of the organism.

reflex (148) a rapid, automatic response to a stimulus.

renal (77) a term relating to any structure or function to do with the kidney.

replication (168) the production of an exact copy of a structure.

reservoirs of infection (216) animals which carry in their bodies organisms pathogenic to man. The pathogens may be transmitted to man by vectors.

respiration (27) the release of energy from food molecules and its transfer to molecules which play a part in the vital chemistry of the cell.

RNA (166) abbreviation for ribonucleic acid, a chemical which helps to convey the 'instructions' from the nucleus to the cytoplasm.

saprophytes (200) organisms, usually fungi or bacteria, which derive nourishment from dead or decaying organic matter.

selective permeability (23) allowing some substances but not others to diffuse through a membrane.

sensory (129) to do with detecting stimuli and sending impulses to the brain.

serum (74) blood plasma from which fibrinogen has been removed. Sometimes it contains antibodies which are used to combat a disease when the serum is injected into the patient.

soluble (20) able to be dissolved in a liquid.

solubility (20) a measure of how much of a substance can be dissolved in a given quantity of liquid.

solution (20) a mixture of a substance in a liquid in which the substance is dissolved and uniformly dispersed throughout the liquid.

specialization (8) the development of a cell or a structure so that it carries out one particular function more efficiently than any others.

spermatogenesis (171) the production of mature sperms from sperm-mother cells.

spore (199) a reproductive or resistant body formed by a micro-organism.

sterilization (207) destruction of bacteria and viruses in food or on medical or experimental equipment.

stimulus (129) a physical or chemical event in an organism or its environment which makes it alter its pattern of activity.

subcutaneous (96) beneath the skin.

thorax (88) the upper part of the trunk, from the neck to the diaphragm.

thoracic (117) a structure or function having something to do with the thorax.

toxin (74) a poisonous chemical produced usually by a bacterium.

urea (92) a chemical compound containing nitrogen. It is a product of metabolism and is excreted by the kidneys, forming one component of urine.

ureter (92) the tube conducting urine from the kidney to the bladder.

urethra (93) the tube conducting urine from the bladder to outside the body.

urine (94) the solution of nitrogenous waste products and salts which is produced in the kidney, stored in the bladder and expelled at intervals through the urethra.

uterus (100) the female organ in which the embryo develops.

utriculus (139) an organ of balance in the inner ear.

vaccination (252) the introduction of a disease-causing antigen into the body, which makes an antibody against the antigen and so develops immunity to an attack of the disease.

vaccine (252) a preparation of an antigen from a disease-causing organism, which is rendered harmless or incapable of reproduction but if introduced to the body will stimulate it to produce antibodies.

vector (214) an animal that carries a disease-causing organism from one host to another.

virus (200) a submicroscopic particle that can reproduce inside living cells and cause disease.

vitamin (34) a chemical taken in with the food that plays an essential part in chemical reactions in cells, but has no energy or body-building value.

zygote (100) the single cell resulting from the joining together of male and female gametes. It can develop into a new individual.

Index

Bold type refers to pages on which topics are given their main treatment.